W0042444

Springer-Verlag London Ltd.

## International Programme Committee:

G. Rabe (*Chairman*) (D)  
R.E. Bloomfield (UK)  
S. Bologna (I)  
G. Cleland (UK)  
P. Daniel (UK)  
G. Dahll (N)  
W. Ehrenberger (D)  
J. Górski (PL)  
W. Halang (D)  

D. Inverso (USA)  
J. Järvi (SF)  
K. Kanoun (F)  
F. Koornneef (NL)  
V. Maggioli (USA)  
A. Pasquini (I)  
G. Picciolo (I)  
J. Rainer (A)  
F. Redmill (UK)  

B. Runge (DK)  
G. Schildt (A)  
E. Schoitsch (A)  
A. Servida (EC)  
I.C. Smith (UK)  
L. Strigini (I)  
A. Weinert (D)  
M. Wilikens (EC)  

## National Organizing Committee:

M. Dowell (*Chairman*), A. Cambon, G. Pirelli, A. Pollicini, G. Rabe, D. Schlittenhardt

# The 14th International Conference on Computer Safety, Reliability and Security

Belgirate, Italy
11–13 October 1995

Edited by
## GERHARD RABE

**Sponsors**

*EuropeanWorkshop on Industrial Computer Systems
Technical Committee 7 (EWICS TC7)*

*European Commission-Joint Research Centre –
Institute for Systems Engineering and Informatics*

**Co-Sponsored by**

*IFIP Technical Committee 5 WG 5.4.
OCG, the Austrian Computer Society*

**Organized by**

*European Commission-Joint Research Centre –
Institute for Systems Engineering and Informatics*

 Springer

Gerhard Rabe
Technischer Überwachungs-Verein Nord e.V.
Große Bahnstr. 31
D-22525 Hamburg, Germany

ISBN 978-3-540-19962-5     ISBN 978-1-4471-3054-3 (eBook)
DOI 10.1007/978-1-4471-3054-3

British Library Cataloguing in Publication Data
A catalogue record for this book is available from the British Library

Apart from any fair dealing for the purposes of research or private study, or criticism or
review, as permitted under the Copyright, Designs and Patents Act 1988, this publication
may only be reproduced, stored or transmitted, in any form, or by any means, with the
prior permission in writing of the publishers, or in the case of reprographic reproduction
in accordance with the terms of licences issued by the Copyright Licensing Agency.
Enquiries concerning reproduction outside those terms should be sent to the publishers.

© Springer-Verlag London 1995
Originally published by Springer-Verlag London Limited in 1995

The use of registered names, trademarks etc. in this publication does not imply, even in the
absence of a specific statement, that such names are exempt from the relevant laws and
regulations and therefore free for general use.

The publisher makes no representation, express or implied, with regard to the accuracy of
the information contained in this book and cannot accept an legal responsibility or liability
for any errors or omissions that may be made.

Typesetting: camera-ready by contributors
34/3830-543210 Printed on acid-free paper

# Preface

Safety-related computer systems are those which may lead to loss of life, injury or plant and environmental damage. Such systems therefore have to be developed and implemented so that they meet strict requirements on safety, reliability and security because their applications cover nearly all areas of daily life and range from controlling and monitoring industrial processes, through robotics and power generation, to transport systems.

Highly reliable electronic systems for safety-related applications represent an area in which industry has been involved for many years and which is now gaining increasing importance in academia. Their relevance also results from an increased perception of safety by society. Therefore, not only are technicians involved in this area, but psychological and sociological aspects also play a major role. Dealing with safety-related systems we have to consider the whole lifecycle of these systems, starting from specification up to implementation, assessment and operation.

All those issues mentioned above are covered in this book, which represents the proceedings of the 14th International Conference on Computer Safety, Reliability and Security, SAFECOMP '95, held in Belgirate, Italy, 11–13 October 1995. The conference continues the series of SAFECOMP conferences which was originated by the European Workshop on Industrial Computer Systems, Technical Committee 7 on Safety, Security and Reliability (EWICS TC7) and reflects the state of the art, experience and new trends in the area of safety-related computer systems. In 35 contributed papers from 13 countries, the reader will get an insight into the subjects of safety analysis and assessment, formal methods, human and legal aspects, design and validation and verification. The contributions on applications and case studies demonstrate the broad range of issues covered in SAFECOMP '95.

As Chairman of the International Programme Committee (IPC) I would like to thank all authors who submitted their work, the presenters, the members of the IPC and the National Organising Committee, the session chairman, and the sponsors and co-sponsors for their efforts and support.

My thanks must also go to Martyn Dowell for the excellent cooperation in preparing the conference and to Felix Redmill for his considerable help and advice.

Finally, I would like to express my thanks on behalf of EWICS TC7 to the European Commission for supporting SAFECOMP '95 and to my company TÜV Nord e.V. for allowing me sufficient time and resources to prepare the conference.

*Gerhard Rabe*                                        *Hamburg, Germany*
                                                    *May 1995*

# Contents

# List of Contributors

T. Anderson
Computing Science Department
University of Newcastle
upon Tyne
NE1 7RU, UK

A. Anselmi
Ansaldo Transporti, Genova, Italy

F. Baranowski
INRESTS, 20, rue Elisee Reclus
F-59650 Villeneuve D'ascq., France

C. Bernardeschi
Dipartimento di Ingegneria
dell'Informazione
Università di Pisa, Pisa, Italy

P.G. Bishop
ADELARD, Coborn House,
3 Coborn Road
London, E3 2DA, UK

J. Blieberger
Department of Automation,
Technical University Vienna
Treitlstr. 3/4, A-1040 Vienna
Austria

R.E. Bloomfield
ADELARD, Coborn House
3 Coborn Road
London, E3 2DA, UK

A. Bondavalli
CNUCE-CNR, Via S. Maria 36
I-56126 Pisa, Italy

J.R. Catmur
Arthur D. Little, Science Park
Milton Road
Cambridge, CB4 4DW, UK

K. Chan
University of Teeside
Middlesbrough
Cleveland, TS1 3BA, UK

G.O. Chandroth
Department of Computer Science
University of Sheffield
Sheffield, S1 4DP, UK

S. Chiaradonna
CNUCE-CNR, Via S. Maria 36
I-56126 Pisa, Italy

J. Christmansson
Laboratory for Dependable
Computing
Department of Computer
Engineering
Chalmers University of
Technology
S-412 96 Göteborg, Sweden

M.F. Chudleigh
Cambridge Consultants Ltd.
Science Park, Milton Road
Cambridge, CB4 4DW, UK

A. Coombes
Department of Computer Science
University of York
York, YO1 5DD, UK

C.M. Crawshaw
Department of Psychology
University of Hull
Hull, HU6 7RX, UK

D. Davis
Eversheds, Hepworth and
Chadwick
Cloth Hall Court, Infirmary Street
Leeds, LS1 2JB, UK

F. Di Giandomenico
IEI-CNR, Via S. Maria 46
I-56126 Pisa, Italy

M. El Koursi
INRESTS, 20, rue Elisee Reclus
F-59650 Villeneuve D'ascq., France

x

F. Engelmann
SYNSPACE AG
Rottmannsbogenstr. 30
CH-4102 Binningen, Switzerland

A. Fantechi
IEI-CNR, Via S. Maria 46
I-56126 Pisa, Italy

C. Fencott
University of Teeside
Middlesbrough
Cleveland, TS1 3BA, UK

P. Fenelon
Department of Computer Science
University of York, Heslington
York, YO1 5DD, UK

P. Gianninò
INNOTEC Engineering
Via Neri di Bicci 12
I-50143 Florence, Italy

S. Gnesi
IEI-CNR, Via S. Maria 46
I-56126 Pisa, Italy

J. Górski
Franco-Polish School of New
Information
ul. Mansfelda 4
60-854 Poznan 6, Poland

W.A. Halang
FernUniversität Hagen
Postfach 940
D-58084 Hagen, Germany

A.J. Harrison
Railtrack PLC, 355 Euston Road
London, NW1 3AG, UK

B. Hebbron
University of Teeside
Middlesbrough
Cleveland, TS1 3BA, UK

J.-P. Heckmann
Aerospatiale Avions
Toulouse, France

M. Heisel
Institut für Angewandte Informatik
Technische Universität Berlin
Franklinstr. 28-29
D-10587 Berlin, Germany

M. Hietikko
VTT Manufacturing Technology
PO Box 17011, F-33101 Tampere
Finland

K.M. Hobley
Safety Critical Computing Group
School of Computer Studies
University of Leeds
Leeds, LS2 9JT, UK

G.R.J. Hockey
Department of Psychology
University of Hull
Hull, HU6 7RX, UK

D. Hughes
School of Computing, University
of Plymouth
Plymouth, PL4 8AA, UK

T. Jennings
EDS Limited, Wavendon Tower
Wavendon
Milton Keynes, MK17 8LX, UK

P.H. Jesty
Safety Critical Computing Group
School of Computer Studies
University of Leeds
Leeds, LS2 9JT, UK

Z. Kalbarczyk
Laboratory for Dependable
Computing
Department of Computer
Engineering
Chalmers University of
Technology
S-412 96 Göteborg, Sweden

K. Karunakar
Aeronautical Development Agency
PB No. 1718, Vimanapura Post
Bangalore 560 017, India

T.P. Kelly
Department of Computer Science
University of York, Heslington
York, YO1 5DD, UK

B.J. Krämer
FernUniversität Hagen
Postfach 940
D-58084 Hagen, Germany

H. Krebs
TÜV Rheinland, Am Grauen Stein
D-51105 Köln, Germany

S. La Torre
IEI-CNR, Via S. Maria 46
I-56126 Pisa, Italy

S. Larosa
IEI-CNR, Via S. Maria 46
I-56126 Pisa, Italy

R. de Lemos
Computing Science Department
University of Newcastle upon Tyne
NE1 7RU, UK

B. Letrung
INRESTS, 20, rue Elisee Reclus
F-59650 Villeneuve D'ascq., France

B. Littlewood
City University
Northampton Square
London, EC1V OHB, UK

J. Magott
Franco-Polish School of
New Information
ul. Mansfelda 4
60-854 Poznan 6, Poland

J.A. McDermid
Department of Computer Science
University of York, Heslington
York, YO1 5DD, UK

K.W. Miller
Department of Computer Science
Sangamon State University
Springfield, IL, USA

J. Moffett
Department of Computer Science
University of York, Heslington
York, YO1 5DD, UK

G. Mongardi
Ansaldo Transporti, Genova, Italy

P. Morris
ISEI, Joint Research Centre
I-21010 Ispra, Italy

K.N. Narahari
Aeronautical Development Agency
PB No. 1718, Vimanapura Post
Bangalore 560 017, India

G. Picciolo
EniChem, Via Taramelli, 26
Milan, Italy

R.H. Pierce
York Software Engineering Ltd.
University of York
York, YO1 5DD, UK

S. Prasad
Aeronautical Development Agency
PB No. 1718, Vimanapura Post
Bangalore 560 017, India

C. Rabéjac
LIS – Matra Marconi Space
LAAS–CNRS
7 Avenue du Colonel Roche
F-31077 Toulouse Cedex
France

F. Redmill
Redmill Consultancy
22 Onslow Gardens
London, N10 3JU, UK

E. Ruiz Morales
Joint Research Centre, TP 210
I-21020 Ispra, Italy

A. Saeed
Computing Science Department
University of Newcastle upon Tyne
NE1 7RU, UK

E. Schoitsch
Austrian Research Centre
Seibersdorf
A-2444 Seibersdorf, Austria

W. Schynoll
SWE, Pestalozzistr. 1
D-74321 Bietigheim-Bissingen
Germany

I.D.R. Shannon
Prism Engineering, London, UK

A.J.C. Sharkey
Department of Computer Science
University of Sheffield
Sheffield, S1 4DP, UK

N.E. Sharkey
Department of Computer Science
University of Sheffield
Sheffield, S1 4DP, UK

S. Shirlaw
GEC Alsthom Signalling
Systems Division
33, rue des Bateliers
F-93400 St. Ouen, France

N.M. Shryane
Department of Psychology
University of Hull, Hull
HU6 7RX, UK

H. Stienen
SYNSPACE AG
Rottmannsbogenstr. 30
CH-4102 Binningen, Switzerland

P. Taylor
EDS Limited, Wavendon Tower
Wavendon
Milton Keynes, MK17 8LX, UK

R.J. Tiezema
GTI, Landdrostlaan 51
7327 GM Apeldoorn
The Netherlands

R. Tiusanen
VTT Manufacturing Technology
PO Box 17011
F-33101 Tampere, Finland

F. Torielli
Ansaldo Transporti, Genova, Italy

J. Torin
Laboratory for Dependable
Computing
Department of Computer
Engineering
Chalmers University of
Technology
S-412 96 Göteborg, Sweden

M. Viola
Ontario Hydro,
700 University Avenue
Toronto, Ontario, M5G 1X6
Canada

J.M. Voas
Reliable Software Technologies
Corporation
21515 Ridgetop Circle
Sterling, VA 20166, USA

N. Völker
FernUniversität Hagen
Postfach 940
D-58084 Hagen, Germany

H. Waeselynck
INRESTS, 20, rue Elisee Reclus
F-59650 Villeneuve D'ascq., France

A. Wardziński
Franco-Polish School of New
Information
ul. Mansfelda 4
60-854 Poznan 6, Poland

S.J. Westerman
Department of Psychology
University of Hull
Hull, HU6 7RX, UK

D. Wright
City University
Northampton Square
London, EC1V OHB, UK

C.W. Wyatt-Millington
Department of Electronic and
Electrical Engineering
University of Bradford, Bradford
BD1 1DP, UK

# Session 1
## General Issues, Guidelines

# Software Best Practices in Dependable Systems

# The European Research Projects ENCRESS, OLOS and ESPITI from a Partners Perspective

Erwin Schoitsch
Austrian Research Centre Seibersdorf
Vienna, Austria

## Abstract

European Research Programmes have a long term tradition in supporting activities in the area of dependability and quality of software. EWICS TC 7, the European Workshop on Industrial Computer Systems, Technical Committee 7, Reliability, Safety and Security, has been supported by the European Commission in producing its guidelines on "Dependability of Critical Computer Systems", which, as an deliverable, have been published in three volumes by Elsevier Science Publishers, long before the first framework programme of ESPRIT has been started. The author and its institution are now involved in three current European projects in this area, ENCRESS (European Network of Clubs for REliability and Safety of Software, an ESSI dissemination action), OLOS (A Holistic Approach to the Dependability Analysis and Evaluation of Control Systems involving Hardware, Software and Human Resources, a HCM scientific cooperation network) and ESPITI (European Software Process Improvement Training Initiative, an ESSI Training Initiative). These projects will be described from a (small) partners perspective.

## 1. Introduction

EWICS TC 7, the European Workshop on Industrial Computer Systems, Technical Committee 7, Reliability, Safety and Security, has a very long tradition in dealing with dependability and quality issues of complex computer control software-intensive systems. The Austrian Research Centre Seibersdorf and its Department of Information Technology, is involved in the development of such systems since 25 years. Examples such as the Fully Electronic Railway Interlocking System ELEKTRA of Alcatel Austria [DAN90], the sCaleable Security control System CSS of Philips [GOR93] have been presented at several SAFECOMP's in the past, as well as work for standardisation bodies in the area of dependability of software-intensive systems [GEN89]. Technology transfer between the scientific community and Austrian industry is another major task of the Research Centre.

Therefore, the Research Centre Seibersdorf is supporting EWICS TC 7 since the days of the CAMAC development, a high speed electronic bus and interface system developed by the European nuclear research community. The author has contributed to the EWICS TC7 books on "Dependability of Critical Computer Systems", Vol. 2 and Vol. 3 [RED89] [BISH90], is active member of several subgroups of EWICS TC 7 and is chairman of the Project Management Subgroup, developing the "Guideline for the Project Management of the Development of Critical Computer Systems", which is almost completed by now and just before the final review and editing [MAGG94].

In performing its task as national research lab and in close cooperation with Austrian associations and institutions in the area of software and systems dependability and quality, the Austrian Research Centre Seibersdorf has taken part in the Second and Third Framework Call for Proposals of the CEC [CEC92] (and in the Fourth Framework Call as well, but this is outside the scope of this presentation) in order to benefit from the European efforts to promote best practices for dependable software-intensive systems. The result of this effort is, that the Austrian Research Centre Seibersdorf became associated partner and Austrian project leader of ENCRESS (project number 10475), OLOS (project number 0577) and ESPITI (project number 11000).

## 2. ENCRESS, OLOS and ESPITI as means of technology transfer

Technology transfer includes two major activities: Acquisition of knowledge from outside sources, and distribution of knowledge among the addressed target groups, mainly Austrian industry and enterprises, most of them belonging to the the group of small and very small enterprises (in the Austrian software industry, more than 50% of the enterprises have less than 6 persons of permanent staff).

The Research Centre Seibersdorf fulfills some other important prerequisites for performing successful technology transfer: As the largest Austrian independent, non-university research institution, it cooperates very closely with Austrian universities, providing possibilities for research, diploma and thesis work, and cooperates with industry, enterprises and governmental agencies on the basis of contract work. Such, our contribution to technology transfer is not only on a theoretical basis, but triggered by practical development work. Additionally, the Research Centre has a very close link to the other Austrian Research Organisations like Joanneum Research (Graz) and the Research and Test Lab Arsenal (Vienna) within the umbrella organisation "Research Austria".

All three projects addressed here are involved in technology transfer. ENCRESS is a network of clubs (working groups, special interest groups) dealing with dissemination and knowledge acquisition of the state of the art of best practices in the area of (highly) dependable systems. OLOS has not only the task of dissemination but also to promote the development of an integrated set of methodologies for analysis and evaluation of critical systems, and such providing a new, holistic view of systems, not only theoretically but also in practice. ESPITI is a training initiative increasing awareness among management and the public in the area of software quality, assessment and certification, supporting training of enterprises and people (especially of SMEs) and such increasing long-term competetiveness of the European software-intensive industry.

## 3. ENCRESS

ENCRESS, the European Network of Clubs for REliability and Safety of Software, is an ESSI Dissemination action. ESSI, the European Software and System Initiative, consists of three lines of actions:

- Application experiments to demonstrate software best practices in a concrete application (only the experiment, not a product development is funded !),
- Dissemination actions to promote dissemination of best practices, know how and experience within the industrial and scientific community by installation of workgroups and regular conferences,
- a training initiative.

The ENCRESS dissemination project aims at establishing a Special Interest Group for people who are concerned with software and systems safety and reliability (and dependability in the broader context). The model for this idea are the British Clubs for Safety Critical Systems and for Reliability and Metrics, which are, after some years of support from the government, now fully operational without funding.

The project started last year, covering UK (University of Newcastle, Centre for Software Reliability), France (Objectif Technologie), Italy (ENEA), Germany (ISTec, Garching) and Denmark (Delta). At the end of 1994, Austria (Austrian Research Centre Seibersdorf) joined ENCRESS, later on followed by Sweden (SP, Boras), Greece (Kyros) and Spain (INTA). Further countries are planned to be included within an ENCRESS II project (proposal submitted by June, 15, 1995).

Regular meetings of the clubs and a conference in September in Bruges, Belgium, are part of the European aspect. Within each country the national club is operating independently. Some links to existing associations and organisations are recommended (especially for smaller countries).

There is a close cooperation between companies involved in application experiments and the dissemination actions expected by the CEC. In Austria, there are only two ESSI application experiments at the moment, ODP (Universal Online Database Manager for better Project Supervisory, FESTO Vienna, involved in the PLC business, project no. 10788) and EMS (Energy Management System, applying the AMI method to achieve software process improvement to a small software processing unit of the VOEST steel works in Linz). Both have already reported their experience (good or bad !) to the interested community taking part in the ENCRESS club of Austria and to ESPITI, the Software Process Improvement Training Initiative, in the context of awareness events and working groups. To enlarge the perspective of Austrian research and industry and to open markets, international and bilateral events are sponsored, namely joint meetings with the German ENCRESS club, reports from other relevant European events are presented, and SAFECOMP '96 will be organized in Vienna (1996 is the millenium year of "Ostarichi", Austria).

Fortunately, there are more enterprises involved in critical computer systems application in Austria than represented among the ESSI application experiments, e.g. process industry, railway engineering, traffic and transportation systems, air traffic control organisation and suppliers, security control systems, military applications, fireguard- and ambulance systems, medical systems, systems with very high availability requirements in control and communications etc. Topics addressed include design, development, test and operation of safety critical and highly dependable systems, human interfaces and user documentation (guidance), certification, standardization, experience reports and research.

Since only very marginal costs (in principle a very small amount of money for labour and subsidence of regional events, and total travel subsistence for the European Encress meetings for transnational experience exchange) are funded, the national clubs are organized in most countries in the framework of some existing professional or scientific organisation to enable long-term existence of the club beyond the duration of the current project (18 months, for Austria 12 months). In Austria, the Austrian Computer Society OCG and the support of the Austrian Electrotechnical Association (OeVE) supply the basic infrastructure, running cost like copying and labour are mainly supplied by the Research Centre Seibersdorf.

The clubs are operating very successfull, most of the goals have been reached (membership, attendance at meetings, satisfaction of members), in all partner countries. In Austria, the membership of the working groups existing before the establishment as ENCRESS club, was not more than 40 members, and has raised since then to 100, thanks to the publicity of ENCRESS. Five meetings have taken place, and a national newsletter has been startedty. Services such as meeting handouts, information about relevant papers, conferences, standards etc. are requested by members, e.g. when club members are not able to attend a specific meeting. For details, please contact Chris Dale (UK), the European project manager.

## 4. OLOS

OLOS is contracted within the framework of the Human Capital and Mobility Programme (HCM) as a scientific cooperation network, including experience exchange, integration of results of current European research programmes in the area of dependability, exchange of (young) researchers to acquaint them with concepts, methodologies, research and project work at other European sites.

OLOS means "A Holistic Approach to the Dependability Analysis and Evaluation of Control Systems involving Hardware, Software and Human Resources". OLOS proposes a significant innovation in the dependability analysis and evaluation of safety-critical systems by taking a holistic approach rather than studying isolated components of a system. This project is complementary to other European Research Programmes which concentrate on certain components or aspects of critical systems only. The primary goals of OLOS are:

- to develope interdisciplinary skills and competencies, especially among young researchers, concerning "Global System Dependability"
- to define and develope the concept of "global system dependability" in order to make visible how various dependability and reliability notions and methodologies contribute to overall dependability
- to promote the developement of an integrated set of methodologies for analysis and evaluation of critical systems in a holistic manner.

The holistic approach proved already useful during the kick-off meeting in Rome in May 1995. Originally, it was planned to have presentations of each partner separated according to hardware, software and human reliability. It very soon was demonstrated, that this wasn't a reasonable approach - many of the partners are active and have experience in some of the key areas, not just in one, and topics had to be discussed in a more holistic manner already at this basic level !!

OLOS is a three years project, and the plans for this period include:

- to acquaint researchers with the concepts and methodologies of related disciplines
- to identify and adapt methodologies for integrated approaches
- to evaluate these methodologies on existing projects
- to create a conceptual framework which combines the methodologies and information of each discipline
- to disseminate the results to the scientific and industrial community (means are workshops, conferences, newsletters, publications, direct involvement of researchers in (industrial) projects)

The list of partners in the network demonstrates the broad view of dependability which is taken by the OLOS group (including hardware, software and human factors people, system researchers and psychologists):

- ENEA (Rome, I),
- Centre for Software Reliability (City University of London, UK),
- Vrije Universiteit (Amsterdam, NL),
- Laboratory for Dependability Engineering (LIS, LAAS, CNRS, Toulouse, F),
- Fault Tolerant Computer Lab (Texas A&M University, USA),
- Universit di Siena (Siena, I),
- Institute of Systems Engineering and Informatics (ISEI, JRC, Ispra, I),
- Istituto per l'Elaborazione dell' Informazione (IEI, CNR, Pisa, I),
- Software Pruefstelle TUeV Nord (Hamburg, D),
- Laboratory for the Foundation of Computer Science (LFCS, Edinburgh, UK),
- Department of Systems and Computer Engineering (Univ. Roma, I),
- Human Reliability Associates Ltd. (HRA, UK),
- Austrian Research Centre  Seibersdorf (A),
- Institut fuer Rechnerentwurf und Fehlertoleranz (Univ. Karlsruhe, D).

The cooperation program is based on four main phases:

- Knowledge acquisition (interdisciplinary, plenary workshops, identification of suitable research projects, exchange of young researchers)
- Definition of Dependability (to find a satisfactory definition of global dependability; assessment of contribution and interdependencies of different factors and components, special and plenary workshops)
- Promotion of Integration (identify methodologies more suited for integration in each domain, direct involvement of young researchers in "pilot" projects to create a conceptual framework)
- Dissemination of results (industrial and academic world, workshops, conferences, newsletters, publications, send "exchanged" young researchers to industry and research)

The beginning of this program looks very promising: The initial meeting in Rome demonstrated already the problem of "semi-holistic" approaches, sector specific terminology and views, and interesting aspects provided from other disciplines (not hardware or software only). Based on my experience with the usefulness of organisations like EWICS TC7, I believe that OLOS in the end could lead to a more common understanding of dependability and end the wars between different disciplines (still to be seen now in IEC standards related to hardware or software only, leading to contradictionary results !!)

For details, ask Alberto Pasquini (ENEA, Rome) or George Cleland (Univ. of Edinburgh, LFCS).

## 5. ESPITI

ESSI is the European Software and System Initiative, and is designated to improve best practices by application experiments, dissemination actions and training. It is very application oriented (the responsible DG is DG III, industry, therefore), and deliberately the set-up was chosen such that the programme started with application experiments and dissemination actions, and in 1994, the training initiative ESPITI (European Software Process Improvement Training Initiative) completed the programme. There has been considerable funded work in the area of IT around for years, and the goal of ESPITI is

"..... to maximise the benefits from European efforts to improve the software process and of its subsequent ISO 9000 certification through information exchange and training"

This has to take into account the experiences gained from earlier projects and actions. One of the experiences was, that many European activities are more supply-driven than demand driven, therefore the assessment of the actual demands and the actual situation is a major part of the initiative.

Since the situation in each European region differs from the other regions in some way or other (structure of industry, training demands and supplies, existing infrastructure for training, sectors of relevant software-intensive industry, etc.), a regional approach according to the principle of subsidiaridy, was chosen by the EC. Within each region, a regional organisation is repsonsible for the ESPITI activities, and is managed by one of two partners, either MARI N.I. (Belfast) or Reserch Centre Karlsruhe (Germany), who are responsible for the associated contracts. With the exception of Scandinavia, a region is identical with a country.

Managed by MARI (N.I.) are:

- Austria (Austrian Research Centre Seibersdorf)
- Belgium (VCK)
- Denmark (DELTA)
- France (AFNOR)
- Germany (FZ Karlsruhe)
- Ireland (CSE)
- Portugal (IPQ)

Managed by Research Centre Karlsruhe are:

- Greece (Intrasoft)
- Italy (Etnoteam)
- Luxembourg (Centre de Recherche Public)
- Netherlands (SERC))
- Scandinavia (CCC: Finland, Sweden, Norway, Iceland)
- Spain (SIP)
- UK (MARI)

The activities include awareness events (normally attendance for free), information events, workshops and trainings (the letter for marginal cost or regular attendance fees), and should lead to more ISO 9000 certifications, more software process assessments and improvement actions, and to post-certification activities (continuous software process improvement, business reengineering, total quality management activities). The message can't be "go for ISO 9000 certfication" only, the idea of continuous improvement is included as well ! Increased competetiveness in the long term has to be the ultimate goal (therefore tayloring of process models and QM-systems to the real requirements, project management issues, requirements engineering and management have been key issues in the user survey too).

The first results in Austria are very promising:

There have been a series of awareness events, workshops and training seminars (25 events from December 94 to July 1995, between 9 and 140 attendants), well distributed over all Austrian (sub-) regions, the list of replies to the user survey (approx. 300) show more than 50% from enterprises with less than 6 permanent staff, more than 80% from enterprises with less than 50 permanent staff, which is representative for Austria according to some national studies about the situation of the

Austrian software industry. Knowledge about the benefits of ISO 9000 and Software Process Improvement has increased significantly. The "train the trainer" concept seems to work, there is considerable interest of software consultants and trainers too in the initiative. There is good cooperation with major players in the area of software and computer- resp. user organisations and associations, including the Chamber of Commerce (the goal of the initiative is NOT to develope new courses, but to achieve a multiplicator effect by including all major players in this area in a planned effort).

For details, ask Kevin Eakin (MARI N.I., Belfast) or Ingward Bey (FZ Karlsruhe, Germany).

# 6. Conclusions

From a partners point of view, especially from a small country, ENCRESS, OLOS and ESPITI are successful activities within the European framework for research and technology transfer to improve the situation in European industry with respect to dependability and quality of software-intensive systems. Of course, this is not a thorough evaluation, which is a task of the respective offices in Brussels responsible for the European Research Programmes, but it is the impression from an involved partner, who can compare the status of these three projects after one years duration.

Although covering different aspects and segments in this area, and with different goals especially with respect to future continuation of these activities, each of these projects provide a significant contribution to awareness, know-how and experience transfer and research on European and national level, such contributing to a common European-wide safety- and quality culture by bringing together the best of all worlds within Europe (and even beyond !) - and by taking into account and utilizing regional and national differences.

The projects are positioned such that Europen industry and research will be able to strengthen its muscles in areas where it is worthwhile doing so, and not competing with US and Japan in areas which are according to studies like the Bangemann report [BAN94] or of ESI [KOCH95], the European Software Institute, out of scope with respect to profit (such as basic hardware and software products). We have to concentrate on the real challenges [CEC93] of the next century !!

# 7. References

[BAN94]      Bangemann et.al. Europe and the Global Information Society. Recommendations to the European Council. Brussels, May 26, 1994.

[BISH90]      P.G. Bishop (Ed.) "Dependability of Critical Computer Systems - Techniques Directory", Vol. 3, Elsevier Science Publ., London, New York, 1990.

[CEC92]        Commission of the European Community "Third Framework Programme", Brussels, 1992.

[CEC93]        European Commission "Growth, Competivenes, Employment - The Challenges and Ways Forward into the 21st Century", White Paper.

[DAN90]        B.K. Daniels (Ed.) "Safety of Computer Control Systems 1990 (SAFECOMP '90) - Proceedings of the IFAC/EWICS/SARS Symposium, Gatwick, UK, 1990", Pergamon Press, Oxford, UK, 1990.

[GEN89]        R. Genser, E. Schoitsch, P. Kopacek (Eds.) "Safety of Computer Control Systems 1989 (SAFECOMP '89) - Proceedings of the IFAC/IFIP Workshop in Vienna, Austria, 1989", Pergamon Press, Oxford, UK, 1989.

[GOR93]        J. Gorski (Ed.) "SAFECOMP '93 - Proceedings of the 12th International Conference on Computer Safety, Reliability and Security", Springer, London, 1993.

[MAGG94]      V. Maggioli (Ed) "SAFECOMP '94 - Proceedings of the 13th International Conference on Computer Safety, Reliability and Security", ISA Publ., USA, 1994.

[KOCH95]      G. Koch, "Software process Improvement - the European Dimension (ESI)", presentation at the Austrian ESPITI Kick-off event, March 15, 1995, Vienna.

[RED88]        F.J. Redmill (Ed.) "Dependability of Critical Computer Systems", Vol. 1, Elsevier Science Publ., London, New York, 1988.

[RED89]        F.J. Redmill (Ed.) "Dependability of Critical Computer Systems", Vol. 2, Elsevier Science Publ., London, New York, 1989.

# Assessment on the Basis of Standards-Gaps and how to Bridge Them

Author

**H. Krebs**

TÜV Rheinland, Institute for Software, Electronics, Railroad Technology,
Am Grauen Stein D-51105 Köln

## Abstract

The standards IEC 65A(Secretariat) 122, 123 and CENELEC prEN 50126, 50128, 50129 can serve as the basis for development and assessment of safety critical systems in the automotive and in the railway sector. The safety measures described in these standards are not described in sufficient detail to be applied properly in the system. Standards which contain detailed descriptions of the safety measures would avoid the mentioned gap but they would grow in size and in complexity. This would decrease the lifespan of the standard and increase the delay between the technology and the appropriate standard. To avoid these disadvantages CASCADE proposes: the standards mentioned should become more generic and less complex. The gaps at the lower detailed levels could be closed by addition of well tried up to date examples. Such a collection of examples placed in an annex could be completed at short intervals without any changes of the actual obligatory standard. CASCADE is going to produce a short collection of examples for this purpose.

## 1 Introduction

This paper is a result from CASCADE[1] project, especially from the activities in CASCADE to formulate a Generalised Assessment Method for the railway and the automotive sector. CASCADE deals with the assessment of safety critical systems, in particular with the assessment of the software of safety critical systems.

## 2 The Probabilistic Approach in Assessment

The theory of the probabilistic approach is simple. A quantitative risk analysis of the complete system has to show that the remaining risk is smaller than the acceptable risk. It could be required for instance that the probability of a travelling passenger

---

[1] The project CASCADE has the full title "Certification and Assessment of Safety-Critical Application Development" and is partly funded by the CEC under the ESPRIT III programme in the area of Information Processing Systems, project No. 9032

being killed should be less than $10^{-8}$ per hour. A transportation system has a great number of subsystems. Each of them will contribute to the passenger's risk. The designer has the freedom to allocate to each subsystem a risk parameter, or a reliability parameter which corresponds to a risk parameter, such that the risk of the total system will not exceed the acceptable risk of $10^{-8}$ fatalities per person per hour.

Let us presume that one result of the risk analysis is that the software of the train protection system shall not exceed $10^{-8}$ failures per hour. Taking into account that only a small subset of software failures will have serious consequences the requirement $10^{-8}$ failures per hour could be plausible.

The problem with this approach is that there is no current method of demonstrating that the software meets such demanding requirements. We cannot even be sure whether such software can be realised at all [LITTL 90]. The difficulties arising from high reliability hardware subsystems are also serious.

We can state: the purely probabilistic approach, from a high level point of view, is very convincing. Great difficulties exist at the lower levels of assessment. The main problem of the probabilistic approach is to decide whether or not the precisely described probabilistic requirements are met by the subsystems and components.

# 3  The Measure Oriented Approach in Assessment

Since the probabilistic approach causes assessment problems at the subsystem and component level, the standards [IEC 65/12x], [EN 5012x], [DIN 0801] which deal with safety related computer applications do not use this approach. The approach used in the standards mentioned above, is called here "measure oriented approach"

One feature of the measure oriented approach is to substitute the probabilistic requirements for appropriate classes. The standard IEC 65A(Secretariat) 123 for instance defines four System Integrity Levels (SIL1 to SIL4) and relates them to the following probabilities of dangerous failures per hour (FPH) for a safety related continuous control system.

| SIL | FPH |
|-----|-----|
| 4 | $\geq 10^{-9}$ to $< 10^{-7}$ |
| 3 | $\geq 10^{-7}$ to $< 10^{-6}$ |
| 2 | $\geq 10^{-6}$ to $< 10^{-5}$ |
| 1 | $\geq 10^{-5}$ to $< 10^{-4}$ |

There are many reasons to do this. One reason is to point out that there is no need to deal with very accurate figures of FPH and to encourage the user of the standard to assess a system even in difficult cases. According to the standard IEC

65A(Secretariat) 123 the accuracy of a system assessment is sufficient as long as can be decided whether the system FPH is smaller than $10^{-7}$. It is unnecessary to know whether the FPH parameter reaches $10^{-9}$ or not.

The second important feature of the measure oriented approach is the set of safety measures determined to meet the requirements of a particular System Integrity Level.

We collect and categorise the threats which can appear in a safety critical system. Assuming that all threats can be avoided by an appropriate safety system the reason for residual threats are failures of this safety system. The reasons for these failures are random failures in the hardware or faults in hardware or software which cause systematic failures. Therefore we can concentrate on the term failure and on failure avoiding and failure controlling measures.

From a high level point of view we will mainly have to deal with the following measures:

- Measures taken to control random failures in the hardware
- Measures taken to avoid systematic failures in the hardware
- Measures taken to control systematic failures in the hardware
- Measures taken to avoid systematic failures in the software
- Measures taken to control systematic failures in the software

This is a structure at the upper description level. Each measure mentioned above represents a class and has to be particularised. This particularisation has to consider the source of failure and its type. Systematic software failures for instance can derive from faults in:

a)   the system requirements specification
b)   the software requirements specification
c)   or the software design phase

Different measures have to be taken to avoid failures according to a, b and c. Figure 1 gives an overview of the main activities in the development and assessment of safety critical systems according to the measure oriented approach. The functional description of the sytem is the starting point for further considerations. The analysis of the risk leads to the System Integrity Level. The System Integrity Level determines a set of safety measures, which have to satisfy the System Integrity Level. If the safety measures chosen from the standard are applied properly, then the system can be considered safe according to the standard used.

On the right hand side of the activity boxes in Figure 1 standards are mentioned which deal with the left hand side activities and which are the basis of these activities. Applying the chosen safety measures in his system, the designer needs much more information about these measures than he can find in the standards. The knowledge of the state of the art in safety technology of the application area in

**Figure 1:** Main activities in development and assessment of safety critical systems

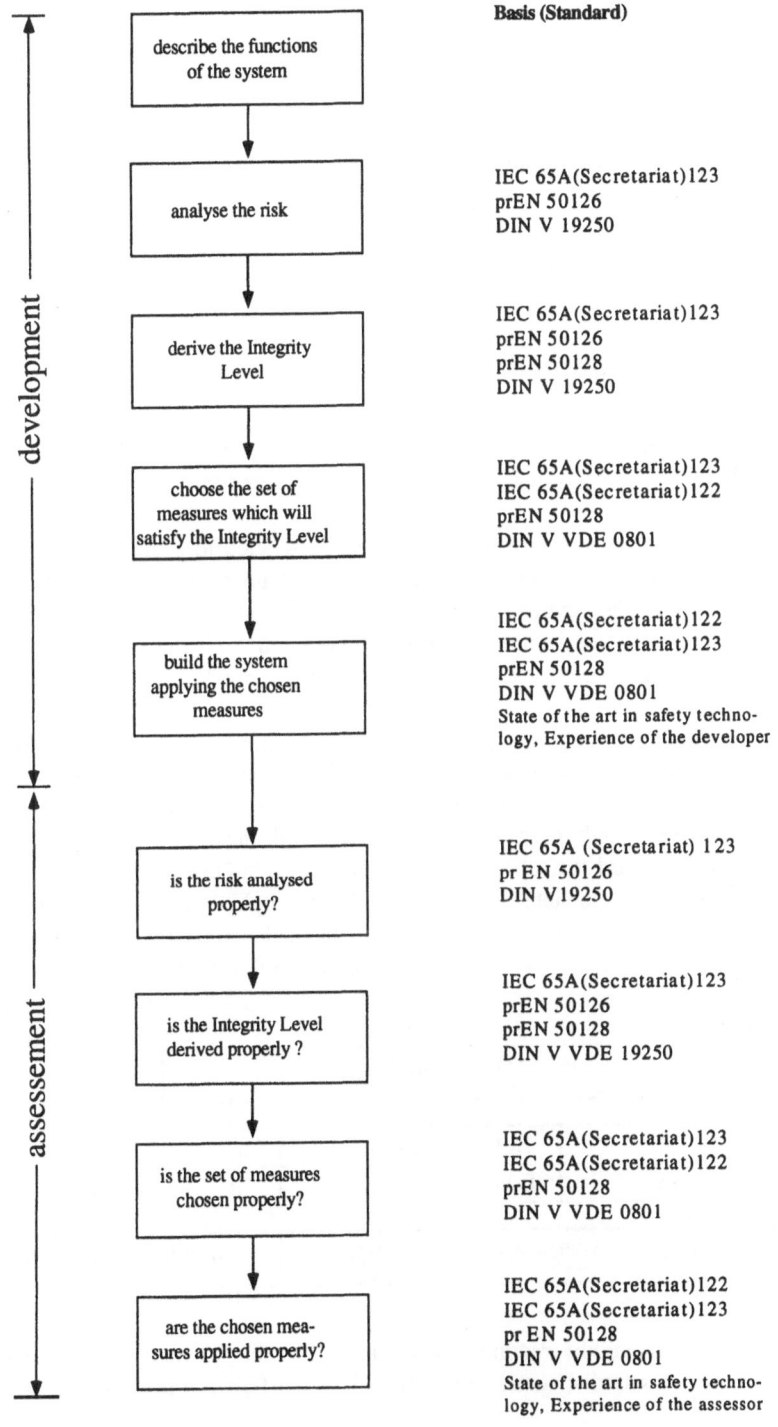

question, and the experience of the designer provide further important information. The assessment process in the measure oriented approach is very similar to the development process. The activities of the designer to construct a safe system form questions or criteria for the assessment process:

- Is the risk analysed properly?
- Is the System Integrity Level derived properly?
- Are the chosen measures applied properly?

We can sum up: The requirements given by the standards to the designer are also the basic assessment procedure prescribed by the standards.

# 4 Gaps in the Standards

The safety measures described in the standards are not described in sufficient detail to be applied properly in the system. This has to be considered the first general gap.

A measure recommended by a given standard can be applied to a greater or lesser extent. As a result it is to be expected that there will be a greater or a smaller contribution to safety. This contribution of a measure is called its effectiveness. A code inspection for instance is one measure to avoid failure in the software. The standard DIN V VDE 0801 requires for the Requirement Classes 5 and 6 (this corresponds to the Integrity Level 3 in IEC 65A) that the code inspection measure is applied with "middle" effectiveness. There are different opinions on what code inspection has to look like. There are even more opinions about the question what a code inspection with the "middle" effectiveness should look like. But designers and assessors need such detailed descriptions of measures with the appropriate degree of effectiveness. The standard IEC 65A(Secretariat) 122 does not explicitly mention the term "effectiveness" with its states "simple, middle and high" but when assessing a system on the basis of this standard we have to take also the effectiveness of a measure into consideration. This has to be demonstrated for clause D.9, "static analysis".

| D.9 Static Analyis | | | | | | |
|---|---|---|---|---|---|---|
| | Ref | IL0 | IL1 | IL2 | IL3 | IL4 |
| 6. Fagan's Inspection | B24 | – | – | R | R | HR |

| effectiveness according to DIN V VDE 0801: | simple | middle | high |
|---|---|---|---|

Fagan's Inspection is highly recommended (HR) for the Integrity Level IL4. According to the standard Fagan's Inspection is a sufficient measure for IL4.

Consequently Fagan's Inspection must be also a sufficient measure for the lower levels IL3 and IL2. The full effectiveness of Fagan's Inspection is not really necessary for the Integrity Levels IL2 and IL3. Fagan's Inspection is therefore only "recommended" and not "highly recommended" for these Integrity Levels. The recommendation for the Integrity Level IL1 is even more feeble.

Fagan's Inspection, as with many other measures, is applicable to a greater or lesser extent. In this way the effectiveness of such safety measure has to be graduated.

Before looking for further gaps we want to evaluate the standards in general concerning the commitment and the completeness of their statements. A standard recommends more or less a precise set of safety measures that have to satisfy the required Integrity Level. The question arises: how far can we rely on this recommendation? The standard IEC 65A(Secretariat) 122 contains lists of measures called "clauses" which are to avoid a special type of failure. Clause 9, for instance, contains measures to avoid systematic failures in the Software Requirements Specification. The measures contained in the clauses of IEC 65A(Secretariat) 122 reflect the opinion of experts who were involved in the formulation of the standard, and those experts who participated in the discussions about the standard. It is a great advantage to have a common opinion from a large number of experts. But we also have to be aware of the weak points of such an approach.

The lists of measures, called clauses, give an idea of how effective several measures are in contributing to the safety of the system. This measure effectiveness is not a result of exact observations on applied measures in real systems. Quantified contributions to the safety of a system are not known at all. One main point in creating the measure list was to highly recommend (HR) the best known measures for systems with the highest Integrity Level. It cannot be excluded that some of these measures are overestimated or underestimated. [Fen 93] states: *"Many of these standards prescribe, recommend, or mandate the use of various technologies which have not themselves been validated objectively"*. The following quotation from IEC 65A(Secretariat) 122 can be considered the standards self assessment. In Section 18 "Guidance on Deriving Application Specific Standards" paragraph 18.2.5 reads: *"This International Standard requires the use of a package of techniques and their correct application. These techniques are required from the tables and detailed in the bibliography. However, the techniques and measures recommended by this International Standard do not guarantee that the required integrity level will be achieved."*

As a result of our considerations, the following two general gaps have been identified:

**G1:** The descriptions of measures proposed in the standard are not detailed enough to be applied properly with the appropriate effectiveness.

**G2:** The set of measures recommended by the standard for a particular Integrity Level cannot be considered to be compulsory such that this set of measures really meets the requirements of the particular Integrity Level.

How can these gaps be bridged? A general answer is: the proposed measures have to be described in detail, and the dependability of the standards has to be improved by further explorations. To find out whether such a general approach is feasible we have to discuss the state of standardisation in safety critical computer based systems and the legal position of standards in the assessment and certification prosess.

# 5 The State of Standardisation in Safety Critical Computer Based Systems

The main objective of CASCADE is to improve the assessment of safety critical systems. One of the most important ways for achieving this objective is to improve the situation in standardisation. We have to look for gaps in the standards and try to bridge them. In doing so, a new generic question arose: What is a real improvement of a standard?

We are most interested in the international standards IEC 65A(Secretariart) 122 and 123 and the CENELEC standards prEN 50126, prEN 50128, prEN 50129. From the 5 standards quoted above, which are highly relevant for our scope of work, two are still Committee Drafts and three are "intended to become a Prestandard".

An "introductory note" on IEC 65A(Secretariat) 123 contains the information that the "meeting held in Chapel Hill in July 1990" took a decision concerning the scope of the standard. The start of the standardisation work on the standards in question may lay long before 1990. We can assume that standardisation work has been going on for 5 years, and it will be another 2 years before the standards IEC 65A(Secretariat) 122, 123 become valid. That means that a total of 7 years will be required to create the valid standard. In the meantime the following note on the front page of the standard has to be observed:

*"This document is still under study and subject to change. It should not be used for reference purposes".*

Therefore it is probable that an unbiased reader comes to the conclusion that the current situation in international standardisation of development and assessment of safety critical systems is poor. The situation in national standardisation is only slightly better. An expert who is familiar with the history of the last 15 years will come to the contrary conclusion: the current situation in standardisation of development and assessment of safety critical systems is of the highest quality since 15 years. The first standard (at least in Germany) which really deals with the safety of computer hardware and software did not exist until 1990. Independent of the existence of such detailed standards computers were applied in safety critical systems and assessed already 15 years ago.

From that and from a more detailed analysis we can give the following estimations:

- The delay time between a technology and its standardisation is about 15 years.
- The meantime between the issue of a standard and its first major change is about 5 years, or at least this is the time after which a standard requires major changes.

The standardisation work on a single standard is in danger of becoming an ongoing process. This provokes the question: how long should we plan for a standard to be valid, for a very limited duration or for eternity? The current versions of standards IEC 65A(Secretariat) 122 and 123 have in total 231 pages. The current versions of the CENELEC standards prEN 50126, prEN 50128, prEN 50129, which also form a unit, have in total 348 pages. Beside the large number of pages, the complexity of these standards is very high.

We consider a 5 years life span as the minimum for such voluminous standards of high complexity. Shorter intervals of major changes would remarkably decrease the acceptance of such standards. The more details a standard contains, the shorter its life span is. The more details a standard should contain, the longer the development time, and the greater the delay between the technology and the appropriate standard will be.

We propose: these standards should become more generic and less complex. These objectives would allow a guaranteed minimum life span of at least 5 years and a reduction in the delay from about 15 years to about 7 years.

A further conclusion drawn from our assessment work is: *standards are not always up to date*. We only expect a standard to be correct and up to date on the date of its issue. In reality even this is much too optimistic. Even a new standard will contain some inconsistencies, and the content of the standard, on its issue date, is a couple of years old.

So we have to sum up: standards are not always up to date, not detailed enough, and not allways correct.

The objectives to make the standards more generic and more topical are in contradiction to the objectives formulated in chapter 4 for bridging the two general gaps.

# 6 Standards Versus State of the Art in Safety Technology

A very tempting idea is the idea of "conformity assessment" of a safety critical system. Such a conformity assessment could be, if possible, much simpler and much cheaper. A good standard is a great help for designers and assessors in the field of safety critical systems. But the peculiarity of a standard being:

> not always up to date;
> not detailed enough;
> not always correct;

does not justify a simple conformity assessment in the area of safety critical systems. The assessment has to take into consideration all information about the state of the art in safety technology. This is the assessment practice but where is the legal basis for this approach?

As a rule legal safety requirements are not related to special standards. According to [Marb 93] safety requirements are related (at least in Germany) to the following terms:

- "Generally Accepted Rules of Technology"
- "State of the Art in Technology"
- "State of the Art in Science and Technology"

Initially the supposition is valid that the standards reflect the state of the art in technology. In most cases the state of the art in technology is only roughly described in standards. It is fixed more precisely by the current literature, but above all by the certified systems in the same application field, and by the certified systems of other application fields with comparable safety requirements.

Standards are not binding on the producer since he may have achieved the safety of the system by another method and convinced the assessor of the safety of his system. However, if the producer follows the recommendations of the standard, then the safety proof becomes easier.

This is the reason why safety relevant systems could be assessed and certified before appropriate standards existed. A prerequisite for assessment and certification of a safety relevant system is the comprehensive knowledge of the state of the art in safety technology. It is the permanent task for the assessor to identify this state of the art. Standard should be updated as quickly as sensible to reflect this state of the art as good as possible. We have to be aware of the fact that standards are following the state of the art in technology and not vice versa.

# 7 How to Bridge Gaps in the Standards

The incomplete descriptions of the measures is one general gap in the standards. This gap cannot be bridged completely in CASCADE, because the standards recommend many hundreds of measures. To describe all of them would require a huge effort, and a deep knowledge of every single measure, and its effectiveness would be necessary. This deep knowledge exists only in special cases.

To limit the effort to a manageable size, the CASCADE team decided to deal with this problem in the following way. Few examples of so called "well tried safety cases" will be described. Only the safety measures contained in these safety cases will be described in detail. A "well tried safety case" is of course only an outline of a real safety case. In the following the benefits of a collection of examples of such well tried safety cases for the development and assessment of safety critical systems will be discussed.

## Examples of Safety Cases are good Illustrations of the Standards

Examples give good illustrations of the standards. The standard DIN V VDE 0801 contains a small number of examples but they cover only the lower Integrity Levels. There are no examples for the Integrity Level IL4, which means the examples in DIN V VDE 0801 are not relevant for railway protection systems. The examples given in CASCADE deal with such systems.

## Detailed Description of the Measures

Although an example of a well tried safety case is only an outline of a real safety case it represents in comparison with the descriptions of the standards a very detailed description for this particular realisation of a safety relevant system.

### High Commitment of the Examples

The high commitment of an example of a "well tried safety case" concerns the attribute "well tried". The examples of "well tried safety cases" are generalisations of one system or a few original systems that have been assessed successfully. To get an idea of the commitment of the standard statements in the field of safety critical computer based systems we want to compare the standardisation process with the assessment process.

Assessing the software of one track guided transport system takes many man years. The standardisation body who produced one of the standards (e.g. IEC 65A(Secretariat) 122) had to create a system which contained the essentials of numerous safety cases in all application fields. To express it in a graphic way, the standardisation body had to assess at least the structure of all safety cases that will be produced in the future, in the whole world, on the basis of today's technology. The question arises as to how a standardisation body could perform such a huge task in a short time? The answer is: the structures described by the standards have to be considered very generic examples of safety cases and are still very flexible. Important items that have a great influence on the safety of the system (e.g. measures) are only described vaguely. By specifying these items in a proper way, real safety systems can be created.

The "well tried safety cases" are of different quality. These are safety cases of systems that are similar to real systems that have been assessed successfully by one assessor. The "well tried safety case" presented by one assessor would be assessed again successfully by him in a very same way. During the progress of CASCADE these examples will be reviewed thoroughly by the remaining assessors, so that finally it can be said that these examples of the well tried safety cases are accepted by the three assessors involved in CASCADE. A designer who would construct his system in compliance with one of these examples could be sure that his system would be assessed successfully at least by the three assessing companies involved in CASCADE.

## Conclusions

A representative collection of well tried safety cases would be a valuable complement of the standard.

Such a collection of examples placed in the annex to the standard could be completed every two years without any changes of the actual obligatory standard.

This collection of well tried safety cases would describe a great number of safety measures in sufficient detail.

The set of safety measures in a particular well tried safety case is of higher commitment than the set of safety measures recommended by the standard.

The above mentioned benefits of the collection of examples would allow to make the standards more generic. This would reduce the size, complexity, and the development time of a standard remarkably. As the examples are independent from each other, the complexity of the annex would be low.

The delay between the technology and its standardisation would be decreased and the lifespan of a standard increased drastically.

# Reference List

[IEC 65/122]   IEC 65A(Secretariat) 122, November 1991. Software for computers in the application of industrial safety-related systems. Draft under study and subject to change.

[IEC 65/123]    IEC 65A(Secretariat) 123, May 1992. Functional safety of electrical/electronic/programmable systems: Generic Aspects, Part 1: General requirements. Draft under study and subject to change.

[EN 50126]    CENELEC prEN 50126, Version 00.06, June 1993. The Specification and Demonstration of Reliabilty, Availability, Maintanability and Safety (RAMS) for Railway Application, Part 0: Dependability.

[EN 50128]    CENELEC prEN 50128, February 1994. Railway Applications: Software for Railway Control and Protection Systems. Draft intended to become a prenorm.

[EN 50129]    CENELEC prEN 50129, August 1993. Railway Applications: Safety Related Electronic Railway Control and Protection Systems. Draft intended to become a prenorm.

[DIN 19250]    DIN V 19250, Mai 1994. Grundlegende Sicherheitsbetrachtungen für MSR-Schutzeinrichtungen.

[VDE 0801]    DIN V VDE 0801, Januar 1990. Grundsätze für Rechner in Systemen mit Sicherheitsaufgaben.
DIN V VDE 0801, January 1990. Principles for Computers in Safety-Related Systems 2nd Proof Copy of English Translation (translated by DKE October 1991).

[VDE 0801A]    DIN V VDE 0801 Änderung A1, April 1994.

[Marb 93]    Marburger, P.: Die rechtliche Bedeutung technischer Normen im Umwelt- und Technikrecht. Vortragsmanuskript, Fachtagung am 02.02.1993. Veranstalter: Bundesverband der Deutschen Gas- und Wasserwirtschaft e. V. BGW & EG-Generaldirektion XII.

[Fen 93]    Norman E. Fenton. How effective are software engineering methods? Journal of Systems & Software, 20:93-100, 1993.

[LITTL 90]    Littlewood, B.: Limits to Evaluation of Software Dependability. Predictably Dependable Computing Systems. ESPRIT project 3092, 1990.

# Session 2
## Safety Analysis

# Safety Analysis for Requirements Specifications: Methods and Techniques

Amer Saeed, Rogério de Lemos and Tom Anderson
Department of Computing Science
University of Newcastle upon Tyne, NE1 7RU, UK

## Abstract

As the complexity of modern safety−critical systems rises it is becoming increasingly evident that there is an intricate coupling between the different domains that make up such systems. The effective management of the interactions between the entities of these domains is essential to obtain high−levels of safety. In this paper, we present a method for conducting safety analysis of requirements specifications that accommodates different domains. The method is based upon providing a common formal basis for the requirements and safety analyses, and an approach for explicitly identifying the potential causal relationships between these domains.

## 1. Introduction

As the complexity of modern safety−critical systems rises it is becoming increasingly evident that there is an intricate coupling between the different domains (i.e. major components) that are part of such systems. These domains include computing systems, physical plants and human operators; providing a way of managing the interactions between entities of these domains is particularly significant for the realisation of high−levels of safety. Several recent incidents have served to illustrate that inadequate consideration of interactions between domains is a significant factor in the occurrence of accidents.

Safety critical systems must be subjected to a safety analysis before deployment, to ensure that the risk associated with the system is acceptable. Traditionally, this analysis is conducted using a number of techniques, each being appropriate for a specific domain and a particular facet of the analysis. However, the increased coupling of modern systems means that local safety analysis of the individual domains must be reinforced by a global safety analysis that is appropriate for the interactions between them. In modern systems, it can be observed that failure behaviours within one domain can impact another domain, and conversely recovery actions (engineered safety features) to cope with the failures of one domain can be effectively defined in another domain.

Too often, in current practice, inter−relationships between the domains are not fully considered until the later stages of system development, during these later stages the absence of a common modelling abstraction can complicate the safety analysis. The focus of the work presented here is to enable an integrated safety analysis to be conducted during the requirements phase of development. This phase is particularly suitable for conducting an effective safety analysis, since the models obtained for each domain are at a sufficiently high–level of abstraction that it is feasible to use a common modelling abstraction. From a pragmatic perspective shifting the emphasis placed upon safety analysis from the later phases to the earlier phases of software development aims to significantly reduce the number of iterations (by increasing the assurance in the results of requirements analysis before proceeding to subsequent phases of development) through the phases of development and the amount of rework involved in each iteration. From the above discussion we can enumerate two basic criteria for an effective approach.

- *Integrated*. Integration should be achieved between the requirements analysis and safety analysis, and between the different domains. In particular, the results should provide guidance on the development of risk reduction strategies and evidence that is suitable for a system safety case.

- *Comprehensive*. The safety analysis should be comprehensive in the sense that examination of all aspects of a domain that are encoded in the associated models and specifications, is supported.

In previous work we have focused on the provision of techniques to support requirements analysis [Saeed 94, de Lemos 95a, de Lemos 95b], and on the integration of activities involved in requirements analysis and safety analysis [de Lemos 94]. This constitutes only a partial foundation for integration; in practice the relationships between the activities of requirements analysis and safety analysis must be supplemented by a consideration of the products of these activities. In the work presented here, we address the following issues: the provision of a *common formal basis* for requirements techniques and safety analysis techniques, in terms of an event/action model; *causal relations* between the entities of a domain; and *effective methods* for safety analysis.

The rest of this paper is organized as follows. In section 2, we discuss the basic aspects of our approach. In section 3, we describe our approach to safety analysis. Section 4 presents a simplified aerospace system that will be used to illustrate our method for safety analysis. Section 5 discusses the relationship between requirements analysis and safety analysis. In section 6, a method for safety analysis is outlined and applied to a simple system. Finally, section 7 presents some concluding remarks.

## 2. Context of Approach

In this section we provide an overview of the basic aspects of our approach to the analysis of safety requirements; these are: *process model*, *information model*, and *modelling abstractions and concepts*.

## 2.1. Process Model

The overall approach to the analysis of safety requirements is divided into phases, each of which matches a domain. The analysis conducted in each phase is based upon two interrelated activities: requirements analysis, which produces safety specifications, and safety analysis, which assesses whether or not the risk associated with a safety specification is acceptable. The aim is to conduct the requirements analysis and the safety analysis in tandem, within each phase of analysis. The safety specifications obtained from the requirements analysis of the domain will feed into the safety analysis, and any defects identified by the safety analysis will be used to guide the modification of the safety specifications before proceeding to the next phase.

## 2.2. Information Model

The notion of a *Safety Specification Graph* (SSG) [Saeed 93] was proposed as the information model for the results of the requirements and safety analyses. The structure provided by the SSG supports the selection of information to be recorded, and records the information in a format that facilitates further analysis. An SSG is represented as a linear graph, in which a *node* represents a safety specification and an *edge* denotes a relationship between a pair of safety specifications.

## 2.3. Modelling Abstractions and Concepts

Modelling the overall system (and its environment) into which the computing system will be embedded is considered to be an integral part of requirements analysis, because such models provide the context in which requirements are expressed and examined. The activity of modelling is concerned with the description of the structure and behaviour of a domain.

### 2.3.1 Modelling Abstraction

The principal modelling abstraction used in the proposed approach is the *interactor* [Duke 93], an object−based concept that supports representation of both structural and behavioural aspects of a system. An interactor corresponds to a class in object−based terminology and is described by a template with the following fields: a *name*, a collection of *components*, declarations of *constants* and *variables*, and a *behaviour specification*. In order to specify the behavioural composition between a group of instances of interactors we employ a mechanism similar to a *contract* [Helm 90]. By analogy with the format of interactor descriptions, contracts are specified by templates. To depict the structural relationships between interactors we adopt the graphical notation of OMT [Rumbaugh 91].

## 2.3.2 Modelling Concepts

An event/action model (E/A model) [de Lemos 95a] is used to describe the behaviour of an interactor. The E/A model is based on the following primitive concepts. A *state* of a system is that information which, together with the system input, determines the subsequent behaviour of the system. A *transition* represents a transformation in the system state. The system state is modified by the occurrence of events and the execution of actions. An *event* is a temporal marker of no duration which causes or marks a transition. An *action* is the basic unit of activity, which implies duration. The primitive concepts of the E/A model can be used to reason about system behaviour in both the value domain and time domain.

The behaviour of an entity is specified in terms of its *standard, failure* and *exceptional* behaviours. By specifying the failure behaviour of the entities we are also able to establish how the behaviour of the domain can be made more robust in the presence of failures of its entities. The specification of the exceptional behaviour of an entity takes into account those situations when it becomes impossible for the entity to deliver the specified standard behaviour, but can signal an exception.

# 3. Safety Analysis

The purpose of the safety analysis of the safety specifications is to increase the assurance that the contribution of a specification to the overall system risk is acceptable. The results of a safety analysis are used during the requirements phase for *risk reduction* and subsequently for *certification* of the system. Our focus here is on qualitative analysis; approaches for quantitative analysis have been discussed elsewhere [Saeed 93].

The qualitative analysis is partitioned into two activities: *preliminary analysis* and *vulnerability analysis*. The preliminary analysis activity must ensure, under clearly defined circumstances, that the safety specifications are adequate for maintaining the absence of hazards. Preliminary analysis aims to increase assurance in the safety specifications by *verification* and *validation*. A detailed example of a preliminary analysis has been presented in previous work [Saeed 94].

The vulnerability analysis is similar in intent to that of a traditional safety analysis, its activity must uncover any defects in the safety specifications and determine their impact on the safety of the system. Vulnerability analysis probes the safety specifications to find circumstances which can affect the safe behaviour of the system; once such circumstances are identified the safety specifications can be modified to reduce their contribution to the system risk. The objective is to develop and propose safety analysis techniques, applicable to the different domains of interest, that capture the intentions of traditional safety analysis techniques.

# 4. Example: An Aircraft Fuel Management System

In this section, we present a case study based upon a Fuel Management System (FMS) for an aircraft to illustrate our approach to safety analysis [de Lemos 95b]. The role of the FMS is to supply fuel to the engines, and to balance the fuel evenly in the tanks of the aircraft; an imbalance of fuel is considered to be the primary cause for the displacing the aircraft centre of gravity (a system hazard). The overall system is the aircraft and modelled by the interactor *Aircraft*, whose structure is represented by the aggregation hierarchy of figure 1.

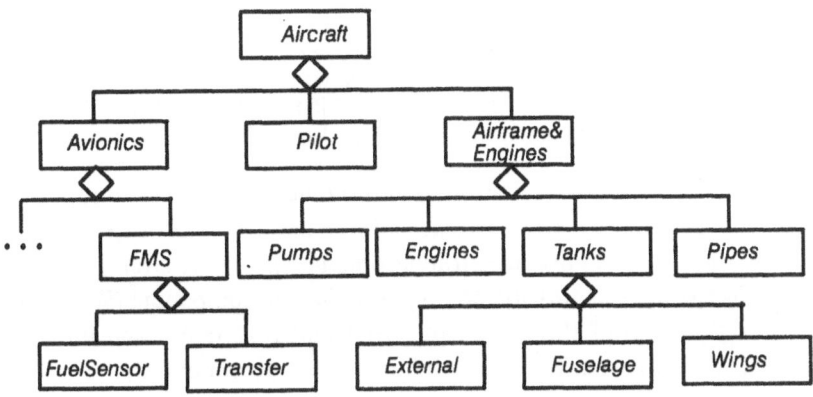

Figure 1. Structure of *Aircraft*.

The failure behaviour of the *Aircraft* that constitutes a *system hazard* is that the centre of gravity deviates from a safe range in the pitch axis. The standard behaviour of the *Airframe&Engines* is expressed as the following *safety constraint*: the mass of the airframe and engines are evenly distributed. The standard behaviour of the *Tanks* is expressed in terms of the following *safety constraint*: the mass of fuel is balanced across the tanks of the aircraft. It is assumed that if the relative masses of the fuel in the front and rear tanks of the fuselage are kept within safe bounds, then the centre of gravity along the pitch axis will remain within a safe range. The generic behaviour of a fuselage tank will be described in the interactor *Tank*, the three variables of interest at the level of *Fuselage* are: *mass* which represents the mass of an instance of the tank (e.g. *front*) the tank, *flow* which represents the flow rate from an instance of the tank (e.g. *front*) to the other fuselage tank (*rear*).

The *safety strategy* that satisfies the safety constraint for *Tanks* has to maintain the mass of fuel balanced between the two tanks of the fuselage, respectively, the *front* and *rear* instances of the interactor *Tank*, by performing the necessary transfer of fuel between the two tanks. This is captured in the definition of the Interactor *Fuselage*, which incorporates two actions *TransferFront* and

*TransferRear* which transfer fuel between the *front* and *rear* tanks and are defined over the variable *flow*.

---

**Interactor** *Fuselage {Fuselage Tanks}*

**Composed – of**

*Tank:*          ⟨*front, rear*⟩

**Behaviour**

**Standard**: There are no fuel losses while transferring fuel between tanks.
*front.flow* = −*rear.flow*.
The safety constraint on the mass distribution is bounded by *safeThresh*.
|*front.mass* − *rear.mass*| < *safeThresh*.                                    [SC]

Within at most *DT* units of time after the relative masses have exceeded
a predefined threshold (thresh), a transfer action must be started.
(*front.mass* − *rear.mass* > *Thresh*)@*t*⇒∃*t'* ∈ [*t, t+DT*]:↑*TransferFront* @*t'* ∧
(*rear.mass* − *front.mass* > *Thresh*)@*t*⇒∃*t'* ∈ [*t, t+DT*]:↑*TransferRear* @*t'*     [SS]

---

In the definition of *SS* the notation |(<*predicate*>)@*t* is used to represent the event that <*predicate*> becomes true at time t, and ↑<*action*>@*t'* that the start event of the <*action*> becomes true at time *t'*. The safety strategy *SS* is to be realised by the interactor *Avionics*, in particular the component *FMS*. The *FMS* measures the mass of the fuel in the front and rear tanks, and initiates fuel transfer in accordance with *SS*. The following interactor definition presumes the existence of an interactor *Sensors* appropriate for measuring the masses in the tanks (represented by the variable of *mass*) and an interactor *Transfer* that contains two actions that will enable the transfer of fuel from front or rear tanks (represented by *Tfront* and *Trear*).

---

**Interactor** *FMS {Fuel Management System}*

**Composed – of**

*FuelSensor:*          ⟨*frontSen, rearSen*⟩

*Transfer:*            ⟨*transfer*⟩

**Behaviour**

**Standard**:    Detects when difference in the masses exceeds *Sen* and initiates
transfer of the fuel in the appropriate direction.

(*frontSen.mass* − *rearSen.mass* > *Sen*)@*t*⇒∃*t'* ∈ [*t, t+DT*]:*transfer.*↑*Tfront*@*t'* ∧
(*rearSen.mass* − *frontSen.mass* > *Sen*) @*t*⇒ ∃*t'* ∈ [*t, t+DT*]:*transfer.*↑*Trear*@*t'*.
[CSS]

---

The relationship between the behaviour of the instances of *FuelSensor, Transfer*, and *Tanks* are captured in the contract *FMSInterface*.

---

**Contract** *FMSInterface*

**Participants:**

| | |
|---|---|
| *front, rear*: | Tank |
| *frontSen, rearSen*: | FuelSens |
| *transfer*: | Transfer |
| *fuselage*: | Fuselage |

**Invariants**

The accuracy of the fuel sensors is bounded by *sensThresh*.
$|frontSen.mass- front.mass| < sensThresh \land$
$|rearSen.mass- rear.mass| < sensThresh$

When the transfer mechanism is executing the action *front* or *rear*, fuel is transferred in the appropriate direction.

$(transfer.Tfront \Rightarrow TransferFront) \land (transfer.Trear \Rightarrow Transfer Rear).$

# 5. Relating Requirements Analysis and Safety Analysis

In this section, we examine the relationships between the techniques employed for requirements analysis and the principles that underpin the activity of safety analysis, in order to guide the integration of these two activities. Firstly, the issue of providing a common formal basis is discussed. Secondly, a model supporting the examination of causal relations between interactors is presented.

## 5.1. Formal Support for Requirements and Safety Analyses

To promote the effective integration of safety analysis and requirements analysis, a common formal basis should be provided for the results of these techniques. With regard to formal models for safety analysis techniques, it is worth emphasising that the concept of an *event* underpins safety analysis; in essence the purpose of safety analysis is to examine the relationship between the *(failure) events* of components and hazards in the system. Hence recent work on the formalization of safety analysis techniques has focused on the development of formal models for events and their inter−relationships, most notably the work on Common Safety Description Model (CSDM) [Gorski 95].

Despite the different motivations for the E/A model (specification of real−time systems) and the development of the CSDM (formalization of safety analysis techniques), the two models are based upon a common set of primitive modelling concepts and relations. In fact, the two models can be considered as alternatives for a formal basis for both the behaviour description of interactors and the formalization of safety analysis techniques, here we adopt the E/A model.

## 5.2. Causal Relationships over Interactors

Central to safety analysis is the examination of causal relationships between the events in an accident sequence. The notion of causality has also been studied as a basis for the expression and analysis of requirements. In an approach proposed by Coombes [Coombes 94], the behaviour of a domain is expressed in terms of explicit causal relationships between its attributes (i.e. variables). In the FOREST approach [Goldsack 91] the causal relationships can be inferred from the actions that the entities of a domain are "permitted" or "obliged" to perform. Our focus on causal relationships is not as the basis of expressing requirements, but rather as a means for projecting a view of a domain that will support subsequent vulnerability analysis.

We propose a graphical notation for making explicit the causal relationship between the interactors of a domain. This will be achieved by introducing the notion of an *impacts relation*. A diagram depicting the impacts relations between the interactors of a domain, will be referred to as its *impact structure*.

### 5.2.1. Impacts Relation

The basic causal relation between two interactors will be defined as the binary relation *impacts*. Informally, the impacts relation holds between two interactors $A$ and $B$ if a change in a variable of $A$ could cause a change in a variable of $B$. The impacts relation is *transitive* (i.e. if A *impacts* B and B *impacts* C then A *impacts* C), and all interactors that can be influenced by an interactor are included in the transitive closure of the *impacts* relation, denoted by $impacts^+$.

### 5.2.2. Derivation of the Impact Structure

The impacts relation for a domain can be derived from the aggregation hierarchy (representing the "part−of" structure of the domain) and the contracts (representing relationships between the behaviour of the interactors). To guide the derivation of an impact structure, the following heuristics are proposed.

- For each each component interactor (i.e. an interactor which is a component of the aggregate) determine, by examining the description of its behaviour, whether it can directly influence its enclosing interactor. If the component interactor has a direct influence an appropriate impacts relation is added.

- For each component interactor determine, by examining its behaviour description and any contracts involving the interactor, whether it can influence other component interactors. If the interactor under analysis can influence another interactor an appropriate impacts relation is added.

A couple of observations can be made on the structure that results from applying the above heuristics. Firstly, an aggregate interactor should normally be in the $impacts^+$ of all its components, other wise the component has no effect on the variables of interest for the aggregate interactor. Secondly, the impact structure for a particular domain may be modified in order to reflect relationships that only become visible in subsequent decompositions of the interactors of that domain.

### 5.2.3. Example Impact Structure Derivation

The impact structure for the interactor *Aircraft*, is depicted in figure 2. The impacts relations are distinguished by either being directly derivable from the information in the specifications (fixed line), or derived from domain knowledge (broken line). As an example the derivation of the relationships involving *Avionics* and its components is discussed. It should be noted, that the depicted structure is a working model, further relationships will be uncovered as a result of subsequent analysis.

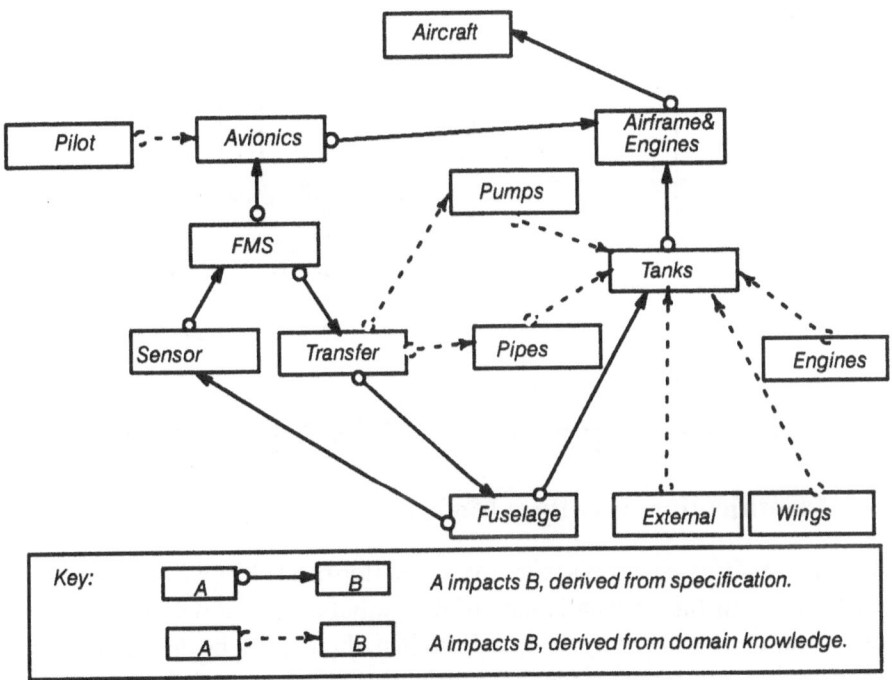

Figure 2. Impact Structure for *Aircraft*.

*Avionics* is a component of *Aircraft*, although it is not directly related to its aggregate object it can have an impact via the *Airframe & Engines*. There is no direct relationship between *Pilot* and *Aircraft*, but there is an indirect relation via the *Avionics*. *FMS* has a direct influence on *Avionics*. From the behavioural descriptions of the *FMS*, *FuelSensor* has a direct influence on *FMS* and *Transfer* is influenced by *FMS*. The relationships between *FuelSensor*, *Transfer* and *Fuselage* are derived from the contract *FMSInterface*.

# 6. A Method for Vulnerability Analysis

In previous work [de Lemos 94] a high—level process model was established for the vulnerability analysis of safety specifications; this model is depicted by the SADT diagram in figure 3. The basic activities of the process model are similar to those followed during a traditional HAZOPS and those proposed for software HAZOPS [Mod 95]. In this section, we decompose the basic activities in the context of the links established in section 5.

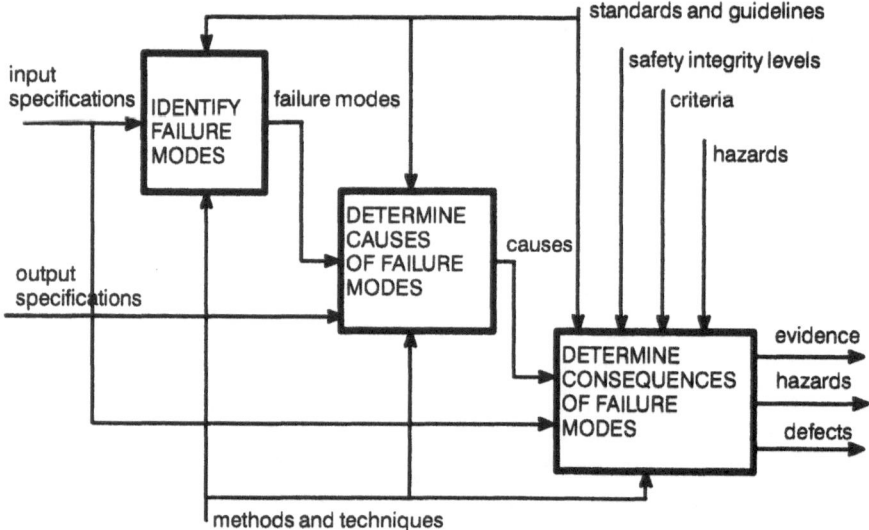

Figure 3. Relationships between the activities of the vulnerability analysis

## 6.1. Identification of Failure Modes

The nature of this activity is similar to the first step of a HAZOPS, and involves an examination of the "design intent" of the output specifications (i.e. that is the behaviour to be realised by the specifications), this is given by the related output specifications. in order to identify possible deviations (i.e. failure modes). It should be recognised that the failure modes identified during this activity are not definitive — rather, on completion of this activity, a set of *anticipated* behaviours are defined that are to be validated by the subsequent activities.

An approach to the systematic derivation of possible failure behaviours is proposed, based upon the notion of guide words that can be used to direct an examination of the design intent. Extending the notion of guide words to cover software specifications has been suggested by several researchers [Chudleigh 93, McDermid 94], and recently an IDS 00—58 [MoD 95] has been proposed on HAZOPS. However, the suggested methods have focused on the design phase, whereas here we suggest an approach suitable for the requirements phase. There are two key themes to the method: firstly a distinction between guide words applicable to the *value domain* and *time domain* derived from the error classification of [Powell 92]; secondly, the interpretation of the guide words in the context of an event/action model. In future work we intend to formalize the guide words in terms of operations that can be applied to the formulae of the event/action model. This formalization will support the analysis of simple deviations, and introduce the possibility of considering combinations of guide words. The following table summarises the interpretation of the guide words with respect to the primitive concept of an action.

| Guide Word | Domain | Interpretation |
|---|---|---|
| Arbitrary | Value | Action does not provide desired changes to variables. |
| Detectable | Value | Action does not provide desired changes to variables, but interactor can detect error. |
| Late | Time | Start event or finish event occur too late. |
| Early | Time | Start event or finish event occur too early. |
| Infinitely Late | Time | Start event or finish event never occur. |

## 6.2. Determination of Causes

This activity involves a deductive analysis of the failure behaviours in order to determine those circumstances that can lead to the postulated failure modes. These circumstances will be a combination of violations in assumptions that are part of the output specifications and the effect of failure modes of the interactors over which the specification is expressed.

The deductive analysis to be conducted during this activity can be guided by a reverse traversal of the impact structure; the analysis would originate at the node that represents the interactors for which the postulated failure mode is defined. The basic steps are similar to those performed during fault tree analysis and are defined below:

- Identify all the interactors causally related to the interactors and contracts in the output specification, this can be obtained from the transitive closures of the *"impacts"* relation of the interactors that constitute the output specification and the participants entry for the contracts.

- Examine each interactor and contract identified in the previous step determine if a behaviour of that interactor (or contract) can lead to the postulated failure behaviour in the absence of other failure behaviours. If so, the interactor is marked as a primary cause of the failure behaviour.

- Examine combinations of the behaviours of related interactors and contracts. Each combination that can lead to the postulated failure behaviour, are marked as causes and the interrelationships recorded for further analysis. The analysis is complete when all combinations have been examined.

- Analyse the potential causes of the failure behaviour to determine if it is credible, and hence worthy of further investigation. If it is not credible, then evidence is provided for a safety case.

The completeness of this activity at a particular level of abstraction is predicated on the knowledge of the failure modes of the output specifications, which will be updated by the vulnerability analysis at the next level of abstraction. The influence of the vulnerability analysis of lower levels of abstraction will provide feedback, when the consequences of the failure modes are considered. Furthermore, the completeness of the behavioural specification of an interactor can be checked by confirming that all combinations of standard, failure and exception behaviours of its components can be mapped back to the interactor.

## 6.3. Determination of Consequences

This activity involves an inductive analysis of the credible failure behaviours to determine their consequences. The examination during this activity will be dominated by an inductive analysis, guided by a forward traversal of the *impact structure*. The analysis would start at the nodes representing the interactors that constitute the output specification, and examine consequences of combinations of failures of these interactors on causally related interactors.

The basic steps are similar to those performed during an event tree analysis and are defined below:

- Identify all the interactors that have a direct causal relation to the interactors over which the failure mode is defined; this can be obtained from the *impact structure*.

- Examine the impact identified in the previous step to determine the consequences of the failure on safe behaviour.

- If the consequences are considered to hazardous a *defect* is identified and some recommendations on risk reduction strategies should be provided. Guidance can be obtained by conducting a reverse analysis of the impact structure starting from the node that represents the interactor over which the hazard is defined; the general policy is to segregate (as much as is feasible) the risk reduction strategy from the causes of the failure behaviour. If the consequences are not considered hazardous, then evidence is provided for the safety case.

## 6.4. Example of Vulnerability Analysis

The vulnerability analysis of the safety strategy *CSS* of the interactor *FMS* is presented below. The "design intent" of *CSS* is to maintain the safety strategy *SS* of the interactor *Fuselage*. It is presumed that a preliminary analysis has been conducted of *CSS* which has confirmed that the *CSS* is indeed a refinement of *SS* under the circumstances defined in *FMSInterface*. This example illustrates how a failure in one domain (the *Avionics*) can lead to a hazard in another domain (the *Airframe&Engines*), and the development of an engineered safety feature to mitigate against the failure can involve another domain (the *Pilot*).

### 6.4.1 Identification of Failure Behaviours

To illustrate the identification of a postulated failure behaviour the guide word *Late* is applied to the safety strategy *SS*. The terms over which *Late* will be applied are the actions *TransferFront* and *TransferRear*, the postulated failure behaviour is that the start event of either action does not occur within its stipulated interval. This postulated failure behaviour is characterized by the following formulae:

$$| \, (front.mass - rear.mass > Thresh)@t \Rightarrow \forall t' \in [t, t+DT]: \neg \uparrow TransferFront @t' \vee$$
$$| \, (rear.mass - front.mass > Thresh)@t \Rightarrow \forall t' \in [t, t+DT]: \neg \uparrow TransferRear@t'.$$

### 6.4.2 Identification of Causes

The causally related interactors and contracts are: *FMS*, *FuelSensor*, *Transfer* and *FMSInterface*. A failure of any of these components can lead to the identified failure behaviour. For example, if the variable *mass* of the *FuelSensor* does not behave in accordance with the contract *FMSInterface* then the postulated failure behaviour can occur even if all the other components are providing their proper service. Following the above examination, the failure mode is considered to be credible.

### 6.4.2 Identification of Consequences

The causally related interactors and contracts are: *Tanks*, *Airframe&Engines* and *Aircraft*. The effect on the *Tanks* is to cause an imbalance in fuel which can lead to an imbalance in mass across the pitch axis in the *Airframe&Engines*, and eventually a deviation in the centre of gravity out of the safe range in the *Aircraft* leading to a system hazard.

Since a consequence of the failure behaviour is hazardous, a defect has been identified and some recommendations on how it should be rectified are needed. Examining the causal chains back from *Aircraft*, potential engineered safety features can be developed involving the *Pilot*, *Avionics* or *Airframe & Engines*. Of these options, it is decided to develop a scheme based on allocating to the *Pilot* the responsibility of transferring the fuel after the *FMS* has failed to perform the transfer.

Firstly the behaviour of *Avionics*, is extended to include the *exceptional* behaviour: "the avionics cannot evenly distribute the fuel and signals an alarm to the pilot". The notion of a warning system is introduced to inform the pilot when a transfer of fuel must be initiated. The behaviour of the interactor

*WarningSystem* is then a refinement of the exceptional behaviour of the *Avionics*. The following definition presumes the existence of a centre of gravity sensor, and a mechanism for signalling an alarm to the pilot.

**Interactor** *Warning System*

**Behaviour**

**Standard**: Raises a warning to signal excessive centre of gravity displacement.

$cogSensor \in alarmRange \Leftrightarrow cogAlarm.$

**Failure**:    Fails to raise a warning.

$cogSensor \in alarmRange \land \neg cogAlarm.$

To complete the analysis of this complementary scheme, the procedures that must be followed by the pilot after the warning system has raised the centre of gravity alarm have to be defined. These procedures would also be subjected to a safety analysis; if this subsequent safety analysis confirmed that the warning system provided an adequate complementary strategy then this would be used as evidence in a safety case. The introduction of the interactor *WarningSystem* modifies the impact structure of the *Aircraft*.

# 7. Concluding Remarks

This paper has presented a method for the safety analysis of requirements specifications, in the the context of an overall approach to the analysis of safety requirements. This work extends previous work on establishing links between the process model for requirements analysis and safety analysis. Emphasis is placed upon providing a common formal basis for the requirements and safety analysis based upon an event/action model, and the exposition of the causal relations between the interactors.

A method for conducting a vulnerability analysis over the requirements specifications and domain models represented by interactors was presented. An example was used to illustrate how the method can be used to examine the interactions between the failure behaviours of three different domains of analysis. In future work we will consider the support provided by exploiting the formal basis provided by the E/A model, in particular the formalization of the guide words (i.e. to give a formal semantics to the guide words enable a more precise generation of possible deviations. Regarding the impact structure, we intend to investigate annotations that can be added to the impact structure to enrich the information on causal relations that is encoded, particularly the role of contracts. The relationship of the proposed method to current standards, such as IDS 00−58 [MoD 95] is currently under investigation.

# Acknowledgements

The authors would like to acknowledge the financial support of British Aerospace (Dependable Computing Systems Centre), the EPSRC (UK) SCHEMA project, and the CEC Copernicus Joint Research Action ISAT.

# References

[Chudleigh 93] M. Chudleigh. "Hazard Analysis Using HAZOP: A Case Study". *Proceedings of SAFECOMP'93*. Springer–Verlag. J. Górski (Ed.). Poznan–Kiekrz, Poland. October 1993. pp. 99–108.

[Coombes 94] A. Coombes, J. A. McDermid, P. Morris. "Causality as a Means for the Expression of Analysis of Requirements for Safety Critical Systems". *Proceedings of the Ninth Annual Conference on Computer Assurance (COMPASS '94)*. Gaithersburg, MD. July 1994. pp. 223–231.

[de Lemos 94] R. de Lemos, A. Saeed, T. Anderson. "On the Safety Analysis of Requirements Specifications". *Proceedings of SAFECOMP'94*. Springer–Verlag. J. Górski (Ed.). Anaheim, California. October 1994. pp. 217–227.

[de Lemos 95a] R. de Lemos, A. Saeed, T. Anderson. "Formal Techniques for Requirements Analysis for Safety–Critical Systems". *Mathematics of Dependable Systems*. Eds. C. Mitchell and V. Stavridou. Oxford University Press. 1995. pp. 63–95. (to appear)

[de Lemos 95b] R. de Lemos, B. Fields, A. Saeed. "Analysis of Safety Requirements in the Context of System Faults and Human Errors". *Proceedings of the IEEE International Symposium and Workshop on Systems Engineering of Computer Based Systems*. Tucson, Arizona. March 1995.

[Duke 93] D. Duke, M. D. Harrison. "Abstract Interaction Objects". *Computer Graphics Forum* Vol. 12 (3). 1993. pp. 25–36.

[Goldsack 91] S.J. Goldsack, A.C.W. Finkelstein. "Requirements engineering for real–time systems". *IEE Software Engineering Journal*. May 1991. pp 101–115.

[Gorski 95] J. Gorski, A. Wardzinski. "Formalising Fault Trees". *Proceedings of the Safety–critical System Symposium*. Springer–Verlag. F. Redmill, T. Anderson (Eds.). Brighton, UK. February 1995. pp. 311–327.

[Helm 90] A.R. Helm, I.M. Holland, D. Gangopadhyay. "Contracts: Specifying Behavioural Compositions in Object–Oriented Systems". *Special Issue of SIGPLAN Notices – Object–Oriented Programming Systems, Languages and Applications Conference*. Ottowa, Canada. 1990. pp. 169–180.

[McDermid 94] J.A. McDermid, D. Pumfrey. "A Development of Hazard Analysis to Aid Software Design". *Proceedings of the Ninth Annual Conference on Computer Assurance (COMPASS '94)*. Gaithersburg, MD. July 1994. pp. 17–25.

[MoD 95] Ministry of Defence. *Draft Interim Defence Standard 00–58: A Guideline for HAZOP Studies on Systems which include a Programmable Electronic System*. UK Ministry of Defence, Glasgow, UK. March 1995.

[Powell 92] D. Powell. "Failure Mode Assumptions and Assumption Coverage". *Proceedings of 22nd IEEE International Symposium on Fault Tolerant Computing (FTCS–22)*. Boston. June 1992. pp. 386–395.

[Rumbaugh 91] J. Rumbaugh, M. Blaha, W. Premerlani, F. Eddy, W. Lorenson. *"Object–Oriented Modelling and Design"*. Prentice–Hall. 1991.

[Saeed 93] A. Saeed, R. de Lemos, T. Anderson. "Robust Requirements Specifications for Safety–Critical Systems". *Proceedings of SAFECOMP'93*. Springer–Verlag. J. Górski (Ed.). Poznan–Kiekrz, Poland. October 1993. pp. 219–229.

[Saeed 94] A. Saeed, R. de Lemos, T. Anderson. "An Approach for the Risk Analysis of Safety Specifications". *Proceedings of COMPASS '94*. Gaithersburg, MD. July 1994. pp. 223–231.

# A Guideline for HAZOP Studies on Systems which include a Programmable Electronic System

M. F. Chudleigh
Cambridge Consultants Ltd
Cambridge, U.K.

J. R. Catmur
Arthur D. Little
Cambridge, U.K.

F. Redmill
Redmill Consultancy
London, U.K.

# 1  Introduction

There is increasing use of programmable electronic systems (PES) in safety-related applications and the task of ensuring that such systems are planned, designed and produced with the appropriate attention to safety is not trivial. In PES, failures due to design mistakes may dominate random hardware failures, so it is crucial to identify hazards early in the design process, then to take appropriate design measures to mitigate the hazards. Hazard analysis is the process of identifying the potential for such undesirable events, and their likely causes and consequences.

Carrying out hazard analysis at various stages of the development process is now mandated in emerging standards produced by the International Electrotechnical Committee (IEC) and the U.K. Ministry of Defence (MOD). The IEC standard has now been published as IEC 1508 [IEC 95]. Interim Defence Standard 00-56 [MOD 91], published by the MOD, is now under revision, but retains the identification of Hazard and Operability Studies (HAZOP) as being an appropriate technique for use in hazard analysis of PES.

HAZOP is a hazard identification technique, developed by ICI in the late 1960s, which has become well established in the petrochemical industry. Over the last five years Arthur D. Little and Cambridge Consultants have extended their experience gained in that industry to the investigation of hazard analysis of systems in other industries, including electro-mechanical systems and PES. Some of that experience is reported in [Chud 92, Chud 93, Chud 94, and Catmur 92].

The MOD recognised that the established HAZOP guidance available, such as [CIA 87, Kletz 92], was targeted towards petrochemical plants and so, after competitive tendering, let a contract to Cambridge Consultants Ltd (working with Arthur D. Little and Redmill Consultancy) to write a guideline for carrying out a HAZOP on a PES.

The first two stages of the work have been carried out: first to establish the feasibility of writing such a guideline, and second to draft it. A draft guideline has now been published by the MOD as draft Interim DEF STAN 00-58 [MOD 95], with the same title as this paper, for public comment. The intent of the guideline is to be a self-contained guide to carrying out HAZOPs and to be generic to all types of systems. This is an important point: contrary to some opinions, the major part of any HAZOP guidance does not vary between application domains. However, what do change are the ways in which designs are represented and deviations from design intent are interpreted. Thus, the examples given in the guideline and in this paper are taken from the PES application domain.

The purpose of this paper is, first, to disseminate the experience gained while preparing the guideline and to explain the particular difficulties in carrying out a HAZOP on a PES; and second, to introduce the guideline to a wider audience than the U.K. defence community.

In carrying out the work we had wide consultation within the U.K. and found that, although HAZOP on PES had been carried out by a number of organisations, a guideline was needed as: first, there were a number of differences of approaches see [Chud 92, Chud 93, Chud 94, Catmur 92] and also [McDer 94, Fenc 94, Burns 93]; and second, there were several problems which had been troubling practitioners.

In the next section we offer an overview of the HAZOP technique, as applied to PES. Then, in the following sections, rather than give a précis of the contents of the guideline, we describe those areas which have caused practitioners the most concern and problems. These are:

- Team Structure and roles. The dynamic and creative interaction between team members is a key factor in the success of the HAZOP process.
- Lifecycle issues of when to carry out a HAZOP. This is an area which has caused problems to many practitioners as HAZOP may not be the most appropriate technique throughout the development process.
- Design representation. A wide variety of ways to represent the design of a PES exist and these must be catered for as it is not realistic to mandate a particular design representation. We believe that a HAZOP may be carried out on almost all forms of design representation, even those which are not pictorial.
- Attributes. The path between components indicates an interaction or relationship between them which possesses properties, or 'attributes'. These affect the correctness and perhaps the safety of the system. A HAZOP examines the possible deviations of attributes from their design values ('design intent'). The choice of attributes is important.
- Guide words and their interpretations. A rigorous and systematic procedure based on the use of 'guide words' is used to investigate the possible deviations of attributes from their design intent. Guide words may be interpreted differently in different circumstances. The key issue is the interpretation of the guidewords in the context of the attributes shown up by a particular design representation.
- Recording of the HAZOP results. This has been a problem for many HAZOP practitioners.

# 2    Overview of the HAZOP Process

A HAZOP is a process of hazard identification. It is carried out by a team, never by an individual. As its purpose is hazard identification, it cannot on its own cover the complete safety-analysis spectrum. It may be considered to consist of four sequential stages:

- Initiating the study;
- Planning the study;
- Holding the study meetings;
- Dealing with follow-up work.

The description below follows the sequence of these activities, considering each in turn.

## 2.1    The Study must be Initiated

If a HAZOP is to be successful, or indeed if it is to be carried out at all, there needs to be someone in the organisation who possesses:

- The awareness that a study is necessary;
- The authority to initiate a study;
- Sufficient understanding of the project in hand to know when a HAZOP is appropriate.

This person may be referred to as the study initiator, and may be a safety officer or some other manager. Within DEF STAN 00-56, the equivalent of the study initiator is the Project Safety Engineer. In too many organisations HAZOPs are not taken seriously mainly because of the absence of a competent, aware and responsible study initiator with the appropriate level of authority to initiate and support the study.

When a HAZOP is appropriate, the study initiator defines the scope and objectives of the study and appoints a study leader (a discussion of scope and objectives is not considered in this paper). The study leader is then responsible for planning the study - with, if necessary, the help and support of the study initiator.

## 2.2    The HAZOP must be Planned

The central activity of a HAZOP is the investigation, by a team, of the design of the system under consideration, in a chosen documented form referred to as the 'design representation'. As the study can be a lengthy process, it often extends over a number of meetings. Thus, the main aspects of the planning of the study (for which the study leader is responsible) are:

- Selecting the study team (see Section 3 below);
- Ensuring the availability of an appropriate design representation (see Section 5);
- Determining how many study meetings will be necessary, scheduling them, and arranging their accommodation and other logistical necessities;
- Briefing the study team members and ensuring their availability;
- Identifying the 'attributes' on the interconnections on the design representation (see Section 6);

- Selecting the 'guide words' for use in the study and determining the interpretations of the guide words when applied to the particular attributes on the design representation (see Sections 2.4 and 7).

## 2.3 Conduct of the Study Meetings is Key

The study leader 'chairs' study meetings and has the responsibility to ensure that they are effective. As the effectiveness of the study depends on using the time available for hazard identification, it is important that extraneous discussion is minimised. The study leader should stop the study at any time if it becomes unproductive and should exclude any team member who is in any way compromising the effectiveness or efficiency of the meeting.

In a HAZOP, it is not the components on the design representation, but the interconnections between them, which are studied. An interconnection may be logical or physical, and represents the flow of some entity (for example a fluid in a chemical plant, data in a computer system) between the components which it connects. Each entity possess attributes according to the circumstance (for example, relevant attributes of a fluid may be temperature and pressure, those of a data stream value, sequence and bit rate), and in the design of every system there are intended design values (the 'design intent') for each attribute. The purpose of the investigation is to identify, in the particular circumstances, what deviations from the design intent might occur, and then to determine their possible causes and hazardous effects. The goal of the study is to identify all possible hazardous conditions.

To commence the study, the study leader selects an interconnection (usually working from left to right, or from inputs to outputs), identifies the flow entities on it (there may be more than one, particularly on data paths) and the attributes appropriate to each entity, and seeks to identify hazardous deviations from design intent.

The identification of deviations is based on the use of predetermined 'guide words' which focus attention on a particular type of possible deviation. An example of a guide word is 'more'. As in some cases this may be interpreted as 'greater' or 'higher', not only the guide words, but also their interpretations in the context of the circumstances and the attribute being studied are important. Thus, the study employs a number of attribute-guide-word-interpretation combinations. As each combination is applied, the team examines whether there could be a deviation from design intent as defined by the combination. If the answer is yes, the team enquire into any possible causes of the deviation. When causes are found, the possible consequences for the system as a whole are investigated. Thus, hazardous consequences are identified.

When the guide words chosen for use in the study have all been applied to the first attribute, the other attributes of the entity under consideration are examined in turn. Then the process is repeated for each other flow entity on the interconnection in question. The study leader then marks off that interconnection and chooses another, and the process is repeated. Thus, the design representation is subjected to a systematic investigation.

Throughout the study, results are recorded according to a predefined procedure (see

Section 8 below). As the purpose of a HAZOP is to identify hazards and not to resolve them, the safe redesign of the system is not within the ambit of the study. The identification of hazards leads to recommendations by the study team. Spending too long on any one issue would not be effective use of the study team's time, so when immediately unresolvable questions arise, actions for follow-up work are raised.

A study meeting ends with the documentation of the activities of the study, identified hazards and their causes and consequences, recommendations resulting from these, and actions for follow-up work. All documentation should be agreed by the team and signed off by the study leader.

## 2.4    Follow-up Work must be dealt with

A HAZOP must end on recommendations, not questions. It cannot be considered complete until the follow-up study of questions raised has been carried out. The answers to questions raised at one meeting should be dealt with at a subsequent meeting, and it is often necessary to schedule an extra meeting beyond the original schedule in order to deal with the final follow-up work and close the study.

# 3    The Team is Essential to the Success of the HAZOP

A HAZOP is a team effort, and it is only successful if the team is well composed and well led. An experienced study leader, with the ability to apply firm chairmanship while still maintaining harmony and motivating good team work, is essential. Many HAZOPs have been unsuccessful because of a lack of care in selecting the team and the study leader, and also, on occasions, because of the selection of inappropriate persons 'because the appropriate persons are too busy'.

Typically, a team comprises the following roles:
- The 'study leader' who is responsible for leading the study;
- The 'recorder' who will document the results of the study;
- A designer with a good understanding of the design representation to be studied;
- A user or intended user of the system being studied;
- Team members selected for their specialist knowledge of the system, aspects of the system, or the system's environment.

There are occasions on which two of these roles may be played by the same person, and, in any case, the optimum team size is between five and seven persons.

While the leader's control of a study meeting is imperative, a HAZOP is a creative activity. It involves processes of:
- Postulation of a possible deviation from design intent;
- Exploration of the deviation's possible causes and consequences;
- Explanation of the behaviour of the system, of problems known with similar systems, of the protection and alarm mechanisms in place or

planned, and of other relevant factors;

- Drawing conclusions about whether a hazard exists;
- Recording the results.

Each stage of this process requires a particular skill, and each skill is necessary to the success of the study and should be possessed by at least one of the team members. While there is no rule for who should possess the various skills, they would normally be expected to be distributed as follows. The study leader would postulate the possible deviation; the expert members, (and perhaps the leader) would be competent explorers of the causes and effects of deviations within the system; the designer too might be a competent explorer, but should certainly be capable of explaining how defined problems could be caused and how the system would behave in the presence of these problems; the user would explain what the operational consequences of a defined deviation could be and the extent to which they may be hazardous. The study leader needs to be adept at summarising the results of the discussion and drawing conclusions from them, and the recorder should be competent at recording, in concise prose, the results of a diverse discussion.

This expected distribution of roles is summarized in Table 1. However, these roles should not be made the responsibilities of particular team members. Rather, it is the study leader's responsibility to ensure, in selecting the team, that the role abilities are present and, in conducting the study, that they are deployed. Having said that, it should be observed that there is no point in selecting a 'designer' who does not possess an understanding of the design, a 'user' without an understanding of the operation of the system, or 'expert members' without relevant expertise. Thus, in selecting the team, the study leader needs to seek not only appropriate knowledge but also the ability to play one or more of the key roles.

| | Postulate | Explore | Explain | Conclude | Record |
|---|---|---|---|---|---|
| **Leader** | Yes | Possibly | Possibly | Yes | |
| **Expert** | | Yes | Yes | | |
| **Designer** | | Possibly | Yes | | |
| **User** | | Possibly | Yes | | |
| **Record** | | | | | Yes |

Table 1: Key roles and their likely possessors

The study leader should determine what expertise is necessary for a successful study and select the expert members accordingly. Typical areas of expertise which may be useful or necessary in PES studies are: supplier, applications software developer, operator, maintainer, independent expert. The team may grow and shrink as experts are called in to answer given questions or for the analysis of different parts of the design, but there should be a balance between allowing change in order to introduce the right expertise, achieving consistency across study meetings or studies by maintaining a core team, and keeping the size of the team to a cost-effective and manageable level (between 5 and 7 members).

# 4 How to Decide when to use HAZOP

Hazard Analysis should be carried out at various stages in the lifecycle of the project. The decision as to which analysis technique is most appropriate at each stage of the lifecycle or of design abstraction is often complex, and no absolute rules can be given. As the project proceeds the aim is to increase the knowledge of each hazard.

## 4.1 Hazard Analysis is Necessary Throughout the Lifecycle

At an early stage in a project the principal aim is to discover whether the system being designed and built has safety implications and, if it has, what hazards it presents. If hazards are identified, these are used to plan future studies, identifying the need for both the studies and their coverage. It should be noted that it is the level of design detail which drives the selection of the technique, not the stage in the design lifecycle. As the design detail increases three fundamental questions are addressed:

- Does the design address adequately the hazards already identified?
- Have any new hazards been introduced, due to design changes or simply found through better knowledge of the system?
- What are the potential causes of the higher-level hazards identified earlier?

Thus, as design detail increases, the increase in knowledge about the hazards and their causes, as well as the repetitive confirmation that the hazard set is complete, will enable a full safety analysis of the design of the system to be produced by the time the design is complete.

The amount of detail available increases as we go through the lifecycle, and the design representation (see Section 5 below) may well change. A suitable hazard analysis process will identify 'components' (of hardware, software or firmware) that have a role in safety at each level of design detail and ensure that they are covered in greater depth at levels of greater design detail. A component which at one level has been shown to have no safety role should not be excluded from any future hazard analysis, but may well receive less attention.

## 4.2 The Choice of Technique is Complex

The choice of technique is complex, as different techniques are useful at different levels of design detail. It is our experience that HAZOP is a useful technique at certain levels of design detail, while at others it may well be of less value or even, in some cases, actually be very difficult or time consuming to carry out. It is therefore critical that the appropriate technique be adopted and that HAZOP only be used when it is the appropriate technique. The knowledge needed to select the appropriate technique will be discovered by analysts as they use the various techniques and build up their experience, but we believe that we can give some basic advice.

When deciding whether to use HAZOP a series of points should be borne in mind:

- HAZOP studies the interactions between components, so meaningful design information on such interactions will be required if a HAZOP is to be successful.
- HAZOP works forward from the guideword/attribute combination to the consequence of a design deviation and backward to the cause. If top level hazards are not reasonably well known it may not work well. It will also not work well if the level of design detail is very close to the lowest level, as the backward search process is then likely to be tedious.

These points tend to imply that HAZOP is most useful at the mid-levels of design detail. When the first hazard analysis is carried out ('Preliminary Hazard Analysis' is the term often used), little design information will exist (possibly not even a block diagram) so a technique based on historical review, expert review and 'what if' review will be far more appropriate. It is, however, possible that even at a very high level of design abstraction (low level of design detail) the design's interfaces with other systems will be defined. In this case, as interactions are being considered, HAZOP would be a good tool for studying the interface hazards. However, the techniques mentioned above should still be used to review the other possible hazards. So, at this low level of design detail, HAZOP may well not be used, and if it is it should be used in conjunction with other techniques.

As more design detail exists, the interactions between components become better defined, as do the interactions with other systems. HAZOP is a very suitable technique for reviewing that the hazard list is complete, that the design copes with hazards already identified and exploring the causes and consequences of hazards that arise from interactions at these mid-levels of design detail. It should be recognised that if hazards can arise from the internal failure of a single component then other techniques such as system Failure Modes and Effects Analysis (FMEA) or modified HAZOP (to include some FMEA) should be used. At these mid-levels of detail, studies such as interface hazard analysis will also be carried out. HAZOP can be a powerful tool for these studies. So, at medium levels of design detail we have found HAZOP to be an ideal tool, although we do need to supplement it with other techniques in certain cases.

At very detailed levels of design detail we have found HAZOP not to be of such great use. At these levels of detail the aim is to identify the ultimate cause of hazards, so the ability to work backward from the guide-word/attribute does not exist. HAZOP at this level can often be a slow and frustrating exercise which may add little to the hazard analysis process. Techniques such as Failure Modes Effects and Criticality Analysis (FMECA), FMEA and Software Errors and Effects Analysis (SEEA) may well be more appropriate. So, at these levels of design detail we would not recommend the use of HAZOP.

When the system is in use, modified or removed from service, there will be a need to carry out hazard analysis studies. The choice of technique will still depend on the level of design detail at which such a study is being carried out, and whether interaction hazards are relevant.

As a final point it is worth noting that if the interactions between components are few and simple at a particular level of design detail, then HAZOP may well be too complex a technique and something simpler may be more appropriate.

# 5    A Wide Variety of Representations may be Used

One of the key features of a HAZOP is its structured examination of a representation of the design of a system. The representation may be of the physical or logical design, and both will need to be examined during the system lifecycle.

## 5.1    A Wide Variety of Design Representations may be the Basis of a HAZOP

A wide variety of PES design representations are in use, and we believe it is possible for any to be the basis of a HAZOP. Examples include: block diagrams, data flow diagrams, state transition diagrams, Object Oriented Design (OOD) diagrams, timing diagrams, circuit diagrams. Not only is no one design methodology with its attendant design representations likely to become dominant, but also different representations are likely to be used at different points in the system lifecycle. Most HAZOPs are carried out with a pictorial representation of the design, but this is not mandatory. No precise and understandable means of representation need be precluded from use in HAZOPs. Indeed, we have recently carried out a HAZOP using the mathematically formal representation of a functional language [Ande 95]. However, there are times when one given design representation is more appropriate than another. For example, at an early stage of development, when overall system hazards are being identified, a simple block diagram may be most appropriate; when timing is a factor, a timing diagram is desirable if not essential.

Each type of representation has its own set of conventions; the symbols on each have predefined meanings. Thus, there is a certain type of information which each presents naturally. For example, a block diagram naturally expresses the logical relationships between components; a data flow diagram naturally expresses the flows of information between components. Neither explicitly expresses attributes such as 'timing', 'bit rate', 'sequence'. The representation used should cover those aspects of a system which could relate to hazards. If a single design representation does not, or cannot, cover all the attributes or credible failures relevant to the study, then one or more other forms of representation should be used.

## 5.2    All the Team Need to Understand the Representation

An important point to note is that the design representation should be understandable to all members of the HAZOP team. This may appear obvious but in practice it can be a constraint on the choice of team members. Our experience is that a dataflow diagram or state transition diagram is very easy for the non-computer expert to understand. An important member of the HAZOP team is the user who will often be someone with an application need but little, if any, knowledge of the design technology behind the system that will fulfil that need. An example might be an avionics system where a high level of integrity is necessary and a decision is made to use a formal mathematical approach as the design representation. Most pilots (or aeronautical engineers) would have problems understanding the functional language without specific training. One approach suggested to overcome this problem is to derive a pictorial representation specifically for use in the HAZOP: we do not recommend that approach because the process of translation can introduce misrepresentations which mask hazards. We believe it is preferable for all team members, including the user, to receive sufficient training to enable them to understand the representation. In our experience, this need not be too time-consuming. With many problems of computer- based systems traced to a poor match between the user and the system, the investment in training necessary to enable users to contribute to hazard analysis would seem to be essential.

The representations used in the HAZOP should cover those aspects of the system which could relate to hazards. In the next section we give some advice about how to check that the design representations used give rise to the attributes necessary to explore the potential hazards.

## 5.3 Compound Interconnections Need to be Handled Carefully

An area that has given rise to problems with some practitioners is that of compound interconnections between components on a design. This is often the case with dataflow diagrams where the flow may be anything from a single item to a group of any number of items. At a low level of design detail, the name is usually descriptive of that whole group, and decomposition into sub-groups and then into primitive single-item flows often takes place as the group passes from a parent to a child diagram. The primitive flows need to be studied during a HAZOP. It is difficult to give definitive advice on when this should be. If only a single level of design is available, then the individual flows must be identified and addressed at that level. On other designs, a hierarchical decomposition may be available and it may be natural to look at the primitive flows when the decomposition reaches that level. On the HAZOPs we have carried out, and in training others in the process, we have found that the team approach aids the process. If the team needs to discuss the primitive flows in order to explore deviations of the group flows, then this indicates that the HAZOP leader should move to looking at the primitive flows. A cautionary note is necessary here: it can be tempting when the team is aware that further design detail is available for them to begin exploring deviations at a lower level than is appropriate at the time. This can lead to an unstructured session and ineffective results, so the HAZOP leader should keep the focus at the level intended.

Another potential problem is when a compound interconnection splits between components, such that one part of the group goes to one receiving component and the remainder to another receiving component. We have found it convenient for the HAZOP to address them as separate interconnections between the sending component and the two receiving components. Similarly, a two-way interconnection between components may be addressed as an interconnection in each direction.

# 6    Use of Correct Attributes is Important

On a design representation, a path between two components indicates an interaction between the components. An interaction consists of a transfer, or flow, from one component to the other. A flow may be tangible (such as a chemical) or intangible (such as a data entity), but in either case it possesses properties, or 'attributes', which affect the correctness of functioning and perhaps the safety of the system. Examples are the 'value' of a data entity and the 'bit rate' of its flow. The adherence of attributes to their design values determines the correctness of operation of the system.

The process of ensuring that all attributes are used in the HAZOP of a particular design depends on the representation that is being used. Each design representation throws up certain attributes of the design. When carrying out a HAZOP two questions should be addressed:

•        What attributes are covered by the representation being used?
•        Are the attributes covered sufficient for identifying the possible hazards?

For a particular representation each interaction shown may directly or indirectly represent certain attributes. In a data flow diagram the attribute 'value' is not represented directly, but can be easily considered for each data flow. In a similar way on a chemical plant P&ID (piping and instrumentation diagram), none of the attributes of flow, pressure, temperature etc are directly shown on each pipe, but they are easy for the HAZOP team to understand as being attributes of the P&ID flow.

The process of ensuring all relevant attributes are addressed should be a two staged process. In the first stage the representation(s) to be used should be reviewed and for each flow type those attributes that can clearly be understood and associated with it should be identified. This should check that all the attributes on each path on each representation have been identified.

Once all the representations that are to be used have been reviewed and all the clear attributes have been identified, a second stage of completeness checking should be carried out. The representations should be reviewed, along with attribute lists from other studies, in order to verify that no other attributes exist which could have an impact on safety. This is best done by considering other attributes that could be used and checking that they are not relevant. Thus 'temperature' is not a relevant attribute of a data flow diagram: however, if temperature is an important factor of the design another representation must be found to ensure completeness.

If further attributes are identified that are relevant but which are not clearly associated or understandable with the design representations chosen, then the need for further representation(s) to cover them must be noted and taken into account.

The final identification of attributes should be a complete list of those which are relevant, given the availability of a suitable representation. It may be useful to review this phase of the HAZOP preparation process with the HAZOP team before commencing the HAZOP.

# 7 Guide Words Must be Interpreted in Context

As stated in Section 6 above, it is the adherence of attributes to their design intent that determines the correctness of operation of the system.

A rigorous and systematic procedure is used to investigate the possible deviations of attributes from their design intent. This investigation is based on the use of 'guide words' and is the core of a HAZOP. A guide word is a word or phrase which expresses and defines a specific type of deviation. The role of the guide word is as a mnemonic, to focus the study and elicit ideas and discussion and, thus, to maximise the chance of identifying and studying all possible hazards.

## 7.1 Standard Guide Words Need Interpretation

Guide words may be interpreted differently in different industries, at different stages of the system life cycle, and when applied to different design representations. When guide words are chosen for a HAZOP, their interpretations should be defined in the given context. Each guide word may have more than one interpretation in the context of its application to the design representation.

The set of guide words used in the petrochemical industry is well established, together with their interpretations. At first sight that set of guide words did not seem to match the needs of HAZOPs on PES and in our early work we invented a number of other guide words. We were able to carry out HAZOPs successfully using those new guide words but gradually came to the realisation that they were not actually new guide words. In fact, they were interpretations of some of the generic guide words in the context of particular attributes of the design. Thus we came to the conclusion that the existing set of guide words is generally adequate, except that new guide words are needed to address timing issues.

As an example of this, some interpretations of guide words for both the chemical industry and PES are given in Table 2. The PES interpretations are for the entities of data flow and control flow taken from a dataflow diagram which both have the attribute 'flow'.

| Guide word | Standard Interpretation for Chemical Industry | Example Interpretation for PES |
|---|---|---|
| no | no part of the intention is achieved | no data or control signal passed |
| more | a quantitative increase | data is passed at a higher rate than intended or more data is passed |
| less | a quantitative decrease | not used here because this is already covered by 'part of' |
| as well as | all design intent achieved but with additional results | not used here because this is already covered by 'more' |

| part of | only some of the intention is achieved | the data or control signals are incomplete |
|---|---|---|
| reverse | covers reverse flow in pipes and reverse chemical reactions | normally not relevant |
| other than | a result other than the original intention is achieved | the data or control signals are complete but incorrect |
| early | not used | the signal arrives too early with reference to clock time |
| late | not used | the signal arrives too late with reference to clock time |
| before | not used | the signal arrives earlier than intended within a sequence |
| after | not used | the signal arrives later than intended within a sequence |

Table 2: Guide word interpretations

It is recommended that interpretations of guidewords are made in the context of the system being studied and its representation. It is important that the set of attributes, guide words and interpretations is:
- understood by the team using it;
- applied consistently;
- sufficient to explore plausible deviations from design intent.

The interpretations given in this section are examples which have been found to work on particular HAZOPs. Other interpretations are not precluded and, in fact, may be more appropriate for particular studies. We are aware that there has been much debate on the choice of the theoretically correct set of guide words for particular design representations. Although this is a valid subject for research, we believe that much of the debate stems from a misplaced focus on the guide words above rather than on the attribute-guide-word-interpretation combinations. Nor need the debate obstruct the carrying out of effective studies. We take the pragmatic view that the objective of the HAZOP process is to identify hazards and that guide words are not ends in themselves but facilitators of the process. The key issue is that the set of guide words and interpretations should be adequate to allow the creative exploration by the team members of deviations on the interconnections between system components - that is the purpose of using guide words.

It is worthwhile noting that certain design constructs are such that the deviations implied by some guide words are impossible. If this is the case, and it can be guaranteed, then the guide word can be ignored. In theory, an ideal representation/implementation would allow no guide word to be applicable, and therefore be inherently safe.

For some design representations, there may be more than one interpretation for a particular guide word. An example is using the attribute 'relationship' on an entity relationship diagram taken from an object oriented design. A possible set of interpretations is given in Table 3.

| Attribute | Guide word | Interpretation |
|---|---|---|
| Relationship | no | relationship will not take place for some reason (ie the relationship is absent) |
| | more/less | wrong cardinality in a relationship (for example it is one to one instead of one to many) |
| | part of | the set of relationship an entity has, shows incompatibilities between individual relations |
| | part of | a needed relationship is missing |
| | other than | the wrong relationship is defined between objects (ie there should be a relationship between the objects, but not the one given) |
| | other than | relationship is wrong (ie. there should not be a relationship between the objects even though one is given) |

Table 3: Guide words and interpretation for entity relationships

There should not be debate during the sessions on what a particular guide word means: those interpretations should have been defined by the HAZOP leader prior to the study and may be repeated by the leader at the beginning of the session so that all the team members have a common understanding. It is possible that during a study it becomes clear that studying a particular deviation needs either another guide word or another interpretation of an existing guide word. We had to add an extra interpretation during our HAZOP of a functional language design. When this happens, it is important to review the interconnections from the earlier part of the study to check whether the application of the new interpretation gives rise to any new hazards.

# 8 The Recording of Results Should Match Objectives

Recording HAZOP results is a non-trivial task, not only because of the quantity of information produced during a meeting, but also because the team members need to check it before agreeing it as a true record of what took place - within the constraints of the recording style.

There are two principal styles of recording in use. The first is to record the results of applying every attribute-guide-word combination throughout the study. This is intended to provide permanent documented proof of a thorough study, both for short-term audit requirements and possible longer-term disputes. This could be decisive in the event of liability legislation, but it imposes a heavy burden during the study. The second style is to record only the discovered hazards and the actions for follow-up work. This throws up more easily manageable documentation without necessarily diminishing the value of the study, but does not demonstrate the thoroughness of the study. In setting the objectives for a HAZOP, the study initiator should define the style to be used, taking into consideration such factors as corporate policy, relevant regulatory requirements, contractual obligations, the need for audit information, and the possible future need for proof of what was done in the study.

Regardless of the style employed, documentation should include:

- The objectives of the study;
- The attributes identified on the interconnections on the design representation, the guide words used, and the attribute-guide-word interpretations;
- Details of the hazards identified;
- Any features existing in the design for the detection or mitigation of the hazards;
- Recommendations, based on the team members' expert knowledge, for the elimination or mitigation of the hazards or their effects;
- Recommendations for the study of uncertainties surrounding the causes or consequences of identified hazards;
- Questions to be answered, as follow-up work, regarding uncertainties as to whether a hazard could occur;
- Operational problems to be investigated for possible further hazards.

The traditional recording medium is paper. However, if a copy of the documentation is produced for each team member, there is not only a large volume of paper but also the difficulty for the members to find time before the last session to read it and give their approval or proposed changes. Modern technology offers other options, in the form of word processors integrated with wall boards for immediate display, and, if the recorder is a competent typist, significant savings can be made if time is set aside at each meeting for approval of the documentation. In any event, there needs to be the process of:

- Recording;
- Review by the study leader, and correction where necessary;
- Review by the team members, discussion of anomalies and corrections where appropriate;
- Signing off by the study leader.

# 9 Conclusions

Draft Interim DEF-STAN 00-58 is a self-contained guide to carrying out HAZOP that is intended to be generic to all types of systems. This is an important point: we believe that the major part of any HAZOP guidance does not vary between application domains. However, what do change are the ways in which designs are represented and in which deviations from design intent are interpreted.

The paper has discussed those areas which have caused practitioners the most concern and we conclude that the following are key points:

- The dynamic and creative interaction between team members is a key factor in the success of the HAZOP process.
- It is important to realise that HAZOP may not be the most appropriate

technique throughout the design process and we have given guidance on when HAZOP is most useful.

- It is not realistic to mandate a particular design representation for PES and we have shown that a HAZOP may be carried out on almost all forms of design representation, even those which are non-pictorial.
- For a PES, the choice of attributes is important because it is their deviations from their design values ('design intent') which guides a HAZOP. Attributes arising from the design representation may need to be supplemented.
- The set of guide words traditionally used in the chemical industry is adequate for HAZOPs of PES, with the addition of others to address timing issues. The key issue is the interpretation of the guidewords in the context of the attributes shown up by a particular design representation. We argue that as long as the interpretations aid the hazard identification process then they are adequate.
- The 'correct' level of recording of the HAZOP results will vary - it is a policy issue, decided when the objectives of the HAZOP are set. Two main alternatives are to record the results of applying every attribute-guide-word combination to all the various interconnections on the design representation (useful when a full audit trail is needed) or to record only the discovered hazards and the actions for follow-up work (more cost-effective but does not document the thoroughness of the study).

# Acknowledgements

The work to draft Interim DEF-STAN 00-58 was carried out under MOD Contract NSM13C/1063. We wish to thank the Project Officer, Kevin Geary of SM836, for his help and support during the work. We would also like to thank the technical reviewers working for Kevin: namely Barry Hebbron of the University of Teesside and Adelard. In addition, we thank a number of members of the Hazard Analysis Interest Group for taking part in workshops and providing useful advice, experience and feedback.

# References

[Ande 95]     Andersen M, Catmur J, Chudleigh M: 'HAZOP of a PES with a non-pictorial Representation' In preparation.

[Burns 93]    Burns D, Pitblado R: 'A Modified Hazop Methodology for Safety Critical System Assessment' In: Redmill and Anderson (eds) Directions in Safety-Critical Systems, Proceedings of the First Safety-critical Systems Symposium, February 1993.

[Catmur 92]   Catmur J, Anderson P: 'Developing a Safety Case for signalling systems which employ safety critical software'.

Computers in Railways III, August 1992.

[Chud 92]        Chudleigh M, Catmur J. Safety Assessment of Computer Systems using HAZOP and Audit Techniques. In: Frey (ed) Safety of Computer Control Systems 1992 (Safecomp '92) pp 285-292.

[Chud 93]        Chudleigh M: 'Hazard Analysis using HAZOP: A Case Study' In: Górski (ed) Proceedings of the 12th International Conference on Computer Safety, Reliability and Security, October 1993 (Safecomp '93) pp 99-108.

[Chud 94]        Chudleigh M: 'Hazard Analysis of a Computer Based Medical Diagnostic System' Computer Methods and Programs in Biomedicine 44 (1994) pp 45-54.

[CIA 87]        A Guide to Hazard and Operability Studies. Chemical Industries Association Limited, 1987.

[Fenc 94]        Fencott C P, Hebbron B D: 'The Application of HAZOP Studies to Integrated Requirements Models for Control Systems' In: Maggioli (ed) Proceedings of the 13th International Conference on Computer Safety, Reliability and Security, October 1994 (Safecomp '94)

[IEC95]        Functional Safety of Electrical /Electronic /Programmable Systems. Generic Aspects. IEC 1508. 1995

[Kletz 92]        Kletz T A HAZOP and HAZAN. Institution of Chemical Engineers, 1992

[McDer 94]        McDermid J A, Pumfrey D J: 'A Development of Hazard Analysis to Aid Software Design' Published in proceedings of Compass '94.

[MOD 91]        Interim DEFSTAN 00-56. Hazard Analysis and Safety Classification of the Computer and Programmable Electronic System Elements of Defence Equipment. U.K. Ministry of Defence 1991.

[MOD 95]        Draft Interim DEF-STAN 00-58. A Guideline for HAZOP Studies on Systems which include a Programmable Electronic System. U.K. Ministry of Defence 1995

# An Automated Code-Based Fault-Tree Mitigation Technique

Jeffrey M. Voas

Reliable Software Technologies Corporation

Sterling, VA 20166 USA

Keith W. Miller

Department of Computer Science, Sangamon State University

Springfield, IL USA

### Abstract

This paper presents a framework for an automated safety methodology that: (1) generates fault-trees from code, and (2) then applies a fault-injection based technique to mitigate the potential for non-root nodes to cause hazardous outputs. This methodology reads in source code and user-defined hazards, builds the fault-tree, and then feeds the fault-tree, code, and user-defined operational profile to a mitigator routine that estimates the frequency with which the event in the root node can occur. Preferably this frequency will be zero, but if not, this methodology will allow a user to quickly assign non-zero probabilities to events that could result in hazards.

## 1 Introduction

There is an increasing use of computing in safety related applications. Ensuring that such systems are conceived and designed with the appropriate attention to safety is not easy. The procedure of identifying undesirable (unsafe) events and their consequences on an entire system is known as *system hazard analysis*. Carrying out a system hazard analysis before a system is designed and built has received widespread support as a mandatory step in developing critical systems. In software engineering, a hazard analysis can also be performed on the software in isolation, where output events from the software are labeled as hazards if they put the physical system into a hazardous state. For instance, new standards produced by the International Electrotechnical Commission, NASA, UK Ministry of Defense, and Underwriter's Laboratory all mention safety and how systems that are controlled by software must contain "safe" software. How much safer systems will be as a result of these process-oriented software standards is questionable.

Software quality is also receiving renewed attention. Corporations and the public sector are beginning to understand that in a world where software is becoming all-pervasive, software quality is too often left to chance. It is a common belief that consumers are at risk from software hazards that have the potential of causing loss of life and tremendous liabilities for software developers [UL94]. To reduce this risk, many have turned to issues associated with the software process. The Software Engineering Institute (and in Europe, the

International Standards Organization) have attempted to describe informally the *way* in which quality software should be developed, and to indicate characteristics of an organization that can develop it.

This paper presents a fresh perspective for software-engineering practice that is aimed towards *safe* software development and assessment. Development ideas and practices such as fault-tree analysis must be made practical through extensive automation if we are to move from theoretical safety development models (state-of-the-art) to wide-spread tools (state-of-the-practice). The process referred to by the phrase "software safety" is a two step process. The first step is to design a system that protects against hazardous outputs. This step includes tasks such as hazard analysis and designing the software *fire-wall* mechanisms necessary to insure that non-critical software functions do not impact critical functions of the code. The second step is to verify that either the design or code is indeed safe.[1] This is done by analyzing the software and/or its design, and mitigating any circumstances that could lead to a hazard. The process of mitigating sources of hazards is the *assessment* of safety, whereas the first step is developmental in nature, i.e., assurance.

This paper presents a technique that addresses safety assessment, i.e., hazard mitigation; an overview of our technique is shown in Figure 1. This automated methodology: (1) generates fault-trees from code, and then (2) applies a fault-injection based technique to mitigate the potential non-root nodes causing catastrophic outputs (or what are frequently termed "hazards"). This technique reads in source code and user-defined hazards, builds the fault-tree, and then feeds the fault-tree, code, and user-defined operational profile to a mitigator routine that estimates the frequency with which the event in the root node, the hazard, could occur. Preferably this frequency will be "never," but if not, the methodology automatically assigns non-zero probabilities to events that could result in hazards.

Our paper is structured as follows: we begin by describing fault-tree analysis and how fault-trees can be generated from source code via language-dependent templates [FRIE95, LEVE91]. We then talk about how you can mitigate potential sources of hazards by either proof-by-contradiction or by the empirical fault-injection-based technique, EPA [VOAS94]. We will also talk briefly about the role of fire-walls in decreasing the size of fault-trees and their benefit as a design-for-safety programming utility.

# 2   The Framework

## 2.1   Fault-trees

*Static fault-tree analysis* is a design/development technique that was developed by Watson in the 1960's at Bell Laboratories for assessing the causal relationship between particular events in a process. Software fault-tree analysis is a technique that has been brought to software from systems safety engineering. Systems safety analysis techniques attempt to ensure safety by using a "backwards" approach; catastrophic output events (hazards) are determined first, and then the designer works backwards from them to either show that those

---

[1] Here, we are not talking about absolute safety, but rather an acceptably low degree of risk.

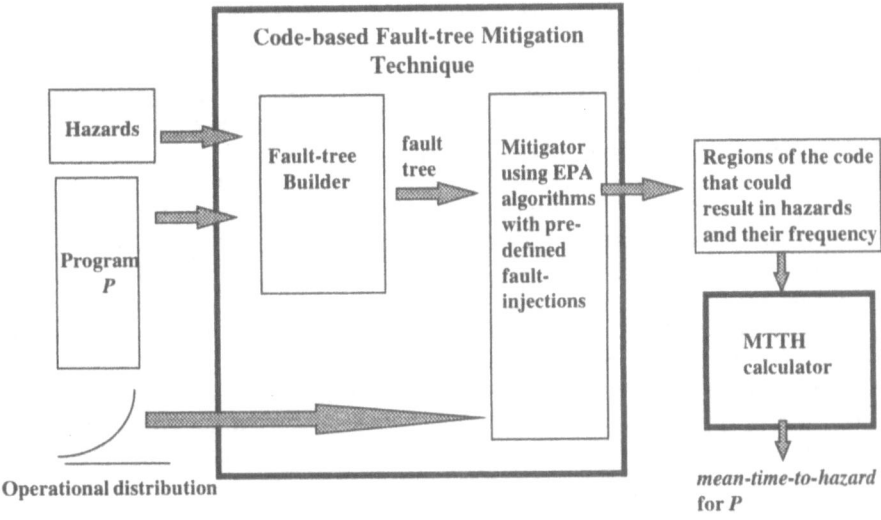

Figure 1: **Overview of Methodology**

output events cannot occur or the probability that they could occur is low. In a "forwards" safety analysis, the designer tries to show that all output events are non-catastrophics, whereas in a backwards analysis, you attempt to show that hazards are not possible or rarely possible. For a backwards approach to be practical, there must be a small set of hazards, otherwise the time involved in performing the backwards analysis will become intractable. For a forwards approach to work, you must show that for any input, the output events are non-hazardous, which for most applications is also intractable.

Software safety can be viewed as one partitioning of the problem of demonstrating *system* safety; hardware fault-tolerance would be an example of a different partitioning. Assume that a physical system has a fixed set of functions that it performs: $\mathcal{F} = \{f_1, f_2, f_3, ...f_N\}$. And assume that for each subset of $\mathcal{F}$, there are zero or more hazards that must be protected against for that subset. If we further assume that the software *directly* or *indirectly* controls some percentage of $\mathcal{F}$'s functionality, the goal of software safety assessment for this physical system is to show that the software cannot affect a subset of $\mathcal{F}$ in a manner that could lead to a physical hazard. Before software safety can be demonstrated in a backwards manner, there must be an enumeration of the system hazards. So system safety (or a lack thereof) must be completely and unambiguously defined before software safety can be demonstrated.

You can think of a fault-tree as a graphical representation of events; the graph is shaped like a tree data structure. The root node, or what is termed the "top node," represents the hazard or class of hazards. This is the final event in the sequence that hopefully cannot occur with respect to the rest of the system. The remainder of the tree represents parallel and sequential events that potentially could cause the hazard to occur. These events could come from a variety of sources: programming faults, design errors, human errors, malicious attacks, or hardware failures. Thus a fault-tree represents all types of events that have the potential to begin the "domino effect," meaning start the first domino falling. In this analogy, if the last standing domino falls, the hazard has occurred. The goal in software fault-tolerance is to ensure that if the domino effect begins, there are some standing dominos in the line that are so large that even if they are hit by a neighbor, they do not fall and hit their next neighbor.

Fault-trees are conceptually very simple; they are composed of events, and logical event connectors: OR-gates and AND-gates. For a given event node, its children are the necessary pre-conditions that could cause that event to occur, and these can be combined in any number of combinations using logical OR-gates and AND-gates. Events in the tree are continually expanded creating subtrees until leaf-events are created for which we can assign a probability for the leaf-event or leaf-events are created that cannot be decomposed into subevents. Events in a fault-tree can describe varying system levels of abstraction. Typically, higher nodes are at a higher level of abstraction, and lower nodes are at a more precise, refined level. For example, a high level event for an auto-pilot system could be: "auto-pilot controls aircraft," and a lower level event for a subcomponent of the auto-pilot, the auto-land, could be: "auto-land controls flaps."

Conventional software safety has typically been performed during the design phase of the software life-cycle for two important reasons:

1. other engineering fields have applied fault-tree analysis to designs long before the expense of building the system was authorized. This is prudent: imagine building a fleet of aircraft and then discovering that the design resulted in a fleet of aircraft that cannot be flown.

2. catching problems in the design phase before coding begins is cost effective.

The downside of applying fault-tree analysis only to designs is the possibility that the code will not reflect the original design, and therefore not provide adequate safety. Although fault-trees can be generated for software designs, this paper focuses on code-based fault-trees. When fault-tree analysis is applied to the code itself, it shows the calling hierarchy and the interrelationships between the actual code modules and the top event of interest. We are not suggesting that basing the fault-tree on the code is better or worse than applying it to the design, but rather that there are several benefits to applying it to software:

1. it will likely be easier to mitigate potential problems, because the code is a more precise definition of exactly what computations are being represented than nodes based on a software design,

2. there are commercial tools that can be applied to code (that cannot be applied to a design) to mitigate potential sources of hazards,

3. potential sources of hazards can be more easily decomposed and analyzed when the code is used instead of the design,

4. a fault-tree that is automatically generated from the code can be compared to previous fault-trees that were generated from previous versions of the code and compared to see if the system is becoming "safer" as the system evolves, and

5. a fault-tree that is automatically generated from code can be compared to a fault-tree that was manually generated from the design to indicate whether the original intuition about what can or cannot lead to a hazard was accurate, substantiating or disproving the original safety hypotheses.

## 2.2  Translating Code to a Fault-Tree

Recall that our goal in this paper is to produce a method for performing safety assessment on critical source code, by first automatically generating the fault-tree from the code, and then mitigating non-root nodes of the tree. The process of automatically reverse engineering a fault-tree from code is a fairly straight-forward process. The key is to define the semantics for what failure means for a particular construct of the language. Translators for generating fault-trees from source code are language dependent. Leveson *et al.* describe the fault-tree templates for the ADA language [LEVE91]. Similar templates could be used for other languages such as C and C++, but we do not know of any tools available for doing so.

## 2.3  Mitigating Tree Nodes that Represent Potential Hazard Triggers

Once a fault-tree is completed, work begins to demonstrate that the subevents under the top node can or cannot infrequently cause the root event to occur. There are two broad classes of techniques for mitigating fault-trees: (1) a formal proof-by-contradiction that a hazard cannot occur, and (2) an empirical approach that is not a formal proof, but rather a demonstration that the likelihood of a hazard occurring is so low that this risk is acceptable. In this section, we will first briefly describe the formal approach, and then describe an empirical technique that we have developed.

The first step of proof-by-contradiction mitigation is to hypothesize that a hazard has occurred. Proof-by-contradiction demonstrates negative behavior, i.e., something cannot occur, as opposed to demonstrating positive behavior, i.e., something can occur. To demonstrate a negative, you hypothesize a positive, and then by contradiction show that the positive cannot occur. For fault-tree mitigation, you begin by building the branches from a parent node until you reach a contradiction while building children nodes. If a contradiction is reached, meaning that some combination of events that are necessary pre-conditions for a parent node cannot occur, then mitigation of the parent node has occurred, and hence that branch is mitigated, i.e., you have a "dead" branch that can be ignored. There is no need to expand this branch further.

If a branch is generated that does not contain a logical contradiction, the fault-tree analysis has revealed the set of data inputs or environmental events that could trigger the hazard. This immediately reveals that the potential is real for the domino effect to occur, and hence either measures to ensure that it does not should be taken or it should be determined that the risk of this sequence of preconditions (for the top event) occurring are so low that there is little reason for concern. Either way, in this situation, the fact that a contradiction was not possible immediately demonstrates the potential for harmful damage.

As you build a fault-tree branch, your goal is to mitigate the branch by generating a contradiction, which is equivalent to killing that branch (total mitigation). But in practice, there will be branches for which this will not be possible, and if you cannot show contradiction for all cases, you should at least show the percentage of cases where propagation to a hazard is likely. This can be accomplished by: (1) estimating mean-time-to-hazard over all defined hazards, or (2) estimating minimum-time-to-hazard for the hazard that that is the most likely to occur. For this, we will look at a technique that can be applied to those regions of the fault-tree that have not been mitigated by proof-by-contradiction to assess how soon it will be that if they fail, a hazard is manifested.

There are a variety of techniques available that suggest whether a system can get into a hazardous state. Recall that the goal is to become sufficiently convinced (confident) that a particular node in the tree cannot cause the top event to occur. Techniques that are options here included Petri-nets, Markov modeling, state diagrams, and formal analysis. Formal analysis can be applied to demonstrate that certain output behavior is not possible. Software testing is an empirical procedure for either demonstrating an absence of faults or a particular level of reliability. Unfortunately, it is not as easy to *test* for software safety as it is to test for reliability. Unlike reliability, in software safety, you must also demonstrate that the program will compute safely (not necessarily correctly) for the *illegal input space*, i.e., inputs outside of the legal input space that the program is expected to receive. To make matters worse, this space is probably larger than the legal input space.

The leaves of a fault tree form the terms of a boolean expression, and the internal nodes contain the boolean operators. In the absence of negation (and negation is rare in fault trees), identifying each leaf as a contradiction is sufficient to identify the entire fault-tree as a contradiction. In practice, it may not even be necessary to do that much work. If a series of leaves are connected using AND-gates, identifying any one of them as a contradiction is sufficient to identify the subtree as a contradiction. This is important because there are *"normal operating mode contributes to root fault"* nodes in fault trees that cannot be identified as contradictions, and there are "event" nodes that might be extremely difficult to identify as contradictions [SHIM95].

For several years, we have has been working on an empirical method to mitigate the possibility of hazardous output states occurring, *extended propagation analysis* [VOAS94, VOAS95a]. This method has been incorporated into our framework, and will be applied to code according to the nodes in the code-based fault-tree that could not be mitigated via a formal proof-of-contradiction. The following subsection discusses extended propagation analysis and how this information can be used to mitigate (and assess risk for) nodes of the code-based fault-tree.

## 2.3.1 Extended Propagation Analysis for Hazard Mitigation

Our fault-tolerance assessment method for mitigating code-based fault-trees is extended propagation analysis (EPA) [VOAS94].[2] Extended propagation analysis is a technique that is automated for C, C++, and Fortran-77 source code in the *PiSCES Safety Analysis Tool* $^{(TM)}$. Extended propagation analysis allows the user to specify hazards; the system then analyzes the propagation capability of the code after hardware failures and software faults are forced into non-root tree nodes. The tool queries the user to specify which class of faults they want to simulate: hardware failures, software faults, or a combination thereof. The tool's output immediately pinpoints for the user *where* hazards will or will not dynamically propagate from. If no regions put the output into a hazardous state, we have a piece of evidence that has partially mitigated the nodes that were forcefully corrupted. If a hazard is observed, then the user knows exactly where in the code the problem originated, and how frequently this hazard was observed. The tool then allows the user to make a *mean-time-to-hazard* (MTTH) calculation or *minimum-time-to-hazard* (MinTTH) calculation for each class of simulated faults that caused the hazard to occur [VOAS95b].[3]

EPA collects *dynamic* information concerning which output variables are affected by a data state value that is altered. By "affected," we mean a variable whose value is distinct enough that we can immediately recognize that some manner of corruption must have existed in the state during computation. In essence, EPA is a measure of the *observability* of the software. EPA is also related to *failure modes and effects analysis* (FMEA) [LAWS83], which provides for the consideration of different types of failures by working forward from the components to the system. FMEA postulates an anomalous event, and studies whether the event can propagate to the output space, which is conceptually similar to the processes of EPA. The HAZOP methodology is another safety assessment scheme that is similar in concept to EPA; it attempts to ensure that both (1) the features that could lead to undesirable outcomes are avoided and (2) that necessary features are incorporated into the design for safe operation. The methodology was developed in the late 1960's and is well established in the petro-chemical sector. In this industry, a plant design is normally described by piping and instrumentation diagrams. A HAZOP study is carried out by engineers who postulate, for each member of the design, what the normal operating modes are for that element, and then assess what they feel might be reasonable deviations from those modes. The consequences of those deviations are then assessed with respect to operability and safety. For each deviation, the engineers ask "can it happen?" and if the answer is "yes", is this likely to

---

[2]EPA is a spin-off of Voas's Sensitivity Analysis technique [VOAS92]; the main difference is that EPA is concerned with incoming hardware failures as opposed to Sensitivity Analysis that is only concerned with resident program faults. Also, EPA differentiates classes of failure; Sensitivity Analysis does not.

[3]Note that in our technique, we will assume that a program execution requires some mean amount of clock time to execute, and that is how we will factor in time to assess mean-time-to-hazard and minimum-time-to-hazard. In fault-tree analysis mitigation, it is common to consider some mean duration time for an event node of the tree, and use that to compute a mean-time-to-failure; that is not how we are assessing those parameters here in our paper. Another interesting but more formal method for assessing the impact of time on hazards is the Merlin and Farber Petri Net [FARB76].

lead to a hazard. The design team will consider any mitigating situations that might control the problem. HAZOP is a methodology which in theory is very similar to EPA, with three very important exceptions:

1. our method for applying this process is *automated*, and hence the number of different "what if" games that we can play at various parts of the code is enormous.

2. our method is applied at the code level and is dynamic, versus at a static design level.

3. our method of mitigating a possible problem is done by actually running the code and seeing if a hazardous output event occurs. This is very different than having a group of people speculate about whether a hazardous output event will occur.

For software to be safe, there are two classes of problems that must be protected against: software faults and hardware failures, i.e., erroneous incoming data to the software. EPA simulates both classes, and thus we know the impact on system output if a hardware sensor were to malfunction or if the software itself were to malfunction (with respect to the specification). By repeatedly executing a program under simulated, adverse conditions, we can ascribe a mathematical confidence to the fault-tolerance of the system based on the hardware and software anomalies that we inject. This provides an experimental means for testing the safety of the system, and thus mitigating potential hazards. EPA provides a means for *testing* the hypothesis:

The software cannot put the physical system into an unsafe state.

If we are unable to contradict this hypothesis, then we have experimental evidence of safety. If we are able to contradict this hypothesis, we know where additional design efforts are needed as well as the level of risk being assumed.

### 2.3.2    The Underlying EPA Fault-Injection Model

A *data state error* is an incorrect variable/value pairing in a data state where correctness is determined by an assertion between locations (statements).[4] We refer to a data state error as an *infection*, and use these two terms interchangeably. If a data state error exists, the data state and variable with the incorrect value at that point are termed *infected*. A data state may have more than one infected variable. *Propagation* of a data state error occurs when a data state error affects the output.

Let $S$ denote a specification, $P$ denote an implementation of $S$, $x$ denote a program input, $\Delta$ denote the set of all possible inputs to $P$, $Q$ denote the probability distribution of $\Delta$, $l$ denote a program location in $P$, and let $i$ denote a particular execution (or what we term an "iteration") of location $l$ caused by input $x$. And let $\mathcal{A}_{lPix}$ represent the data state produced after executing location $l$ on the $i^{th}$ execution from input $x$.

---

[4]Admittedly, this is tenuous, since for any specification, there is an infinite number of correct programs that implement that specification, and therefore for each statement in a program version there must be an assertion if we are to expose data state errors, which as a practical matter will never happen.

It is important to group data states into sets with similar properties. For instance, assume that location $l$ is executed $n_{xl}$ times by input $x$. Now consider all of the data states that are created by this input immediately before $l$ is executed or immediately after $l$ is executed. The following set allows us to do so:

$$\mathcal{A}_{lPx} = \{\mathcal{A}_{lPix} \mid 0 \le i \le n_{xl}\}$$

We further group these sets for all $x \in \Delta$:

$$\alpha_{lP\Delta} = \{\mathcal{A}_{lPx} \mid x \in \Delta\}$$

A *simulated infection* is a modified value forced into the value of some variable (that already had a different value) in a data state. As we have already stated, $\mathcal{A}_{lPix}$ denotes the data state created after the $i^{th}$ iteration of location $l$ on input $x$; $\breve{\mathcal{A}}_{lPix}$ denotes this same data state after a simulated infection is injected into $\mathcal{A}_{lPix}$. A simulated infection usually affects a single live variable.

It important at this point to explain the relationship between simulated infections and the potentially disastrous states that a system can get into that can lead to a hazard. When a system gets into a "bad state" during execution, the next event that we would like to see occur is recovery from that state back to an acceptable state. Simulated infections are the mechanisms that are employed in EPA to allow observation of the impact of different classes of "bad states"; they mimic the effect of both programmer faults and hardware failures (coming into a system). Simulated infections are created by "perturbation functions." The process of injecting a simulated infection into an executing program is termed *perturbing*. A *perturbation function* is a mathematical function that takes in a data state as an incoming parameter, changes it according to certain parameters that are either input to the function or hard-wired, and produces as output a different data state. A data state that has had a value changed by a perturbation function is said to have been *perturbed*.

We consider that program $P$ has a fixed set of output variables: $\{v_1, v_2, v_3, ..., v_n\}$. An *observable variable* is an output variable whose value differs after a simulated infection is forced into the program and the program execution is resumed and termination occurs. For instance, if after a simulated infection is forced into the program, program execution is resumed, termination occurs, and output variable $v_3$ contains a different value, then $v_3$ is an observable variable.

The following EPA algorithm creates sets of observable variables that occur after some variable $a$ is perturbed on all iterations at location $l$. Because we are interested in the program's fault tolerance under circumstances with both software faults and incoming data corruptions, this algorithm will be applied to every program variable, including input variables.

**Algorithm 1:**

1. Set **k** to 0.
2. Set **variable_set** = $\emptyset$.
3. Increment **k**.

4. Randomly select an input $x$ according to $Q$, and if $P$ halts on $x$ in a fixed period of time, find the corresponding $\mathcal{A}_{lPx}$ in $\alpha_{lP\Delta}$. Set $\mathcal{Z}$ to $\mathcal{A}_{lP1x}$.

5. Alter the sampled value of variable $a$ found in $\mathcal{Z}$ creating $\check{\mathcal{Z}}$, and execute the succeeding code on both $\check{\mathcal{Z}}$ and $\mathcal{Z}$. If $l$ is executed more than once for $x$, i.e., $\mathcal{A}_{lP2x}, ... \mathcal{A}_{lPmx}$, alter $a$ in each $\mathcal{A}_{lPix}$, $2 \leq i \leq m$.

6. For each output variable that contains a different value (after comparing $P$'s output using $\check{\mathcal{Z}}$ to the output $P$ regularly produces), add it as a member to the set **variable_set**.

7. Set $\Pi_{alPQk} = $ **variable_set**. ($\Pi_{alPQk}$ represents the set of output variables that have different values given execution of $P$ occurs with $\check{\mathcal{A}}_{lPx}$. $\Pi_{alPQk}$ is the empty set if no output variables are affected by the injection of $\check{\mathcal{A}}_{lPx}$.) When this set of empty, the software has low observability.

8. Repeat steps 2-7 $n$ times.

There will be instances, however, were we are not necessarily concerned with whether variable corruption occurred, but more specifically whether a particular output event occurred, which we will denote by predicate $PRED$. To determine this, we do not need to run the unperturbed version of program $P$. $PRED$ will represent a predicate expression that relates specific variables to values ranges or combinations of variables and ranges. Also, $PRED$ may contain certain restrictions on the input that was used during an execution. The following algorithm provides this information; it determines the proportion of outputs that satisfy $PRED$:

## Algorithm 2:

1. Set **count** to 0.

2. Randomly select an input $x$ according to $Q$, and if $P$ halts on $x$ in a fixed period of time, find the corresponding $\mathcal{A}_{lPx}$ in $\alpha_{lP\Delta}$. Set $\mathcal{Z}$ to $\mathcal{A}_{lP1x}$.

3. Alter the sampled value of variable $a$ found in $\mathcal{Z}$ creating $\check{\mathcal{Z}}$, and execute the succeeding code on $\check{\mathcal{Z}}$. If $l$ is executed more than once for $x$, i.e., $\mathcal{A}_{lP2x}, ... \mathcal{A}_{lPmx}$, alter $a$ in each $\mathcal{A}_{lPix}$, $2 \leq i \leq m$.

4. If the output satisfies $PRED$, increment **count**.

5. Repeat steps 2-4 $n$ times.

6. Divide **count** by $n$ yielding $\hat{\psi}_{alPQ}$; $(1 - \hat{\psi}_{alPQ}$ is the degree of *fault-tolerance*). When $\hat{\psi}_{alPQ}$ is small, the software has low observability, which translates into high fault-tolerance.

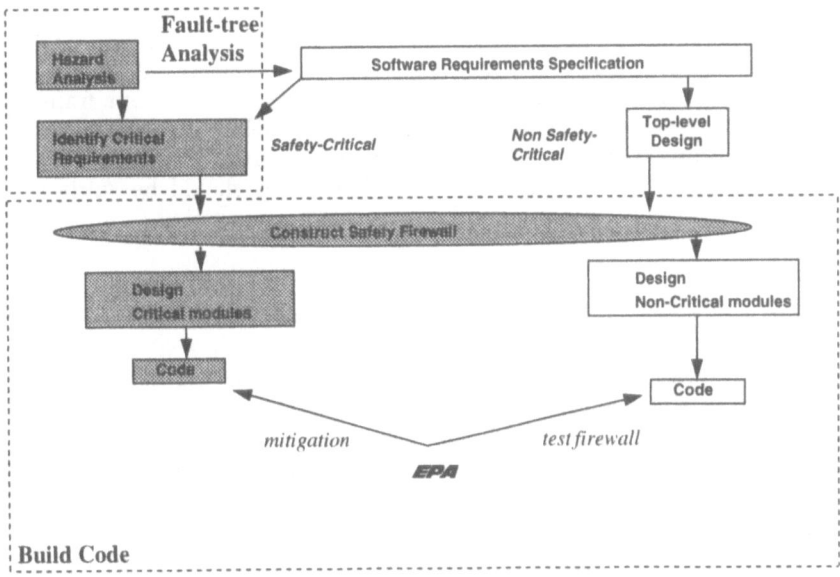

Figure 2: **The Firewall Concept**

## 2.4  Firewalls

Software firewalls are one means for designing in safety [FRIE95]. Figure 2 shows a development methodology in which fault-tree analysis is applied in the early stages of development before code exists. Then coding occurs according to the results of hazard and fault-tree analysis and a firewall is constructed. Next, EPA is applied both to the critical side and to test the fire-walls; this tests for the level of risk given the code in its current form.

The general hypothesis behind employing fire-walls is that critical requirements must have the modules that compute them isolated from modules that do not compute critical requirements. Non-critical software components must not have an unpredictable effect on the critical components. This integrity must be maintained under abnormal operation (i.e., when faults manifest themselves) as well as under normal operation. One approach to ensuring integrity is to analyze all possible interactions between modules, which is infeasible. With fire-walls, data flows and control flows are restricted between critical and non-critical modules. This is handled via programming language and system architecture features. For example, Ada packages, C++ objects, and separate address spaces are each alternatives for building a firewall [FRIE95]. Non-critical data can only flow from a critical module to a non-critical module, not the reverse case. In languages that allow for pointer manipulation, such firewalls are less effective unless the use of pointers is disallowed.

Fire-walls have an important role in separating critical functionality from non-critical functionality. And the use of fault-trees during the design phase can aide in determining how to apply fire-walls in the code. Our framework addresses an additional issue in risk mitigation, namely critical functions that could cause other critical functions to result in hazards. Hence even systems with fire-walls that were based on design fault-trees should be subjected to code-based fault-tree generation and the associated mitigation.

## 2.5  Applications and Examples

Step 5 of Algorithm 1 and similarly Step 3 of Algorithm 2 are critical to the value of the information produced by these algorithms; the alterations of $a$ must reflect either a hardware failure class or a class of programmer faults. For example, Underwriter's Laboratory's standard, "Safety-Related Software," [UL94] defines the following classes of events that would need to be simulated in Step 5 to be used for their standard:

> "These requirements [UL1998] address risks that may occur as a result of faults caused by software errors, such as the following: a) design errors such as incorrect algorithms or interfaces b) Coding errors, including syntax, incorrect signs, endless loops, and the like; c) Timing errors that can cause program execution to occur prematurely or late; d) Induced errors caused by hardware failure; e) Latent errors that are not detectable until a given set of conditions occur;......"

Step 5 of Algorithm 1 (and similarly Step 3 of Algorithm 2) can simulate the anomalies mentioned in b), c), and d) of UL1998. We have shown elsewhere how using inverted operational distributions in Step 4 of Algorithm 1 and Step 2 of Algorithm 2 can demonstrate the risks mentioned in e) [VOAS95a]. As another example, NASA's new interim software safety standard (due to be approved in 1995) requires demonstration of software fault-tolerance for problems involving timing and hardware failure sensitivities [NASA94]. Timing and hardware failure anomalies can also be simulated in Algorithm 1 and Algorithm 2.

An example of how this technique can automatically mitigate nodes with the fault-tree system developed under this program is discussed within the following example: Suppose our goal is to answer the: "if a plane is in landing mode and the altimeter reports to the flight-control system that the plane is 50 feet from touchdown when the plane is really 150 feet from touchdown, will their be any impact on either the control of the engines or control of the wing flaps?" Here we are concerned with the class of hardware failures associated with a broken altimeter, and we are concerned with the output commands from the flight control system to the flaps and the engine.

By defining a broken altimeter as a hazard, a fault-tree can be generated from source code. When the code receives EPA, EPA can simulate this hazard to determine the impact that a broken altimeter has on the control of the airplane. Preliminary experiments with NASA software have shown that EPA works well. We performed analysis on the yawdamp function of an autopilot system. The yawdamp module has six functions: **YAWDAMP, WOUT,**

**FOLAG**, lim, **FCAS**, and **LINTERP**.[5] After defining hazards for **yaw.c**, and applying hardware and software fault simulation, we found that function **YAWDAMP**, when injected with problems, was safe 57% of the time, but 43% of the executions caused the system to enter into hazardous output states. lim was able to cause an unsafe output state nearly 60% of the time. But for **LINTERP**, we were never able to get it to put the system into a unsafe output state. So even though the code based fault-tree suggested that several functions were potential dangers, we were able to empirically mitigate one of these functions.

# 3 Summary

Our framework allows a developer to demonstrate via fault-tree analysis that the code is safe (given that a certain degree of risk is always assumed). Our framework will allow for any number of different techniques (Petri nets, Markov models, state diagrams, EPA, data flow and control flow graphs, formal analysis, etc.) to be incorporated to mitigate possible risks. This framework provides a development technique that will help developers/testers visualize what portions of the code have a static relationship to a hazard, and better yet, code-based fault-trees can be automatically built and displayed by a tool. This paper has concentrated on applying one measure of software observability, EPA, to fault-tree mitigation.

Unfortunately, state-of-the-practice software safety tools are manual and static. To perform extensive, practical, cost effective safety analysis on next-generation software systems, advanced software safety tools are needed that automate the process, change dynamically over time, maintain auditing information, and work well together. This technique provides a means for developing safe software via a highly automated code-based fault-tree generation tool. This provides auditing for all employed risk mitigation procedures.

### Acknowledgements

We thank Jeffery Payne for his comments on an earlier draft. This work was partially funded under NASA-Langley contract: NAS1-20388.

### Contact Address

Voas may be reached at RST Corporation, Suite 250, 21515 Ridgetop Circle, Sterling, VA 20166, USA, phone: (703) 404-9293, jmvoas@rstcorp.com.

# References

[FARB76]    P. MERLIN AND D. FARBER. Recoverability of Communication Protocol–Implications of a Theoretical Study. *IEEE Transactions on Communications*, COM-24:1036–1043, 1976.

---

[5]This code was supplied to us by NASA-Langley; it was generated with a CASE tool and is part of a research avionics flight control system.

[FRIE95]    M. FRIEDMAN AND J. VOAS. *Software Assessment: Reliability, Safety, Testability.* to be published by John Wiley and Sons, New York, 1995.

[LAWS83]    D. J. LAWSON. Failure Mode, Effect, and Criticality Analysis. In J. K. Skwirzynski, editor, *Electronic Systems Effectiveness and Life Cycle Costing*, pages 55–74, NATO ASI Series, F3, Springer-Verlag, Heidelberg, 1983.

[LEVE91]    N.G. LEVESON, S. S. CHA AND T. J. SHIMEALL. Safety Verification of ADA Programs Using Software Fault Trees. *IEEE Software*, pages 48–59, July 1991.

[NASA94]    NASA. NASA Software Safety Standard. Office of Safety and Mission Assurance, June 1994. Interim Report 1740.13.

[SHIM95]    T. J. SHIMEALL. Personal communications.

[VOAS92]    J. VOAS. *PIE*: A Dynamic Failure-Based Technique. *IEEE Trans. on Software Engineering*, 18(8):717–727, August 1992.

[UL94]      UNDERWRITERS LABORATORY INC. Safety Related Software, January 1994. Standard for Safety UL1998, First Edition.

[VOAS94]    J.VOAS AND K. MILLER. Dynamic Testability Analysis for Assessing Fault Tolerance. *High Integrity Systems Journal*, 1(2):171–178, 1994.

[VOAS95a]   J. VOAS AND K. MILLER. Examining Software Quality (Fault-tolerance) Using Unlikely Inputs: Turning the Test Distribution Up-side Down. In *Proc. of Eighth Annual Conference on Computer Assurance.*, National Institute of Standards and Technology, Gaithersburg, MD, June 1995.

[VOAS95b]   J. VOAS AND K. MILLER. Predicting Software's Minimum-time-to-hazard and Mean-time-to-hazard for Rare Input Events. In *Proc. of the International Symposium on Software Reliability Engineering*, (Submitted) Toulouse France, October 1995.

# Session 3
## Formal Methods

# Formal Support for the Safety Analysis of Requirement Models

Ken Chan        Clive Fencott        Barry Hebbron

Centre for Modelling and Simulation

University of Teesside

Middlesbrough

Cleveland TS1 3BA

k.chan@tees.ac.uk        p.c.fencott@tees.ac.uk        bdh@tees.ac.uk

June 9, 1995

### Abstract

This paper demonstrates the use of an established integrated method
(that is Hazard and Operability Studies, Ward and Mellor Essential Models and the Synchronous Calculus of Communicating Systems) to model
and analyse control systems. In particular, we discuss the interplay between traditional hazard analysis techniques and formal methods and
their associated analyses in the context of an integrated model. Also a
process model with tool supports is proposed for developing of a safety-
critical software for real-time control systems. Our approach is illustrated
by a small but realistic industrial case study.

## 1   Introduction

The means for achieving dependability [Lap89] are the methods, tools, and
solutions that enable us to deliver a service on which reliance can be placed
and in which we have the appropriate level of assurance [McD90].

Currently, no single method or technique alone can be used to design a safety
critical system for real-time software. One possible approach is by using integrated methods [Kro93, FKV94] which enhance the assurance of the software
development process. Integrated methods increase both the comprehension
of the system under review and diversity of its construction and verification,
thereby increasing the system's integrity.

Formal methods should be able to contribute to this process and, recently,
researchers have attempted to integrate structured and formal methods [LFT92,
SFD92] to produce an integrated method to facilitate the specification and
verification of the system. This type of integrated method is still inadequate
for producing high integrity software systems because it lacks the analysis of
the safety issues involved.

With this in mind, an important area of current research, is the integration
of safety analysis, structured methods, and formal methods to yield a single
method, capable of capturing a holistic view of a real-time system, for the
production of high integrity real-time software system. This approach aims to
make use of the strengths of each method to compensate for their individual

weaknesses.

The Methods Integration Group, within the Centre for Modelling and Simulation at Teesside, is concerned with research based on integrating three methodologies:

HAZOPS

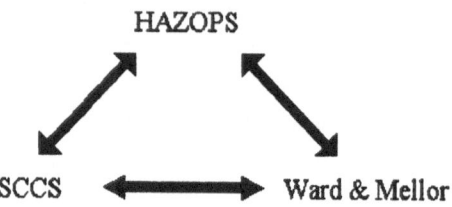

SCCS            Ward & Mellor

Figure 1: The Integrated Method.

- Ward and Mellor Structured Analysis for Real-time systems (Ward & Mellor) [WM85];
- Synchronous Calculus of Communicating Systems process algebra (SCCS) [Fenng, Mil89];
- Hazard and Operability Studies (HAZOPS) [CIS77].

The integration of SCCS and Ward & Mellor has been intensively researched and at present greater attention is placed on the integration of Safety and Hazard Analysis with the rest of the method. In an earlier paper [FH94a] we presented the application of HAZOP studies to Ward and Mellor models and in this paper we document work which seeks to add the formal dimension to the integration. Our use of formal methods is very much applied and can be incorporated readily into traditional process models for safety related systems development. We use the term applied because the mathematics is used primarily as an analysis technique with which to support the investigation of a systems model already developed within a more traditional framework.

In the next section we present a process model for the three-way integration. Section three gives an overview of the formalisms used. In section 4 we show how deviations, identified by a HAZOP study, may be investigated both by testing them against the formal view of the structured model and by formally expressing and verifying them by means of temporal logic. A particular interest is the use of proof tree analysis [Cha94] to identify the error states which give rise to the deviations that have been shown to be properties of the model.

The ASCENT CASE tool[LG95] is used for modelling Ward & Mellor Diagrams while The Edinburgh Concurrency Workbench (CWB) [CPS89, Mol92] is used for the formal verification activities such as automating the proof procedure. Prototype software support is used to generate the formal view of the Ward & Mellor model, on which CWB operates, from data held by ASCENT.

## 2 The Integrated Process Model

In this section we briefly outline our approach in terms of a proposed life cycle which incorporates all three methods HAZOP analysis, Ward & Mellor SA/RT and CCS formal method. Figure 2 presents the life cycle in terms of a Ward & Mellor *style* data flow diagram (DFD).

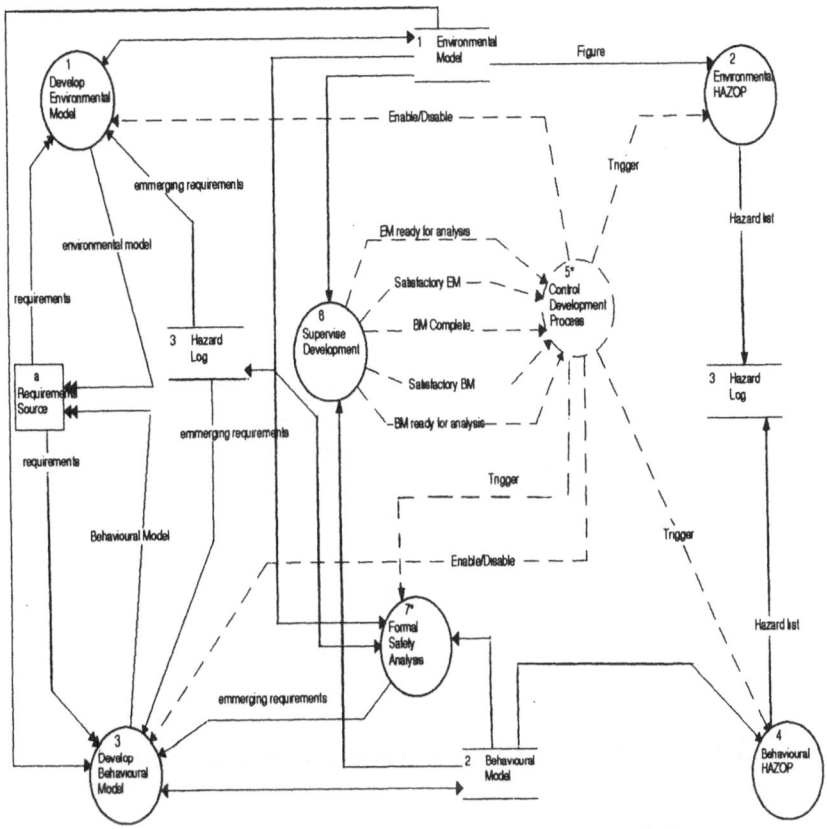

Figure 2: Overview of the process model.

The process has 6 activities:

- Developing the Environmental Model
- Undertaking an Environmental HAZOP
- Developing the Behavioural Model
- Undertaking a Behavioural HAZOP
- Undertaking a Formal Safety Analysis
- Supervision of the development process.

The dynamic behaviour of the activities is described within the specification of the control transformation "control development process"

## 5 Control Development Process

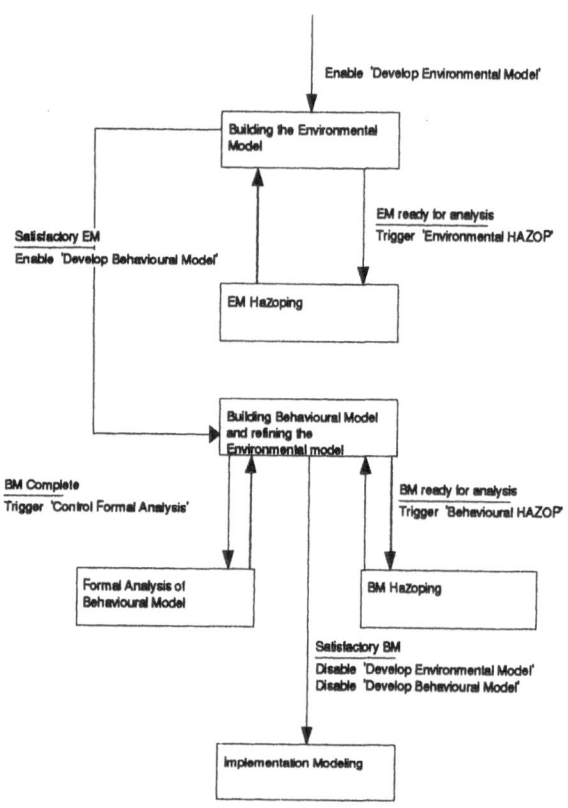

Figure 3: The control development process behaviour.

When commissioned the development of the Environmental model will be initiated (enable 'development of environmental model'). This activity is described in [Heb95]. Once the supervisory activity judges the Environmental model to be sufficiently mature an Environmental model HAZOP study is undertaken ('Trigger Environmental HAZOP'). The resulting Hazard Log then provides emerging requirements that will subsequently influence modification of the model under construction. Normally, once these modifications have been adopted the supervisory activity will indicate satisfactory state of the Environmental model and development of Behavioural model will begin ('Satisfactory EM Enable Develop Behavioural model'). Exceptionally, if the modifications are judged to be substantial, the supervisory activity may initiate a subsequent Environment HAZOP and also, running parallel, the activity Formal Analysis of Behaviour model can be conducted (Trigger 'formal safety analysis'). The behaviour of the data transformation 'Formal Safety Analysis', as follows:

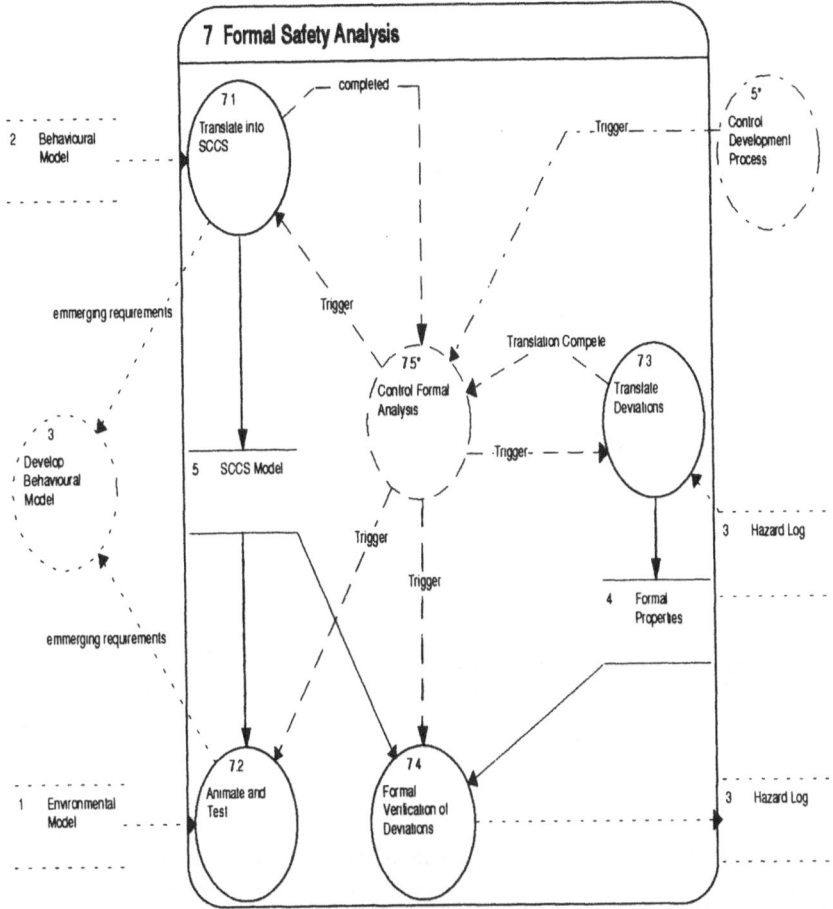

Figure 4: The Formal Safety Analysis process.

This process has 4 activities:

- Translating Ward & Mellor into SCCS
- Verifying the SCCS model by Animation and checking Required System's Properties
- Translating Deviations into Formal notation
- Conducting Formal Verification of the deviations

The dynamic behaviour of the activities is described within the specification of the control transformation "control formal analysis process"

7.5 Control Formal Analysis

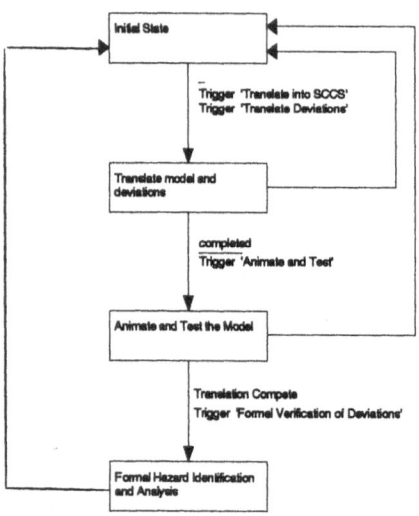

Figure 5: The control formal analysis process behaviour.

The process of translating a Ward & Mellor Behavioural model into SCCS follows a set procedure described in [LFT92]. This process is well understood and can be automatically translated using the ASCENT tool, but translation of the systems' properties (i.e. the event response list from the environmental model) and the hazardous deviations into temporal logic takes a little more effort by using human intuition. These two activate may run in parallel. Once the SCCS model is developed the functionality of the system can be validated against particular deviations by using the CWB tool to animate the model and to check the model has the required systems' properties (Trigger "animate and test"). Eventually the animation of the SCCS model will be deemed to have achieved all it can under constraints of time etc. and the formal verification of the deviations can be conducted (Trigger "formal verification of deviation"). Deviations are formally expressed in temporal logic and CWB is again used to show whether or not the model does indeed satisfy the properties. If the model does satisfy the formally expressed deviation further analysis of the proof tree can be conducted to identify the faulty state(s) within the SCCS model and ultimately these states can be traced back to the Ward & Mellor models.

# 3    Formalisms Employed

## 3.1    Synchronous Calculus of Communicating Systems

The Synchronous Calculus of Communicating Systems (SCCS) is a truly concurrent extension to CCS which allows transitions consisting of multiple atomic actions. In other words For example, consider a simple vending machine,

$$V \stackrel{\text{def}}{=} m : (b\#\overline{c} : V + l\#\overline{c} : V)$$

where m is money; b is big bar; l is little; and c is collect. After some money $m$ is deposited the customer can select a big bar $b$ or a little bar $l$ and at the same moment receive a chocolate bar $c$. The expression $b\#\overline{c}$ stands for the instantaneous performance of the actions $b$ and $\overline{c}$. SCCS is used in our work because true concurrency, rather than interleaving, is necessary in order to fully express the semantics of Ward& Mellor Essential Models. For a fuller discussion of SCCS see [Mil89] while for a discussion of SCCS as a semantic domain for Ward & Mellor Essential models see [FGL+94].

## 3.2 Formal Properties

Formal properties embody design decisions about the nature of the behaviour of the system under development. It must be the designer's intention that the properties will be satisfied for any state which the system may evolve to. Some of the properties which are of interest to a system designer are liveness, deadlock, security, safety etc. The safety property amounts to nothing bad ever happens while a liveness property expresses something good does eventually happen.

The notion of 'necessity' and 'possibility' can be defined in terms of modal logic and the notion of 'always' and 'sometimes' can defined in terms of a branching time temporal logic [ES89]. With these notions interesting properties can be formally expressed. First of all some examples of properties involving 'necessity' and 'possibility', where $tt$ stands for true and $ff$ for false.

| | |
|---|---|
| $[K]$ **P** | $P$ holds after every action in $K$; |
| $\langle K \rangle$ **P** | $P$ holds after any actions in $K$; $\langle K \rangle$ $P$ is dual of $[K]$ $P$, that is $\langle \mathbf{K} \rangle$ **P** $\stackrel{\text{def}}{=} \neg[\mathbf{K}] \neg\mathbf{P}$; |
| $[K]ff$ | expresses an inability to perform any action in $K$; |
| $\langle K \rangle tt$ | expresses capacity to perform any actions in $K$; |
| $[-K]ff$ | expresses that every action but any action in $K$ is impossible; |
| $\langle -K \rangle tt$ | expresses that some action can happen except actions in $K$; |
| $[-]ff$ | expresses deadlock, an inability to perform any action; |
| $\langle - \rangle tt$ | expresses that some action can happen. |

Some useful temporal operators defined in modal mu-calculus are:

| | |
|---|---|
| **AG** $P$ | strong always: $P$ is always true in every trace; |
| **EG** $P$ | weak always: at least one trace in which $P$ is always true; |
| **AF** $P$ | strong eventually: in all traces $P$ becomes possible; $AF$ is the dual of $EG$, that is **AF P** $\stackrel{\text{def}}{=} \neg$**EG** $\neg$**P**; |
| **EF** $P$ | weak eventually: at least one trace in which $P$ become possible; $EF$ is the dual of $AG$, that is **EF P** $\stackrel{\text{def}}{=} \neg$**AG** $\neg$**P**; |
| $P$ **U** $Q$ | strong until: there is one trace in which $Q$ becomes true and $P$ is always true in that until $Q$ becomes true; |
| $P$ **U**$^*$ $Q$ | weak until: as for strong until but $Q$ might never become true; |

where $A$ and $E$ are trace or path quantifiers, for all traces and there exists a trace respectively. For an introduction to the application of branching time temporal logic to CCS/SCCS see [Fenng] and See [Sti91] or a detailed explanation of the underlying theory of the mu-calculus in which the temporal operators are defined.

Using these brief definitions complex formal properties may be represented. For example, consider a home heating system with the property *it is always when the heat system is* **switched off** *then the furnace will* **shut down**". This can be expressed formally as $\mathbf{AG\langle switchedoff \# shutdown\rangle tt}$ assuming the action name **switchoff** and **shutdown** are used in the SCCS formal specification. NB. properties of this form apply to SCCS code generated from Ward & Mellor models Essential Models and would in general be more complex for SCCS models generated in other ways.

## 3.3 Safety Properties

Safety properties may be derived from the hazardous deviations identified from a HAZOP Study. Having identified the hazards within a system, a safety property would express that the hazardous state can not be reached. Considering the home heating system example again and if a hazardous deviation was identified as *when the heating system is* **switched off** *the furnace cannot* **shut down** this may be represented in temporal logic by the formula

$$\mathbf{EF[switchoff \# shutdown]ff}$$

and the safety property can be derived by placing a negation in front of deviation, that is

$$\mathbf{\neg(EF[switchoff \# shutdown]ff}$$

this may be simplified, using its dual modal and temporal operators, to

$$\mathbf{AG\langle switchoff \# shutdown\rangle tt}$$

which translates as "it is always possible to have the actions switch off and shut down."

The formalised deviation can be checked to show if it can become true sometimes in the future of the system, indicating the system may have a potential hazardous consequences and would require further investigation. So the ideal HAZOP study outcome is when all the deviation properties, say $D$, satisfies the system under review, say **System**$X$, where **SystemX** is a SCCS design specification and **D** is a formalised deviation property, that is

$$SystemX \models \neg D$$

which means that **SystemX** satisfies the property saying "always the deviation cannot be performed". When the outcome is

$$SystemX \not\models \neg D$$

which says **SystemX** does not satisfies the safety property and needs further investigation into fault(s) of the design specifications with the proof tree analysis technique. The severity of the fault and the recommendations are left for the HAZOPS team to decide. This provides a powerful tool for adding to the assurance in safety critical software.

# 4 Proof Tree Analysis

Proof trees are generated while checking the required properties, that is the formalised safety properties, against the SCCS design specification. They are based on a set of tableau rules [SW89]. When checking the properties within a system there can be two types of outcomes, the property is either satisfied or not satisfied, that is $\mathbf{E} \models \mathbf{P}$ or $\mathbf{E} \not\models \mathbf{P}$ where $E$ is the SCCS agent expression and $P$ is the temporal property. The former outcome is not so important because it just means the system satisfies the requirement specification, but the latter is of interest because it means the system has a fault in the design and needs further investigation to find the cause of the fault. The tableau proof system can be automated for finite transition systems.

A tree like structure is produced, meaning that proofs are conducted in top-down fashion, when checking whether or not an SCCS agent/state satisfies a particular property. Each step of the proof a state is checked against each part of the property until all of the states are examined and if the outcome does not satisfies the property the concerning failed state(s) can be deduced. We wish to demonstrate the correctness of the assertion that the simple vending machine $V$, introduced above satisfies the property that after putting in some money both a big bar and a small bar can be selected. We wish to demonstrate the correctness of the assertion that $V \models \langle m \rangle tt \wedge [m](\langle b\#\bar{c}\rangle tt \wedge \langle l\#\bar{c}\rangle tt$,

$$
\cfrac{
\cfrac{V \vdash \langle m \rangle tt}{b\#\bar{c}:V + l\#\bar{c}:V \vdash tt}
\quad
\cfrac{V \vdash \langle m \rangle tt \wedge [m]\langle b\#\bar{c}\rangle tt \wedge \langle l\#\bar{c}\rangle tt}{
\cfrac{V \vdash [m]\langle b\#\bar{c}\rangle tt \wedge \langle l\#\bar{c}\rangle tt}{
\cfrac{b\#\bar{c}:V + l\#\bar{c}:V \vdash \langle b\#\bar{c}\rangle tt \wedge \langle l\#\bar{c}\rangle tt}{
\cfrac{b\#\bar{c}:V + l\#\bar{c}:V \vdash \langle b\#\bar{c}\rangle tt}{V \vdash tt}
\quad
\cfrac{b\#\bar{c}:V + l\#\bar{c}:V \vdash \langle l\#\bar{c}\rangle tt}{V \vdash tt}
}
}
}
}{}
$$

All leaves of the proof tree are of the form $F \vdash tt$ and so we have proved the correctness of the sequent $V \vdash \langle m \rangle tt \wedge [m]\langle b\#\bar{c}\rangle tt \wedge \langle l\#\bar{c}\rangle tt$ at the root of the tree. We will now show that the vending machine does not possess the property $\langle m \rangle tt \wedge [m][l\#\bar{c}]ff$, as follows:

$$
\cfrac{V \vdash \langle m \rangle tt}{b\#\bar{c}:V + l\#\bar{c}:V \vdash tt}
\quad
\cfrac{V \vdash \langle m \rangle tt \wedge [m][l\#\bar{c}]ff}{
\cfrac{V \vdash [m][l\#\bar{c}]ff}{
\cfrac{b\#\bar{c}:V + l\#\bar{c}:V \vdash [l\#\bar{c}]ff}{b\#\bar{c}:V + l\#\bar{c}:V \vdash tt \quad l\#\bar{c}:V \vdash ff}
}
}
$$

The unsuccessful leaf is $l\#\bar{c}:V \vdash ff$ which does not meets the requirements laid down for a successful leaf node and we conclude that $V \not\models \langle m \rangle tt \wedge [m][l\#\bar{c}]ff$.

## 4.1 Fault Modes

The analysis of proof trees can be base-on a chain of causality fault to catastrophe [FH94b]. It is possible to distinguish the interesting stages, in an unsuccessful proof tree, leading to a hazardous system which can be roughly characterised as follows:

**Fault:** A fault occur at the beginning of the error SCCS state, that is at the proof tree stage when the system commit itself from a norm state to an error state.

**Failure:** A failure happen at the stage when the SCCS specification does not satisfy the safety property.

**Hazard:** At this stage it indicates the state(s) where in SCCS specification is dis-functioning. This is the end result of the proof tree, that is the leaf nodes of the proof tree

The fault and failure are the events and the hazard is the state caused by the failure event. An accident event cannot be defined by the proof tree because it is to do with the effects of the system's environment, e.g. software indirectly cause deaths.

A proof tree shows the stages leading to a hazard can be identified and hence the proof tree analysis technique can be performed to locate the faults, i.e. the causes of the fault, within the SCCS design specifications. Consider for example the failed prof tree above, $V \vdash \langle m \rangle tt \wedge [m][l \# \bar{c}] ff$, the fault modes be:

- **Fault** $\quad V \vdash \langle m \rangle tt \wedge [m][l \# \bar{c}] ff$
- **Failure** $\quad V \vdash \langle m \rangle tt \wedge [m][l \# \bar{c}] ff$
- **Hazard** $\quad l \# \bar{c} : V \vdash ff$

# 5 The Silly Mixer Case Study

The silly mixer, has been created to provide a small yet effective illustration of the method developed during this research. The requirements for the Silly Mixer are as follows:

> "A liquid mixing system mixes three liquids, A, B and Bulking agent. When the operator presses a start button, liquid A are added to a mixing vat, once 10 litres have been input, the heater is switched on and 2 litres of liquid B are added. At this point the mixer motor is turned on, then a further 3 litres of liquid B are added. When all of liquid B has been mixed in, the bulking agent is added half a litre at a time until the required 1.0 specific gravity value is reached. The heating system is then switch off and a message is sent to the operator. The operator then requests the system to turn off the mixer motor and empty the vat."

> Environment release of liquid A, B and the bulking agent in isolation does not constitute a serious hazard. However, the intermediate product, an A & B Mixture does pose a serious threat. If released, it is to be considered a major hazard. Once the bulking agent is added the resulting A/B/Bulking Agent Mixture is safe.

The partial environment model used is composed of an Event - Response list and Context Diagram, the event/response is as follows:

| No. | Event Name | Response (signals to be sent) |
|-----|-----------|------------------------------|
| 1. | operator start manufacturing | close vat and open liquid A |
| 2. | 10 litres of liquid A added | close liquid A, open liquid B and heat on |
| 3. | 2 litres of liquid B added | mixer on |
| 4. | a further 3 litres of liquid B added | close liquid B and add bulking agent |
| 5. | $\frac{1}{2}$ litres of agent added and specific gravity not reached yet | add $\frac{1}{2}$ litres of bulking agent |
| 6. | $\frac{1}{2}$ litres of agent added and specific gravity reached | inform operator batch is complete and heater off |
| 7. | operator empty vat | mixer off and open vat |

Table 1: Event Response List.

The following diagram represent a partial Ward & Mellor Behavioural Model of these requirement.

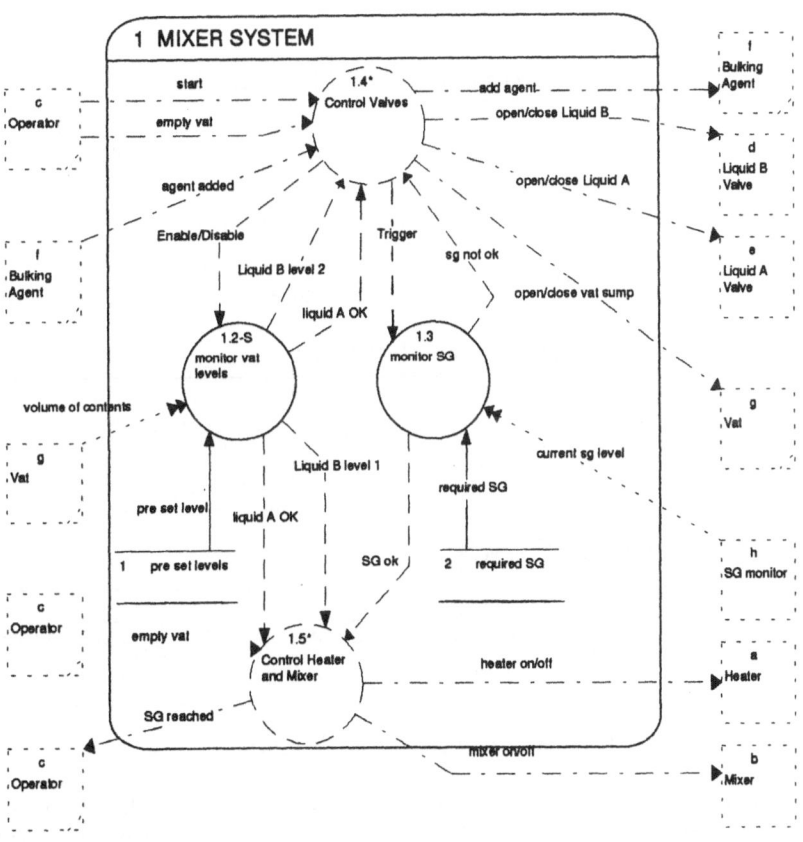

Figure 6: The Silly Mixer Behavioural Model (top level).

The HAZOPS seeks to identify the hazardous deviations within the Behavioural model. Some of the deviations found are shown in Table 2.

| No. | Hazard | Event No. | Deviation |
|---|---|---|---|
| 1. | Overheating an A/B mixture | 6 | No Response: heater off and SG reached signal not sent |
| 2. | Heating an Empty Vat | 5 | Early Event: The system perceived a request to empty the Vat before the required SG is reached |
| 3. | Potentially dangerous liquid A/B mixture | 7 | Part of Response: Open Vat signal not sent |
| 4. | The operator prompted to empty a spoiled batch and the heating remains on | 4 | Early Event: Final liquid B level perceived before the actual level is reached |

Table 2: Some of the Hazardous Deviations Identified.

The Behavioural model can be translated into SCCS automatically, the Silly Mixer top level agent definition is:

$$Silly\_Mixer \stackrel{\text{def}}{=} (ControlValves \mid MonitorSG \mid MonitorVatLevels \mid ControlHeaterMixer) \, / \, L$$

where the set $L$ is the action permission set. This SCCS model would be verify by animation and testings against the hazardous deviations. The deviation from table 2 are formalised, as show in the table 3.

| No. | Formalised Deviation |
|---|---|
| 1. | $EF \, \langle AgentAdded\#CurrentSGLevel_{1.0}\rangle tt$ |
| 2. | $\bigvee EF \, \langle AgentAdded\#CurrentSGLevel_i\#EmptyVat\#\overline{OpenVat}\#\overline{MixerOff}\rangle tt$ where $i \in \{r\!:\mathbf{R} \mid r \geq 0 \wedge r \leq 1.5 \wedge r \neq 1.0\}$ |
| 3. | $EF \, \langle EmptyVat\#\overline{MixerOff}\rangle tt$ |
| 4. | $EF \, \overline{\langle VolumeOfContents_{10}\#\overline{CloseLiquidA}\#\overline{OpenLiquidB}\#HeatOn\rangle} \, \langle VolumeOfContents_{15}\#\overline{CloseLiquidB}\#\overline{AddAgent}\rangle tt$ |

Table 3: The Formalised Deviations.

These deviations are potential hazards, although some may be not hazards at all. Safety properties are derived using the formalised deviations from table 3, that is $\neg EF \, D$ where $D$ is the formalised deviation, this is discussed in the previous section.

Now using Proof Tree Analysis to verify the SCCS model against the safety properties deduced from table 3. Considering for example, the second deviation identified in table 2 and we wish to verifying against the formalised safety property derived from table 3, that is we want to show the correctness of the assertion that

$$SillyMixer \vdash \neg EF \bigvee \langle AgentAdded\#CurrentSGLevel_i\#EmptyVat\#$$
$$\overline{OpenVat\#}\ \overline{MixerOff}\rangle tt$$
$$\text{where } i \in \{r\!:\mathbf{R} \mid r \geq 0 \wedge r < 1.5 \wedge r \neq 1.0\}$$

The proof tree generated would too large to be included in this paper but results indicates the Silly Mixer does satisfies the safety property, that is

$$SillyMixer \models \neg EF \bigvee \langle AgentAdded\#CurrentSGLevel_i\#EmptyVat\#$$
$$\overline{OpenVat\#}\ \overline{MixerOff}\rangle tt$$
$$\text{where } i \in \{r\!:\mathbf{R} \mid r \geq 0 \wedge r < 1.5 \wedge r \neq 1.0\}$$

We can also demonstrate the result by animation, that is the actions sequence:

1. $Start\#\overline{CloseVat}\#\overline{OpenLiquidA}\#VolumeOfContents_0$

2. $VolumeOfContents_{10}\#\overline{CloseLiquidA}\#\overline{HeatOn}\#\overline{OpenLiquidB}$

3. $VolumeOfContents_{12}\#\overline{MixerOn}$

4. $VolumeOfContents_{15}\#\overline{CloseLiquidB}\#\overline{AddAgent}$

5. $AgentAdded\#CurrentSGLevel_{0.5}\#\overline{AddAgent}\#$
   $VolumeeOfContents_{15}$

6. $AgentAdded\#CurrentSGLevel_{1.0}\#\overline{HeaterOff}\#\overline{SGReached}$
   $\#VolumeOfContents_{15}$

7. $emptyVat\#\overline{MixerOff}\#\overline{OpenVat}\#VolumeOfContents_{15}$

Lets take another example but this time with a known error in the Silly Mixer system, in particular the fourth hazardous deviation identified in table 2. Using the associated formalised deviation from table 3, we can verify the Silly Mixer against the derived safety property from formalised deviation, that is we want to show

$$SillyMixer \vdash EF \langle VolumeOfContents_{10}\#\overline{CloseLiquidA}\#\overline{OpenLiquidB}\#$$
$$\overline{HeatOn}\rangle\langle VolumeOfContents_{15}\#\overline{CloseLiquidB}\#\overline{AddAgent}\rangle tt$$

As expected using the proof analyser, the Silly Mixer system does not satisfy the safety property and by animation we can demonstrate this, that is the actions sequence:

1. $Start\#\overline{CloseVat}\#\overline{OpenLiquidA}\#VolumeOfContents_0$

2. $VolumeOfContents_{10}\#\overline{CloseLiquidA}\#\overline{HeatOn}\#\overline{OpenLiquidB}$

3. $VolumeOfContents_{15}\#\overline{CloseLiquidB}\#\overline{AddAgent}$

4. $AgentAdded\#CurrentSGLevel_{0.5}\#\overline{AddAgent}\#$
   $VolumeeOfContents_{15}$

5. $AgentAdded\#CurrentSGLevel_{1.0}\#VolumeOfContents_{15}$

6. $emptyVat\#\overline{OpenVat}\#VolumeOfContents_{15}$

The action $VolumeOfContents_{12}\#\overline{MixerOn}$ is bypassed placing the control transform "Control Valves" into a deadlocked consequently the heater is not turned off and the mixer is never turned on, this concludes that

$$SillyMixer \not\models EF \langle VolumeOfContents_{10}\#\overline{CloseLiquidA}\#\overline{OpenLiquidB}\#$$
$$\overline{HeatOn}\rangle\langle VolumeOfContents_{15}\#\overline{CloseLiquidB}\#\overline{AddAgent}\rangle tt$$

The fault indicates the Silly Mixer model has an error in the specification model. This model must be corrected by modifying the Ward & Mellor Essential Models and then the whole process of HAZOPing the Essential models; the translation of the Behavioural Model into a modified SCCS model; the translation of deviation and systems' properties into temporal logic; and the verification of the SCCS model is conducted.

# 6 Conclusion

This work has demonstrated the feasibility of providing a formal basis for safety analysis. We have successfully shown firm links between SCCS and HAZOPS with the use of proof tree analysis and a proposed process model which incorporates Ward & Mellor, HAZOP Study, and SCCS.

Future work intends to investigate the automation of formal safety analysis. Formal safety analysis can be automated in two ways: firstly by verifying the system under review has the required systems' properties defined by the Event-Response List; and secondly by examining all the deviation, causes and consequences, these deviation can be automatically generated by following the HAZOPS procedure using the property words and guide words.

# 7 Acknowledgement

We would particular like to thank Simon Hooker for his tool supports and his insights for the translation of Ward & Mellor to SCCS.

# References

[Cha94]   K. Chan. The investigation into proof tree analysis for concurrent systems and its practical applications. Master's thesis, University of Teesside, Middlesbrough, 1994.

[CIS77]    CISHEC. A guide to Hazard and Operability Studies. Technical report, The Chemical Industry Safety and Health Council of the Chemical Industries Association Ltd., 1977.

[CPS89]    R. Cleaveland, J. Parrow, and B. Steffen. The concurrency workbench: A semantics based tool for verification of concurrent systems. Technical report, University of Edinburgh, 1989.

[ES89]     E. Emmerson and J. Srinivasan. Branching time temporal logic. *Lecture Notes in Computer Science*, 354, 1989.

[Fenng]    P.C. Fencott. *Formal Methods for Concurrency*. Chapman and Hall, 1995 Forthcoming.

[FGL⁺94]   P. C. Fencott, A. J. Galloway, M. A. Lockyer, S. J. O'Brien, and S. Pearson. Formalising the semantics of ward-mellor sa/rt essential model using a process algebra. *Lecture Notes in Computer Science*, 873, 1994.

[FH94a]    C. Fencott and B. Hebbron. The application of HAZOP studies to integrated requirements models for control systems. In *Proceedings of SAFECOMP 94*, pages 83–92, Anaheim, USA, 1994.

[FH94b]    C. P. Fenelon and B. D. Hebbron. Applying HAZOP to software engineering model. In *Risk Management and Critical Protective Systems*, Altrincham, UK, 1994.

[FKV94]    M. D. Fraser, K. Kumar, and V. K. Vaishnavi. Strategies for incorporating formal specification. *Communication of the ACM*, 37 No.10:74–86, 1994.

[Heb95]    B. Hebbron. Applying HAZOP to a Ward and Mellor essential models. Technical Report TEES-BDH-001, University of Teesside, 1995.

[Kro93]    K. Kronlof. *Methods Integration: Concepts and case studies*. John Wiley and sons, Nokia Research Center, Finland, 1993.

[Lap89]    J. C. Laprie. Dependability: A unifying concept for reliable computing and fault tolerance, LAAS report (1986). In T. Anderson, editor, *Dependability of resilient computers*, chapter 1, pages 1–28. BSP Professional Books, 1989.

[LFT92]    M. Lockyer, P.C. Fencott, and P. Taylor. The integration of structured and formal methods for real-time systems specification. In *5th Conference on Putting into Paractice Methods and Tools for Information Systems Design*. University of Nantes, 1992.

[LG95]     M. Lockyer and G. Griffiths. *ASCENT: Automated Strict CASE Environment at Teesside*. University of Teesside, Cleveland, England, 1995. Technical Manual.

[McD90]    J. A. McDermid. Issues in Developing Software for Safety Critical Systems. Technical report, University of York, England, 1990.

[Mil89]    R. Milner. *Communication and Concurrency*. Prentice-Hall, Hemel Hempstead, 1989.

[Mol92]    Faron Moller. The Edinburgh Concurrency Workbench (version 6.1). Technical report, University of Edinburgh, 1992.

[SFD92]    L. T. Semmens, R. B. France, and T. W. G. Docker. Integrated structured analysis and formal specification techniques. *The Computer Journal*, 35(6):600–610, 1992.

[Sti91]    C. Stirling. An introduction to modal and temporal logics for CCS. *Lecture Notes in Computer Science*, 491, 1991.

[SW89]     C. Stirling and D. J. Walker. Local model checking in the modal mu-calculus. *Lecture Notes in Computer Science*, 351:369–382, 1989.

[WM85]     P.T. Ward and S.J. Mellor. *Structured Development for Real-Time Systems*, volume 1,2,3. Prentice-Hall, 1985.

# Modeling Fault Trees Using Petri Nets

Janusz Górski, Jan Magott, Andrzej Wardziński
Franco-Polish School of New Information and Communication Technologies
(EFP)
Poznań, Poland

## Abstract

The paper presents an approach to safety analysis with the use of Fault Trees. The aim is to provide for more precise analysis of timing dependencies between the events of a tree. A Fault Tree is first represented formally and then converted into a time Petri net. The reachability analysis of the net provides the answer if the hazard can actually occur. The approach is illustrated by an example.

## 1 Introduction

Safety Analysis can benefit from providing it with more formal foundations. Fault Tree Analysis is an important Safety Analysis technique which deals with identification of possible scenarios of system hazards. This technique is widely used in practice with respect to such applications as e.g. nuclear power plants, medical systems and process control systems. A sample Fault Tree is given in Fig.1. The tree relates to a gas burner and provides the scenario of the major hazard related to this system: a leak of gas which may lead to the explosion.

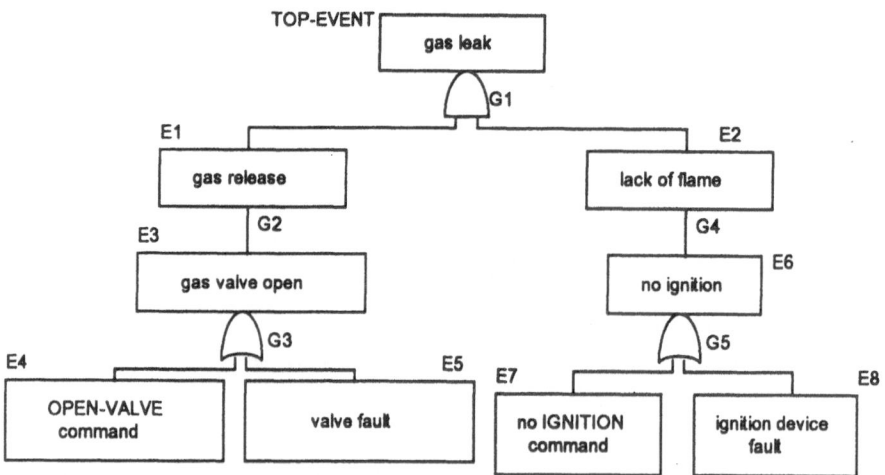

Fig.1. A Fault Tree of the gas burner.

The most common analysis performed with respect to Fault Trees involves assigning probabilities to the events of a tree and then calculating the resulting probability of the system hazard (the top event in the tree) [Vesely'81]. This article proposes an alternative approach based on the development of a formal model of the tree and subsequent analysis of this model. In [BCG'91, Górski'94] it has been proposed that the semantics of a Fault Tree can be based on a formal model which includes events and the causal and time relations between them. In [Górski'94, GW'95] it has been shown how the gates and events from a Fault Tree can be represented by expressions which are interpreted in terms of this formal model. The advantage of such approach is that the semantics of a Fault Tree is given precisely and unambiguously even if the tree is large and the events are related one to another in time, sometimes in complex ways. The paper extends this approach by proposing a method through which a formally specified Fault Tree can be analysed. The key idea is that the formal definition can be used as the specification of an executable model of the tree, expressed in terms of time Petri nets (TPN). The net is then analysed with respect to reachability of the state which corresponds to the hazardous event. The analysis provides a precise answer if the hazard can actually occur. This kind of analysis seems to be particularly useful while introducing to the system additional controls which aim at preventing the hazard occurrence. Then, the same tree (after its appropriate modification) can be analysed again to verify if the introduced modifications guarantee the hazard exclusion. The proposed approach is illustrated in Fig.2.

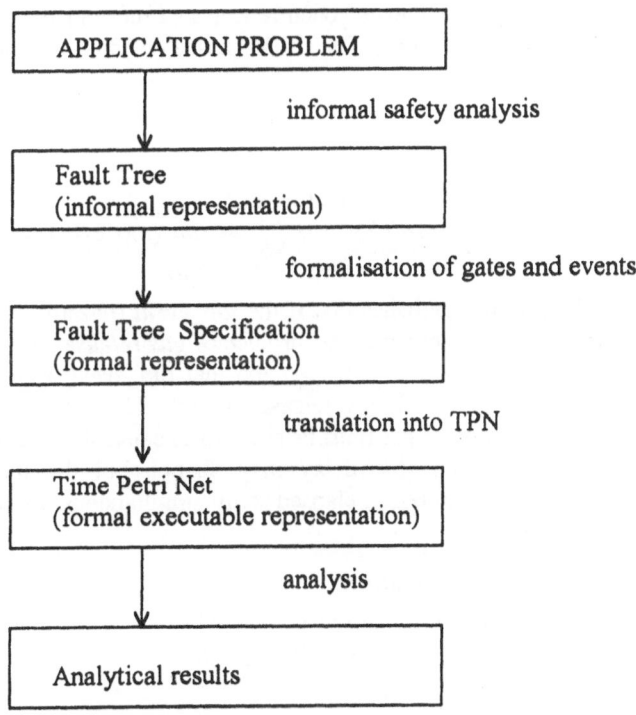

Fig.2. The method of Fault Tree modeling and analysis.

The paper is organised as follows. In the next section we introduce, following [Górski'94, GW'95] the formal definition of the Fault Tree of Fig.1.

given in the CSDM notation. The subsequent section introduces time Petri nets and presents a method of reachability analysis which decides if a particular state of the net is reachable. In order to pass from the formal specification of Fault Tree to its time Petri net representation a set of *translation rules* has been defined in the Section 4 which forms the *CSDM-to-TPN dictionary*. The dictionary consists of entries giving formal specifications of the basic *gate types* that occur in Fault Trees together with their Petri net counterparts. Section 5 gives a TPN representation of the example Fault Tree and discusses possible analyses which can be performed with respect to it. Section 6 concludes the paper.

# 2 Formalisation

The Fault Tree in Fig.1. is specified informally and leaves room for many ambiguities. As an example let us observe that the events E1:*gas release* and E2:*lack of flame* have to be related in time to cause the hazard. Short period of gas release in the presence of the lack of flame is perfectly normal and always occurs, before the ignition takes place. To provide for removal of such (and many other) ambiguities it has been proposed to extend the FTA method with formal semantics. In [BCG'92, Górski'94] such an extension was introduced under the name CSDM (*Common Safety Description Model*). The meaning of gates and events of a Fault Tree is given in terms of event occurrences in time and their relationships are defined by logical expressions. The CSDM based definition of the tree from Fig.1. is given below.

**G1 - Causal AND:**

$occur(\text{top\_event}) \Rightarrow$
$$occur(e1) \wedge occur(e2) \wedge duration(e1 \wedge e2) > t_{Gmin} \wedge$$
$$\max(\; start(e1),\; start(e2)) + t_{Gmin} <= start(\text{top\_event}) <=$$
$$\max(\; start(e1),\; start(e2)) + t_{Gmax}.$$

The above definition establishes that if the *top_event* (hazard) occurs then it must have been caused by the common occurrence of *gas release* and the *lack of flame* (the overlap period greater than $t_{Gmin}$) and that *top_event* occurs at least $t_{Gmin}$ seconds after the unflamed gas release started. This delay represents that a dangerous concentration of gas must be reached to cause the hazard. Depending on the ventilation conditions the delay may change but the hazard must have happened before the time $t_{Gmax}$ elapsed (only simultaneous occurrence of e1 and e2 could give rise to it).

**G2, G4 - identity transformations:**
$$occur(e1) \Rightarrow occur(e3) \wedge e1 = e3$$
$$occur(e2) \Rightarrow occur(e6) \wedge e2 = e6$$

**G3 - causal OR:**

$occur(e3) \Rightarrow$
$$(\; \exists\; e4 \;\cdot( duration(e4) > t_1 \;\wedge\; start(e4) + t_1 \leq start(e3)) \vee$$
$$(\; \exists\; e5 \;\cdot\; start(e5) < start(e3)\; ).$$

This definition states that the gas valve becomes open as the result of the OPEN VALVE command (with the delay $t_1$) or in effect of the valve fault (the valve remains open disregarding the commands). Here, the delay $t_1$ represents the causal dependency between e4 and e3.

**G5 - generalization OR:**

$$occur(e6) \Rightarrow occur(e7) \wedge e6 = e7 \vee occur(e8) \wedge e6 = e8$$

*No ignition* means that either there is a fault of the ignition device or no "ignition_on" command has been given.

## 3 Time Petri Nets

Let **N** be a set of natural numbers $N=\{0,1,2,...\}$, and **Q** be a set of non-negative rational numbers. A *Time Petri Net* (TPN) [MF'76] is a tuple $N = <P, T, B, F, M_0, SI>$ where $P$ is a finite set of *places*, $T$ is a finite set of *transitions*, $B$ is the *backward incidence function* $B: P \times T \rightarrow \{0,1\}$, $F$ is the *forward incidence function* $F: T \times P \rightarrow \{0,1\}$, $M_0$ is the *initial marking* $M_0: P \rightarrow N$, and *SI* is the *static interval* function $SI: T \rightarrow Q \times (Q \cup \{\infty\})$.

We illustrate the above definition using a time Petri net from Fig.3.

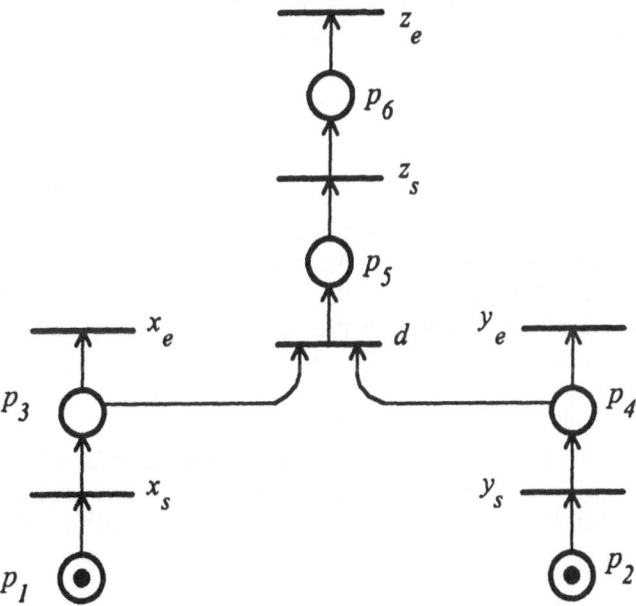

Fig.3. A time Petri net.

Places are represented by circles, transitions are represented by bars. The backward incidence function is expressed by arcs directed from places to transitions, whereas the forward incidence function is illustrated by arcs directed from transitions to places. A directed arc from the place $p$ to the transition $t$ exists iff $B(p,t) = 1$. The initial marking expresses the number of tokens in each place.

The transition $t_i$ is *enabled* by the marking $M$ if there exists at least one token in each input place of this transition, i. e. $(\forall p \in P)(M(p) \geq B(p,t))$.

The *static firing interval* function is interpreted as follows: $SI(t_i) = <SEFT_i, SLFT_i>$, where $SEFT_i(SLFT_i)$ is the earliest (latest) firing time of the transition $t_i$, and $SEFT_i \leq SLFT_i$. The times $SEFT_i$, $SLFT_i$ are relative to the absolute time instance $\tau \in Q$ at which $t_i$ became enabled. The transition $t_i$ cannot be fired before the time $\tau + SEFT_i$, can be fired at or after $\tau + SEFT_i$, and if it remains enabled during the time interval $[\tau, \tau + SLFT_i)$ then it has to be fired at the time $\tau + SLFT_i$.

Let us assume the following values of the $SI$ function for the net from Fig. 3.:

$SI(x_s) = SI(y_s) = <0,\infty>$,

$SI(z_s) = <0,0>$,

$SI(x_e) = <t_{Xmin}, t_{Xmax}>$, $SI(y_e) = <t_{Ymin}, t_{Ymax}>$, $SI(z_e) = <t_{Zmin}, t_{Zmax}>$,

$SI(d) = <t_{Gmin}, t_{Gmax}>$.

*States* of a TPN are represented by pairs $S = <M,I>$ where $M: P \rightarrow N$ is a marking and $I$ is the set of *dynamic firing intervals* (one interval for each transition). For each transition $t_i$ enabled by $M$ the corresponding interval has the form $<DEFT_i, DLFT_i>$ where $DEFT_i(DLFT_i)$ denotes the dynamic earliest (latest) firing time of the transition. The intervals depend on current time. At the time $\tau$ when the transition $t_i$ becomes enabled, $<DEFT_i, DLFT_i> = <SEFT_i, SLFT_i>$ holds. If the transition remains enabled at the time instant $\tau + \tau'$, the dynamic interval of $t_i$ is given as follows:

$$<DEFT_i, DLFT_i> = < \max(0, SEFT_i - \tau'), SLFT_i - \tau'>.$$

Because the time domain is dense, so the state space of TPN can be infinite. In order to get a finite representation of TPN behaviour, the notion of the *state class* has been introduced [BM'82]. A state class can represent infinitely many states of TPN. A state class $C$ is given by a pair $C = <M, D>$ where $M$ is a marking (common for all states $S = <M, I> \in C$ and $D$ is the *firing domain* of the class. $D$ is expressed by a set of linear inequalities. The inequalities give the time intervals for the firing of each transition enabled by $M$.

To demonstrate how the state classes for TPN are defined let us consider the net from Fig. 3. The first class to be considered is that related to the initial marking $C_0 = <M_0, D_0>$. The initial firing domain $D_0$ is defined by the following inequalities:

$0 \leq \tau_{x_s} < \infty$,

$0 \leq \tau_{y_s} < \infty$

where $\tau_{x_s}, \tau_{y_s}$ denote the relative times of firing the corresponding transitions $x_s, y_s$. The transitions can be fired at any time instant within the limits given by their static intervals. Then we examine all markings which are reachable from the initial class $C_0$, by firing one of the enabled transitions. The corresponding set of inequalities is then determined for each of the new markings. Only those markings correspond to reachable reachable state classes, for which the associated inequalities do not lead to contradiction.

For the net of Fig.3. two possible transitions have to be examined: $x_s$ and $y_s$. Let us suppose that the transition $x_s$ is fired at the time instant $0 \le \tau_{x_s}^f < \infty$. Firing $x_s$ results in a new marking $M_l$, where $M_l(p_3) = M_l(p_2) = 1$, and $M_0(p) = 0$ for the remaining places. The dynamic firing interval for the transition $y_s$ is computed according to the formula:

$$< DEFT_{y_s}, DLFT_{y_s} > \; = \; < \max(0, 0 - \tau_{x_s}^f), \infty - \tau_{x_s}^f >\; = \;<0, \infty>.$$

For $x_e$, as it became enabled at time instant $\tau_{x_s}^f$, we have that the corresponding dynamic firing interval is

$$< DEFT_{x_e}, DLFT_{x_e} > \; = \; < SEFT_{x_e}, SLFT_{x_e} > \; = \; < t_{Xmin}, t_{Xmax} >.$$

Consequently, the resulting state class $C_l = <M_l, D_l>$ is given by the following inequalities:

$$0 \le \tau_{ys} < \infty$$

$$t_{X \, min} \le \tau_{Xe} \le t_{X \, max}$$

Now, let us suppose that $y_s$ fires for the state class $C_0$. This results in the state class $C_2 = <M_2, D_2>$, where $M_2(p_1) = M_2(p_4) = 1$, $M_2(p)$ for the other places. The firing domain $D_2$ is described by:

$$0 \le \tau_{Xs} < \infty$$

$$t_{Y \, min} \le \tau_{Ye} \le t_{Y \, max}$$

For the class $C_1$ two transitions are enabled, namely $y_s$ and $x_e$. The transition $x_e$ has to fire before relative time txmax. If the transition $y_s$ fires first then it must have happened before txmax (otherwise, according to the TPN semantics, $x_e$ would have fired first). Consequently we have that the time instant $x$ of firing $x_e$ or $y_s$ (whichever happens first) has to satisfy $0 \le x \le t_{ymax}$. In general, if the transition $t_i$ is to be fired for the marking $M$, then the relative firing time $\tau_i$ of this transition has to satisfy the relation

$$DEFT_i \le \tau_i \le min \; \{ DLFT_j \mid t_j \text{ enabled by } M \}. \tag{*}$$

If this requirement contradicts with the dynamic firing interval of $t_i$ then the firing $t_i$ is impossible and the corresponding state is unreachable.

The reachability graph for the state classes of the example is given in Fig.4. Each node of the graph characterizes the marking and the dynamic firing intervals of the transitions enabled in a given state class. Let us observe that firing of d while being in the state class $C_4$ is possible only if

$$t_{Gmin} \; <= \; t_{Xmax} \quad \text{and} \quad t_{Gmin} \; <= \; t_{ymax} \tag{**}$$

holds. If this condition is not fulfilled, the condition of firing of d would lead to contradiction with the condition (*). Thus, the reachability of the state class $C_6$ depends on the condition (**).

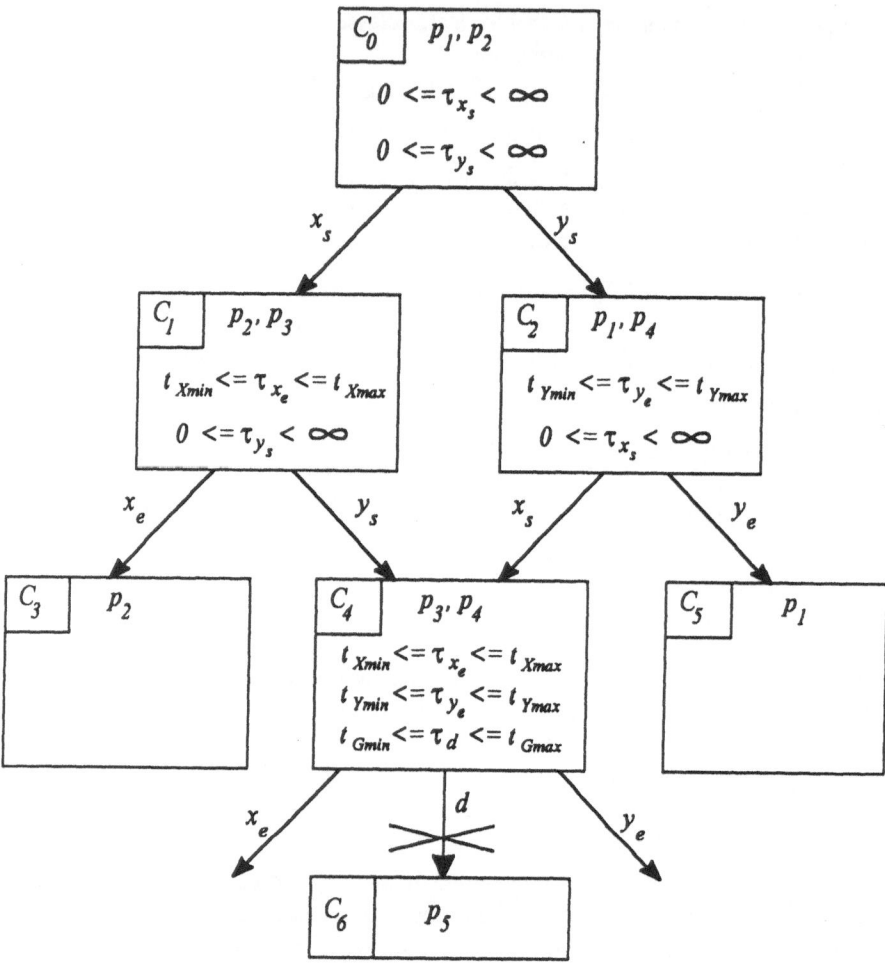

Fig.4. The reachability graph for the net from Fig. 3.

## 4 Dictionary

The CSDM based definition of a fault tree gate can serve as the specification of the TPN construct which simulates the gate behaviour. This results in the *CSDM to TPN dictionary* which for each type of gate gives its CSDM specification and the corresponding TPN structure. Examples of the entries of such a dictionary are given below.

**Causal AND:**
This gate relates the input events x and y with the output event z. It is required that x and y overlap on the interval greater than $t_{Gmin}$ and that z must occur within the time period $< t_{Gmin}, t_{Gmax} >$ after the common occurrence of x and y.

$$occur(z) \Rightarrow$$
$$occur(x) \wedge occur(y) \wedge duration(x \wedge y) > t_{Gmin} \wedge$$
$$max(\ start(x),\ start(y)\ ) + t_{Gmin} <= start(z) <= max(\ start(x),\ start(y)\ ) + t_{Gmax}$$

The corresponding TPN is given below.

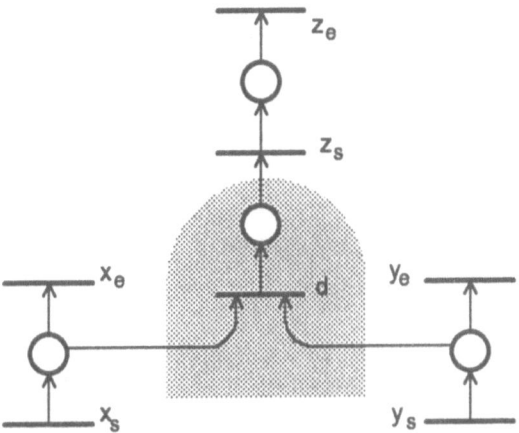

where:
- $x_s$ and $x_e$ represent the start and the end transitions of the event x (and similarly for the events y and z),
- $SI(x_s) = <0,0>$, $SI(y_s) = <0,0>$, $SI(z_s) = <0,0>$,
- $SI(x_e) = <t_{Xmin}, t_{Xmax}>$, $SI(y_e) = <t_{Ymin}, t_{Ymax}>$, $SI(z_e) = <t_{Zmin}, t_{Zmax}>$. The time intervals given here specify duration of the events x, y and z.
- d is the transition which represents the delay (time difference between the occurrence of the cause events and their consequence). $SI(d)=< t_{Gmin}, t_{Gmax}>$.

**Causal OR:**

$occur(z) \Rightarrow$
$( \exists x \cdot ( duration(x) > t_1 \wedge start(x) + t_1 \leq start(z) ) ) \vee$
$( \exists y \cdot ( duration(x) > t_2 \wedge start(y) + t_2 \leq start(z) ) ).$

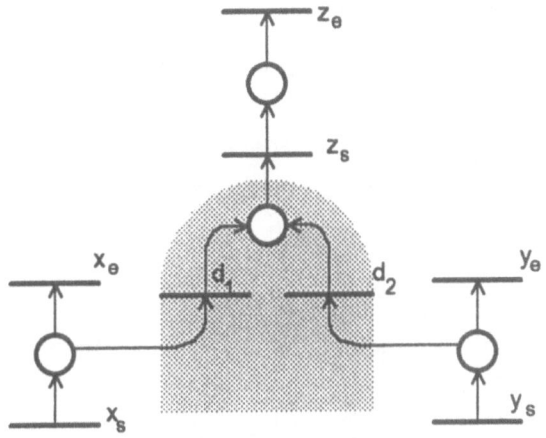

- $SI(x_s) = <0,0>$, $SI(y_s) = <0,0>$, $SI(z_s) = <0,0>$,

98

- $SI(x_e) = <t_{Xmin}, t_{Xmax}>$, $SI(y_e) = <t_{ymin}, t_{ymax}>$, $SI(z_e) = <t_{Zmin}, t_{Zmax}>$. The time intervals given here specify duration of the events x, y and z.
- $SI(d_1) = <t_1, \infty>$,  $SI(d_2) = <t_2, \infty>$. Transitions $d_1$ and $d_2$ represent the delay between the cause and its effect.

**Generalisation OR:**

$$occur(z) \Rightarrow ( occur(x) \wedge x = z ) \vee$$
$$( occur(y) \wedge y = z )$$

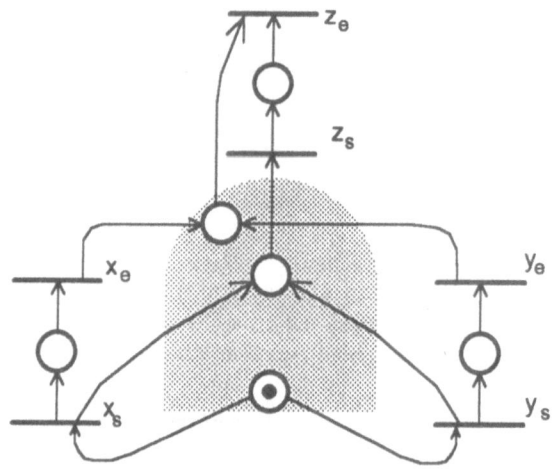

- $SI(x_s) = <0,0>$,  $SI(y_s) = <0,0>$, $SI(z_s) = <0,0>$,
- $SI(x_e) = <t_{Xmin}, t_{Xmax}>$,  $SI(y_e) = <t_{ymin}, t_{ymax}>$, $SI(z_e) = <0,0>$.

**Generalisation AND:**

$$occur(z) \Rightarrow occur(x) \wedge occur(y) \wedge overlap(x, y ) \wedge$$
$$max( start(x), start(y) ) = start(z) \wedge$$
$$min( end(x), end(y) ) = end(z)$$

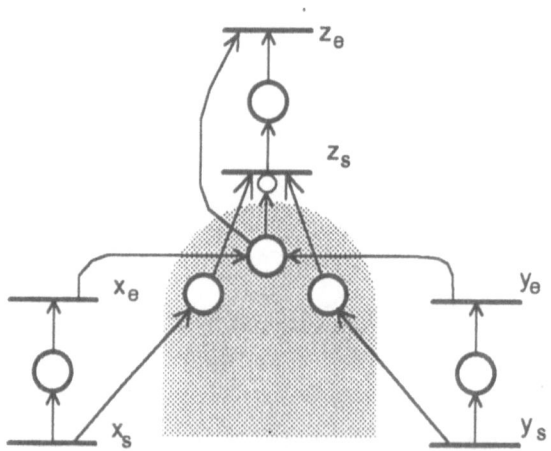

- $SI(x_s) = <0,0>$, $SI(y_s) = <0,0>$, $SI(z_s) = <0,0>$,
- $SI(x_e) = <t_{Xmin}, t_{Xmax}>$, $SI(y_e) = <t_{ymin}, t_{ymax}>$, $SI(z_e) = <0,0>$.

# 5 Analysis

The CSDM-to-TPN dictionary can be used for systematic conversion of a (formally specified) Fault Tree into its TPN counterpart. Then, through the reachability analysis of the state classes we can verify if, starting from the initial marking of the net, a state class which includes in its marking the place representing the hazard is reachable. The net corresponding to the Fault Tree from Fig. 2. is shown below. The reachability analysis shows that depending on the values of the times $t_{Gmin}$, $t_{Gmax}$ and $t_2$ the hazardous state is reachable or not. This gives the hints on how the system could be modified to avoid the hazard.

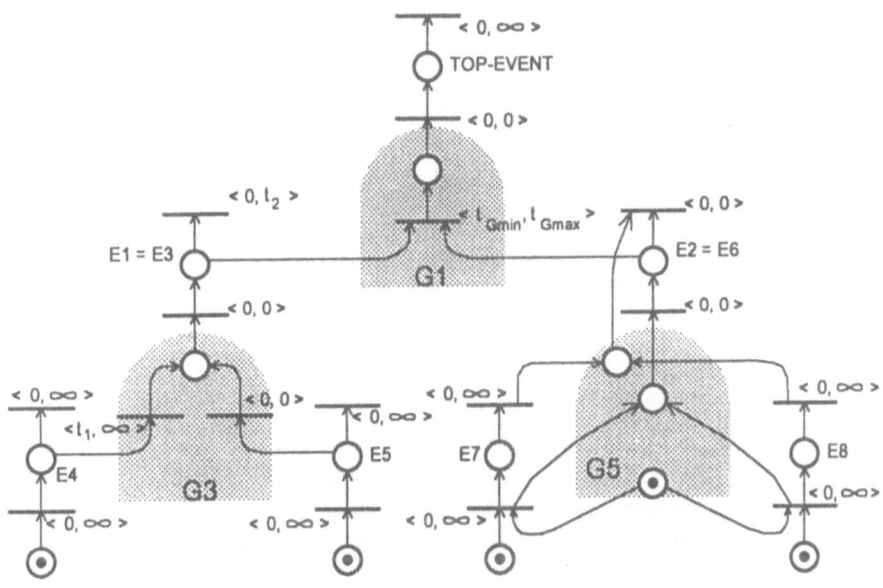

Fig. 5. Time Petri net model of the example Fault Tree.

# 6 Conclusion

The approach presented in the paper provides a method which can be used to analyse Fault Trees with complex time dependencies. The method gives a precise answer if the hazard is reachable. The analysis provides some hints on how the system should be modified to prevent the hazard occurrence. The advantage of this approach is that the existing tools for TPN reachability analysis can be exploited. To prove its practicallity the method should be validated through more case studies.

We compare our method with the other Petri net approaches to the Safety Analysis. In [LS'87] a Petri net model is built for the system to be studied. Then the reachability graph analysis is performed. This graph covers the hazardous states and many normal states as well. The number of normal states could be very large. In our method, the reachability graph is generated for hazardous states only. In the approaches presented in [SMG'91], a system is modelled by time Petri net and then the fault tree is constructed. In our approach a Fault Tree is developed first and then after its formalisation, a time Petri net is built and analysed.

# References

[BCG'91]     R. E. Bloomfield, J. H. Cheng, J. Górski, *Towards A Common Safety Description Model,* Proceedings of Safecomp'91, Pergamon Press, 1991

[BM'82]      B. Berthomieu, M. Menashe, *A State Enumeration approach for analyzing time Petri Nets,* Proceedings of 3rd European Workshop on Applicatrions and Theory of Petri Nets, Varenna, Italy, Sept. 1982

[Górski'94]  J. Górski, *Extending Safety Analysis Techniques With Formal Semantics,* In Technology and Assessment of Safety Critical Systems, (F.J. Redmill and T. Anderson, Eds.), Springer-Verlag, 1994

[GW'95]      J. Górski, A. Wardziński, *Formalizing Fault Trees,* Proceedings of SCSS'95, Brighton, UK, (F.J. Redmill and T. Anderson, Eds.), Springer-Verlag, 1995.

[LS'87]      N. G. Leveson, J. L. Stolzy, *Safety Analysis Using Petri Nets,* IEEE Transactions on Software Engineering, Vol. 13, No .3, March 1987, pp. 386-397

[MF'76]      P. Merlin, D. J. Farber, *Recoverability of Communication Protocols - Implications of a Theoretical Study,* IEEE Transactions on Communications, Vol. 24, No. 9, Sept. 1976, pp. 1036-1043

[SMG'91]     T. J. Shimeall, R. J. McGraw, Jr., J. A. Gill, *Software Safety Analysis in Heterogenous Multiprocessor Control System,* Proceedings of IEEE Annual Reliability and Maintainability Symposiom, 1991, pp. 290-294

[Vesely'81]  W. E. Vesely et el., *Fault Tree Handbook,* Nureg 0492, US Nuclear Regulatory Commission, 1981

# The Application of Formal Methods to Railway Signalling Systems Specification and the Esprit III Project CASCADE

A J Harrison
Railtrack PLC
London, UK

I D R Shannon
Prism Engineering
London, UK

## Abstract

This paper discusses the preparation of a Requirements Specification for a railway signalling interlocking system using the Z notation. It then reviews the use of the specification project as a case study within the Esprit III project, CASCADE, which aims to develop a generic assessment method for safety critical software-based systems.

# 1    Introduction

The structure of the railway industry within the UK is undergoing fundamental changes. On the 1st April 1994, responsibility for infrastructure ownership and train regulation management transferred from British Rail, the UK national railway authority since 1948, to a new company, Railtrack PLC. British Rail is being divided into a number of train operating companies, along with other railway-support businesses. This re-structuring is occurring as a precursor to the transfer of ownership from the public to the private sector.

As the infrastructure controller for the UK mainland railway network operating under a licence issued by the Secretary of State for Transport, Railtrack generates revenue through the effective exploitation of train paths over its infrastructure. By purchasing train paths from Railtrack, train operating companies are provided with a managed routing for a railway service between two points on the railway network. The successful provision of train paths depends upon the implementation of cost-effective, functional and reliable infrastructure elements, such as railway track, electrical distribution systems and signalling.

Effective railway signalling systems are fundamental to the safe provision of train paths. Consequently, signalling systems have the potential to contribute to the success of Railtrack's business. Initiatives such as the European Train Control System project, along with work being carried out under the auspices of

CEN/CENELEC aimed at harmonising railway standards between member states, are catalysts to the dissolution of the boundaries which have traditionally existed in the railway signalling marketplace.

Effective exploitation of the wider competitive market for signalling systems is a component of Railtrack's business philosophy. To realise the benefits of an open market, Railtrack would like to be able to procure both bespoke and commercial-off-the-shelf (COTS) systems which meet Railtrack requirements in terms of both functional and non-functional attributes.

On vesting, Railtrack inherited, among others, two related projects started by the Network SouthEast Division (NSE) of British Rail in late 1993, both of which support the company's railway signalling systems procurement philosophy. The first project aimed to capture, in an implementation-independent form, current UK railway signalling practice as a specification of requirements. The second project was the NSE involvement in a European Community collaborative project, undertaken within the ESPRIT III Framework, known as CASCADE, the acronym for "Certification and Assessment of Safety Critical Application Development".

This paper introduces the background to both projects and discusses the use of the specification project as a case study within CASCADE. The work undertaken by both projects is reviewed and the intermediate results and forward plans outlined.

# 2 Railway Signalling

## 2.1 Evolution

In the UK, railways evolved from the use of horse-drawn wagons to carry coal and ore from mines during the 17th and 18th century. Hard-surfaced, guided ways eventually replaced the original rubble roads and, following the invention of the steam engine, this power source replaced the horse.

From its early days, the railway had to cope with two fundamental problems [Heard93]. Low friction at the rail/wheel interface meant that, once moving at speed, vehicles required long distances in which to stop. The second problem was that, in order for trains to divert from one route to another, a moveable portion of track was required. To ensure the safety of passengers and prevent damage to both track and trains, it is vital that these moveable portions of track, known as points, do not move while a train is passing over them.

The continued resolution of these two problems remains the objective of the railway signal engineer, who must ensure that:

- there is always sufficient distance between any two trains so that the rear train can stop short of the train in front of it; and
- points are prevented from moving while a train is passing over them.

To meet these objectives, the principles of UK signalling practice have evolved over a period of 150 years. Essentially, these principles involve:

- the use of pre-defined track spacings, "block sections", to keep trains apart;
- the locking of signal operation with both points and level crossings; and
- the provision of information to train drivers on divergence to left or right from the straight-through route at track junctions, with train speed regulation assisted by signals released on the approach of trains.

The term "interlocking" refers to the equipment which prevents points and signals from operating in an unsafe situation. The term "controls" refers to the logic which the interlocking has to satisfy.

## 2.2   UK Regulatory Framework

The UK government has regulated the operation of the railways since 1840. Between 1869 and 1875, three basic principles were established [Holden93] which govern the relationship between a UK railway company and the government:

- the responsibility for the safety of a railway operation rests with the railway company;
- once a railway has been opened, the railway company is responsible for maintaining it to the standard necessary to ensure public safety; and
- the government is not responsible for the safety of systems designed and built by the railway company.

In support of these principles, numerous regulations have been published over the years. Most notable for railway signal engineers are:

- the Regulation of Railways Act (1889), which mandated the interlocking of points and signals and gave powers for making statutory orders which could force any railway to adopt and use the block system for passenger railways;
- the Department of Transport Requirements for Passenger Railways and Recommendations for Goods Lines (1950), which stated the minimum standards for a railway signalling system. A new version of these requirements is currently being prepared;
- the Transport and Works Act (1992), which changed the way that authorisation for the construction of a railway is obtained. The Approval of

Works, Plant and Equipment Regulations prescribes the way in which approval to take any new or altered works, plant or equipment into use on the railway may be obtained; and

- the Railways Safety Case Regulations (1994), which aim to ensure that the management structures, operating procedures and provision of equipment necessary for railway safety have been set up. Railtrack is required to produce a Railway Safety Case for validation by the Health and Safety Executive and all train operators must prepare a Railway Safety Case, for validation by Railtrack, covering their operation over Railtrack infrastructure.

It is essential that Railtrack can demonstrate compliance with these statutory responsibilities, which cover both the operation of the business and also the broad functionality of systems used within the business, together with the mechanisms to be used for the approval of such systems.

## 2.3   Railway Signalling Systems

Against this background of evolving legal and technical requirements, signal engineers have devised systems, based on the technology of the day, to safely regulate train movement.

Early systems used mechanical levers and locking to enable the control of a small area of track. As electromagnetic technology developed, relays were used in combination to provide systems capable of controlling much larger areas of track. The development of the microprocessor enabled a new generation of signalling systems to evolve, and British Rail commissioned its first processor-based interlocking system at Leamington Spa in 1985. Processor-based systems offered significant cost savings over traditional electromagnetic systems in the areas of installation, testing, commissioning and long term support. Since then, processor-based systems have been widely used in the UK for interlocking, train detection and signalling system display.

The application of these diverse engineering technologies shared a series of common threads:

- the fundamental architecture of signalling systems did not change, with new technology directly replacing the functionality of the old;
- each implementation adopted an approach based essentially on the concept of "intrinsic fail-safety"; and
- for each implementation, the railway authority worked alongside the supply industry to develop systems and equipment over a lengthy period.

For modern programmable systems, used within a commercial railway environment, the market is fundamentally different from that which previously existed for traditional technologies. Rapid advances in the technology are being matched by a desire for shorter times to market. The fail-safe engineering approach is not as appropriate to processor-based systems as it was for more traditional systems. The potential offered by processor-based systems means that railway authorities are initiating a re-appraisal of the traditional architecture of signalling systems with the aim of producing a more integrated approach to train control.

Consequently, suppliers are now more likely to take the initiative in developing generic systems capable of specific configuration and application within a variety of different railway networks.

Railtrack's aspiration is to be able to benefit from the commercial advantages offered by processor-based systems in terms of increased functionality, improved control capacity and lower costs for train control whilst not importing unacceptable risk onto the railway network. This aspiration, fuelled by the need to react positively to changes in the market and to demonstrate compliance with statutory requirements, was behind the decision to undertake work on the two NSE projects.

# 3 Aims and Objectives of the Specification Project

The aim of this project was to develop a specification which encapsulates current UK railway signalling practice. The key objectives of the specification were:

- to specify requirements for UK railway signalling systems;
- to be technology-independent;
- to be system architecture-independent; and
- to be a complete, unambiguous, consistent and traceable description of signalling practice.

The purpose of the specification is to provide a basis for:

- the assessment of COTS systems developed by industry against Railtrack requirements;
- guiding the development of new systems to Railtrack requirements by industry; and
- controlling the enhancement and modification of existing systems.

A contract for the production of the requirements specification was awarded in September 1993.

# 4 Formal Methods

'Formal methods' is a generic term given to the use of discrete mathematics to describe the properties of a system. The use of a mathematical notation allows the system requirements to be specified without detailing how they will be achieved.

This abstraction provides several benefits including the ability to reason confidently about the specification without the ambiguity of natural language.

There are a number of excellent books and papers that the reader may consult for a more detailed treatment of the subject, including [Denvir86] and [Spivey89].

It was decided for the purposes of the NSE project to utilise the Z notation in the definition of a conceptual model of the current signalling system employed on Railtrack infrastructure. This notation was chosen following a wide-ranging review of specification methodologies, as it seemed to offer the best combination of industry-acceptance, language stability and information analysis capability. The use of a notation, such as CSP, to support the Z model was considered, although it was concluded that the actual complexity of the functionality of an interlocking did not merit its additional use. An important point is that this project represented the first application of formal methods within British Rail, and that signal engineers are not trained in the use of techniques such as formal methods. As such, we were very aware of both the need to "walk before running"; and that other fundamental principle of safety critical systems development, "simplicity in all things". In the event, a regular expression notation, based on the FUSION design method, has been used within the specification to represent some sequences of operations.

The use of mathematics removes the ambiguity of natural language from the description of the properties of a system, although it should be noted that an English commentary was also produced in parallel with the mathematics. Being based on mathematics, this commentary is more precise and less ambiguous than if it had been written without first developing the mathematical specification of the system properties. The use of the English commentary will be necessary when verifying the specification of UK signalling practice as it will provide the interface between signalling and software engineers.

It was thought that the increased clarity of the English commentary that would result from the specification of UK signalling practices in mathematics would alone justify the effort of adopting a formal notation within the project.

# 5 CASCADE

The CASCADE project focuses on the assessment and certification of software intensive safety critical systems. The need for assessment is driven, in CASCADE's view, by the increasingly sophisticated uses of computer based control systems and the economic and technical advantages of systems incorporating such components. Where computers are used in applications with safety implications, there is an emerging requirement either to develop systems which comply with the safety regulations and internationally recognised standards, or to purchase systems that comply with the regulations and standards. In particular, procurers require independent assessment of conformance that is cost effective, objective and which constrains liability in the event of failure. The suppliers of assessment services must ensure that their offerings be objective, repeatable, cost effective, technically sound, generic (in so far as different technical solutions chosen by implementors can be accommodated) and risk-limiting in both a technical and legal sense.

The CASCADE project views assessment and certification as an integral part of the system safety case. The safety case is required to present the evidence for the justification of those safety functions that the system must exhibit. The technical work of the project has so far focused on how to articulate these properties and on determining how current methods, techniques, tools and management practices provide evidence to support assessment judgement.

A number of interesting technical questions have surfaced as the project has tried to address this problem:

- how do we classify systems, ie. how do we determine their safety functional requirements and safety integrity requirements, and decide what development and assessment methods are appropriate to different classes of systems?
- what is the role of safety critical standards in a safety case? what is needed to show compliance with an industrial standard, such as CENELEC prEN50128?
- how important are the evidential contributions from product and process assessment?
- what constitutes an adequate audit trail for the development of safety related systems and what data should be collected to support assessment judgements?
- how can standards and technical guidelines be written so that compliance can be objectively established?
- how should we approach the development and assessment of tools which are themselves used for the development and assessment of safety related systems?
- how can formal and semi-formal methods be effectively deployed in the development and assessment of safety critical systems?

Solutions to these questions will be derived from the existing practices employed by the members of the project and from the results of current research being undertaken by the partners. This research is based on a series of case studies of which the interlocking specification project is but one.

CASCADE will produce a generalised framework for the assessment of systems that can be deployed by the assessors and trusted by the users. From the experience gained in the applications, the project will also produce guidelines for the development and procurement of safety critical systems, to be deployed by the users.

To date, the CASCADE project has developed a Provisional Generalised Assessment Methodology (PGAM). Central to the methodology are the twin concepts of standards compliance and development of a system safety case. The PGAM has been influenced by the results of the first phase case studies carried out by the partners, including the interlocking specification project, and this has resulted in a very practical and useable framework which will contribute significantly to the confidence of users of safety critical systems. The PGAM will be refined in later phases of the project and further case studies will be used to assess its effectiveness.

The results of the project will be exploited immediately by both the users and the assessors. Assessor partners will migrate the findings of the project directly into their commercial assessment work. User partners will be in a position to deploy the development guidelines in their practical development work and will be able to have their own, or third party systems assessed in accordance with the framework. Results from the project will be made available to standards organisations and the individual members of the project already represented on standards bodies will be able to make use of the work through those bodies.

# 6    The Specification Project

## 6.1    Project Approach

The signalling interlocking system implements a core safety function within a railway operation. A systematic approach to the specification of such a system is essential if confidence is to be placed in the final requirements document. The approach adopted by the specification project involved a series of linked tasks, undertaken within a management system, covering the following key project phases:

- definition of the scope of the system under consideration;
- definition of the scope and structure of the requirements capture and analysis activity;
- recommendation of a specification methodology, consistent with the safety criticality of the interlocking system, suitable for later verification and validation and acceptable for the proposed future use of the specification within industry;
- definition of a cascading approach to the specification process;
- capture of system requirements; and
- analysis and documenting of system requirements.

## 6.2   Project Control

It is essential that the management methods used to control a project of this type are robust and adequate for the development of safety systems.

NSE spent some time carefully analysing proposals from those contractors considered suitable to carry out work of this type. The contract was ultimately placed with a systems engineering house with considerable expertise in the specification and development of software-based safety systems, and with a track record of successful work in the railway industry. Following contract award, project control was based on the use of rigorous project management techniques supported by complimentary quality and safety management systems.

Both the contractor and ourselves carried out the project within comprehensive Quality Management Systems (QMS), both certified to ISO 9001, and both of which addressed the development of safety critical systems within their certification scope. In addition, the contractor's certification included TickIT, a rapidly growing UK-based initiative which supports ISO 9000 specifically in the area of software engineering. A project-specific Quality Plan was produced, maintained and implemented throughout the project, with both internal and external quality audits adding confidence during the development of the specification.

In addition to the QMS, a Safety Management System was implemented within the project. A project-specific Safety Plan was prepared which addressed the safety management controls to be placed on the project and  the safety engineering and analysis techniques, including hazard analysis and risk assessment tasks, to be employed in support of the specification work. The Safety Plan also included requirements for a Safety Case to be produced at the conclusion of the project, along with details of the approval mechanism for both the Safety Case and the final specification. The purpose of the Safety Case was to justify in safety terms, the adequacy of the final document for its intended purpose. The Safety Plan was prepared,  independently reviewed and finally approved by NSE. The plan was then implemented  throughout the project, with the Safety Case being produced at the end of the work. A series of internal and external audits and assessments were carried out in support of the Safety Plan and the Safety Case was independently assessed and approved by NSE on conclusion of the work.

These control mechanisms enabled NSE to have sufficient confidence in both the quality of the final requirements specification produced by the project, and the processes used to generate the specification.

## 6.3   The Requirements Specification

The Requirements Specification project began in September 1993. The requirements form the foundation for a set of specifications. They define the business requirements, both mandatory and desirable, to be met by an interlocking system and the technical requirements which derive from the business requirements. Both functional and non-functional requirements are included. Requirements cascade, at decreasing levels of abstraction, through the specification. A core set of generic requirements are defined which are equally applicable to the interlocking requirements of any railway system. These requirements are refined to generate requirements specific to a UK implementation. The requirements may be applied to systems and applications at all levels of abstraction.

The data capture exercise was based on a wide-ranging series of interviews with key players in the railway signalling industry, both within and outside British Rail. In addition, analysis of existing documentation covering the principles of UK signalling and their implementation in both electromagnetic and processor-based technologies supported the core requirements identified during the interviews.

The scope of the specification can be summarised in three main areas, breadth, depth and generality.

The breadth of requirements covers non-functional areas, for example system support, design, system interfaces and installation and commissioning requirements. In addition, functional requirements cover:

- network topology and static behaviour;
- network dynamic behaviour;
- safety constraints on static and dynamic characteristics;
- signalling operations;
- enforcement of signalling principles;
- personnel track safety and possession management;
- operation under failure conditions; and
- system diagnostics.

An example of a functional requirement written in the Z notation is shown in Figure 1. This is part of the specification of the route protection system, specifically the reservation of paths within the route protection system.

The depth of the specification is such that the document concentrates on requirement rather than solution. Whilst the specific effect of this distinction varies between different sections of the specification, all requirements are given at a level which ensures that they are free of implementation choices. All non-functional requirements are at a level which is free of application-specific requirements.

```
┌─ PathReservations ─────────────────────────────────
│ reservationPath : RESERVATION ⇸ Path
│ permissiveReservations, absoluteReservations : P RESERVATION
│ reservations : P RESERVATION
│ reservedPaths : P Path
├────────────────────────────────────────────────────
│ reservations = dom reservationPath
│ reservedPaths = ran reservationPath
│ ⟨permissiveReservations, absoluteReservations⟩ partition reservations
│
│ ∀ p : reservedPaths • ¬ (∃ l : Location •
│     l.offset = 0 ∧ (p.start isSameLocAs l ∨ p.end isSameLocAs l))
│
│ ∀ p₁, p₂ : reservedPaths •
│     ∀ n : dom(requiredConnectionConfig p₁) ∩ dom(requiredConnectionConfig p₂) •
│         requiredConnectionConfig p₁ n = requiredConnectionConfig p₂ n
│
│ ∀ ra : absoluteReservations; pa, p : reservedPaths |
│     pa = reservationPath ra ∧ p interferesWith pa •
│     p isNDContiguousWith pa
└────────────────────────────────────────────────────
```

**Figure 1: Specification Extract**

For generality, the specification is intended to cover all types of interlocking system which may be used in the UK, from traditional block working to speed signalling and moving block systems. The requirements are presented at a level of abstraction which is capable of covering all types of foreseen rail networks and signalling systems. Individual applications would generate application-specific requirements specifications based on the generic requirements contained within the document. A series of optional, forward-looking requirements are also included.

The specification, which runs to some 280 pages, half of which is written in the Z notation, was completed in March 1994.

# 7    Further Work

The next stage in the CASCADE project is the refinement of the PGAM. The Railtrack input to this phase will be the verification of the interlocking specification which will have two main thrusts. The first will be the mathematical verification of the specification, including a check on the consistency between the mathematics and the English narrative. The second will be the verification that the signalling principles contained within the specification are correct, consistent and an accurate representation of current UK signalling practice.

There are several interesting aspects to the verification process, one of which is the synergy between the verification of the mathematics and the verification of the signalling principles, as two different types of engineers will have to interact in two very different problem domains in order to complete the project.

Following the verification of the signalling interlocking requirements specification, Railtrack will utilise the specification to facilitate the procurement of new systems and equipment.

# 8 Conclusion

This paper has introduced a Railtrack project which uses formal methods as a means of improving the specification of the UK railway signalling system. It has also introduced the Esprit III project CASCADE, and discussed the aims and objectives of the work and the possible long-term benefits of the project.

The specification represents the first attempt to record a complete and unambiguous representation of current UK railway signalling practice. It focuses on the functional requirements of signalling systems whilst also addressing the benefits that interlocking systems can bring to areas such as staff trackside safety.

The first draft of the specification has been assessed through the CASCADE project. This assessment has highlighted a number of areas where the specification can be improved. The assessment focused on the process used to develop the specification and included a review of project safety management.

It is hoped that the CASCADE project will further the acceptance of computer based safety critical systems through the provision of a systematic and consistent approach to certification involving the developers, assessors and users of such systems. The use of real case studies within the CASCADE project gives greater confidence that the resultant assessment methodology and certification scheme will be of practical benefit to the safety critical systems community.

**Date**: 31st May 1995

## References

[Denvir86]     Denvir T. Introduction to Discrete Mathematics for Software Engineering, Macmillan, 1986

[Heard93]      Heard B D. Philosophy and Principles of British Signalling, Third Railway Signalling and Telecommunications Course, RIA 1993

[Holden93]     Holden Major C B. The Regulation of Railway Safety in Great Britain, Third Railway Signalling and Telecommunications Course, RIA 1993

[Spivey89]     Spivey J M. The Z Notation A Reference Manual, Prentice Hall, 1989

# Accessible Formal Method Support for PLC Software Development

J. A. McDermid
R. H. Pierce
York Software Engineering Limited
York YO1 5DD, UK

**Abstract**

This paper describes a formal notation for the design of PLC software based upon the use of graphical and tabular notations, and indicates how this notation is translated into Z so that the design model can be checked for consistency and determinism. The facilities of a tool to support this notation are also described.

## 1 Introduction

Programmable Logic Controllers or PLCs are widely used in many industrial safety related systems such as chemical reactor control, emergency plant shutdown, and fire detection and suppression. PLCs have been developed as low-cost, rugged, stored program replacements for a previous generation of analogue or relay logic control systems, and are capable of handling large numbers of analogue and digital inputs and outputs. PLC programs are frequently developed by engineers with no software background, and the most frequent programming notation used (at least in the UK and USA) is ladder logic, which is a diagrammatic notation mimicking the coils and contacts of a relay system. The principles of sound software engineering, which it is desirable to use in developing demonstrably safe systems, are frequently lacking in organisations which perform PLC development.

The SEMSPLC project, partially funded by the UK Department of Trade and Industry, is concerned with the introduction of an improved methods for the development of PLC software which is to be used in safety related applications. The work of the project is focused on the production of a Code of Practice [CLA 95] which is suitable for safe PLC development and conforms to the draft IEC standard 1508, based on the work of TC65A. Formal methods are "highly recommended" in 1508 for high integrity applications, but the cultural background of PLC engineers, coupled with market pressure to keep PLC software development costs low, seems to make introduction of formal methods particularly difficult in this context. One work package in SEMSPLC is to study how formal methods can help in the development of safe PLC software, and this work has resulted in an approach to formal methods, described in this paper, which is intended to make them more acceptable to the PLC community. In particular, the aim of our work has been to show how to gain the benefits of formal methods with a notation suitable for PLC applications. This means achieving a precision of specification and ability to mechanically check consistency, so that specification errors can be found early, and not by testing. To this end we are working on a "proof of concept" tool, as well as defining the method. We describe the tool in terms of its intended functions at the end of the project, and report on the status of the tool in section 5.

## 2 A Graphical Method and Tool

The principle underlying the work is that it is necessary to provide methods which are overtly acceptable to systems engineers working in the PLC business, yet which have the underlying rigour of a formal development method. The objective of the work has therefore been to define a combined graphical/tabular method which has an underlying formal basis. Note that the requirement of being "overtly acceptable" is meant to imply that the notations used are not excessively mathematical, not that all knowledge of the underlying formalism can be removed. In particular, it is unlikely that the details of formal analysis, such as proof, can be hidden in their entirety. For some parts of a system, for example in the definition of critical safety properties, it may be preferable to use "classical" formal specifications rather than the graphical notation outlined here.

### 2.1 Technical Basis for the Method and Tool

The technical basis for the method and tool developed in SEMSPLC is that it is appropriate to specify PLC software as a set of state machines, using a variant of classical state-machine mechanisms for dealing with the control structure of the system, and tables to show the actions on the transitions. Thus engineers will be able to see the overall control structure of the system in one or a small number of diagrams, and look in more detail at actions of interest by inspecting relevant tables. This will provide a well-defined method and tool support for it, as well as facilitating acceptability to engineers.

The method is based on a "safety critical" subset[1] of the Statecharts notation [HAR 88], allowing only simple transition labels, and it restricts the relationships of levels in the hierarchy. Further, the notation is restricted to synchronous or polling systems, i.e. transitions cannot be triggered by interrupts, and broadcast events are not permitted. This fits in well with the normal scan cycle concept of a PLC (input-computation-output) which is most clearly seen in ladder logic.

---

[1] While Statecharts provide a powerful notation for system modelling, they contain a number of facilities whose use is questionable in a safety critical system. Discussion of such issues is outside the scope of this paper.

A Statechart is a more elaborate form of the state transition diagram familiar to many PLC engineers in the Sequential Function Charts (SFC) notation. It differs mainly in that states can be decomposed into simpler states, and that the state machine can be in several "and" states simultaneously. Such additional features can of course be ignored and simple, single-level models produced.

For the actions performed on each transition, a simple table using a precondition-postcondition structure is used, with labels for each pair of conditions for ease of reference.

## 2.2  Tool Facilities

The method and the associated tool provide three main facilities:

- specification development;
- specification analysis;
- code generation.

Specification development involves graphical and tabular editors for the Statechart and tabular notations. The tool provides "normal" editing facilities, enabling diagrams and tables to be introduced and modified in the style of other CASE tools. The "look and feel" of the interface is that of Microsoft Windows[2].

### 2.2.1  Notation

The following is an outline of the notation, which uses Z [SPI 92] as a basis.

System 'data dictionary'. A data dictionary is provided by the tool for each graphical model. The dictionary is presented in tabular form and contains the following.
- a collection of Z type definitions to be used in variable declarations, held as Z text.
- lists of input and output variables, together with their Z types and optionally the target programming language types, and the corresponding PLC I/O identifiers.
- list of state variables, together with their types.

The type definitions are global, but in order to provide a form of modularity in the definition the input, output and state variables are local to a module; this encourages the designer to think in an "object based" fashion. Externally visible variables (those which are accessible from other modules) are shown separately from internal state variables, and an "import" notation is used to identify variables which are defined in other modules.

---

[2] Windows is a trademark of Microsoft Corporation.

A PLC I/O identifier is generally in some cryptic form, but is useful if either executable code is generated or to allow the formal model to be related to other specifications concerning the PLC in use (for example, those provided by the end user such as the plant operator).

Statecharts. The method is centred on the Statechart notation, with all states named and defined by means of a predicate over the state variables (a possibly quantified Z predicate). It has been found useful to allow the state predicate to include the output variables, which are assumed to be "latched", in other words, once set to a particular value, an output variable retains that value until it is reset. A unique start state must be defined, and transitions must be labelled with a condition name and an action name (except for transitions out of a start state).

It is possible to define a number of Statecharts in any given model. Each Statechart represents one "module" in the formal definition, and as noted above has it own input, output and state variables defined. State and action names must be unique across all the Statecharts in the model, and each Statechart is given a unique name. Each transition name identifies an entry in the transition table; this entry is a predicate over the state variables and inputs (a non-quantified Z predicate). Each action definition consists of a named table with numbered rows, each giving a precondition-postcondition pair, over the state, input and output variables (possibly quantified Z predicates).

When nested states are used, the inner states naturally inherit the state predicate of the outer state. State variables which are only used in nested states need not be considered in the state predicates if the system is not in the state containing the nested states.

In the "standard" Statechart notation, as described by Harel [HAR 88], the labels attached to transitions are written as <transition_name> [condition]/<action_name>, where the condition and the action may be omitted if they are null. In this form, the <transition_name> is intended to represent an event, for example, an interrupt or the arrival of a message. The transition is made when the event occurs, provided that the associated condition is TRUE at that time. In the method described here, events can only be recognised by a condition becoming TRUE, since there are no interrupts or broadcast events. It seems simplest, therefore, to combine both the "event recognition" and any other enabling conditions in one predicate, and for simplicity of tool construction and avoidance of clutter on the diagram, to detach the transition predicate and move it into the separate table of transitions. In other words, the transition name is just used as a convenient (and, hopefully, meaningful) label which identifies the transition in the associated table.

In Z, it is the rule that if a variable is not mentioned in the postcondition of a schema it is undefined. We feel that to apply this rule to the action postconditions would be irksome to the average engineer. We therefore introduce the rule that state and output variables which are not mentioned in an action postcondition remain unchanged.

## 2.2.2 *Dynamic Behaviour of the Model*

The dynamic behaviour of a PLC program defined by one or more Statecharts constructed according to the above rules is informally defined in this section. A key aspect of the dynamic semantics concerns the relationship of the possible state transitions to the scan cycle of a typical PLC.

On program start or reset, the system enters the start state of each Statechart in the model. The predicate defining the start state defines the initial values of the state variables in that state. A transition out of the start state takes place after the initial input scan is completed.

At the start any given scan cycle, and after the inputs have been read, the system will be in some state or states. For each such initial state, the predicates on the transitions out of that state are considered in some arbitrary order. For whichever predicate is TRUE, the action associated with the corresponding transition is performed and the system enters the resulting state. When more than one Statechart exists for the system, each such Statechart is considered in turn in some arbitrary order. When multiple simultaneous states exist in a Statechart, due to the presence of "or" states, each such state is considered in turn in some arbitrary order.

The actions are performed by evaluating the precondition on each pre/post pair in the action; whenever a precondition is TRUE, the postcondition defines the resulting state of the internal variables and output variables.

When the postcondition of a transition action refers to an *output variable*, the system will not make any further transition out of the resulting state during the current scan cycle. However, if the postcondition does not refer to an output variable then a further transition out of that state is considered. Thus, a series of transitions can be made within the same scan cycle but such a series will stop whenever any action causes the value of an output variable to be potentially changed. This rule is motivated by efficiency considerations; it would be simpler to allow only one transition to be taken (in each Statechart) in one scan, but this may be too inefficient for practical use. This rule could be made optional.

When all transitions which can be taken according to the above rules have been taken, the current values of the output variables are sent to the output lines and the system waits until the end of the current scan cycle. It is erroneous to reach the end of the scan cycle period without having completed all the transitions which can be taken according to the above rules, since this would indicate a time overrun.

The model must be capable of dealing with real time, and in particular to be able to specify time-out transitions. The way we have chosen is to define a "countdown timer" type. When a variables of that type is set to a positive integer value, it is decremented (automatically) by 1 on every scan until it reaches zero; a time-out can then be defined as the condition when the timer variable has the value zero. PLC languages tend to provide such timers. In the example in section 3, the name TIMER is used for this special type. A built-in function TIME_NOW to obtain the date and time would also be useful.

Obviously, the resolution of the timer can be no better than the period of the PLC scan cycle.

### 2.2.3 Checking the Model

The graphical model is translated into Z for input to the PC version of the CADiZ tool [TOY 95], for checking of scope and type correctness. Local checks, such as the uniqueness and well-formedness of names in the model, can be checked either on entry of such names or prior to creation of the Z specification.

In addition to the "classical Z" checks, checking of consistency within the Statechart, and between the Statechart and action definitions is required. In addition certain "determinism" checks can be made.

The primary consistency checks are as follows:

- the condition on a transition is compatible with the state definition, i.e. it can arise in that state;

- the action pre-conditions are compatible with the corresponding transition condition, i.e. they can arise in that condition;

- the action post-conditions are compatible with the ensuing state definition.

The determinism checks are as follows.

- Each state is uniquely defined, i.e. the set of states partitions the space of state variables.

- The pre-conditions in an action table are disjoint. The exception to this rule occurs where an action is unconditional. In this case, a number of such unconditional actions can be written with a precondition of TRUE; this helps both the reader and the automatic code generator.

- The pre-conditions in an action table cover the space of possibilities admitted by the condition and the state from which the transition starts.

A useful optional check is that all the predicates on transitions out of a given state are disjoint.

In addition, it may be possible to check the reachability of states automatically, by linking to a model checking tool.

The consistency and the determinism checks will need to be represented by generating them as conjectures in Z, and then proving them (proof is likely to be beyond the skills of PLC engineers, and for the most critical projects a proof specialist would have to be engaged). The Z generation tool will produce these conjectures on request. In general, such conjectures are not mechanically decidable. However, in practice, many of the consistency conjectures will be very simple, as illustrated below, and many will be proven automatically, using decision procedures, e.g. for Presburger arithmetic. The extent to which automated proofs are feasible can only be demonstrated experimentally, and this is one of the reasons for developing the "proof of concept" tool.

### 2.2.4  Generation of Z

The rules for mapping the graphical notation into Z are outlined in this section. It is important that the generated Z has an obvious relationship to the diagrammatic and tabular model, since errors are reported by the CADiZ tool in terms of the generated Z. This implies that names as used in the model must be preserved and that the Z has a clear structure.

Firstly, a free type definition is introduced to represent the states in the Statechart machine:

STATE  ::=  S1 | S2 | S3.... | Sn

The literals in this type are the state names themselves. If there are decomposed states, inner state names are generated in the form <Outer_State>_<Inner_State >. A schema to represent the state of the Statechart machine is then generated (the name STC_... given to the various schemas would of course be replaced by the name of the actual Statechart in question).

```
┌─ STC_State ─────┐
│ State : State   │
└─────────────────┘
```

followed by a schema representing the state and output variables of the Statechart.

```
┌─ STC_Internals ──────┐
│ Sv_N      : <type>   │
│ Output_N! : <type>   │
└──────────────────────┘
```

The state space of the Statechart machine is then captured by the following schema, which contains the state invariants.

```
┌─ STC ──────────────────────────┐
│ STC_State                       │
│ STC_Internals                   │
├─────────────────────────────────
│ state = S1 ⇔ S1_predicate       │
│ state = S2 ⇔ S2_predicate       │
│ ..........                      │
└─────────────────────────────────┘
```

The inputs which are used by the Statechart are then represented:

```
┌─ STC_inputs ──────┐
│ input_1 : <type>   │
│ input_2 : <type>   │
│ .........          │
└────────────────────┘
```

As defined previously, a transition condition is a predicate over the inputs and the initial values of state variables in the Statechart:

$$\text{STC\_Condition} \triangleq \text{STC\_Inputs} \wedge \text{STC\_Internals}$$

Actions change the internal state of the Statechart machine:

$$\text{STC\_Action} \triangleq \Delta\text{STC\_Internals}$$

and each action is defined using a schema of the form:

$$\text{Action\_N} \triangleq [\text{STC\_Action} \mid \text{precondition\_1} \Rightarrow \text{postcondition\_1} \wedge$$
$$\text{precondition\_2} \Rightarrow \text{postcondition\_2} \wedge \text{postcondition3}$$
$$\wedge ... \text{postcondition\_n}]$$

where each implication corresponds to one precondition-postcondition pair in the action tables (if the precondition is true the implication is omitted). Although the classical Z rule is that variables whose values are not explicitly defined in the postcondition of a schema are undefined, rule is relaxed in the tabular representation (section 2.2.1). The Z generator, therefore, automatically introduce extras predicates VAR' = VAR for all state and output variables which do not appear dashed in the action postconditions.

Finally, a general schema for transitions is generated; a transition will clearly alter the state of the state machine:

$$\text{STC\_Transition} \triangleq [\Delta\text{STC\_State} ; \text{STC\_Condition}; \text{STC\_Action} \mid \Delta\text{STC}]$$

Each transition is now represented as a schema which allows the transition if the state machine is in the correct state and the transition predicate is true, changes the state to the new state and performs the action:

Transition_N
     $\hat{=}$ [STC_Transition | State = Sn $\wedge$ Condition_N $\wedge$ State' = Sn_successor
     $\wedge$Action_N]

Having now generated all the necessary schemas, the complete description of the behaviour of the state machine represented by the Statechart is given as the disjunction of the transition schemas.

STC_Behaviour $\hat{=}$ Transition_1 $\vee$ Transition_2 $\vee$ .... $\vee$ Transition_n

# 3  Example of the Graphical Method in Use

The in this section example is drawn from an actual PLC system, developed by one of the SEMSPLC partners, which is used as a practical case study within the project. The example shows a (small) part of a Fire and Gas Protection System, which is a classical application for a PLC.

The example chosen is the inert gas release system, which floods the plant control room with inert gas, after a suitable delay, on receipt of a confirmed fire signal (2 out of 2 smoke detectors signal that smoke is present), or on manual request from a pushbutton. The control room door must be closed before the inert gas can be released. Complications arise because manual release can take place even if the inert gas release is inhibited by a zone inhibit, while automatic release cannot.

The inert gas system contains two banks of extinguishant, Bank A and Bank B. These are selected by a keyswitch, which also has a third "INHIBIT" position. The zone and bank inhibit signals are intended to allow the detection system to be tested without actually releasing extinguishant.

The system tests to see if extinguishant has indeed been released; if the release has failed, the bank selector switch may be changed and another attempt may be made using the manual pushbutton, but only if the button has been released and then pushed again. Automatic release cannot be retried. An audible alarm is sounded when release of extinguishant has been initiated; a 20 second period is allowed for staff to evacuate the control room before the inert gas is released; an "extinguishant released" light flashes during this period, and then shows a steady indication.

Figures 1 gives the Statechart representing the behaviour of the inert gas release component. The following text shows the tabular part of the notation, as it might be produced by a document generation tool. The smoke detector module signals that a fire has been detected by setting the interface variable CR_Confirmed_Fire to TRUE.

122

For the sake of brevity, the state Failed_to_Release and its associated transitions and internal states are omitted from the tables, as are the initial and reset actions. Internal variables concerned with timing are omitted from the state predicates. The clock symbol on certain transitions in the inert gas Statechart indicates that the transition is triggered by a time-out (this is a local extension to the notation, and has the status of a comment - it is not checked against the transition predicate).

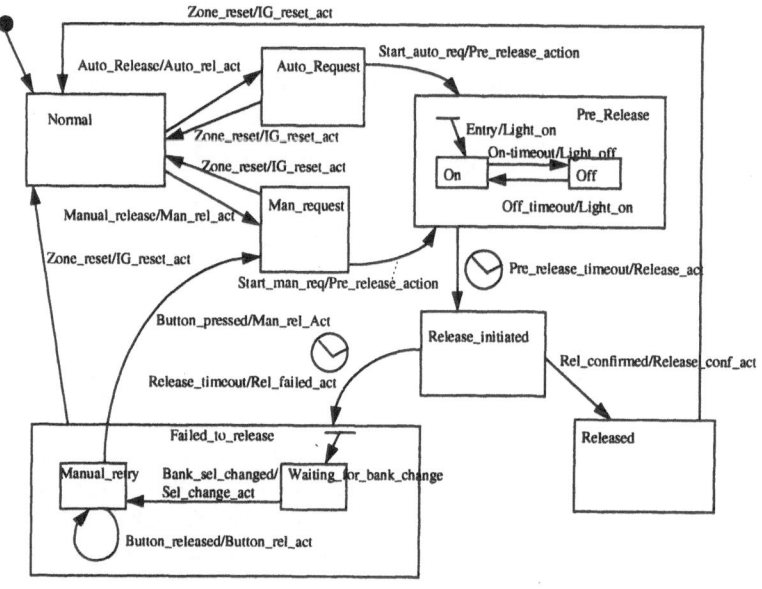

**Figure 1 - Statechart for control room inert gas release**

## Type Declarations

| | |
|---|---|
| BANK_SEL_STAT | ::= BANK_A I BANK_B I INHIBIT |
| LIGHT_STATUS | ::= OFF I ON |
| BOOLEAN | ::= FALSE I TRUE |

**Module:** INERT_GAS_SYSTEM

**Import** CR_Confirmed_Fire **from** Smoke_Detectors
**Import** Cont_Area_Alm_Req **from** Alarm_System

## Inputs

| Name | Z type | Plant Tag | Comment |
|---|---|---|---|
| Bank_Selector? | BANK_SEL_STAT | | Keyswitch |
| Ext_Release? | BOOLEAN | HS-8760-1 | Manual p/button |
| Inert_Gas_Released? | BOOLEAN | PSH-8760-A | Pressure sensor |
| Door_Shut? | BOOLEAN | ZSL-5701-A | Microswitch |
| Control_Room_Inhibit? | BOOLEAN | | Zone inhibit |
| Control_Room_Reset? | BOOLEAN | | Zone reset |

## Outputs

| Name | Z type | Plant Tag |
|---|---|---|
| Release_Bank_A! | BOOLEAN | XSOV-8760-XA |
| Release_Bank_B! | BOOLEAN | XSOV-8760-XB |
| Ext_Light! | LIGHT_STATUS | XL-8760-C |
| Avail_Light! | LIGHT_STATUS | XL-8760-A |
| IG_CR_Alarm! | BOOLEAN | XA-8760-A |

## Internal Variables

| Name | Z type | Comment |
|---|---|---|
| Release_Start_Timer | TIMER | For 20 second delay on release of inert gas |
| Release_Check_Timer | TIMER | For 3 second delay while waiting to confirm IG release |
| Flash_Timer | TIMER | For flashing light |

## States

| Name | Predicate | Comment |
|---|---|---|
| Normal | Release_Bank_A! = FALSE<br>Release_Bank_B! = FALSE<br>Avail_Light! = ON<br>Ext_Light! = OFF<br>IG_CR_Alarm! = FALSE | |
| Auto_Request | Release_Bank_A! = FALSE<br>Release_Bank_B! = FALSE<br>Avail_Light! = ON<br>Ext_Light! = OFF<br>IG_CR_Alarm! = TRUE | |
| Man_Request | Release_Bank_A! = FALSE<br>Release_Bank_B! = FALSE<br>Avail_Light! = ON<br>Ext_Light! = OFF<br>IG_CR_Alarm! = TRUE | |

| Name | Predicate | Comment |
|------|-----------|---------|
| Pre_Release | TIME_NOW? < Release_Start_T<br>Release_Bank_A! = FALSE<br>Release_Bank_B! = FALSE<br>Avail_Light! = ON<br>IG_CR_Alarm! = TRUE | |
| On | Ext_Light! = ON | |
| Off | Ext_Light! = OFF | |
| Release_Initiated | (Release_Bank_A! = TRUE ∨ Release_Bank_B! = TRUE)<br>Ext_Light! = ON<br>Avail_Light! = OFF<br>IG_CR_Alarm! = TRUE<br>Release_Check_Timer > 0 | |
| Released | Release_Bank_A! = FALSE<br>Release_Bank_B! = FALSE<br>Ext_LIght! = ON<br>Avail_Light! = OFF<br>IG_CR_Alarm! = TRUE | |

## Transitions

| Name | Condition | Comment |
|------|-----------|---------|
| Auto_Release | CR_Confirmed_Fire = TRUE | |
| Manual_Release | Ext_Release? = TRUE | |
| Start_Auto_Req | Door_Shut? = TRUE ∧<br>Bank_Selector? ≠ INHIBIT ∧<br>Control_Room_Inhibit? = FALSE | |
| Start_Man_Req | Door_Shut? = TRUE ∧<br>Bank_Selector? ≠ INHIBIT | |
| Entry | TRUE | |
| On_timeout | Flash_Timer = 0 | All times are measured in milliseconds |
| Off_timeout | Flash_Timer = 0 | |
| Pre_Release_Timeout | Release_Start_Timer = 0 | |
| Rel_Confirmed | Inert_Gas_Released? = TRUE | |
| Release_Timeout | Release_Check_Timer = 0 | |
| Zone_reset | Control_Room_Reset? = TRUE | |

## Actions

| Auto_Rel_Act | Precondition | Postcondition | Comment |
|------|-----------|---------|---------|
| 1 | TRUE | IG_CR_Alarm!' = TRUE | |

| Man_Rel_act | Precondition | Postcondition | Comment |
|---|---|---|---|
| 1 | TRUE | Cont_Area_Alm_Req' = TRUE<br>IG_CR_Alarm!' = TRUE | |
| Pre_Release_action | Precondition | Postcondition | Comment |
| 1 | TRUE | Release_Start_Timer' = 20000 | Start 20s timer |
| Light_On | Precondition | Postcondition | Comment |
| 1 | TRUE | Ext_Light!' = ON | |
| 2 | TRUE | Flash_Timer' = 1000 | Start 1s timer |
| Light_off | Precondition | Postcondition | |
| 1 | TRUE | Ext_Light! =OFF | |
| 2 | TRUE | Flash_Timer' = 1000 | Start 1s timer |
| Release_act | Precondition | Postcondition | Comment |
| 1 | Bank_Selector?= BANK_A | Release_Bank_A!' = TRUE | |
| 2 | Bank_Selector? = BANK_B | Release_Bank_B!' = TRUE | |
| 3 | Bank_Selector? = INHIBIT | | Null action |
| 4 | TRUE | Release_Check_Timer' = 3000 | Start 3s timer |
| Release_conf_act | Precondition | Postcondition | Comment |
| 1 | TRUE | Ext_Light!' = ON | "Released" light |
| 2 | TRUE | Avail_Light!' = OFF | "Available" light |
| 3 | TRUE | Release_Bank_A!' = FALSE | Release signals cleared |
| 4 | TRUE | Release_Bank_B!' = FALSE | |

The consistency checks mentioned in section 2.2.3 can be illustrated simply. For example, in the transition Manual_Release from state Normal we require that the condition can arise in that state, and this is expressed as the conjecture:

Normal ∧ Manual_Release

⊢

True

Proof of this conjecture is in this case mechanical.

# 4 Code Generation from the Graphical Notation

Generating the control structure code for a Statechart is straightforward and there are many commercially available tools which do just this for a variety of programming languages. By contrast, generating the code which implements the transition conditions and actions involves translating Z into a programming language, which is infeasible in the general case. However, by limiting the use of Z to simple constructs, it will be possible for the tool to generate the complete executable code directly. In fact, the worked example in the previous section uses only very simple predicates, with no quantifiers, and we believe that this is quite typical of PLC software in general. In this example, the transition conditions can be converted into simple Boolean expressions and the action postconditions can be transformed directly into simple assignment statements.

The executable code should be in a language which a substantial number of PLCs would accept, but since there are many dialects of such a language a practical tool would currently have to cope with several of the best known ones. The advent of the IEC 1131-3 standard, however, should allow the choice of a single language as a target for code generation, with the expectation that a substantial number of PLC manufacturers will in the future support products which conform to the standard. Of the languages defined in IEC 1131-3, Structured Text (ST) is the closest to conventional high-order programming languages and would be the most straightforward target for code generation. Structured Text, in addition to providing a fairly conventional high level language, also provides a textual representation for Sequential Function Charts, a form of state transition diagram, and Function Block Diagrams. However, only the high level language aspects of ST are currently being investigated as a target for code generation, since there are some obvious (and possibly some subtle) differences between the semantics of Statecharts and SFCs.

As with any automatic code generation, there is a risk of errors in the code generator introducing errors in the final system. This hazard can be mitigated by various means, but further discussion of this topic is beyond the scope of the paper.

## 5 Current Status of the Work

At the time of writing, the PC version of the CADiZ tool is complete enough to be used for checking the output of the Z generation system. Work is proceeding on editors for the graphical and tabular notations, on a Z generator and on a code generator. In order to demonstrate proof of concept within the timescale of the SEMSPLC project, which ends in September 1995, a prototype with restricted functionality is being developed, concentrating on Z generation and checking. Executable code will be generated in Ada, but conversion of this code generator to produce Structured Text would be straightforward for a production quality version of the tool. The diagram editing facilities in the prototype will be rudimentary, since nothing will be proved by implementing yet another CASE tool.

## 6 Conclusions

The method and tool described in this document should provide a sound basis for formally specifying and verifying PLC software in a manner which may be acceptable to the PLC engineers after a modest amount of training has been given.

We have had only limited experience of applying the method, but the results have been encouraging. The PLC engineers on the project find the notation quite acceptable, and certainly preferable to Z. Our formalism has detected many areas of inconsistency and incompleteness in the fire and gas specification. This is not surprising, since the specification was entirely informal. However, it shows that there is value in the rigour of formal methods, even for relatively simple systems such as those produced on PLCs. Even manual application of the consistency and determinism rules has proved valuable; for example, errors in the definition of the Normal and Auto_Request states were found by inspection. Mechanisation would aid the application of such checks.

In the long run, we believe that approaches such as the method we have outlined above will become cost-effective for developing PLC software. However, for the potential of such methods to be realised will require cultural changes in both the purchasers and suppliers of PLC systems, as well as technical advances. In many ways these cultural shifts may be the more challenging.

# 7 Acknowledgements

Particular thanks are due to our colleagues Fen-gang Shi, whose doctoral research on Statecharts and Z has contributed to the work described here, and to Jon Hall, Jeremy Jacob and Ian Toyn, whose work on formalisation of an industrial press controller has provided the rules for mapping the notation to Z. We are very grateful to our partners in the SEMSPLC project – ERA Technology Limited (the prime contractor), Servelec Limited (who provided the fire and gas system example, and helped us to understand the world of PLCs), British Gas plc and LDRA Limited. We are also grateful to some of the project sponsors, notably the UK Health and Safety Executive, Nuclear Electric Limited and CEGELEC Projects Limited (who helped us greatly with IEC language issues).

# 8 References

[CLA 95]   Clarke, S. et al, *A code of practice for the development of safe PLC software*, Proc. Safety Critical Systems Symposium, Brighton, UK, Feb. 1995.

[HAR 88]   Harel, D. *On visual formalisms*, CACM Vol.31 No.5, p. 514, May 1988.

[SPI 92]   Spivey, J. M., *The Z notation - A Reference Manual*, second edition. New York: Prentice Hall, 1992.

[TOY 95]   Toyn, I. & McDermid, J. A., *An architecture for Z tools and its implementation*, Software - Practice and Experience, Vol. 25 No.3, pp 305-330, March 1995.

# Session 4
## Human and Legal Aspects

# Eliminating the Unexpected

R.J. Tiezema
GTI Industrial Automation bv
Apeldoorn, The Netherlands

# Impact of Human Factors

## Abstract

Many causes of both casualties and large production losses in the process industries result from human failures.

Especially errors such as "mistakes" are a matter of great concern.

This paper describes the human factor in operational safety and availability.

Attention will be paid to errors in the "thinking process" and also to maintenance induced errors.

# 1 Introduction

In general process safety and availability both will be effected by failures in the following areas:

- Hardware

- Software

- Management

- External

- Human

Almost 40% of all large losses [HPI 95] in the hydrocarbon-chemical industries are caused by operational errors and/or are unknown.

An analysis of 216 accidents [Pietersen 86] found that approximately 30% were caused by human failures.

To improve process reliability, reduction of the human factor is obviously of prime importance.

In principle one might say that 100% of all incidents in the industry are caused by human failures, as every component, plan, construction, etc. has been originated by people. Conversely there are people saying that it might be possible to reduce the human factor to zero by fully automated process control.

It is just a question of the definitions used and of the system specifications.

The following section provides more information about the human factor and will emphasize the rule of "thinking errors" in the operational sector and the reduction of execution failures. Also a number of possible solutions will be suggested to decrease the effect of human failures.

# 2 Basic Considerations

The effects of human errors in the process industry are divided in two main groups, depending on the life cycle of the project.

    a.     System failures

    b.     Operational failures

Concerning the human factor, the process industry usually views system failures (figure 1) as not being human failures.

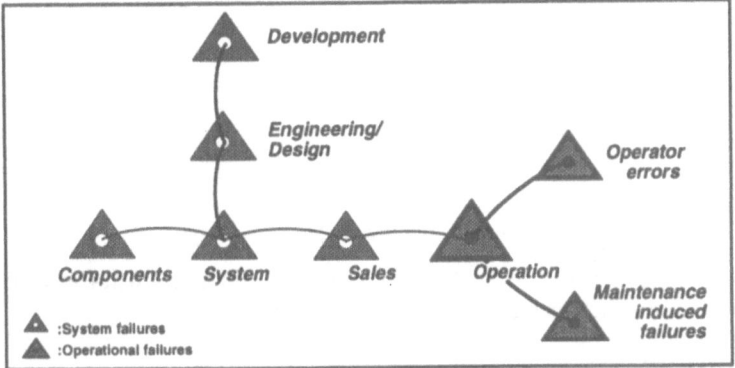

**Fig. 1 Failure effect areas**

In this paper, operation failures are split into two types of failures. The first are defined as operator errors, the second are collectively defined as maintenance induced failures. In turn, and for the sake of clarity, the divisions of these two items are also simply divided into two main parts each.

Figure 2 show the two main parts of operator errors, whilst figure 3 reflects the separation of maintenance induced failures into internal and external failures.

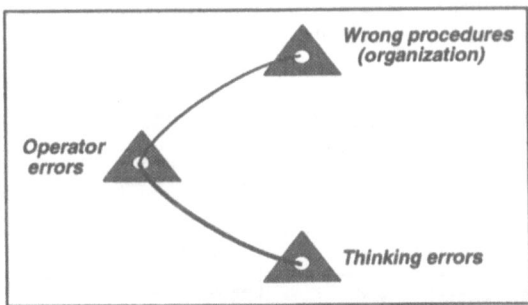

**Fig. 2 Operator errors**

From a control-system point of view, the chosen approach is justified, because of the prime contribution of the thinking errors and the internal maintenance induced failures in the total system failure rate.

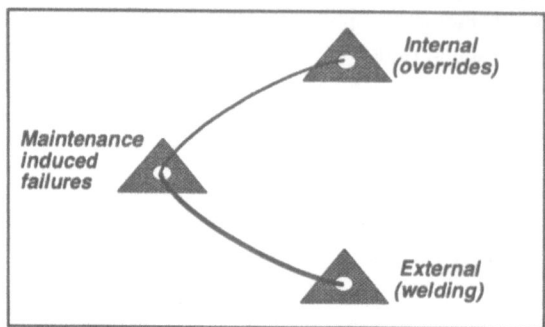

**Fig. 3 Maintenance induced failures**

Therefore section 3 focusses on the thinking errors and section 4 will especially go into the matter of maintenance overrides (MOS).

# 3 Thinking Errors

Since the enormous increase in automation, the ratio between thinking and execution errors has also changed. Three independent investigators analysed 500 events and found that 30% are caused by thinking errors [Gerdes 95]. To protect systems against these "mistakes" and even violations (figure 4) is rather difficult, because human beings are very "inventive" in their efforts to follow their own convictions.

However, there are a number of measures to reduce the probability of accidents and to control the process under unforeseen situations:

| Error: | Intention: | Execution: | Remarks: |
|---|---|---|---|
| Slip | good | wrong | execution failure, unintended action |
| Lapse | good | wrong | memory failure, omission |
| Mistake | bad (unaware) | well | diagnosis failure, thinking error |
| Violation | bad (aware) | well | calculated action, unintended cause of damage |
| Sabotage | bad (aware) | well | no failure! calculated action to cause damage |

**Fig:4 Possible operator errors**

1. Maximum training, with advanced simulators and a careful, ergonomic design of the Man-Machine-Interface (MMI).

2. The separation of the safety part and non-safety related part of a system: The safety part represents the "fuse" of the system.

3. The identification and analysis of thinking errors

## 3.1 Training

The book: "Normal accidents" [Perrow 84] describes the so-called "interactive complexity" of the system as the source of increasing unexpected interactions among inevitable failures. It is even possible that an operator action may make it worse since it is not known what the situation really is.

Sintef (Norway) carried out an analysis of shutdowns on offshore installations and concluded that misinterpretation of the state of the system and the process causes uncontrolled events.

Consequently the question arises: "Can operators be trained to handle unforeseen situations?"

The answer: "In principle no, but adequate training will considerably reduce the effect of the human factor."

What does adequate training mean in practice?

First of all, it has to be recognised that much training is concentrating on maintenance issues, such as trouble shooting and repair. In relation to the investment in maintenance training, it is incredible that investments in real operator training are so low. Currently this situation is improving slightly. The number of extended test stands, where all the equipment is built and integrated together, is increasing. All kind of tests (not limited to just the test protocols) are executed in the presence of operators.

Expecially this includes test, where engineers might say: "Why? This can never happen."

Also, the use of advanced simulators is becoming an accepted way of giving operators more experience to handle calamities. True, these simulators are quite expensive, but what are the costs of an uncontrolled event?

Finally, an effective solution to train operators is to involve future operators from the onset of a project by participating in the project team.

## 3.2 The protection part as system "fuse"

There are manufacturers of MMI's who are improving the Risk Management Function of their systems. These systems are interactive and provide the operator a dynamic forecast of the process. They present their system with the slogan: "The operator decides".

However, because of the "realistic" and fast presentation of alternative measures, the question is raised of who should be making the decision: the system or the operator.

Is it always wise in case of an emergency to delegate the final responsibility to an operator? Especially when complex processes are involved? To answer this question, it is necessary to evaluate the range of the process parameters.

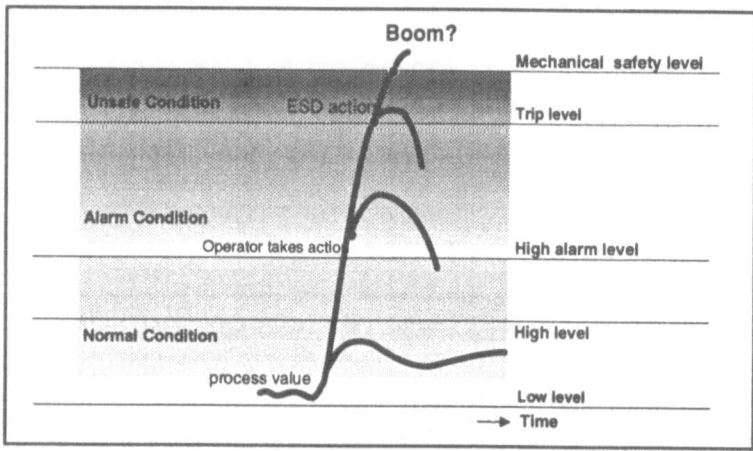

**Fig: 5  Process parameter range**

Figure 5 gives a simplified example of an input parameter for a safety system. Up to and including the alarm condition, the operator will control the process succesfully. But what should be done, if the process attains the unsafe condition?

The operator, of course (probably under a certain stress, ref. Normal Accidents [4]), is too busy to handle a number of other events.

In this situation the use of an emergency system should be considered (Eliminating the Unexpected, part 1 [Lippevelde 90]).

These type of safety system combine high availability with an extremely high safety degree. For this reason these systems are also defined as: "High Integrity Protection Systems" (HIPS).

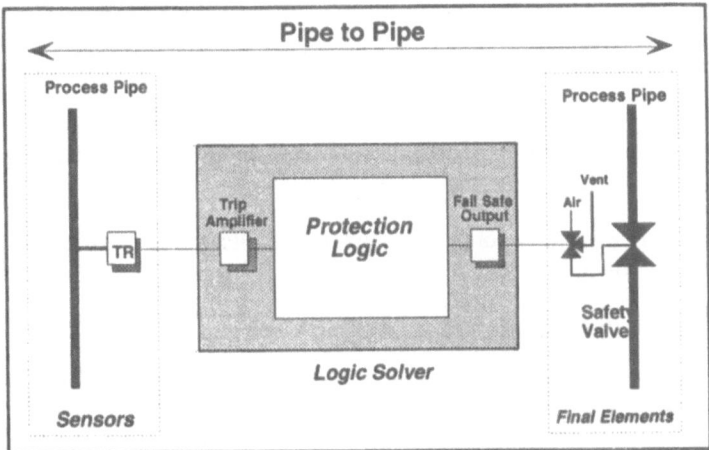

**Fig: 6  High Integrity Protection System (HIPS)**

It is evident that the whole safety loop (sensor, transmitter, logic circuitry, actuator, etc. figure 6) should have the same Safety Integrity Level (SIL, ref. IEC 65A). Besides the high SIL, HIPS can not be effected by either the operator nor by the control part of a control system (figure 7).

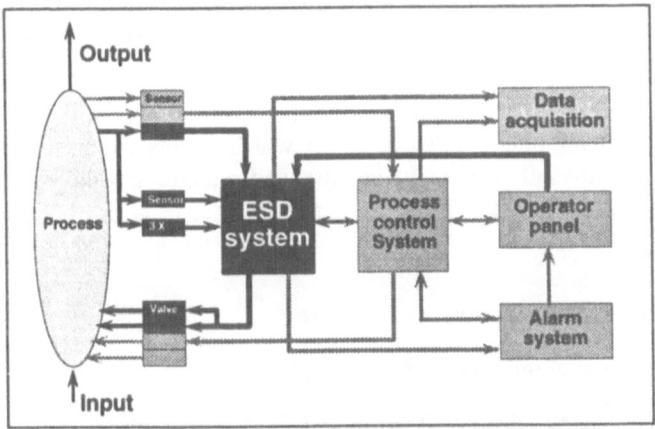

**Fig: 7  The safety system  interfacing**

## 3.3. The identification of thinking errors

Thinking errors are also referred to as "cognitive errors" and defined here as: Errors in the intention-formation process. At this moment the methods available to identify and analyze such errors are scarce and premature [Dougherty 92]. The intention-formation process comprises everything between noticing that there is a need for a task (detection) and the actual performance of the physical interactions with the plant to achieve this task (execution).

A Method to Identify Cognitive Errors (MICE) is developed by N.V. Kema (NL) [Gerdes 95].

MICE essentially consists of two main products;

-    a four-level cognitive error classification that contains 43 different errors, built from 11 basic error types, and

-    a flowchart to identify cognitive errors that results in a decomposed representation of, possibly multiple, errors.

The other products that underlie MICE;

-    a four-level cognitive task decomposition combined with a cognitive process model, and

-    six error evaluation diagrams that evaluate erroneous results at all levels of the cognitive task.

The results of a MICE analysis can be interpreted with these two underlying products. This more fundamental insight gives direction to an effective root-cause analysis.

Obviously, classifying the errors is not enough to analyze them. To obtain insight into the reasons why a particular error occurs, the circumstances that provoke this error should be evaluated.

For every cognitive error a specific set of influencing factors will be designed, presented in a mapping diagram that describes which factors are most relevant for which errors. The overall method to identify and analyze cognitive errors then consists of three units [Gerdes 95];

-    a classification of cognitive errors (figure 8)

-    a classification of influencing factors, and

-    a mapping diagram that combines these two classification.

MICE has been applied to several real cases. The results of this pilot study are very promising.

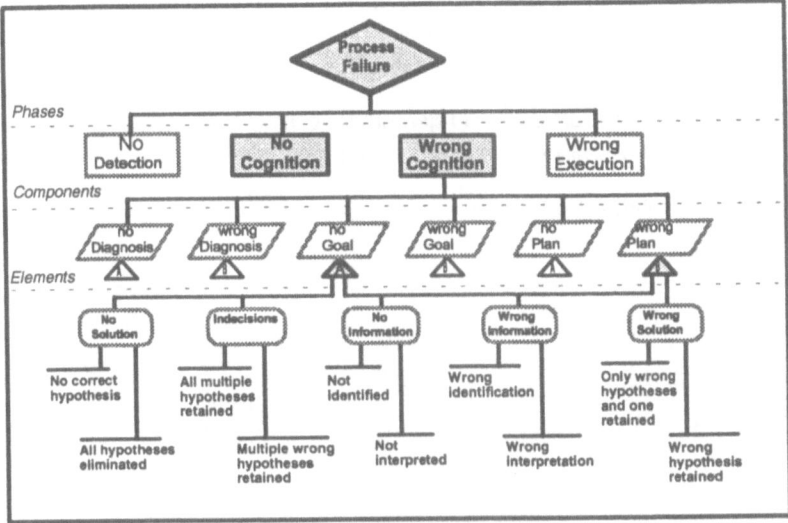

**Fig: 8 Cognitive error classification**

# 4 Induced Maintenance Errors

An analysis of a process automation system, shows that the four significant system requirements are:-

-	Availability

-	Safety

-	Maintainability

-	Flexibility

A combination of all items being at a maximum level will be excluded by technical limitations. Which one (or combination) has to be considered as the most important, depends on the process, environment, laws, etc.

Experienced system designers will also be aware of the fact that the requirements above might effect each other in a negative way.

For example:

-	"Nice to have" flexibility shall reduce the integrity of the safety requirement

-	On-line / Off-line maintenance will effect the safety / availability

A serious incident occured recently [Pearson 93] on a chemical plant when a shutdown function was overriden. This mistake caused an overpressure and a release of materials on a plant handling high toxic substances.

There are more known incidents (Tsjernobyl) caused by wrongly operated maintenance overrides. A reason to reconsider the function and goals of overrides.

Figure 9 shows a typical override circuit.

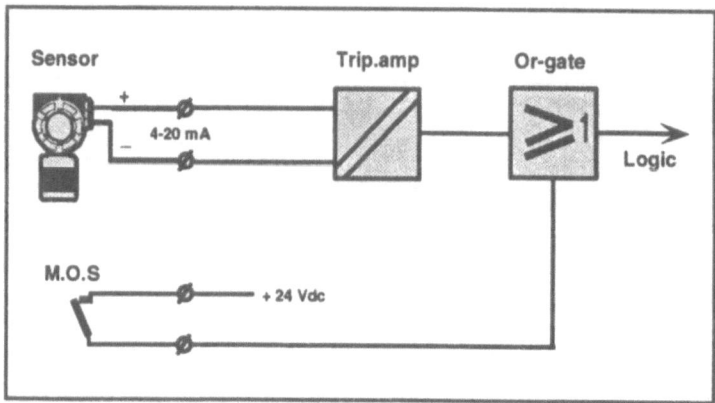

**Fig: 9 Typical Override circuitry**

The principle is rather simple and obviously designed for on-line maintenance (calibration, test, etc.). If the input circuitry (sensor, cabling, termination and tripamplifier) needs maintenance, the MOS (Maintenance Override Switch) will be switched on. The OR-gate will keep the logic in operation independent from the input status.

From a maintenance point of view there will not be any objection. But what about safety? The safety now just depends on procedures and operator capabilities. Reviewing the impact of the human factor in safety the override circuitry should be redesigned by introducing other technics and were possible be eliminated.

## 4.1 Alternative circuitries.

With the circuitry of fig. 9 as a starting-point, possible solutions to reduce the use of MOS are as follows:

- Off-Line maintenance
- Fail-safe Timerprotection
- Redundancy

The right choice depends on the system requirements.

# 5 Summary and Perspectives

The effect of the human factor on process Safety and Availability is quite serious. It would be of great interest to reduce the consequences of the frequent occurrence of operating errors and the unnecessary use of overrides in high integrity systems. Training objectives, test- and maintenance procedures should be carefully reconsidered based on the analysis of thinking errors and a system design where requirements on the area of safety, availability and maintenance are quite clear.

A major problem for the development of a pipe-to-pipe system (ref. figure 6) is the choice of the right sensor.

The use of "smart" transmittors is increasing, but requires careful safety considerations.
On the other hand, to use its "intelligence" for sensor validation could be a useful feature to reduce testfrequencies.
Expecially in combination with neural networks [Spiker 92] "smart" devices, both sensors as well actuators might improve a system to operate in a reliable and maintainable way.

In summary, the probability of being caught by the unexpected will be substantically decreased if the impact of human factors for a particular plant has been studied and analyses. Obviously the results of this analysis should be used to define the correct system requirements.

## References

[Dougherty 92] Dougherty E.M.: Context and Human Reliability Analysis - 1992
Elsevier Science Publishers

[Gerdes 95] Gerdes V.: Method to identify cognitive errors - June 1995 KEMA. NL

[HPI 95] HPI Market Data - 1995   GULF Publishing Company   USA

[Lippevelde 90] Lippevelde, R.I.L. van: Eliminating the Unexpected, part 1
June 1990 GTI IA. NL

[Pearson 93] Pearson J.: Safety- under control? - London 1993        HSA. GB

[Perrow 84] Charles Perrow: Normal Accidents        Basic Books - 1984   USA

[Pietersen 86] Pietersen, C.M.: Impact van recente grote ongevallen
TNO 86-272 TNO. NL

[Spiker 92] Spiker, R.Th.E.: Safety System provided with Neural Circuit
September 1992   GTI IA. NL

# Cognitive Diversity: A Structured Approach to Trapping Human Error

S.J. Westerman, N.M. Shryane, C.M. Crawshaw, G.R.J. Hockey
Department of Psychology, University of Hull, Hull, HU6 7RX, England.
&
C.W. Wyatt-Millington
Department of Electronic and Electrical Engineering, University of Bradford, Bradford, BD1 1DP, England.

## Abstract

This paper considers cognitive diversity as a method of avoiding common-mode error when performing a fault detection task. Factors influencing cognitive diversity are examined, and supporting data from field and laboratory studies are presented.

## 1 Introduction

The use of component redundancy as a means of achieving fault tolerance within hardware systems is well established. However, this does not guard against common-mode errors, within the design of the system, to which identical components are equally susceptible. This problem can be surmounted, in both hardware and software, by means of diversity. Fault tolerance is achieved by virtue of the fact that design errors which are present in one component are unlikely to be replicated. Consequently, by using a voting system, a higher level of reliability can be achieved than could be attributed to any individual component.

Within the context of human system components similar concepts can be applied. The benefits of redundancy are widely understood, as illustrated by the familiar phrase 'two heads are better than one'. Beneficial effects have been demonstrated, with respect to software design, when work is checked by another person, or teams engage in problem solving (e.g. [BiL89]; [WNH92]). The implication of this approach for the design of safety-critical systems is that if one individual 'fails' at a particular point within the task process then another will be likely to detect the error. Unfortunately, as with hardware and software components, redundancy is susceptible to common-mode errors. "Redundancy does not always work to reduce human error... The use of multiple humans in the way multiple inaccurate components are used is not a reliable way to enhance human-machine system reliability" ([SeM91], pp. 118-120). Failures in the performance capacities or strategies employed by one individual may well be shared by others, and, as a result, all individuals may make common errors. These errors are necessarily much more problematic for the human-machine or human-human system to detect. Once again a solution may be found in diversity; in this case 'cognitive diversity'.

Examples of cognitive diversity, within the context of software development, can be found in the case of inspection teams, comprising individuals with a (specified) range of task perspectives ([Fag76], [Fag86]). In such instances, each participant may bring a different approach to the problem and thereby reduce the possibility of 'common-mode' errors. However, in contrast, it should be noted that there are many organisational and task-related factors which serve to reduce diversity. For example, uniform methods of selection and training within organisations will restrict the range of individual abilities and task performance strategies. Similarly, many forms of computer-based task support encourage uniformity in the mental model (see [GeS83]) which is applied to the task ([Hoc88]; [CBG94]; although see [Gar93]).

The potential of cognitive diversity as a tool that can be used during the development and operation of safety-critical systems was recognised by [Sag94]. In a paper which discusses wider issues relating to system diversity, the author identifies oversight, unskilled personnel, thinking traps, and uncommon situations as being causal origins of human faults which can be improved with the application of diversity. We will contend that the first of these (oversight) may not be appropriately included. The remaining areas are in broad agreement with [FlH91] who suggest that diversity arises from individual differences in prior experience and access to task relevant information. In addition they highlight the potential importance of the functional role which individuals adopt in relation to the task. A more extensive model of cognitive diversity is provided by [Hol93], who describes Human Dependent Failures (HDFs) as resulting from a Root Cause ("...a susceptibility of humans to fail in a set of related or different actions"), and a Coupling Mechanism that "creates the conditions for multiple human actions to be affected by the same root cause". He goes on to suggest that "Defensive techniques, aimed at reducing HDFs, can be directed towards the potential root causes (i.e. quality) and/or towards the coupling mechanisms (i.e. diversity). It is the strength of the coupling mechanisms and the effectiveness of the defences that will determine the assessed level of dependence".

As far as we are aware there has been no empirical investigation of the mechanisms involved in cognitive diversity. Consequently, we would like to extend some of these considerations within the context of an applied safety-critical design task. In particular our concern is with the process of fault-detection. However, we believe that the principles which are discussed below could be applied to a wider range of tasks. In order to avoid confusion the term 'fault' will be used to refer to an incorrect or missing component within a system under examination. The term 'error' within this context will refer to a failure to detect a fault or a false detection on the part of an individual performing the fault-finding task.

## 1.1 Facets of Diversity

The aim of any study of cognitive diversity within fault-finding must be to predict systematic variance in performance. In contrast to previous studies of fault-finding, we are not concerned with predicting maximal performance; what is at issue is whether the detection of specific 'types' of faults are associated with factors characteristic of the individual performing the task or the manner of performance. Diversity will be apparent if different predictors are associated with the location of different 'types' of faults (see Figure 1).

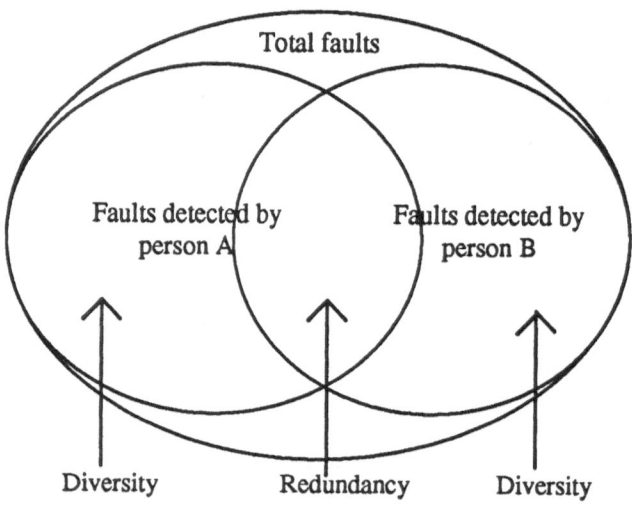

Figure 1 : Diversity and redundancy in fault detection

Following from this, there are a number of assumptions which must hold true before cognitive diversity can make a contribution to fault tolerance.

*Assumption 1 : Individuals are not capable of consistently achieving a perfect solution.*

If the task were sufficiently simple that it were possible for an individual to detect all faults there would be no need to devise a diverse system. Errors may still occur in this instance, but they will tend to be 'slips' [Rea90], where the discrepancy between action and outcome is obvious and detection may be achieved by redundancy. Cognitive diversity will only be of benefit if the task is either (a) complex or (b) sufficiently arduous that vigilance becomes an issue.

*Assumption 2 : There is some 'constancy' in error.*

There must be an association between the circumstances that will promote error and the nature of that error (see [Rea90]). We return to this issue later and it forms the basis of a method for determining the occurrence of diversity.

*Assumption 3 : No single individual/strategy consistently locates a superset of the errors located by other individuals/strategies.*

If it were possible for an individual/strategy to locate all errors detected by other individuals/strategies then, once again, there would be no need for diversity. In the context of diversity, the important question is not whether one individual/strategy is more efficient than another, it is whether individuals/strategies differ in the types of errors they detect (see Assumption 2).

*Assumption 4 : Diversity can be operationalised in a psychologically meaningful way.*

For cognitive diversity to occur it must be possible to represent alternatives in a manner which is psychologically meaningful. Even though it may be possible to determine a number of diverse solutions to a problem, and to implement diverse solutions within computer-based expert systems (e.g. [LeK93]), unless the each component represents a psychologically meaningful entity cognitive diversity will not be possible.

## 1.2 Potential Predictors of Diversity

### 1.2.1 Domain Knowledge

One factor which will necessarily contribute to diversity is domain knowledge. To take an extreme, and perhaps implausible, example, if individuals have completely different domain knowledge relating to a fault detection task, then one could predict a completely diverse solution in which there was no overlap between the faults detected by each individual. In this respect diversity can be seen as specialism, with individuals contributing to the fault detection process within specific areas of expertise. The benefits associated with such a system relate not only to the avoidance of common mode errors, but also to increased training efficiency. However, it should be noted within this context that it may also be the case that there are costs associated with overly diverse human systems. Typical symptoms of too much diversity might be difficulties in communication because people are no longer 'talking the same language' (e.g., see [FlH91]). Anecdotal examples of this situation abound in the case of the interface between subject area specialists and computer specialists in the design of software. Further to this, extreme diversity may have implications for human redundancy. It is possible that specialists will be unable to detect errors in the work of others unless if falls within their area of expertise (see Figure 1).

## 1.2.2 Strategy

A more contentious predictor of diverse fault-finding is performance strategy. Previous studies have identified a number of alternative strategies which may be applied to fault finding (e.g., [MoR85]; [MoD88]; [Pat93]; [Ras86]; [Rou83]). Many of these can be related to the Abstraction Hierarchy described by [Ras86], with a common broad theme relating to a distinction between, on the one hand, a 'bottom-up' strategy in which performance is concerned with the topographical features of the task environment, and a 'top-down' strategy, on the other hand, where performance is concerned with a functional view of the task environment. A further distinction which has received less attention in this regard concerns the use of imagistic and linguistic information [Rou83]. Given the wealth of evidence relating this distinction to cognitive architecture (e.g. [Bad86], [Wic92]), and the facility for interface design support in this area, this may well be a direction worth pursuing in the present context. Previous research has focused upon the comparative efficiency of performance strategies. For example, there is evidence to suggest that the 'top-down' strategy is more cognitively demanding than a 'bottom-up' strategy but potentially more successful ([Ras86]; [MoD88]). However, the question of whether these strategies are reliably detecting different 'types' of faults has not been addressed. Of great importance in this respect are task environmental factors (including interface design) which exert a large influence upon the selection ([Cas94]; [Rou83]; [Wic92]) and efficiency [VCP95] of performance strategies.

A more comprehensive taxonomy of fault-finding strategy, and one which might prove a valuable research tool with respect to cognitive diversity, is proposed by [Pat93], who classifies strategy along to two dimensions. The first dimension concerns the level of cognitive processing, and comprises three categories: pattern-matching; the application of rules and heuristics; and reasoning. These categories are obviously very similar to the model of Skill-, Rule-, and Knowledge-based behaviour developed by [Ras86]. The second dimension of this taxonomy relates to the type of information used, which is classified as: structural; values and relationships; functional; probabilistic; and temporal.

## 1.2.3 Individual Characteristics

A potential source of cognitive diversity relates to individual characteristics (e.g., see [Hoc90]). If through differences in ability or preference specific individuals consistently perform in ways more likely to detect specific (predictable) types of errors then diversity would be apparent. One of the intuitively most promising areas in this respect is that of cognitive style. Whereas measures of cognitive ability are concerned with assessing maximal performance, the study of cognitive style "...implies the measurement of propensities in terms of typical performance with the emphasis on a predominant or customary processing mode" ([Tie89], p. 263). However, previous research, both in relation to fault-finding performance ([MoR85]; [Mor88]) and more general computer-based performance (see [Wes93] for review) suggests that although measures of cognitive style sometimes predict performance, they are more akin to measures of cognitive ability ([McK84]; [Rob85]; [Tie89]). Nevertheless, there is some evidence to indicate reliable individual differences do exist in the cognitive processing strategies as applied to problem solving situations, e.g. abstract vs concrete [EgG82], model vs rule-based ([GBS86]; [MaJ82]), and that the selection of strategy [MHM78] or the efficiency with which different strategies are applied [StW80] may be related to cognitive abilities.

# 2.0 Empirical evidence of cognitive diversity

Our initial interest in the issue of diversity arose from the study of the task of preparing the geographic data which controls the safe movement of trains within a section of railway. This is a complex design task which requires design engineers to express the necessary signalling principles for sections of railway in the form of data, which can then be processed by signalling software and hardware. In order to ensure that data is fault free it undergoes a rigorous process of checking and testing. There are three distinct design phases in this respect. First, the data are written. Second, the data are checked. This involves an experienced design engineer checking printouts of the data for faults. Third, the data are tested. This involves the use of computer-based simulations of the operation of the railway, which permit the data to be functionally tested. Intuitively it would appear that there is a good deal of diversity built in to this verification process, with differences particularly apparent between the writing and checking components on the one hand, and the testing component on the other hand. For example, the strategy employed at the writing and checking stages tend to be of the 'bottom-up' nature, whereas testing strategy tends to be 'top-down'; there is a greater emphasis upon the textual presentation of information for writing and checking whereas the information presented to testers has greater spatial content; and, there are some differences in domain knowledge, with engineers often specialising in either checking or testing tasks.

In this paper we will focus upon the checking and testing components of the data design process. The following sections report some preliminary results from a field study and a laboratory study.

## 2.1 Field Study

One of the primary areas of concern in our current research work has been to determine whether certain types of faults are resistant to detection at particular stages of the data design process. In order to examine the relative efficiency of the different task components (writing, checking, and testing) a range of data has been collected from the field. Engineers were asked to complete additional fault logs at each stage of the data design process, recording the stage at which each fault was detected and the signalling principles which had been violated.

Data relating to 580 faults are presented in Table 1. Included is a figure representing the ratio of checking/testing faults detected. The larger this figure, the greater the comparative frequency of fault detection at the testing stage of the data design process. A shortcoming of this data is that it only relates to those faults which have been detected. Furthermore, if faults are detected at the checking stage of the data preparation process they will be corrected and therefore cannot be detected at the testing stage. However, this data does give some indication of the capacity of each of these methods to detect different types of faults.

Table 1: Breakdown of faults detected at the checking and testing stages of data preparation by signalling principle contravened

| Principle | Checking | Testing | C/T Ratio |
|---|---|---|---|
| None (no action required : false alarm) | 46 | 26 | .57 |
| Identity | 26 | 30 | 1.15 |
| Route | 74 | 120 | 1.62 |
| Signal aspect control | 14 | 32 | 2.29 |
| Approach locking | 11 | 24 | 2.18 |
| Opposing locking | 9 | 39 | 4.33 |
| Aspect sequence | 4 | 12 | 3.00 |
| Other | 106 | 7 | 0.07 |
| Total | 290 | 290 | 1.00 |

A qualitative analysis suggests that certain faults are resistant to detection by the checking process, and are more readily detected by the testing process. Of particular note in this respect are 'opposing locking' faults. These concern situations where the geographic data allows a route to be set without making all the necessary checks to ensure that a conflicting route (i.e. one which requires some of the same sections of track) has not already been set. The ratio of detection frequency for the two task components (checking : testing) suggests that 'opposing locking' faults are detected much more frequently during testing than checking. A further breakdown of the data revealed that this was particularly the case for those opposing routes which either do not cross any points or which require points in the same lie (direction) as the route to be set (see Table 2).

Table 2: Breakdown of opposing locking errors, detected at the testing stage, by points conflict.

| | |
|---|---|
| Opposing route does not include any points | 17 |
| Opposing route requires points in the same lie as the route to be set | 17 |
| Opposing route requires points in the opposite lie to the route to be set | 5 |
| Total | 39 |

In contrast, the checking task component appears to be particularly sensitive to faults categorised as 'other'. A further breakdown revealed that this category comprised largely non safety-critical errors relating to the format of data. As mentioned above, no conclusions can be drawn from this data relating to the sensitivity of the testing task component to 'other' faults, in that, if they are being detected at the checking stage, no opportunity for them to be tested arises. However, the nature of these faults makes it apparent that many would not be detectable at the testing stage.

In summary, these data provide some qualitative support for the existence of cognitive diversity within the design process. It would appear that there are differences in the sensitivity of the checking and testing phases of the data design process to different types of faults. However this must be qualified, in that the absolute frequencies of faults types within this data are unknown, and only those faults which are not detected during checking are available for testing.

## 2.2 Laboratory Experiment

In order to examine these factors in more controlled conditions a laboratory experiment was conducted in which components of the checking and testing tasks were simulated. This simulation related only to the signalling requirements for setting a route and obtaining a proceed aspect (a green signal in this case). Consequently, participants were required to determine: points availability, that opposing routes were not already set, and that track circuits within the route were not occupied. Of primary interest was task (checking vs testing) differences in the detection of opposing route faults.

In addition the relative ease with which faults of commission (additional or changed data) and faults of omission (missing data) could be detected was also examined. The basis of this was reports by task experts that faults of omission are more difficult to detect during checking than faults of commission.

### 2.2.1 Method

Twenty seven participants with an engineering or computer science background were recruited from the student population and randomly allocated to two between-participant conditions. One condition (13 participants: mean age 24.00 yrs.) simulated the task of checking geographic data, and the other (14 participants: mean age 22.78 yrs.) simulated the task of testing the data. The data from one participant in the checking condition were excluded from the analysis due to a misinterpretation of experimental instructions. All participants completed tests of spatial and verbal ability prior to training and experimental task performance.

In order to standardise training as far as possible, all participants initially completed a 'core' component which related to general signalling principles. This was followed by specific training in which each experimental group was taught the mechanics of either checking or testing. All materials for these tasks were computer presented (see Figures 2 and 3). In both task conditions the upper portion of the display included a representation of a railway layout. However, in the checking condition participants were able to display the contents of text files containing the geographic data in the lower portion of the screen (see Figure 2); whilst in the testing condition buttons corresponding to the components of the railway were displayed (see Figure 3). These buttons could be selected, using a computer mouse, in order to establish test scenarios.

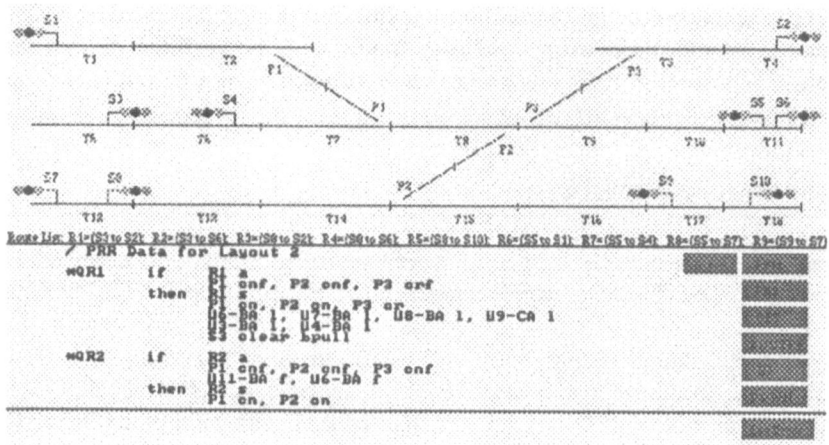

Figure 2 : Layout of checking screen

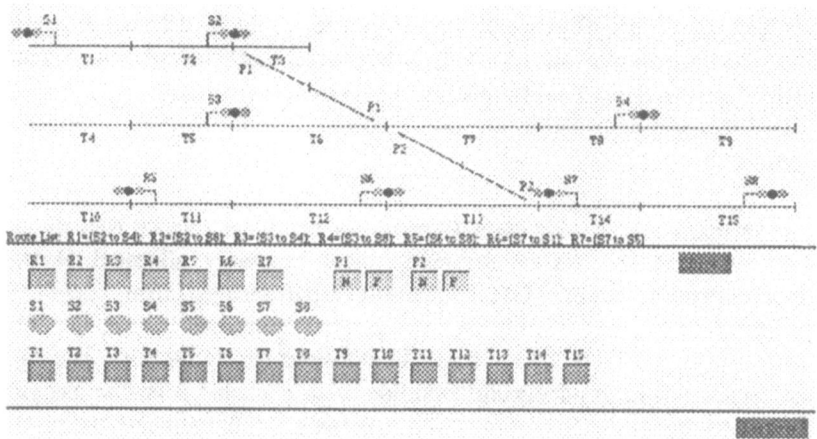

Figure 3 : Layout of testing screen

Data was gathered from performance whilst checking or testing two track layouts (see upper portions of Figures 2 and 3). Faults contravening four signalling principles (SPs) were introduced into the data for each layout. These faults related to (i) setting an opposing route over a different lie of points (ORDL); (ii) setting an opposing route over the same lie of points (ORSL); (iii) data in the OPT file failing to ensure that all sections of track are clear before the signal shows a proceed aspect (OPT); and (iv) other miscellaneous route setting errors (OTHER). For each SP there were faults of commission (additional or changed data), and faults of omission (missing data). This factor of faults of commission vs omission will be referred to as CO. Upon the basis of the field study, it was hypothesised that common-mode error would be apparent in the checking task for opposing route faults, and in particular for those which use the same lie of points (ORSL).

## 2.2.2 Results and Discussion

There was no significant difference between checking and testing groups in spatial ability. However, the testing group scored significantly higher, $t(24) = 2.24$, $p<.05$, in verbal ability. Correlations between cognitive ability and fault detection performance (examined separately for each task group) were non-significant and provided no indication of an association between cognitive ability and cognitive diversity.

A 2 (task type - checking vs testing) x 4 (SP) x 2 (CO) ANOVA (with repeated measures for the latter two factors) was used to analyse the probabilities of detecting errors. The mean proportion of faults detected in each of the experimental conditions are shown in Table 3.

Table 3 : Proportion of faults detected by task type, signalling principle, and omission vs commission.

|  | Checking | | Testing | |
|---|---|---|---|---|
|  | Omi | Com | Omi | Com |
| ORDL | .667 | .708 | 1.000 | 1.000 |
| ORSL | .917 | .750 | .929 | .857 |
| OPT | .958 | .917 | .857 | .405 |
| OTHER | .958 | .833 | .893 | 1.000 |

There was no significant main effect of task type, $F(1,24) = .64$, $p > .05$, or SP, $F(3,72) = 2.08$, $p > .05$. However, errors of presence were detected less frequently than errors of absence, $F(1,24)=10.01$, $p<.01$. There was a highly significant interaction between task type and SP, $F(3,72)=10.58$, $p<.001$. This appeared to be attributable to errors made during the performance of the checking task relating to the detection of ORDL faults, and errors made during the testing task relating to the detection of OPT faults. This latter effect was clarified by a significant three-way interaction between task type, SP, and CO, $F(3,72)=5.59$, $p<.005$, which appeared to be attributable to the comparatively low probability of participants in the testing condition detecting OPT faults of commission. There was also a significant interaction between SP and CO, $F(3,72)=4.32$, $p<.01$, such that faults of commission in data relating to ORDL were detected more frequently than faults of omission. However, the reverse was true in the case of data relating to ORSL and particularly OPT.

These results indicate diversity of fault detection across task components. Participants in the checking condition appeared to find it more difficult to locate ORDL faults than participants in the testing condition; and participants in the testing condition found OPT faults more difficult to locate than did participants in the checking condition. The three way interaction between task type, SP and CO indicated that this latter effect was stronger in the case of faults of commission. The fact that participants in the checking condition did not find ORSL faults the most difficult to detect, as hypothesised, may be attributable to the nature of the simulation. Although this warrants further investigation it need not concern us here.

## 2.3 Conclusions

These data demonstrate that cognitive diversity is an achievable method of improving fault detection in a safety-critical design task. With respect to the mechanics of this process, each of the assumptions stated earlier can be seen to apply. The task (in the field and in the laboratory) was sufficiently complex that perfect performance was not possible; there was constancy in the association between task type (checking vs testing) and fault type; neither checking nor testing located a superset of faults; and the very performance of these tasks with diverse results attests to the fact that differences were expressed in a psychologically meaningful way. The importance of different factors in achieving this diversity is more difficult to determine. As described earlier, there is undoubtedly some domain knowledge diversity between the checking and testing tasks. However, this was minimised in the laboratory study, with all faults being potentially detectable using either method. Individual differences in cognitive ability do not appear to have contributed to diversity in this instance. This is an important fact, in that it implies that selection need not be a primary consideration in the creation of diverse systems. Apparently of greatest importance are strategy and task environment factors. The strategy differences between the checking and testing tasks were described above. However, the nature of these tasks is such that it is not possible to identify the effects of particular strategies with precision. Consequently we are in the process of establishing a series of experimental studies which will use a more generic (less applied) task environment to examine this factor in greater detail. Further to this, the importance of task environmental factors cannot be underestimated. The support provided for each task varied along a number of dimensions (e.g. spatial content; functional information). Once again, more controlled (although less realistic) research is required.

The present position may be summarised by highlighting task structure as a key to this problem; with specific combinations of performance strategy and task environmental support being required to achieve optimal cognitive diversity. Further work is now required to identify the associations between these factors more precisely.

## Acknowledgements

The work described here forms part of the project "Human Factors in the Design of Safety-Critical Systems" (Grant no. GR/J17319) from the UK Safety-Critical Systems Research Program, supported by the DTI and EPSRC. It was completed with the collaboration of Interlogic, GEC Alsthom, and Westinghouse Signals.

# References:

[Bad86]   Baddeley, A. *Working Memory*. Oxford: Clarendon Press. 1986

[BiL89]   Bisant, D.B. & Lyle, J.R. "A two-person inspection method to improve programming productivity". *IEEE Transactions on Software Engineering, 15*, 1294-1304, 1989.

[BrD81]   Brooke, J.B. & Duncan, K.D. "Effects of system display format on performance in a fault finding location task". Ergonomics, 24, 175-189, 1981.

[CBG94]   Canas, J.J., Bajo, M.T., & Gonzalvo, P. "Mental models and computer programming". *International Journal of Human-Computer Studies, 40*, 795-811, 1994.

[Cas94]   Casner, S.M. "Understanding the determinants of problem-solving behaviour in a complex environment". Human Factors, 36, 580-596, 1994.

[EgG82]   Egan, D.E. & Grimes-Farrow, D.D. "Differences in mental representation spontaneously adopted for reasoning". *Memory and Cognition, 10*, 297-307, 1982.

[Fag76]   Fagan, M.E. "Design and code inspections to reduce errors in program development". *IBM Systems Journal*, 3, 182-211, 1976.

[Fag86]   Fagan, M.E. "Advances in software inspections". *IEEE Transactions on Software Engineering, 12*, 744-751, 1986.

[FlH91]   Flor, N.V. & Hutchins, E.L. "Analysing distributed cognition in software teams: A case study of team programming during perfective software maintenance". In J. Koenemann-Belliveau, T.G. Moher, & S.P. Robertson (Eds.) *Empirical Studies of Programmers: Fourth Workshop*. Norwood, NJ: Ablex. 1991.

[Gar93]   Garland, W.J. "Dealing with the dilemma of disparate mental models". In G. Salvendy & M.J. Smith (Eds.) *Human-Computer Interaction: Software and Hardware Interfaces*. Amsterdam: Elsevier. 1993.

[GBS86]   Galotti, K.M., Baron, J. & Sabini, J.P. "Individual differences in syllogistic reasoning: Deduction rules or mental models?" *Journal of Experimental Psychology: General, 115*, 16-25, 1986.

[GeS83]   Gentner, D. & Stevens, A.L. (Eds.). *Mental Models*. Hillsdale, NJ: Lawrence Erlbaum Associates. 1983

[Hoc88]   Hoc, J-M. "Towards effective computer aids top planning in computer programming: Theoretical concerns and empirical evidence drawn from assessment of a prototype". In G.C. van der Veer, T.R.G. Green, J-M Hoc, & D.M. Murray (Eds.) *Working with Computers: Theory Versus Outcome*. London: Academic Press. 1988.

[Hoc90]   Hockey, G.R.J. "Styles, skills, and strategies: Cognitive variability and its implications for the role of mental models in HCI". In D. Ackerman & M.J. Tauber (Eds.) *Mental Models and Human-Computer Interaction I*. Amsterdam: North Holland. 1990.

[Hol93]   Hollywell, P.D. "Human dependent failures: A schema and taxonomy of behaviour". In EJ Lovesey (Ed.). *Contemporary Ergonomics*. London: Taylor & Francis. 1993.

[LeK93]   Lee, J.M. & Kim, J.H. "An integration of heuristic and model-based reasoning in fault diagnosis". *Engineering Applications of Artificial Intelligence, 6*, 345-356, 1993.

[MaJ82]   Mani, K. & Johnson-Laird, P.N. "The mental representation of spatial descriptions". *Memory and Cognition, 10*, 181-187, 1982.

[McK84]   McKenna, F.P. "Measures of filed dependence: Cognitive style or cognitive ability?" *Journal of Personality and Social Psychology, 47*, 593-603, 1984.

[MHM78] McLeod, C.M., Hunt, E.B., & Matthews, N.N. "Individual differences in the verification of sentence-picture relationships." *Journal of Verbal Learning and Verbal Behaviour, 5*, 493-508.

[MoR85]   Morris & Rouse, W.B. "Review and evaluation of empirical research in troubleshooting." *Human Factors, 27*, 503-530, 1985.

[Mor88]   Morrison, D.L. "Predicting diagnosis performance with measures of cognitive style." *Current Psychology: Research & Reviews, 7*, 136-156, 1988.

[MoD88]   Morrison, D.L. & Duncan, K.D. "Strategies and tactics in fault diagnosis." *Ergonomics, 31*, 761-784, 1988.

[Pat93]   Patrick, J. "Cognitive aspects of fault-finding training and transfer." *Le Travail Humain, 56*, 187-209, 1993.

[Ras86]   Rasmussen, J. *Information Processing and Human-Machine Interaction*. Amsterdam: North Holland. 1986.

[Rea90]   Reason, J. *Human Error*. Cambridge: Cambridge University Press. 1990.

[Rob85]   Robertson, I.T. "Human information-processing strategy and style." *Behaviour and Information technology, 18*, 199-214, 1985.

[Rou83]   Rouse, W.B. "Models of human problem solving: Detection, diagnosis, and compensation for system failures. *Automatica, 19*, 613-625, 1983.

[Sag94]   Saglietti, F. "Fault tolerance by software diversity: How and when?" In V. Maggioli (Ed.) *SafeComp '94. Proceedings of the 13th International Conference on Computer Safety, Reliability, and Security. 1994*

[SeM91]   Senders, J.W. & Moray, N.P. *Human Error: Cause, Prediction, and Reduction.* Hillsdale, NJ: Lawrence Erlbaum. 1991.

[StW80]   Sternberg, R.J. & Weil, E.M. "An aptitude X strategy interaction in linear syllogistic reasoning." *Journal of Educational Psychology, 72,* 226-239, 1980.

[Tie89]   Tiedeman, J. "Measures of cognitive styles: A critical review." *Educational Psychologist, 24,* 261-275, 1989.

[VCP95]   Vicente, K.J., Christoffersen, K., & Pereklita, A. "Supporting operator problem solving through Ecological Interface Design". *IEEE Transactions on Systems, Man, and Cybernetics, 25,* 529-545, 1995.

[Wes93]   Westerman, S.J. "Individual differences in human-computer interaction." Unpublished Doctoral Dissertation. University of Aston, England. 1993.

[Wic92]   Wickens, C.D. *Engineering Psychology.* NY: Harper Collins. 1992.

[WNH92]   Wilson, J.D., Nosek, J.T., Hoskin, N., Liou, L.L. "The effect of collaboration on problem-solving performance among programmers." In J. van Leeuwen (Ed.) *Information Processing '92: Algorithms, Software, Architecture.* North Holland: Elsevier Science. 1992.

# Legal Aspects of Safety Critical Systems

Dai Davis

Eversheds, Solicitors

Leeds, London and Manchester, England.

## 1 Introduction

Society is increasingly dependent upon computers and, in particular, computer controlled systems. Unfortunately, the growth of computer controlled systems in the last three decades has not been matched by any systematic reappraisal by legislative bodies of the legal liability attaching to suppliers of those computer controlled systems.

To this haphazard application of the law must be coupled a fundamental difficulty of the supplier of a computer controlled system : the relative ease of supplying a system which inadvertently contains an error even where the system has been tested, validated, verified and assessed.

This paper examines the range of specific liabilities placed upon manufacturers and suppliers of computer systems with particular reference to safety-critical systems. This paper deals with (1) the "classical" remedies available (2) product liability legislation and (3) the newer burdens imposed by European Union legislation.

## 2 Classical Remedies

### 2.1 Contract

In order to bring a claim, there must first be a contract between the parties in dispute. This is by no means a trivial test since typically, there will not be a contract between the injured party and the manufacturer who has introduced the defect into the product.

An example of where a contract claim could be brought would be where a computer control manufacturer supplies a component to a motor car manufacturer who is building an anti-lock brake device. If, as a result of a fault in one of the computer units, an accident occurs the car manufacturer could bring a claim for breach of contract. That claim could be very great, since it could include damages for the loss of business suffered by the car manufacturer where he had needed to undertake a product recall exercise. In practice the computer component manufacturer would seek to limit his liability by excluding claims of the same type as consequential loss.

A claim for a breach of contract could be based on an express term of a contract, for example, that goods will comply with a certain British Standard. Additionally, a claim could be based on a term implied into the contract. Among the most important terms implied into most contracts for the supply of goods are that goods will:-

- comply with their specification and description;
- be of satisfactory quality; and
- be fit for any specific purpose for which they have been supplied.

## 2.2    Law of Negligence

Under the law of negligence, a manufacturer of goods owes a so-called "duty of care" to ensure that the goods he supplies are not likely to cause personal injury. Essentially, the duty is to take "reasonable care".

Historically, claims for defective computer systems have had to be brought under this law - particularly where there has been no contractual relationship between the claimant and the manufacturer. The manufacturer's liability under the law of negligence extends to anyone the manufacturer ought reasonably to believe would be adversely affected by his negligence. This is far wider than under the law of contract where the person bringing the claim must have a direct contractual relationship with the defendant.

## 2.3　　　Health and Safety at Work etc Act

Legislation places additional obligations on a manufacturer, for example the Health and Safety at Work etc Act 1974 which imposes a general duty on suppliers of machinery for use at the workplace as well as on employers. In particular, a manufacturer is required to use reasonable efforts to ensure that machinery is "safe". The remedies for breach of this Act are both criminal and civil.

In general, however, the obligations imposed under this Act are not as onerous as under the newer legislation which is discussed below; if a manufacturer complies with his obligations under the Product Liability law discussed below, the manufacturer is more than likely to have complied with his other legal obligations. Other European Directives may, in the long run, impose even higher duties on manufacturers, however.

## 2.4　　　Reasons for Change

The European Union has taken a lead in introducing legislation into the area of product safety. Traditional remedies have been difficult to enforce, for example defective goods may still have been of "satisfactory quality" under the law of contract. Much of this European legislation has far reaching implications for computer controlled systems. Most importantly, the Council of the European Union (which is made up of the Prime Ministers of each of the Member States) adopted a Directive on product liability on 25 July 1985 [Council 85]. This Directive introduced a standard law on product liability throughout the fifteen states of the European Union. The discussion below on this and other Directives, therefore, gives an outline of European rather than just UK legislation.

# 3  The "New" Remedy : Product Liability

## 3.1       The Five Requirements

Product liability law can be concisely expressed as follows : Where DAMAGE is CAUSED by a DEFECT in a PRODUCT then CERTAIN PERSONS are liable to compensate for that damage.

*3.1.1*       *The Damage.*  Two types of damage may be the subject of a claim under the legislation.  The first type of damage is death or personal injury.  The next of kin can sue when a person has died.  The second type of damage is property damage.  The property damaged must be of a type ordinarily intended for private use or consumption, and the item must be used mainly for private use, occupation or consumption, and the total damages claimed by a person in respect of loss of property (excluding any claim of interest) must exceed 500 ECU (European Currency Units). The value of an ECU as at 2 June 1995 was about 83 pence.  The damage claimed for must also exclude damage to the defective product itself!

*3.1.2*       *The Cause.*  The legislation provides that the injured party needs to prove the damage, the defect <u>and</u> the causal relationship between the damage and the defect.  In certain circumstances, this may be a substantial hurdle, for instance where the injured party is trying to claim against the manufacturer of a component within a computer controlled device, the injured party may not know exactly which component has been defective.

*3.1.3*       *The Defect.*  The legislation defines the defect in terms of the safety which a person is entitled to expect.  Where the safety is that which people (i.e. the public) are entitled to expect, then the product will not be defective.  In determining this test of safety all the circumstances must be taken into account.  These circumstances will include the presentation of the product, whether any instructions or warnings were given in relation to the product), and whether the use to which the product is being put is a reasonable use. The question of safety must be decided with reference to the time at which the product was supplied. So the mere fact that a newer product is safer than an older product will not necessarily mean that the older product was not, at the time it was supplied, as safe as people are generally entitled to expect.

Comparing this test to the previous prevailing law where it was necessary to show that the manufacturer had not taken reasonable care in producing the product, it can be seen how the obligations have increased. It is relatively easy for a software manufacturer to show that it has taken reasonable care under the law of negligence and contract, even though a defect was introduced into the software. Under the product liability law, however, the manufacturer must show that the product is as safe as people are entitled to expect. Although there are no reported cases on this issue, it is believed that this test is very onerous from the point of view of the manufacturer of a computer controlled system.

For instance, in areas of software manufacture where adopted standards exist, manufacturers must follow those standards to comply with the Directive. In addition, manufacturers should have regard to developing and non-mandatory standards, as indeed the public would expect them to do. If a software manufacturer has not used best practice "techniques", the manufacturer may well be liable for a defect thereby introduced into the software.

Care must be taken when giving instructions or warnings. In the case of a complex control system, it may not be sufficient merely to put a warning of a significant danger in the instruction manual. A prominent warning should also be placed in a permanent form next to the operator.

*3.1.4     The Product.* The legislation only applies to products and not services. The definitions contained in the legislation do not really help in determining the dividing line between what is a product and what is a service. As may be anticipated, debates have ensued as to whether items such as computer software are a product or a service. A detailed discussion of this topic is outside the scope of this article. It is now generally accepted, however, that software falls within the scope of this legislation - especially since it is often supplied "bundled" on hardware (such as a floppy disc).

*3.1.5* *The Persons Liable.* The primary person who is liable for a defective product is the "producer". The legislation specifically provides that a product can be comprised within another product. It is therefore possible for more than one producer to be liable and hence to be obliged to pay compensation.

For example, suppose an engine contains a micro-processor which dynamically determines the fuel/air mixture pumped into the engine. A failure in the software could cause the engine to cut out at a critical moment, causing damage and injury. There could be three separate "producers" liable under the legislation in this example the supplier of the programmed control unit; the supplier of the engine who has incorporated the control unit into the pump; the supplier of the engine itself which contains that engine. In the case of an engine imported into the European Union, the importer would also be liable.

In addition, a person who puts his name, trade mark or other distinguishing feature on a product and thereby represents himself as being a producer will also be liable. Furthermore, if the producer of the product cannot be identified then the supplier of the product can be treated as the producer unless the supplier informs the injured party of the identity of the actual manufacturer. The legislation is designed so that the injured party will always be able to find someone to sue within the European Union.

## 3.2 Defences and Limitations

Although the legislation generally provides six defences, only two of them are of any real practical use. The burden of proof in respect of all the defences is placed firmly on the producer. The two more useful defences are the "Development Risks" Defence and the "Sub-Component Manufacturer's" Defence.

*3.2.1      The Development Risks Defence.* This defence is more widely-known as the "state of the art" defence. A producer is not liable where he can show that the state of scientific and technical knowledge at the time when he put the product into circulation was not such as to enable the existence of the defect to be discovered. This defence is quite narrow.

To avail himself of this defence, a producer must use every scientific and technically known method. It is not sufficient if the manufacturer only has regard to typical procedures undertaken by the average manufacturer. Only if he has taken <u>every</u> step which a manufacturer was able to take, but nevertheless he did not discover the defect, can he use this defence. Not only must a manufacturer comply with mandatory standards but, to be certain that the defence will apply, he must also have regard to developing and non-mandatory standards. Similarly, if best practice techniques are not followed, the manufacturer will find himself unable to use this defence successfully. In the context of product liability, "good manufacturing practice" techniques are not good enough!

*3.2.2      The Sub-Component Manufacturer's Defence.* This defence is available to the manufacturer of a component where he can show that the defect is attributable (i) to the design of the product into which the component is fitted or (ii) to the instructions given by the manufacturer of the product into which the component is fitted. This defence only applies to a sub-contractor and only where the sub-contractor can show that the defect is <u>wholly</u> attributable to the design or instructions given by the manufacturer of the overall product.

It is not likely that this defence will arise in the computer control industry in respect of software. This is because it is highly unlikely that a software manufacturer will ever receive instructions that are detailed enough to be able to say that, through compliance with those instructions, the software manufacturer introduced a defect into the product.

An example of where the defence could arise would be where a wiring manufacturer was told to supply wiring with certain characteristics (such as being capable of carrying a current of certain strength). If in fact the wiring in a machine caught fire because a larger current was passed through it, then the wiring supplier could escape from liability using this defence.

# 4  Other European Law

European law is becoming increasingly important, not least because of its sheer volume and also because it imposes higher safety standards upon the manufacturers of products. No European Directives are specific to computer controlled systems. Indeed, by dealing with computer hardware and software amongst other features of products, the legislation fails to recognise the peculiarities of these items. The liability which is introduced may therefore appear to be unreasonable, at least on a superficial examination.

## 4.1  Machine Safety Directive

*4.1.1*  *General.* A Directive of the European Union of 14 June 1989 imposed a duty on Member States to enact safety legislation relating to machinery [Council 89]. The Directive required Member States to bring the Directive into force by 1 January 1993 although certain transitional arrangements were provided. Most products did not need to comply with the Directive until 1 July 1995. In some cases, the important date is 1 January 1996.

The Directive applies to almost all machinery; the main categories of machinery which are exempt are lifting machinery, machinery for medical use, pressure vessels, fire arms, vehicles, storage tanks and machines primarily designed for use by the military or police. The definition of "machinery" is very wide and covers almost any article which contains a moving part. The definition encompasses anything from a microwave cooker to an industrial computer controlled system, as well as an ordinary personal computer (at least one with a disc drive!).

The Machine Safety Directive imposes a duty on the manufacturer or importer into the European Union to assemble a technical file in respect of the machine. In addition, that person must then state that the machine conforms with the safety requirements of the Directive. In certain circumstances, the machine must be examined by an independent body to show compliance.

There are two aspects of those safety requirements which particularly refer to software. The first is that the control systems of machinery must be so designed that they are safe and reliable and in particular that "errors in logic do not lead to dangerous situations".

In particular, the control circuit logic must be designed so that :

-   the machinery does not start unexpectedly;
-   the machinery cannot be prevented from stopping if the command to stop the machinery has already been given;
-   automatic or manual stopping of the machinery must not be impeded.

Secondly, any interactive software which the operator uses to control the machine must be "user friendly".

*4.1.2    Analysis.* Unfortunately, neither "dangerous situation" nor "user friendly" are defined in the Directive. The legislation which enacts the Directive in the United Kingdom, the Supply of Machinery (Safety) Regulation 1992 (SI 1992 No. 3073) uses identical (undefined) terminology. In other circumstances, a dangerous situation has been held by a Court to arise where machinery could injure anybody in a way which might reasonably have been expected to occur. It is possible that the same approach will be given to interpreting the requirements of this legislation.

The term "user friendly" is very much more ambiguous. Probably, the context in which the machine was intended to be used would be very important. What is clear is that a machine would not be "user friendly" if the controls of the machine were such that it was possible for an operator to override the normal and safe functioning of the machine. In these circumstances, the clearest possible warnings need to be given to the operator that the results of those overrides could potentially be dangerous.

The regulations which introduce the Directive into the United Kingdom impose sanctions on companies which do not comply. Also, a director or manager of a company will be criminally responsible where a breach of the regulations can be shown to have been permitted with his (or her) consent, connivance or neglect. A defence is, however, provided where the individual proceeded against can show that he took "all reasonable steps" **and** exercised "all due diligence" to comply with the regulations. In Great Britain, the maximum penalty which can be imposed for a breach of these regulations is a fine of £5,000 (£2,000 in Northern Ireland) and imprisonment for a period not exceeding three months.

As well as being a criminal offence, a civil action will also lie against a manufacturer who fails to comply with the Directive. Such an action is based, in the United Kingdom, upon the legal wrong known as "breach of statutory duty".

## 4.2    General Product Safety Directive

This Directive, passed in 1992, is now in force by the General Product Safety Regulations 1994 [Council 92/A]. Essentially, it is a "catch all" Directive which applies to a product which is put on the market where there is no other specific European Directive applying to the safety of that product. The Directive states that provided a product is safe as judged by the national law of any Member State, it will be deemed to be safe for all Member States. The product can then be circulated throughout the European Union.

There is now a general duty on all manufacturers and in most case on distributors to ensure that the products they place on the market are safe. If they are not safe not only can the company be fined, but also individual directors can be fined (up to £5,000) and sent to prison (for up to three months).

The standard of safety is not as high as the standard of "no defects" required under product liability legislation. In the context of product safety, "safety" means that the risks associated with the use of the product must only be those minimum risks compatible with a high level protection or the health and safety of people (not damage to the environment for example). Only normal and foreseeable use needs to be considered.

Under the new Product Safety Legislation, distributors also have a fairly heavy responsibility. For instance, if a distributor possesses an unsafe product and intends to supply it, he will commit a criminal offence. However, any other business which is involved in the supply chain can also be liable. For example, if a transport company fails to ensure that products remain safe whilst they are in its possession, the transport company will also commit an offence under the Regulations. This could occur, for example, where a transport company fails to keep a component for a silicon chip stored at the correct temperature during transportation.

There are defences available where the company or person charged took all reasonable steps and exercised all due diligence to avoid committing the offence or where the defendant can show that he relied on information from another as to the safety of the product. However, the defendant must show that it was reasonable in all the circumstances for him to rely upon the information supplied by the other party. Whilst in many circumstances, the distributor will be able to take advantage of this defence he will not be able to do so where, for example, he can see that the packaging of the product has been damaged. In these circumstances, he is obliged to examine the product inside to make sure that the product has not been damaged.

Further (civil law) obligations deal with the provision of information by the manufacturer and distributor and perhaps the most radical is that the manufacturer must have plans for product recall where this proves necessary. For example, the manufacturer is required to plan in advance and must mark products or batches of products in such a way that they can be identified. He must test samples of the products he places on the market. He also needs to investigate complaints and to inform distributors should a product recall be necessary. Of course, all good manufacturers have been taking these steps for a long time. However, distributors are now under a duty to assist manufacturers in order to ensure that unsafe products are not placed on the market. For example, distributors must inform manufacturers of any risks they see in relation to a product. Distributors are also now under a positive obligation to assist manufacturers in dealing with any product recall.

Additional burdens on manufacturers include a duty to ensure that second hand goods are safe (within the meaning of the regulations) before they are placed on the market. In some cases, it may be impossible to achieve the safety standard required. This will mean that it will make it almost impossible for those used products to be sold as "safe" within the meaning of the legislation. This has a clear implication for persons who sell second hand cars after trying to improve the performance of the car by interfering with the computer-controlled engine management system. Similarly, refurbishers and reconditioners of products are now liable to ensure that the refurbished and reconditioned products that they supply are safe.

## 4.3    Electromagnetic Compatibility Legislation

The basis of the law in the United Kingdom is the implementation of the EMC Directive. This Directive was introduced on 3 May 1989, its purpose being to harmonise the laws of the Member States relating to this subject. [Council 89/A]. That Directive has been amended four times :

- on 29 April 1991 - regarding telecommunications equipment [Council 91];
- on 28 April 1992 - transitional period extended [Council 92/B];
- on 22 July 1993 - CE Mark amended [Council 93/A];
- on 29 October 1993 - regarding satellite earth station equipment [Council 93/B].

In the United Kingdom the law is now to be found in the Electromagnetic Compatibility Regulations of 1992, as amended in 1994. Whilst the UK Regulations are already in force, they do not require any immediate change in companies' procedures. At present, companies have the choice either to comply with the provisions of the Directive or to comply with the "older" law which was in force as at 20 June 1992. This will continue to be the position until 1 January 1996, from which time companies will have no choice but to comply with the provisions of the Directive. The legislation will apply to all electrical and electronic equipment and systems supplied after that date.

The legislation imposes three basic obligations:-

*4.3.1     The electrical and electronic apparatus must be constructed so that it "conforms".* This means that the equipment must be constructed so that any electro-magnetic disturbance generated by the equipment does not prevent other apparatus from operating as intended. In addition, the equipment must be so constructed that it has an adequate level of immunity from external electromagnetic disturbance, so that the equipment can carry on operating as was originally intended.

*4.3.2     The equipment must be constructed so that it "complies" with the legislation.* This means that the equipment must either be built in accordance with the relevant standards (of the European Union where they exist, otherwise national standards) or else that a so-called technical construction file is created. Where a technical construction file is created a competent body must be consulted to approve the file.

*4.3.3     Certain types of telecommunication equipment must be sent to a notified body for examination.* They may not be sold unless a type examination certificate has been issued for the equipment by the notified body.

The basic obligation is imposed upon suppliers of "single commercial units". This is intended to catch the supplier of the end-product rather than component suppliers. However, the user must ensure also that the apparatus conforms. So, for example, in a factory where the owner of the factory buys various components but constructs an automated assembly line in situ, it is the owner who must ensure that the apparatus "conforms".

## 4.4      Other Directives

Various other Directives affect the legal framework in which computer systems are supplied. Limitations of space mean that they cannot be fully discussed here. However they include the following:-

| Directive | Reference |
|---|---|
| Health & Safety (Display Screen Equipment) Directive | [Council 90] |
| Extractive Industries (Boreholes) Directive | [Council 92/C] |
| Personal Protective Equipment Directive | [Council 89/B] |
| Work Equipment Directive | [Council 89/C] |

# References

[Council 85] - Council Directive 85/374 of 25 July 1985 on the Approximation of the Laws, Regulations and Administrative Provisions of the Member States Concerning Liability for Defective Products. [OJ 1985, L 210/29].

[Council 89/A] - Council Directive 89/392 of 14 June 1989 on the Approximation of the Laws of the Member States relating to Machinery. [OJ 1989, L 183/9].

[Council 92/A] - Council Directive 92/59 of 29 June 1992 on General Product Safety. [OJ 1992, L 228/24]. Enacted in the United Kingdom by the General Product Safety Regulations 1994.

[Council 89/A] - Council Directive 89/336 of 3 May 1989 on the Approximation of the Laws of Member States relating to Electromagnetic Compatibility. [OJ 1989, L139/19].

[Council 91] - Council Directive 91/263 of 29 April 1991, on the Approximation of the Laws of the Member States Concerning Telecommunications Terminal Equipment,

Including the Mutual Recognition of their Conformity. [OJ 1991, L128/1].

[Council 92/B] - Council Directive 92/31 of 28 April 1992, Amending Directive 89/336 on the Approximation of the Laws of the Member States Relating to Electromagnetic Compatibility. [OJ 1992, L126/11].

[Council 93/A] - Council Directive 93/68 of 22 July 1993 Amending ...... Directive 89/336 (Electromagnetic Compatibility) ...... [OJ 1993, L220/1].

[Council 93/B] - Council Directive 93/97 of 29 October 1993 Supplementing Directive 91/263 in Respect of Satellite Earth Station Equipment. [OJ 1993, L290/1].

[Council 90] - Council Directive 90/270 of 29 May 1990 on the Minimum Safety and Health Requirements for Work with Display Screen Equipment. Commonly referred to as the "VDU Directive". [OJ 1990, L156/14].

[Council 92/C] - Council Directive 92/91 of 3 November 1992 Concerning the Minimum Requirements for Improving the Safety and Health Protection of Workers in the Mineral-Extracting Industries Through Drilling (Eleventh Individual Directive Within the Meaning of Article 16(1) of Directive 89/391). [OJ 1992, L348/9].

[Council 89/B] - Council Directive 89/686 of 21 December 1989 on the Approximation of the Laws of the Member States Relating to Personal Protective Equipment. [OJ 1989. L399/18]. Enacted in Great Britain by the Personal Protective Equipment at Work Regulations 1992 [SI 1992 No 2966].

[Council 89/C] - Council Directive 89/655 of 30 November 1989 Concerning the Minimum Safety and Health Requirements for the Use of Work Equipment by Workers at Work (Second Individual Directive Within the Meaning of Article 16(1) of Directive 89/391. [OJ 1989, L393/13]. Enacted in Great Britain by the Provision and Use of Work Equipment Regulations 1992 [SI 1992 No 2932].


Dai Davis, MSc, MA, C Eng, MIEE is partner in charge of Computer and IT law at Eversheds which is a national firm with offices throughout England and Wales. He is based at Cloth Hall Court, Infirmary Street, Leeds, LS1 2JB, England and also at London, England and Manchester, England.

Telephone number 0113 243 0391 (international +44 113 243 0391).

Facsimile number 0113 245 6188 (international +44 113 245 6188).

# Invited Paper

# A Bayesian Model that Combines Disparate Evidence for the Quantitative Assessment of System Dependability

Bev Littlewood

David Wright

Centre for Software Reliability, City University,

Northampton Square, London EC1V 0HB, ENGLAND

### Abstract

For safety-critical systems, the required reliability (or safety) is often extremely high. Assessing the system, to gain confidence that the requirement has been achieved, is correspondingly hard, particularly when the system depends critically upon extensive software. In practice, such an assessment is often carried out rather informally, taking account of many different types of evidence—experience of previous, similar systems; evidence of the efficacy of the development process; testing; expert judgement, etc. Ideally, the assessment would allow all such evidence to be combined into a final numerical measure of reliability in a scientifically rigorous way. In this paper we address one part of this problem: we present a means whereby our confidence in a new product can be augmented beyond what we would believe merely from testing that product, by using evidence of the high dependability in operation of previous products. We present some illustrative numerical results that seem to suggest that such experience of previous products, even when these have shown very high dependability in operational use, can improve our confidence in a new product only modestly.

## 1 Introduction

Critical systems are coming to depend more and more upon the correct functioning of software to ensure their safe operation. At the same time, the size and complexity of these software subsystems is increasing as designers take advantage of the extensive functionality that software makes possible—functionality that sometimes enhances different aspects of safety.

There are important unresolved questions concerning how one might go about designing such systems so that they will be sufficiently safe in operation. In this paper, however, we shall concentrate upon the difficult problems of *evaluation* that they pose. In particular, we shall be concerned with the problem of how to measure the reliability of such a software system when that reliability is likely to be very high.

In several recent papers different authors have pointed out some of the basic difficulties here, [Butler93, Littlewood93]. They show that, if we are only going to use the evidence obtained from operational testing of the software, we shall only be able to make quite modest claims for its reliability. For example, Littlewood and Strigini show that even in the most favourable situation of all, that of a system that has not failed during $x$ hours of statistically representative operational testing, we can draw only the weak conclusion that there is a 50:50 chance that it will survive failure-free for the same time $x$ in the future.

The limitations here seem intrinsic: they arise from the relative paucity of evidence (when compared with the stringency of the reliability level that needs to be demonstrated) and will not be ameliorated significantly by better statistical models. To make a very strong claim—that a particular system is ultra-reliable—needs a great deal of evidence. If that evidence comprises only observation of failure-free behaviour, then the length of time over which such behaviour is observed needs to be very great. To assure the reliability goals of certain proposed and existing systems, for example the $10^{-9}$ probability of failure per hour for the 'fly-by-wire' computer systems in civil aircraft [Rouquet86], would clearly require the systems to be observed *and show no failures* for lengths of time that are many orders of magnitude greater than is practicable.

Faced with these limitations to what can be claimed from merely observing the system in operation, it has been suggested that we should instead base our evaluations upon *all* the disparate kinds of evidence that are available. These include, in addition to the operational data discussed above, evidence of the efficacy of the development methods utilised, experience in building similar systems in the past, competence of the development team, architectural details of the design, etc. Most of these other sources of evidence about the dependability of a system will involve a certain amount of engineering judgement in the evaluator, which might itself introduce further uncertainty and potentiality for error. In addition, there are serious unresolved difficulties in *combining* such disparate evidence in order to make a single evaluation of the overall dependability and thus to make a judgement of acceptability.

In this paper we shall consider only a small part of this problem. We shall treat in detail the situation where we wish to augment the evidence that can be gained from the operational testing of a particular product, by also taking into account the success (or not) in building 'similar' products in the past. An important special case, of course, is that where there is unreserved good news from these previous products—i.e. none of them has failed during operational use up till the present time.

It should be emphasised that the goal in all this work is to obtain a *quantification* of the reliability of a product. The model that is proposed in the next sections, therefore, requires us to make certain assumptions about the failure process, and about how we represent our beliefs about certain model parameters. We acknowledge that these assumptions can be questioned, and are certainly very difficult to validate. However, we believe that they are reasonably plausible. More importantly, our main aim is to demonstrate that this kind of evidence can only improve our confidence in the reliability of a product quite modestly. Thus, we would regard a critique of our results on the grounds that they are not sufficiently conservative as being in the spirit of our own aims; suggestions, on the other hand, that the assumptions here can be modified in order to arrive at much higher confidence in product reliability we would regard with suspicion. It seems to us that, particularly in the case of safety-critical applications, it is safest to adopt a conservative view of the informativeness of evidence unless there are scientifically valid reasons to believe the contrary.

# 2 Modelling Approach

When we use evidence we have obtained from building and operating previous products in order to try to improve the accuracy of the predictions that we can make about the reliability of a novel product, we must take account of two kinds of uncertainty. In the first place, there will be uncertainty concerning the actual reliabilities that have been achieved by these earlier products. Even in those cases where there is extensive operating experience, we shall never know the true reliability of a product and will have to use an estimate based upon the data collected during its operation. In those situations where we are dealing with products that are likely to be very reliable, we shall probably only see a small number (or even none at all) of failures even in quite extensive periods of operation.

The second source of uncertainty will concern the 'similarity' of the products that have been observed in the past, and the 'similarity' of the one under study to these past products. In what follows, we shall assume that the probabilities of failure of the different products, past and present, can be assumed to be realisations of independent and identically distributed random variables. This assumption, although an idealisation, captures the essentials of what we mean by 'similarity'. Thus, it means that the actual reliabilities of the different products will be different, as is clearly the case in reality. We would not expect the reliabilities of, say, two versions of a software-based telephone switch to be identical, even though we might be prepared to agree that the problems posed, and the quality of the processes deployed in their solution, were similar. The notion of 'similarity' in the eye of an observer here seems to be equivalent to a kind of 'indifference'. You might agree that two different products were similar for the purposes of the current exercise if you were indifferent between them in reliability terms: if you were asked to predict which would be the most reliable, before seeing them in operation, you would have no preference. This is represented by their probabilities of failure being identically distributed random variables: any probability statements you would make about the reliabilities of products $A$ and $B$ will be identical. The important point here is that this interpretation of 'similarity' in terms of indifference does not mean that you believe that the two products will have identical reliabilities [Laprie92] - indeed you will know that the actual reliabilities of the products will differ.

The two sources of uncertainty here are both important. However, it is the nature of the uncertainty concerning 'how similar' the products actually are that will be most difficult to estimate in practice, since this requires us to see as many different products as possible. It is far more likely that we have large quantities of information about a few products, than that we have information on many products.

Consider first the failure process of a *single* software product $\mathcal{A}$. Assume a Bernoulli trials process model of the failures of this product in a sequence of 'demands' with neither debugging, maintenance, nor significant variation in the 'stressfulness' of the software's operational environment. Thus, in the first $n$ trials of product $\mathcal{A}$, let $R$ be the random number of failures occurring and $p$ be the probability of failure on demand. Then the distribution of $R$ for fixed $n$ and $p$ is

$$R \,|\, n, p \sim \binom{n}{r} p^r \, (1-p)^{n-r} \tag{1}$$

Now think of $p$ as unknown and construct a Bayesian model by assuming that $p$ is a realisation of a random variable $P$ having a parametric distribution

$$P \,|\, \theta \sim f_p(p|\theta)$$

with parameter $\theta$. Here we can think of this distribution for $P$ as representing the general reliability of products in a particular *product family*, perhaps produced by a single development team, using a common development method, and for similar applications. For example, a family of products known to have highly variable reliability levels would correspond to a distribution $f_p(p|\theta)$ with a large variance, whereas for another product family a high 'average' product reliability would correspond to a small mean for $f_p(p|\theta)$. If we fully understood the true variation in reliabilities of the products in each of these two product families then we could describe the two families by specifying two different $P$-distributions having the required characteristics and index these $P$-distributions with two different $\theta$-values, $\theta_1$ and $\theta_2$, say. More generally, our parameter space $\mathcal{S}$, say, for $\theta$, could be said to represent a set of different conceivable reliability characteristics each of which potentially characterises a different *family of similar products*. I.e., given sufficient data on the reliability variation amongst the products of a particular family, a value of $\theta$ (and hence a particular distribution $f_p(p|\theta)$) could in principle be assigned as descriptive of that variation. In this way, we have defined a model in which $\theta$ can be thought of as a product-family-characterising parameter. For a product chosen at random from those of a particular family of similar products (i.e. particular $\theta$) and observed for a sequence of $n$ demands, it follows that $(R, P)$ has joint distribution

$$(R, P) \,|\, n, \theta \sim \binom{n}{r} p^r \, (1-p)^{n-r} \, f_p(p|\theta), \tag{2}$$

given $n$ and $\theta$. Integrating (2) over $p$ gives the conditional distribution of $R$ given $n$ and $\theta$ as

$$R \,|\, n, \theta \sim \binom{n}{r} \int_0^1 p^r \, (1-p)^{n-r} \, f_p(p|\theta) \, dp \tag{3}$$

or, expressed in terms of moments of $f_p(\cdot|\theta)$,

$$R|n,\theta \sim \binom{n}{r} \mathbf{E}\big(P^r(1-P)^{n-r}|\theta\big) \,. \tag{4}$$

If we *observe* that $R = r$ failures actually occur during $n$ demands, then we can condition on this data by normalising (2) to give the updated distribution

$$P|r,n,\theta \sim \frac{p^r(1-p)^{n-r} f_p(p|\theta)}{\int_0^1 p^r(1-p)^{n-r} f_p(p|\theta)\,dp} \tag{5}$$

of the probability of failure on demand for this program, given $\theta$, $n$ and the observation $r$.

The last three equations describe properties of a general mixture of Bernoulli trials processes, where $f_p(\cdot|\theta)$ is the mixing distribution. Note that although exchangeability[1] of the original Bernoulli trials process has not been lost by mixing the processes, the property that non-intersecting sections of the process are independently distributed does not hold in general for the resulting mixed process. In fact the number $R'$ of failures in a subsequent set of $n'$ demands on the same product now has an updated distribution obtainable from (5) as

$$
\begin{aligned}
R'|r,n,n',\theta &\sim \binom{n'}{r'} \frac{\int_0^1 p^{r+r'}(1-p)^{n+n'-r-r'} f_p(p|\theta)\,dp}{\int_0^1 p^r(1-p)^{n-r} f_p(p|\theta)\,dp}\,, \\
&= \binom{n'}{r'} \frac{\mathbf{E}\big(P^{r+r'}(1-P)^{n+n'-r-r'}|\theta\big)}{\mathbf{E}(P^r(1-P)^{n-r}|\theta)}
\end{aligned}
\tag{6}
$$

given $n$, $r$.

The distributions which we have considered up till this point are parameterised by $\theta$. We now adopt a Bayesian approach to handling this parameterisation by supposing a prior distribution

$$\Theta \sim \mathrm{Prior}_\theta(\theta)\,,$$

with support set $\theta \in S$. If we plan to observe and predict reliability only of a single software product, this extension adds very little to the model as so far described, since, by integrating over $\theta$, the model is reduced to a degenerate ($|S| = 1$) case of the assumptions described earlier. (Simply replace $f_p(p|\theta)$ by $\int_{\theta \in S} f_p(p|\theta)\mathrm{Prior}_\theta(\theta)\,d\theta$ in the distributions above.)

The idea of a prior distribution for $\theta$ becomes a useful concept, however, if we wish to address the problem of *learning* about a *distribution* of product reliabilities by observing the failure behaviour of *multiple* software products from a single family $\langle \mathcal{A}_i \rangle$, say, of similar products. We can then represent a

---

[1]i.e., the property that any permutation of a portion of the binary success-failure sequence has the same probability as the unpermuted sequence. Equivalently, we can say that the probability of a precise sequence of successes and failures during a specified interval of discrete time (say from the 10th to the 20th demand, inclusive) can be expressed as a function of the *number*, only, of successes during that interval.

conservative[2] version of a *process* concept for the trend of their reliabilities, from one product to the next, by modelling these products' individual failure processes as above with *different* $p_i$, and an assumption that each of these $p_i$ arises *independently given* $\theta$ for some *unknown, common* parameter value $\theta$ characterising the entire family of products.

Thus $\theta$ and $p$ now play distinct roles in terms of the model concepts: Whereas each $p_i$ still captures a property of a single software product, $\theta$ now represents a common unknown characteristic of the whole family of similar products. To obtain the value of $\theta$ would be to capture the reliability-relevant characteristic which these software products all have in common. For this *multi*-product model, there is now a real purpose behind including separate distributional assumptions for firstly $\theta$, and secondly $p_i$ given $\theta$. In the following, we do not in fact assume that $\theta$ can ever be known[3]. However, we assume that we hold *probabilistic prior beliefs about* $\theta$ (i.e. beliefs about the possible distributions $f_p(\cdot|\theta)$ of reliabilities of products belonging to the family $\langle \mathcal{A}_i \rangle$). Then, any observation of failure behaviour of any subset of the sequence $\langle \mathcal{A}_i \rangle$ can be regarded as information about $\theta$ which we will use in order to learn about $\theta$ by the usual Bayesian learning mechanisms. Thus the second stage of our doubly stochastic model is to represent our prior beliefs about a subjective random variable $\Theta$ of which the true value $\theta$ for our particular product family is a single unknown realisation.

Observe now that, conditionally given $\theta$ and $\langle n_i \rangle_{i=1}^{k}$, our independence assumption for the $\langle P_i \rangle$ tell us that the first $k$ terms of our $\langle R_i \rangle$ sequence are jointly distributed

$$\langle R_i \rangle_{i=1}^{k} \big| (\langle n_i \rangle_{i=1}^{k}, \theta) \sim \prod_{i=1}^{k} \binom{n_i}{r_i} \int_0^1 p^{r_i} (1-p)^{n_i - r_i} f_p(p|\theta)\, dp. \tag{7}$$

Once we have executed these $k$ software products and observed their failure behaviour (i.e., $r_i$ failures out of $n_i$ trials for each product $\mathcal{A}_i$) then we can regard (7) as the likelihood function $L\big(\theta; \langle n_i, r_i \rangle_{i=1}^{k}\big)$ of the parameter $\theta$ given this failure data. $L\big(\theta; \langle n_i, r_i \rangle_{i=1}^{k}\big)$ is a product involving combinatorial terms together with moments of the parametric distribution $f_p(\cdot|\theta)$

$$\langle R_i \rangle_{i=1}^{k} \big| (\langle n_i \rangle_{i=1}^{k}, \theta) \sim \prod_{i=1}^{k} \binom{n_i}{r_i} \mathbf{E}\big(P^{r_i}(1-P)^{n_i - r_i} \big| \theta\big). \tag{8}$$

In §3 we make use of the factor of this likelihood which depends on $\theta$,

$$L_k(\theta) \;=\; \prod_{i=1}^{k} \mathbf{E}\big(P^{r_i}(1-P)^{n_i - r_i} \big| \theta\big)$$

---

[2]in the sense that we desist from making any stronger assumption of any kind of systematic development of reliability from one product to the next. For example, we do not assume an increasing trend in reliabilities of different products in the family.

[3]Loosely, we can say that in order to *know* the value of $\theta$ characterising a family $\langle \mathcal{A}_i \rangle$ of products, we would require a very large amount of operational failure data on *each* of a very large number of products belonging to that family.—So that we could accurately describe from empirical data the shape of the distribution $f_p(\cdot|\theta)$.

$$= \prod_{i=1}^{k} \int_0^1 p^{r_i} (1-p)^{n_i - r_i} f_p(p|\theta) \, dp \tag{9}$$

# 3 Bayesian Updating of Distributions in the General Case

To implement the Bayesian learning about $\Theta$ given observation of $\langle r_i \rangle_{r=1}^k$ we would like to calculate the posterior distribution of $\Theta$. Recalling that the prior for $\Theta$ is denoted $\text{Prior}_\theta$, for $\theta$ lying in $S$, then the required posterior distribution is proportional to the product of the prior distribution for $\Theta$ and the likelihood function evaluated as (7)

$$\Theta \,|\, \langle n_i, r_i \rangle_{i=1}^k \sim c L_k(\theta) \, \text{Prior}_\theta(\theta)$$

where $c$ is a function of $\langle r_i, n_i \rangle$ not involving $\theta$, i.e.

$$\Theta \,|\, \langle n_i, r_i \rangle_{i=1}^k \sim \frac{\left[ \prod_{i=1}^{k} \int_0^1 p^{r_i} (1-p)^{n_i - r_i} f_p(p|\theta) \, dp \right] \text{Prior}_\theta(\theta)}{\int_{\theta \in S} \left[ \prod_{i=1}^{k} \int_0^1 p^{r_i} (1-p)^{n_i - r_i} f_p(p|\theta) \, dp \right] \text{Prior}_\theta(\theta) \, d\theta} \tag{10}$$

Equation (10) moves the focus of attention away from failure probabilities $P_i$ of products $\mathcal{A}_i$ by the integrations over $p$. It is now of great interest to know an up-to-date distribution for $P$ given what has been observed (in order to make predictions about a particular new product, for example). Then our learning could be expressed directly in terms of the changing nature of the current uncertainty about a failure probability of some particular product. At this stage it is instructive to distinguish between four different stages in our learning about one of the failure probabilities, say $P_k$. The first of these is the prior marginal distribution of $P_k$

$$P_k \sim \int_{\theta \in S} f_p(p_k|\theta) \text{Prior}_\theta(\theta) \, d\theta \,, \tag{11}$$

which represents our initial state of uncertainty concerning the reliability of any given product, $\mathcal{A}_k$, prior to any observation either of that or of any other product's behaviour.

The second most trivial case—observing only the past failure behaviour of the specific product of interest—has effectively already been covered by (5). Substituting $\int_{\theta \in S} f_p(p|\theta) \text{Prior}_\theta(\theta) \, d\theta$ for $f_p(p|\theta)$ in (5) gives a conditional distribution

$$P_k | n_k, r_k \sim \frac{p_k^{r_k} (1 - p_k)^{n_k - r_k} \int_{\theta \in S} f_p(p_k|\theta) \text{Prior}_\theta(\theta) \, d\theta}{\int_{\theta \in S} \int_0^1 p^{r_k} (1-p)^{n_k - r_k} f_p(p|\theta) \, dp \, \text{Prior}_\theta(\theta) \, d\theta} \tag{12}$$

for $P_k$ given $n_k$ and $r_k$.

Thirdly, replacing $k$ by $k-1$ in (10) and then substituting this distribution in place of $\text{Prior}_\theta(\theta)$ in (11) (or, alternatively, directly substituting $n_k = r_k = 0$ in (14) below) gives the distribution

$$P_k \,\big|\, \langle n_i, r_i \rangle_{i=1}^{k-1} \sim$$

$$\frac{\displaystyle\int_{\theta \in S} f_p(p_k|\theta) \left[ \prod_{i=1}^{k-1} \int_0^1 p^{r_i}(1-p)^{n_i-r_i} f_p(p|\theta)\, dp \right] \text{Prior}_\theta(\theta)\, d\theta}{\displaystyle\int_{\theta \in S} \left[ \prod_{i=1}^{k-1} \int_0^1 p^{r_i}(1-p)^{n_i-r_i} f_p(p|\theta)\, dp \right] \text{Prior}_\theta(\theta)\, d\theta} \tag{13}$$

of $P_k$ given observation of the failure behaviour $\langle n_i, r_i \rangle_{i=1}^{k-1}$ only of *other* products $\langle \mathcal{A}_i \rangle_{i=1}^{k-1}$.

Finally, replacing $k$ by $k-1$ in (10) and then substituting this distribution in place of $\text{Prior}_\theta(\theta)$ in (12) gives the distribution

$$P_k \,\big|\, \langle n_i, r_i \rangle_{i=1}^{k} \sim$$

$$p_k^{r_k}(1-p_k)^{n_k-r_k} \frac{\displaystyle\int_{\theta \in S} f_p(p_k|\theta) \left[ \prod_{i=1}^{k-1} \int_0^1 p^{r_i}(1-p)^{n_i-r_i} f_p(p|\theta)\, dp \right] \text{Prior}_\theta(\theta)\, d\theta}{\displaystyle\int_{\theta \in S} \left[ \prod_{i=1}^{k} \int_0^1 p^{r_i}(1-p)^{n_i-r_i} f_p(p|\theta)\, dp \right] \text{Prior}_\theta(\theta)\, d\theta} \tag{14}$$

for $P_k$ given observation both of the failure behaviour $\langle n_k, r_k \rangle$ of the product $\mathcal{A}_k$ itself and *also* the failures $\langle n_i, r_i \rangle_{i=1}^{k-1}$ of other products $\langle \mathcal{A}_i \rangle_{i=1}^{k-1}$.

# 4   The No-Failures Case

Consider the special case in which no failures at all have been observed—neither of the product for which we wish to predict reliability, nor of other products within the same product family. This case is of particular importance since it provides an upper limit for the reliability levels which can be objectively measured in a given amount of observation time purely from observation of failure behaviour. Specialising the equations of §3 to this case is simply a matter of substituting $\langle r_i \rangle = \langle 0 \rangle$. If we similarly specialise the form of our *predictions* by considering the Bayesian predictive probability of a *further* period of failure-free operation, we find that these predictions can be expressed in rather a simple form as the expectations of products of higher non-central moments of a particular conditional distribution. So, conclusions about the reliability levels measurable using this model turn out to depend crucially on our decision about what may be considered realistic model assumptions for these moments. Thinking in terms of the probability $Q_i = 1 - P_i$ of successful completion of an individual demand, and assuming that we do believe that our product family is highly reliable, then the conditional distribution of $Q_i$ given $\theta$ will be concentrated very close to 1 (for all except, perhaps, some values of the product-family parameter $\theta$ which we consider to be highly unlikely, i.e. that are assigned small probability (density) values $\text{Prior}_\theta(\theta)$ by our prior for $\theta$). Defining $\mu'_m$ to be the $m^{\text{th}}$ non-central moment of this conditional distribution of $Q_i$ given $\theta$ makes $\mu'_m$ a deterministic function of $\theta$

$$\mu'_m = \int_0^1 (1-p)^m f_p(p|\theta)\, dp. \tag{15}$$

We now take the expectation of $Q_k^n$ with respect to each of the three updated distributions (12–14) for $P_k$. This yields three expressions representing the Bayesian predictive probability that the next $n$ demands on $\mathcal{A}_k$ will be failure-free given previous observation of failure-free execution of respectively: $\mathcal{A}_k$ only; $\langle \mathcal{A}_i \rangle_{i=1}^{k-1}$; or, lastly, all of $\langle \mathcal{A}_i \rangle_{i=1}^{k}$:

$$\mathbf{E}(Q_k^n \,|\, R_k = 0) = \frac{\mathbf{E}(\mu'_{n_k+n})}{\mathbf{E}(\mu'_{n_k})}, \tag{16}$$

$$\mathbf{E}\left(Q_k^n \,\middle|\, \langle R_i \rangle_{i=1}^{k-1} = \langle 0 \rangle\right) = \frac{\mathbf{E}\left(\mu'_n \displaystyle\prod_{i=1}^{k-1} \mu'_{n_i}\right)}{\mathbf{E}\left(\displaystyle\prod_{i=1}^{k-1} \mu'_{n_i}\right)}, \tag{17}$$

$$\mathbf{E}\left(Q_k^n \,\middle|\, \langle R_i \rangle_{i=1}^{k} = \langle 0 \rangle\right) = \frac{\mathbf{E}\left(\mu'_{n_k+n} \displaystyle\prod_{i=1}^{k-1} \mu'_{n_i}\right)}{\mathbf{E}\left(\displaystyle\prod_{i=1}^{k} \mu'_{n_i}\right)}. \tag{18}$$

These predictive probabilities of $n$ consecutive successful demands on $\mathcal{A}_k$ should be compared with the unconditional

$$\mathbf{E}(Q_k^n) = \mathbf{E}(\mu'_n) \tag{19}$$

which is the probability that the next $n$ demands on $\mathcal{A}_k$ will be failure-free given *no* conditioning observation of either $\mathcal{A}_k$ or any other products—i.e. based solely upon the prior belief.

## 5    An Example of a Particular Choice of Prior Distributions for $P$ given $\Theta$, and for $\Theta$

We shall retain throughout what follows our original assumptions that each product $\mathcal{A}_i$ fails as a Bernoulli trials process with unknown parameter $P_i$, and that the $\langle P_i \rangle$ sequence is i.i.d. conditionally given an unknown product-sequence-characterising parameter $\theta$. To generate particular cases of our model we are then left with the tasks of choosing the distribution family $\{f_p(\cdot|\theta)\,;\;\theta \in \mathcal{S}\}$ and the single prior distribution $\text{Prior}_\theta$ over this family.

The beta-family of distributions

$$f_p(p|\theta) = \frac{p^{a-1}(1-p)^{b-1}}{\beta(a,b)}, \qquad \theta = \langle a,b \rangle, \quad a,b > 0$$

is conjugate [DeGroot70] to both the binomial and the negative binomial (including geometric) distributions, and is thus in some sense a 'natural' choice. If we use this as our $f_p$ distribution family, we obtain a mixed process for the failures of a single product for which the probability of $r$ failures in $n$ demands is given from equation (4) to be

$$R|n,a,b \sim \frac{\binom{n}{r}\beta(r+a,n-r+b)}{\beta(a,b)},$$

obtained by integrating over $p$ the joint distribution of equation (2) which would be

$$(R,P)|n,a,b \sim \frac{\binom{n}{r}p^{r+a-1}(1-p)^{n-r+b-1}}{\beta(a,b)}$$

in this case.

The likelihood (8) resulting from observation of $k$ products in operation is

$$\langle R_i \rangle_{i=1}^{k} \big| (\langle n_i \rangle_{i=1}^{k}, a, b) \sim \prod_{i=1}^{k} \binom{n_i}{r_i} \frac{\beta(a+r_i, b+n_i-r_i)}{\beta(a,b)}$$

with

$$L_k(a,b) = \prod_{i=1}^{k} \frac{\beta(a+r_i, b+n_i-r_i)}{\beta(a,b)}$$

as defined in equation (9).

Having decided to investigate the beta $f_p$, the choice of $\text{Prior}_\theta$ over $\mathcal{S}$, the positive quadrant[4], remains problematic. In real life there would be an 'expert' from whom we would wish to elicit the distribution that truly reflects his a priori belief. This is not an easy task in such a complex model, and the expert may find it difficult to represent his beliefs in a distribution for $\langle a,b \rangle$. A way out of this difficulty is to assume that the expert is 'ignorant', and use that prior distribution which represents ignorance. Even this is a non-trivial task. As an example we consider the simple case of distributions uniform on some finite rectangle with sides parallel to the $a$ and $b$ axes,

$$\text{Prior}_\theta(a,b) = \begin{cases} \frac{1}{(a_2-a_1)(b_2-b_1)}, & \text{if } a_1 < a < a_2,\ b_1 < b < b_2 \\ 0, & \text{elsewhere.} \end{cases}$$

---

[4]possibly extended to include points representing $a,b \to \infty$ with $a/b$ constant, and $a,b \to 0$ with $a/b$ constant, to include all the limiting cases of the beta family

Firstly we can examine characteristics of the prior distribution (11) for $P_k$ implied by these model assumptions,

$$P_k \sim \int_{a_1}^{a_2} \int_{b_1}^{b_2} \frac{p^{a-1}(1-p)^{b-1}}{\beta(a,b)} \frac{db\,da}{(a_2-a_1)(b_2-b_1)}.$$

The first and second non-central moments of $P|a,b$ are $\frac{a}{a+b}$ and $\frac{a(a+1)}{(a+b)(a+b+1)}$. These may be integrated analytically with respect to $f_p(p|a,b)$ (first expanding in partial fractions with respect to $b$ in the case of the second moment) to give the expressions,

$$\mathbf{E}(P) = \tfrac{1}{2} + \frac{(a_1^2-b_1^2)\log(a_1+b_1)-(a_2^2-b_1^2)\log(a_2+b_1)-(a_1^2-b_2^2)\log(a_1+b_2)+(a_2^2-b_2^2)\log(a_2+b_2)}{2(a_2-a_1)(b_2-b_1)}$$

and

$$\mathbf{E}(P^2) = \frac{2}{3} + \frac{t(a_1,b_1)-t(a_2,b_1)-t(a_1,b_2)+t(a_2,b_2)}{6(a_2-a_1)(b_2-b_1)}$$

where $t(a,b) = s(a,b) - s(a,b+1)$, where

$$s(a,b) = (2a^2 - 2ab + 2b^2 + 3a - 3b)(a+b)\log(a+b).$$

The prior reliability function is given from equations (15) and (19) by

$$\mathbf{P}(X_k > n) = \mathbf{E}(\mu'_n)$$
$$= \int_{a_1}^{a_2} \int_{b_1}^{b_2} \frac{\beta(a,b+n)}{\beta(a,b)} \frac{db\,da}{(a_2-a_1)(b_2-b_1)}$$
$$= \int_{a_1}^{a_2} \int_{b_1}^{b_2} \frac{b(b+1)\ldots(b+n-1)}{(a+b)(a+b+1)\ldots(a+b+n-1)} \frac{db\,da}{(a_2-a_1)(b_2-b_1)},$$

where the *first* failure of $\mathcal{A}_k$ occurs on the $X_k^{\text{th}}$ demand.

These expressions can be thought of as different ways of expressing *a priori* belief about the reliability of a product. Now we explore the effects on these beliefs of learning from observation. We examine the realisations under these particular distributional assumptions of both the posterior distributions for $P_k$ given by equations (12–14), and the predictions of $X_k$, the time to next failure of $\mathcal{A}_k$ using equations (16–18). In the most general case of arbitrary periods of observation of some finite number of previous products, each of the probabilities entailed by these questions takes the form of the ratio of a pair of integrals (over the chosen rectangle in the $\langle a,b \rangle$-plane), where the integrands in the numerator and denominator are each equal to some product of terms of the form

$$\mathbf{E}(P^r(1-P)^{n-r}|a,b) = \int_0^1 p^r(1-p)^{n-r}\frac{p^{a-1}(1-p)^{b-1}}{\beta(a,b)}\,dp$$
$$= \frac{\beta(a+r,b+n-r)}{\beta(a,b)} = \frac{a(a+1)\ldots(a+r-1)b(b+1)\ldots(b+n-r-1)}{(a+b)(a+b+1)\ldots\ldots\ldots\ldots(a+b+n-1)}.$$

In practice, since this kind of inference is most likely to be called for in dealing with very high reliability systems, the values $n_i$ of $n$ in these products are likely to be rather large, and the values of $r$ are likely to be small, and ideally zero. So some very large products will be involved in the above term. We shall report elsewhere on the mathematical difficulties that arise as a result of this. Here we show only some illustrative numerical results based upon the observation of three previous products, each of which has been exposed to $10^7$ demands without a single failure. In Table 1 we can see how various different assumptions for Prior$_\theta$ affect the strength of the inferences concerning a fourth product in the same family which can be drawn from this sort of evidence of high reliability of previous, similar products.

All the results in the Table involve assuming uniform distributions over different regions of the $\langle a, b \rangle$–space. We have excluded values of $b$ smaller than one, since these entail beta distributions with infinite density at 1; but we have allowed values of $a$ smaller than one, since infinite density at the origin seems plausible. The region in the positive quadrant where $a$ and $b$ are both large can also be ruled out, since any point here corresponds to a beta distribution with very small variance—i.e. it implies that different products will have essentially identical probabilities of failure upon demand, which runs counter to the spirit of this whole exercise.

The first nine rows of the Table involve several rectangles of the kind described above. The ninth row shows a small rectangle, effectively approximating to a known point value for $\langle a, b \rangle$. Rows 10 to 12 show thin 'wedges' adjacent to the $b$-axis. The informal reasoning here is that it may be reasonable to believe a priori that the mean $\mathbf{E}(P \mid a, b)$ of the distribution of probability of failure on demand does not exceed a certain value $0 \le \mathbf{E}(P \mid a, b) \le M < 1$, say, and this is equivalent to the restriction to $\frac{a}{b} \le \frac{M}{1-M}$. We used $M = 10^{-3}$, $10^{-5}$, and $10^{-7}$. Once again, all points in the wedge are given equal weight.

In the Table we show how 'the reliability' of a product changes as a result of the different types of evidence that could be available. For brevity here we have chosen to present the mean of the distribution of $P_4$, and the reliability function evaluated at $10^7$ demands (i.e. the probability of surviving this number of demands), in each of the four cases: given no data; given only evidence of failure-free operation of this product; given only evidence of failure-free working of earlier products; and given both these latter items of evidence.

The most interesting and important results concern the different predictions of future operational behaviour, expressed as the probability $R(10^7)$ of surviving $10^7$ further demands without failure: the information from the perfect working of previous products makes only a modest contribution to our confidence in the current product when compared with actual evidence of failure-free working on that product itself (compare columns 8 and 10). Thus when we only have evidence from the previous products, although this is of extensive perfect working for each, it only allows us to claim, in the case of the rectangular priors, about 0.75 probability of similarly extensive perfect working (i.e. surviving $10^7$ demands) for the new product[5].

---

[5] We conjecture that some limiting result may be indicated here : perhaps the probability that product $A_k$ will survive its first $X$ demands, given that $k-1$ previous products have done so, tends to $(k-1)/k$ as $X \to \infty$.

| Region of Uniform Prior | | | | Given no Data | | Given no failure of this product | | Given no failure of previous 3 products | | Given failure neither of this nor of previous 3 products | |
|---|---|---|---|---|---|---|---|---|---|---|---|
| $a_1$ | $a_2$ | $b_1$ | $b_2$ | $E(P_4)$ | $R(10^7)$ | $E(P_4)$ | $R(10^7)$ | $E(P_4)$ | $R(10^7)$ | $E(P_4)$ | $R(10^7)$ |
| 0 | 1 | 1 | 2 | .2384 | .6229E-1 | .3966E-1 | .9585 | .1388E-1 | .7498 | .1047E-1 | .9893 |
| 0 | 1 | 1 | 10 | .1037 | .6828E-1 | .1577E-1 | .9547 | .5398E-2 | .7499 | .4062E-2 | .9883 |
| 0 | 1 | 1 | 100 | .2077E-1 | .8048E-1 | .3020E-2 | .9469 | .1019E-2 | .7500 | .7655E-3 | .9862 |
| 0 | 1 | 1 | 1000 | .3207E-2 | .9877E-1 | .4636E-3 | .9355 | .1556E-3 | .7500 | .1168E-3 | .9831 |
| 0 | 2 | 1 | 2 | .3692 | .3114E-1 | .3966E-1 | .9585 | .1388E-1 | .7498 | .1047E-1 | .9893 |
| 0 | 2 | 1 | 10 | .1781 | .3414E-1 | .1578E-1 | .9547 | .5398E-2 | .7499 | .4062E-2 | .9883 |
| 0 | 2 | 1 | 100 | .3833E-1 | .4024E-1 | .3020E-2 | .9469 | .1019E-2 | .7500 | .7655E-3 | .9862 |
| 0 | 2 | 1 | 1000 | .6091E-2 | .4939E-1 | .4637E-3 | .9355 | .1556E-3 | .7500 | .1168E-3 | .9831 |
| .01 | .0101 | 10 | 10.1 | .9990E-3 | .8700 | .9990E-3 | .9931 | .9990E-3 | .8700 | .9990E-3 | .9931 |
| 0 | b/999 | 1 | 1000 | .5002E-3 | .1824 | .2056E-3 | .9401 | .9494E-4 | .7545 | .7593E-4 | .9832 |
| 0 | b/99999 | 1 | 1000 | .5000E-5 | .9689 | .4947E-5 | .9977 | .4843E-5 | .9703 | .4791E-5 | .9978 |
| 0 | b/9999999 | 1 | 1000 | .5000E-7 | .99968 | .4999E-7 | .999977 | .4998E-7 | .99968 | .4998E-7 | .999977 |

.XXXXE-$n$ means $0.XXXX \times 10^{-n}$

Table 1: Effect on Reliability Predictions of Observation of Non-Failure of Previous Products

The evidence from previous perfect working of the *same* product, however, is more informative. It allows us to be much more confident that the product will work perfectly in the future: the probability of it surviving $10^7$ demands, given that it has already survived $10^7$ demands, exceeds 0.9 in all cases.

On the other hand, the small increase in confidence that comes from experience of other products may be useful in the case of safety-critical systems, especially as it is likely to come with little or no cost to developers of the new product. Thus, in the first row of the Table, the *a priori* belief of the $10^7$ demand survival is .062, this increases to .96 after we have actually seen the product survive $10^7$ demands, and to .99 when we are told, in addition, that three other products have also survived $10^7$ demands. Putting it another way, this evidence of previous product survival has reduced the chance of a failure in the next $10^7$ demands by a factor of 4 (from .04 to .01) compared with the result based only on the evidence from operational experience of this product.

We have shown the columns for the means of the various distributions for $P_4$ mainly as a warning that these can be misleading if used to represent 'the reliability' of a product. Thus the mean probability of failure on demand can be quite large (0.24 in the first line prior distribution), but still the chance of surviving $10^7$ demands may be non-negligible (0.063 in this case). The informal reason is that the distribution is such that the mean is not a good summary statistic, and in particular cannot be used in a geometric distribution to approximate to the more complex model that applies here.

In fact, decreasing values of $\mathbf{E}(P_4)$ do not necessarily imply increasing chance of surviving $10^7$ demands, as might naively be expected: see, for example, columns 7 and 8 of rows 1 to 4. Imagine that we have two experts, let us call them James and Peter, represented by two different prior distributions (rows of the Table), who observe the system to survive for $10^7$ demands. They are then asked to tell us how reliable the system is. If the question is posed as 'what is the mean of $P_4$?', then James is more optimistic than Peter; if, however, the question is posed as 'what is the chance of surviving a further $10^7$ demands', Peter is more optimistic than James. Such (only apparent) paradoxes underline the importance of using the right formulation for our purposes when we ask questions about the reliability of a system.

# 6   Conclusions and future work

A major motivation for research of this kind is to make the process of assessing safety-critical systems more open to analysis. Currently, particularly in those cases where complex software is involved, such assessments have a high degree of informality and rely a great deal upon expert judgement. Whilst this process is usually carried out responsibly, and with great rigour, it is difficult for an outsider to analyse how the final judgement has been reached, and much has to be taken on trust. Since there is some evidence of experts being unduly optimistic about their judgemental abilities [Henrion86], simply checking their honesty is insufficient. What is needed is a more formal means of argumentation, where the assumptions and reasoning processes are visible and can be questioned. This new model treats a small part of this problem by providing a representation, and means of composition, of two important types of evidence that are commonly used to make claims for the reliability of a product: evidence from testing of the product itself and evidence from previous experience of 'similar' products.

Whilst we make no great claims for the realism of the example we have used, it does indicate the way in which a formal model of this kind could be used to question whether an optimistic conclusion drawn from past experience might be ill-founded. Essentially, if you were to claim that great trust could be placed in a particular system because of past experience of other systems, you would have to justify this by trying to claim that your prior distribution is reasonable within the model. It is clear that some of the examples of prior distributions we have used could be said to be 'unreasonable' in the sense that they represent beliefs about the reliability, prior to seeing any evidence, that are very strong.

The particular numerical examples used here are meant only to be illustrative. Clearly further work is needed to identify classes of 'plausible' prior distributions, even for the case in which the expert professes 'complete prior ignorance'. For example, rather than addressing the raw $\langle a, b \rangle$ parameters, it may be easier for the subject to think in terms of a reparameterisation - the mean and coefficient of variation are possibilities. Another area of future work concerns the impact of different kinds of evidence upon the conclusions. For example, the case here of complete perfection of operation of the previous products is the best news that it is possible to have, and it would be interesting to investigate the case where there have been failures in the earlier products.

The possibility that conclusions about the reliability of a system can be highly dependent upon the precise way in which they are formulated is somewhat surprising and needs further investigation. However, the results here support those obtained in a different context, concerning stopping rules for software testing [Littlewood94].

Finally, all this modelling depends upon the reasonableness of notions of 'similarity' between different products. In this we are merely making more formal the extremely informal claims that experts make when they argue that the behaviour of one product can be used as a means of inferring the likely behaviour of another. Justification of such assumptions of similarity in particular cases is, of course, outside the direct scope of our studies—presumably it will come, in the case of software, from knowledge of the application domain (the problems being solved were similar), the development process (the methods used were similar), the design teams (they were the same or of comparable competence), etc. However, we believe that our model can be used to provide a curb on the enthusiasm of experts: specifically, the use of 'similarity' arguments to make stronger claims than would be warranted via the model should be treated with suspicion.

# Acknowledgement

This work was supported by the ESPRIT PDCS2 project 6362, the DTI/ EPSERC Safety Critical Systems Research Programme's DATUM Project, and the CEC Environment Programme's SHIP Project. It has benefited considerably from numerous critical comments and suggested improvements by colleagues working on these projects and colleagues at the Centre for Software Reliability.

# References

[Butler93] R. W. Butler and G. B. Finelli. The infeasibility of quantifying the reliability of life-critical real-time software. *IEEE Transactions on Software Engineering*, 19(1):3–12, 1993.

[DeGroot70] M. H. DeGroot. *Optimal Statistical Decisions*. McGraw-Hill, New York, 1970.

[Henrion86] M. Henrion and B. Fischhoff. Assessing Uncertainty in Physical Constants. *American Journal of Physics*, 54(9):791–8, 1986.

[Laprie92] J. C. Laprie. For a Product-in-a-Process Approach to Software Reliability Evaluation. In *Proc. 3rd International Symposium on Software Reliability Engineering (ISSRE92)*, pages 134–9, Research-Triangle Park, USA, 1992. Invited Paper.

[Littlewood93] B. Littlewood and L. Strigini. Validation of Ultra-High Dependability for Software-Based Systems. *Comm. Assoc. Computing Machinery*, 36(11), November 1993.

[Littlewood95] B. Littlewood and D. R. Wright. On a Stopping Rule for the Operational Testing of Safety-Critical Software. In *Proc. 25$^{th}$ Fault Tolerant Computing Symposium*, Pasadena, June 1995. IEEE.

[Rouquet86] J. C. Rouquet and Z. Z. Traverse. Safe and Reliable Computing on board the Airbus and ATR aircraft. In W. J. Quirk, Editor, *Proc. Fifth IFAC Worshop on Safety of Computer Control Systems*, pages 93–97, Oxford, 1986. Pergamon Press.

# Session 5
## Design

Session 3

Paper 3?

# Six Steps Towards Provably Safe Software

Maritta Heisel

Technische Universität Berlin

FB Informatik – FG Softwaretechnik

Franklinstr. 28-29, Sekr. FR 5-6

D-10587 Berlin

heisel@cs.tu-berlin.de

fax: (+49-30) 314-73488

### Abstract

We present an approach to the specification and implementation of provably safe software. It uses well-established tools and techniques that are usually employed to ensure correctness, rather than safety, of software. The approach comprises six steps, each of which is complemented by some proof obligations. For each step, the safety-related aspects are clearly elaborated. Thus, designers of safety-critical systems are given guidance that helps to avoid potentially dangerous gaps in the specification of the system's safety properties.

## 1  The General Setting

The aim of this work is to support the development of provably safe software. Since a safety proof cannot be obtained by conventional software engineering techniques, we use formal methods to achieve this goal. Formal methods as they are used today mostly have the sole purpose of guaranteeing the *correctness* of software. This means, the software is to implement a certain functionality. Software *safety*, on the other hand, is not so much concerned with the implemented functionality. Instead, it must be guaranteed that certain undesirable states are *not* entered. Moreover, the interaction of the system with its environment plays a crucial role.

Our approach covers the specification as well as the implementation of safety-critical software. As a specification language, we have chosen the model-based language Z [Spi92b]. In Z, system states are modeled explicitly. This is in accordance with the fact that most embedded safety-critical systems have a state. For the implementation of specifications, the program synthesis system IOSS (Integrated Open Synthesis System) designed by the author [HSZ95b] is used. IOSS supports the implementation of imperative programs and thus matches well with Z.

The choice of these formalisms, however, imposes some limitations on our approach: distributed systems, parallelism and real-time requirements cannot

192

be treated in full generality[1] because Z has no means to express the corresponding notions.

The approach consists of six steps to be performed, each of which comes with some proof obligations. For each step, its safety-related aspects are highlighted. Their description can serve as a checklist, thus providing guidance for the designers of safety-critical systems.

In the next section, Z and IOSS are introduced. Then the steps of our approach are explained in some detail, followed by an example. Finally, related work is discussed and an assessment of the approach is given.

# 2    Z and IOSS

We have chosen the specification language Z because it has gained considerable popularity in industry and comes equipped not only with a methodology [PST91] but also with some tool support, e.g. for type checking [Spi92a] and theorem proving [BG94]. Z is designed to specify state-based systems which is in good accordance with the reality of safety-critical systems. An undeniable deficiency of Z is the fact that neither time nor complex control structures can be specified.

The author's synthesis system IOSS supports the development of imperative programs using so-called *strategies*, [Hei94, HSZ95b]. Strategies describe possible steps during the synthesis process. Their purpose is to find a suitable solution to some *programming problem*. A strategy works by problem reduction. For a given problem, it determines a number of subproblems. From their solutions, it produces a solution to the initial problem. Finally, it checks if that solution is acceptable. The solutions to subproblems are also obtained by applications of strategies. In general, the subproblems produced by a strategy are not independent of each other or of the solutions to other subproblems. This restricts the order in which the various subproblems can be set up and solved. A strategy describes how exactly the subproblems are constructed, how the final solution is assembled, and how to check whether this solution is acceptable.

Programming problems are basically specifications, expressed as pre- and postconditions of first-order predicate logic. A complete definition is given in Section 4.2.1. Solutions are basically programs in a Pascal-like language. A solution is *acceptable* if and only if the program is totally correct with respect to the specification and additionally fulfills some variable conditions (see Section 4.2.1). For each developed program a formal proof in dynamic logic [Gol82] is constructed. This is a logic designed to prove properties of imperative programs. The proofs are represented as tree structures that can be inspected at any time during development.

Program synthesis with IOSS consists of a loop of strategy applications. The intermediate states of the development are represented by a data structure called *development tree*. Its nodes contain a problem and its solution (once it

---

[1]Simple time constraints can be modeled with timers and thus be specified in Z, see Section 4.

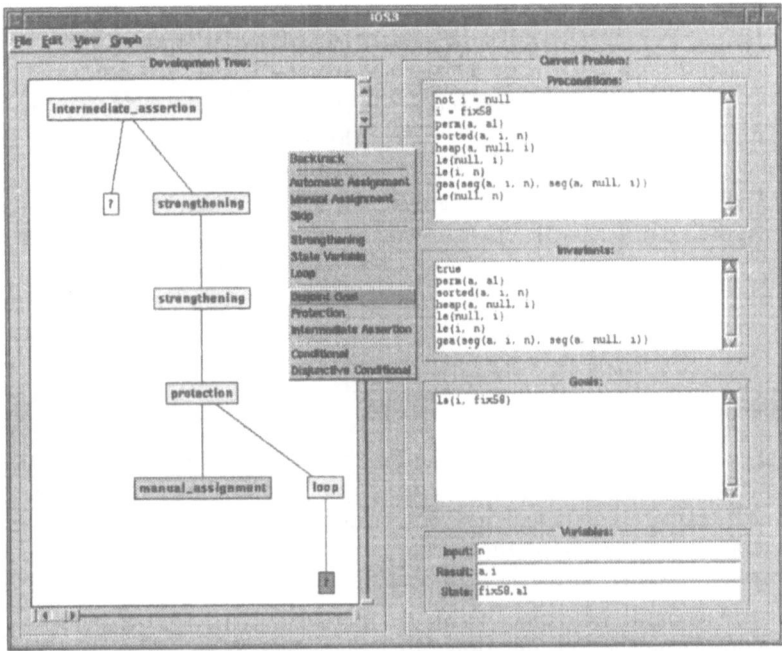

Figure 1: The IOSS interface

has been found). Representing the state of development as a data structure
makes it possible to obtain an overview of the development at any time. Each
new strategy application causes the development tree to be extended, if the
strategy reduces the problem to a number of subproblems. Otherwise, the
problem is solved immediately, and the solution is recorded in the respective
node. When all subproblems of a problem have been solved, its solution can be
assembled from the solutions to the subproblems. The development is finished
when all problems have been solved. The result of the development process is
the final development tree.

The strategy base of IOSS contains formalized development knowledge in
form of strategy modules. A number of interactive, semi-automatic and fully
automatic strategies have been implemented. In the current version, they are
oriented on programming language constructs. In the near future, higher level
strategies, e.g. for the development of divide-and-conquer algorithms or re-
usable procedures, will be built in. A complete description of the available
strategies can be found in [Hei94].

Figure 1 shows the general interface of IOSS. The main window displays
the development task, represented by the development tree – on the left-hand
side of the window – and the specification of the current problem – on the
right-hand side of the window. The tree visualizes the process and the state
of development. Each node is labeled with the name of the strategy applied
to it. The state of the node is color coded, showing at a glance whether it
is reducible, or solved, etc. The strategy menu is shown in the center of the

| No. | Step | Proof Obligations |
|---|---|---|
| 1 | Define the legal states of the system. | Show that the initial state is legal. |
| 2 | Define the actions the system can perform. | Analyze the conditions under which the actions transform legal states into legal states. |
| 3 | Define the interface of the system to the outside world. | Show that the internal system operations are only invoked if their preconditions are satisfied. Show that for each combination of sensor values exactly one internal operation is invoked. Show that – if the sensors work correctly – the system faithfully represents the state of its environment. |
| 4 | Refine the data and operations of the specification until data and control structures of the target programming language can be used. | Show the correctness of the refinements. |
| 5 | Transform the specification obtained in Step 5 into a form suitable for the program synthesis system. | Show the correctness of the algorithm performing this task. |
| 6 | Use the synthesis system to obtain a proven correct implementation of the specified system. | Proof obligations are generated by the synthesis system. |

Table 1: Steps and Proof Obligations

window. Applications of strategies, inspection of nodes or the proof tree and graph manipulations like scaling are performed via mouse clicks or pull-down menus. For a more complete description of IOSS, the reader is referred to [HSZ95a, HSZ95b].

The combination of Z and IOSS can be achieved easily: since both formalisms allow for states and have concepts to deal with changing values of variables, Z specifications can mechanically be translated into IOSS programming problems. The translation mechanism as well as the synthesis process resembles the approach of the refinement calculus [Woo91b] and are described in more detail in Section 4.2.

# 3  The Six Steps in Detail

Table 1 gives an overview of the proposed procedure. The first three steps give a guideline how to set up the specification of a system, where special attention is devoted to the safety requirements. In general it will not be possible to carry out these steps independently of each other and without iteration. Instead, a process resembling the spiral model of software development will have to be employed. The last three steps describe how to perform the transition from a mere specification to a correct (and thus safe) program.

**Step 1**   The definition of the legal states must comprise the safety requirements as well as other properties of the legal states. We do not deal with the question how this specification is obtained. It can be set up by one party, treating functional as well as safety requirements. Another possibility is to set up two specifications, a functional and a safety specification, by different parties and then show that the safety requirements are entailed by the functional specification. The latter approach can be used to double-check the safety requirements, or it may be enforced by certification procedures or safety standards.

Once the legal states are defined, an initial state should be given. This is not only in accordance with the recommended Z methodology but also with other formalisms like finite state machines or statecharts where one has to define start states or default states. In showing that the initial state is legal, we also demonstrate that the requirements for legal states are satisfiable.

**Step 2**   The actions of the system can be triggered either by outside events or by the system itself. In Z, they are defined by operations that may change the system state. The analysis of the conditions under which the actions transform legal states into legal states is done by precondition analysis. This analysis yields the condition that must hold if the state reached after execution of the operation is legal, provided the state before execution of the operation is. If the precondition is not trivial, it must be taken care that the operation is only executed when its precondition holds.

Analyzing preconditions also helps to detect design errors. If the precondition of an operation turns out to be *false*, the operation cannot be executed at all (or it would lead to an illegal state). This clearly shows that something is wrong with the design of the operation or even the whole system.

So far, we have applied standard Z methodology. The next step deals with the peculiarities of safe software. For software safety, the environment in which the software operates has to be taken into account. This is achieved by modeling the environment using sensors and by performing consistency checks on sensor values.

**Step 3**   In order to define the interface between the system and the outside world, sensors must be modeled that enable the system to detect situations to which it must react. It must also be specified how the system reacts to possible sensor values and/or failures. We advocate to model the system so as to provide exactly one internal operation for each combination of sensor values. This guarantees that each situation is taken care of and yields a clear and comprehensible interface. It is not strictly necessary to show that for each combination of sensor values exactly one internal operation is invoked. We introduce this proof obligation to encourage developers to design their systems as clear and simple as possible. The other two proof obligations, however, are necessary to ensure the system's safety.

Once step 3 is performed, it is guaranteed that the state internally maintained by the software always fulfills the safety requirements and that this state is consistent with the state of the environment, under the condition that failure of sensors can be detected. It follows that (under the same condition) also the "real" system state is safe, provided the implementation of the software is correct.

**Remark concerning proof obligations.** The proofs that have to be carried out are standard and fairly simple. However, there are a lot of them to do. Until now, specialized tool support for this purpose with a sufficient degree of automation is not yet available. Full-fledged first-order theorem provers are not necessary because the proof obligations often have the form of existentially quantified statements, with equations for the existentially quantified variables. We believe that the construction of mostly automatic, specialized provers for the proof obligations occurring in this context poses no severe problems.

The steps presented so far only dealt with the *specification* of safe software. A *model* of the system has been defined, and it has been shown that this model behaves safely. The following steps are concerned with the correct implementation of this model. They are not presented in so much detail because they follow a methodology that is common for the application of formal methods.

**Step 4** What refinement means and how it is performed is described in the literature, e.g. [Woo91a]. This step is not necessary if the data structures involved are available in the target programming language. On the other hand, it is also possible that several refinement steps are necessary.

**Step 5** The Z specifications are transformed into IOSS programming problems, as described in Section 4.2.

**Step 6** The program synthesis guarantees that the concrete states of the implementation are always safe, provided the abstract states of the system model are.

# 4 Example: A Microwave Oven

We exemplify our approach with a simple microwave oven. The description of the oven (which is taken from [SM92]) is as follows:

1. There is a single control button available for the user of the oven. If the oven door is closed and you push the button, the oven will cook (that is, energize the power tube) for 1 minute.
2. If you push the button at any time when the oven is cooking, you get an additional minute of cooking time.
3. Pushing the button when the door is open has no effect.
4. There is a light inside the oven. Any time the oven is cooking, the light must be turned on. Any time the door is open, the light must be on.

5. You can stop the cooking by opening the door.

6. If you close the door, the light goes out. This is the normal configuration when someone has just placed food inside the oven but has not yet pushed the control button.

7. If the oven times out (cooks until the desired preset time), it turns off both the power tube and the light. It then emits a warning beep to tell you that the food is ready.

## 4.1 Specification

Two hazardous situations can be identified for the microwave oven. (i) If the power tube is on while the door is open, there is a severe risk of human injury. (ii) If the light is off while the power tube is on, a boiling over of food may remain unnoticed and can cause a damage of the oven or even set it on fire. Requirement (i) is certainly more important than (ii). We will come back to this in Section 5.3.

**Step 1: Define the legal states of the system.** The interesting components of the microwave oven can take on two possible states.

$MICROWAVE\_STATE ::= energized \mid de\_energized$

$LIGHT\_STATE ::= on \mid off$

$DOOR\_STATE ::= open \mid closed$

$TIMER\_STATE ::= running \mid halted$

$BEEPER\_STATE ::= silent \mid beeping$

The global state of the oven must reflect the safety requirements which are expressed in the first two lines of the state invariant. The other predicates of the following state schema reflect the natural language description given above.

```
__MicrowaveOven_____
  power_tube : MICROWAVE_STATE
  light : LIGHT_STATE
  door : DOOR_STATE
  timer : TIMER_STATE
  timer_value : N
  beeper : BEEPER_STATE
  _____
  door = open ⇒ power_tube = de_energized
  power_tube = energized ⇒ light = on
  door = open ⇒ light = on
  door = closed ∧ power_tube = de_energized ⇒ light = off
  power_tube = energized ⇔ timer = running
  timer_value ≠ 0 ⇒ beeper = silent
```

The initial state describes the microwave oven as you can buy it in a store. It fulfills the state invariant. The decoration "'" of variable names means that they describe the state *after* an operation is completed. Plain variables describe the state in which an operation is started.

```
┌─ MicrowaveOvenInit ─────────────────────────────────────
│ MicrowaveOven'
├──────────────────────────────────────────────────
│ power_tube' = de_energized ∧ beeper' = silent
│ light' = off ∧ door' = closed
│ timer' = halted ∧ timer_value' = 0
└──────────────────────────────────────────────────
```

**Step 2: Define the actions the system can perform.** As a user of the oven, you can open and close its door and push the control button. Moreover, there are state-changing operations that are only indirectly invoked by the user. These have to do with the behavior of the timer.

```
┌─ OpenDoor ─────────────────────────────────────────────
│ Δ MicrowaveOven
├──────────────────────────────────────────────────
│ door = closed
│ power_tube' = de_energized ∧ beeper' = silent
│ light' = on ∧ door' = open
│ timer' = halted ∧ timer_value' = 0
└──────────────────────────────────────────────────
```

This operation may only be invoked when the door is closed (precondition *door = closed*). It then leads to a legal state[2]. "Δ*MicrowaveOven*" means that the state of the oven may change. The operation *CloseDoor* is defined analogously, with precondition *door = open*.

For the control button, we have to distinguish whether it is pushed when the door is open or when the door is closed.

```
┌─ PressButtonDoorClosed ──────────
│ Δ MicrowaveOven
├───────────────────────────
│ door = closed
│ power_tube' = energized
│ light' = on ∧ door' = door
│ timer' = running
│ timer_value' = timer_value + 60
│ beeper' = silent
└───────────────────────────
```

```
┌─ PressButtonDoorOpen ──────────
│ Ξ MicrowaveOven
├───────────────────────────
│ door = open
└───────────────────────────
```

The schema on the right-hand side specifies that the state of the oven does not change when the button is pushed while the door is open. When the button is pressed, one of the two above operations will be invoked:

$$PressButton \mathrel{\widehat{=}} PressButtonDoorClosed \lor PressButtonDoorOpen$$

---

[2]This holds for the other operations, too. Therefore, we will not mention this any more in the following.

The combined operation has the precondition *true*, because the door must be either closed or open, according to the definition of *DOOR_STATE*.

The timer can either be running or halted or get a timeout. In the first case, just the time value is decreased (precondition: $timer = running \land timer\_value > 0$). In the second case, nothing happens (precondition: $timer = halted$). The third case occurs when the timer is running and reaches the value 0 (precondition: $timer = running \land timer\_value = 0$).

```
__ TimerRuns _____
  Δ MicrowaveOven
 _____
  timer = running
  timer_value > 0
  power_tube' = power_tube
  light' = light ∧ door' = door
  timer' = timer
  timer_value' = timer_value − 1
  beeper' = beeper
```

```
__ TimeOut _____
  Δ MicrowaveOven
 _____
  timer = running
  timer_value = 0
  power_tube' = de_energized
  light' = off ∧ door' = closed
  timer' = halted
  timer_value' = timer_value
  beeper' = beeping
```

```
__ TimerHalted _____
  Ξ MicrowaveOven
 _____
  timer = halted
```

When the timer is timed out, the beeper starts beeping. In the natural language description, nothing was said about how long the beeper should beep. For simplicity, we decide not to define an extra operation that switches off the beeper but make use of the fact that opening the door does the job.

Again, these cases are combined to form the operation *Timer* with precondition *true*.

$$Timer \cong TimerRuns \lor TimeOut \lor TimerHalted$$

The next operation is only needed because not only the correct functioning but also the safety of the microwave oven are of interest: when something unforeseen happens, the oven must enter a safe state. Of course, this operation has no precondition.

```
__ EmergencyShutdown _____
  Δ MicrowaveOven
 _____
  power_tube' = de_energized
  light' = off ∧ door' = door
  timer' = halted ∧ timer_value' = 0
  beeper' = silent
```

**Step 3: Define the interface of the system to the outside world.** The connection of the internal system state and the environment is modeled by sensors telling if the door is open or closed and if the button is pressed or not. We assume that a failure of the door sensor is detectable.

$$DOOR\_SENSOR ::= door\_open \mid door\_closed \mid failed$$

$$BUTTON\_SENSOR ::= pressed \mid released$$

The sensor values are connected to internal operations via the following schema that has the sensor values as input parameters:

```
┌─ ExternalEvents ──────────────────────────────────────
│ Δ MicrowaveOven
│ ds? : DOOR_SENSOR
│ bs? : BUTTON_SENSOR
├───────────────────────────────────────────────────────
│ ds? = failed ⇒ EmergencyShutdown
│ ds? = door_open ∧ door = closed ⇒ OpenDoor
│ ds? = door_closed ∧ door = open ⇒ CloseDoor
│ bs? = pressed ∧ (ds? = door_open ∧ door = open
│        ∨ ds? = door_closed ∧ door = closed) ⇒ PressButton
│ bs? = released ∧ (ds? = door_open ∧ door = open
│        ∨ ds? = door_closed ∧ door = closed) ⇒ Timer
└───────────────────────────────────────────────────────
```

This means that a pressed button is ignored if at the same time the door is moved. A door movement is sensed by comparing the sensor value with the internal variable storing the door state. Only if those two are equal the internal operation *PressButton* is invoked. If neither the door is moved nor the button is pressed, the *Timer* operation is invoked. This operation is deterministic since the preconditions of the disjuncts exclude each other. Hence *ExternalEvents* is also deterministic. For each constellation of the sensors exactly one internal operation is invoked.

## 4.2  Implementation

Step 4 is not necessary for the microwave oven because the specification does not make use of any non-trivial data structures.

### 4.2.1  Step 5: Translation into IOSS Format.

Problems to be solved with IOSS are specifications of programs, expressed as pre- and postconditions that are formulas of first-order predicate logic. To aid focusing on the relevant parts of the task, the postcondition is divided into two parts, *invariant* and *goal*. In addition to these it has to be specified which variables may be changed by the program (result variables), which ones may only be read (input variables), and which variables must not occur in the program (state variables). The latter are used to store the value of variables before execution of the program for reference of this value in its postcondition.

The translation of a Z schema into an IOSS programming problem proceeds as follows:

- Each input variable (decorated with "?") of the Z schema becomes an input variable of the corresponding problem.
- Each output variable (decorated with "!") of the Z schema becomes a result variable.
- Each variable $x$ of the Z state schema becomes an input variable if the schema predicate entails $x = x'$.
- Otherwise $x$ becomes a result variable, and a new state variable $x_0$ is generated for $x$ if $x$ occurs in the schema predicate.
- The precondition of the IOSS problem is the precondition of the Z schema plus an equation $x = x_0$ for each introduced state variable $x_0$.
- The invariant of the IOSS problem is the invariant of the Z schema defining the system state.
- The goal of the IOSS problem consists of those conjuncts of the schema predicate that depend on result variables of the IOSS problem, where dashed variables have to be replaced by plain variables and plain variables have to be replaced by their corresponding state variables.

As an example, we consider the implementation of the schema *PressButton-DoorClosed*. The above algorithm yields:

| | |
|---|---|
| input variables: | *door* |
| result variables: | *power_tube, light, timer, timer_value, beeper* |
| state variables: | *timer_value$_0$* |
| precondition: | *door = closed* $\wedge$ *timer_value = timer_value$_0$* |
| invariant: | see *MicrowaveOven* |
| goal: | *power_tube = energized* $\wedge$ *light = on* $\wedge$ *timer = running* $\wedge$ *timer_value = timer_value$_0$ + 60* $\wedge$ *beeper = silent* |

*4.2.2  Step 6: Synthesis of a Sample Program.*

We assume that the the light, the timer and the beeper can be switched on and off by setting the corresponding variables accordingly[3]. The synthesis of a procedure `press_button_door_closed` can then be performed completely automatically, using the *Automatic Assignment* strategy shown in Figure 1, because for each result variable we have an equation in the goal that can be transformed into an assignment statement. Note that *PressButtonDoorClosed* is embedded in the schema *PressButton*. To implement this schema, one develops a conditional (motivated by the "$\vee$", using the *Disjunctive Conditional* strategy): **if door = closed then press_button_door_closed else skip fi**, where **skip** is the program that does nothing.

# 5  Discussion

Now that our approach is presented in some detail, we can relate it to other work in the field, compare software safety with correctness and reliability, and finally discuss its merits as well as its drawbacks.

---

[3]If more sophisticated procedures are needed, the right-hand sides of the assignments can be replaced by calls to the respective procedures.

## 5.1 Related Work

Our choice of Z for the specification of safety-critical systems is not completely out of the way, as a look at the literature shows. Several case studies have been performed using the specification language VDM [Jon90], e.g. the British government regulations for storing explosives [MS93], a railway interlocking system [Han94], and a water-level monitoring system [Wil94]. VDM and Z are based on similar concepts and have the same expressive power (and weaknesses). Mukherjee's and Stavridou's as well as Hansen's work, however, place the focus on the adequate modeling of safety requirements, independently of the fact if software is employed or not. Consequently, they do not discuss issues specific to the construction of safe software.

Williams [Wil94] assesses safety specifications. His conclusions are:

1. " Methods used for the development of safety-critical systems should have well-defined criteria for ensuring the specification's completeness and consistency."
2. "The use of theorem proving is not limited to the verification of refinement steps. ..."
3. "Reviews can be an effective means of detecting errors in formal specifications."
4. "A formal statement of the safety requirements should be a part of the formal system specification. ..."
5. "The use of CASE tools can help eliminate simple syntactic errors in model-based specifications. ..."

Our approach fulfills most of these requirements. The completeness criterion is expressed in the proof obligation to show that for each combination of sensor values exactly one internal operation is invoked. Consistency is taken care of by the first three proof obligations shown in Table 1. The proof obligations introduced by our approach exceed the ones occurring in refinement steps. Reviews are not an explicit part of our process model but of course they are encouraged. According to Step 1, the fourth requirement is also fulfilled. Finally, we used the fuzz checker [Spi92a] to check all of the specifications contained in this paper, in order to eliminate simple syntactic errors.

The goals pursued by Halang and Krämer [HK94] are similar to ours. They present a development process, from the formalization of requirements to the testing of the constructed program. Their focus is on programmable logic controllers. As formalisms they use the specification language Obj and the Hoare calculus, where their choice is motivated by the tool support available. Both of these formalisms are weaker than the ones we chose. Obj only allows to state conditional equations, and the Hoare calculus is a proper subset of dynamic logic.

Like our work, Moser's and Melliar-Smith's approach to the formal verification of safety-critical systems, [MMS90], comprises the specification, design and implementation phases. The transition from an abstract top-level specification to a detailed specification suitable as a basis for program development is done by stepwise refinement. This activity is covered by Step 4 of our approach.

Moser and Melliar-Smith use a reliability model for the processors that execute the program. This enables them to take computer failures into account, an aspect not covered by this work. On the other hand, they do not consider the validation of the top-level specification, an issue that is of much importance for us, see the proof obligations of Steps 1–3.

## 5.2   Relation to Correctness and Reliability

In general, safety, correctness and reliability share the goal to make software more dependable. In detail, however, they have to be distinguished carefully.

**Safety vs. Correctness.**   One might consider safety a weaker requirement than correctness. Leveson [Lev86] states "We assume that, by definition, the correct states are safe." The example of the microwave oven, however, shows that safety concerns have an influence on what is considered a correct state. To ensure its safety, we defined the schema *EmergencyShutdown* that switches off the microwave as soon as a failure of the door sensor is detected. This situation is not taken into account when only the correctness of the software is of interest because correctness is a relation solely between a specification and a program. Failures of technical equipment are of no interest in correctness considerations. Hence, we think that the development of safe software has to proceed differently: the environment in which the software operates must explicitly be modeled. This difference is not of a technical, but of a pragmatic nature.

**Safety vs. Reliability.**   Our example study shows that reliability and safety can be conflicting goals (see also [Lev86]). Of the safety requirements for the microwave oven, the requirement that the power tube is de-energized when the door is open is certainly more important than the requirement that the light must be on when the power tube is energized. If the light bulb breaks down, it is a reasonable decision not to invoke the "emergency shutdown" but to sacrifice the less important safety requirement to increase availability (and thus reliability) of the oven.

## 5.3   Assessment of the Approach

We conclude with a summary of the merits and drawbacks of this work.

**Limitations.**   The approach presented here concentrates on the software aspects of safety-critical systems. Nothing can be guaranteed about the hardware. For instance, if the sensors yield false values, the system can enter a non-safe state because the software controls the system according to the sensor values. This limitation cannot be overcome by means concerning the software alone. Instead, fault tolerance methods like redundancy and consistency checks have to be applied.

Moreover, it is not possible to deal with absolute time measures in the formalisms we have chosen. If it is, e.g., necessary that a component reacts within 2 ms, then this cannot be guaranteed with our approach. The maximum execution time of the specified operations cannot be specified in Z, and we are not aware of any formal methods that allow one to *prove* maximum execution time of programs in higher-level languages[4]. Finally, our formalisms are not suitable to develop distributed or parallel systems.

As a result, the kind of safety our approach can guarantee is relative. Since we can only guarantee that the states before and after execution of an operation are safe, the execution must be sufficiently fast, because safety cannot be guaranteed in the intermediate states that occur during execution. It is up to the system designers and implementors to judge if this is the case. Here, traditional methods like testing are indispensable.

**Enhancing the Applicability of the Approach.** In contrast to hardware or power failure which are beyond our capabilities, the problem of unsafe intermediate states can be treated under the condition that sequences of assignments are considered as sufficiently fast. In this case, we can require a "safety invariant" to hold before and after each sequence of assignments. Then the system can be in an unsafe state only for the time that is needed to execute the longest assignment sequence occurring in the implementation. With little effort, IOSS can be extended to deal with such safety invariants.

For relatively small systems like a microwave oven, a complete formal treatment certainly can be recommended because the control software is relatively simple. The cost for a formal safety proof would be much less than potential damages. For larger systems, however, a complete formal treatment might not be feasible. In this case, our approach can be applied nevertheless. It is possible to formalize and prove only selected properties of the system and treat the other requirements with traditional techniques (*partial verification*, [Lev91]). When this approach is taken, still all of the software modules have to be considered. To further reduce cost, one might exclude those parts of the software from the verification process that can be guaranteed to be of no importance for safety.

**Contributions.** Our approach provides a process model for the development of provably safe software. Its contributions are the following:

- A detailed guidance for developers of safe software is provided, complemented by clear and explicit proof obligations.
- The approach can easily be introduced and applied in an organization because it relies on well established techniques and tools.
- The steps of the approach concerned with safety are clearly identified.
- Not only the specification but also the implementation of safety-critical systems is covered.

**Acknowledgment.** Many thanks to Thomas Santen and Jan Peleska for stimulating discussions on the topic and comments on this work.

---

[4]This is true even for formalisms designed to deal with time, like temporal logic or the duration calculus; again, these limitations come from the fact that the formalisms do not consider the hardware on which the programs are executed.

# References

[BG94]    J. Bowen and M. Gordon. Z and HOL. In *Z User Workshop*, Workshops in Computing, pages 141–167. Springer-Verlag, 1994.

[Gol82]   R. Goldblatt. *Axiomatising the Logic of Computer Programming*. LNCS 130. Springer-Verlag, 1982.

[Han94]   Kirsten Mark Hansen. Modelling railway interlocking systems. Available via ftp from ftp.ifad.dk, directory /pub/vdm/examples, 1994.

[Hei94]   Maritta Heisel. A formal notion of strategy for software development. Technical Report 94–28, TU Berlin, 1994.

[HK94]    Wolfgang Halang and Bernd Krämer. Safety assurance in process control. *IEEE Software*, 11(1):61–67, January 1994.

[HSZ95a]  Maritta Heisel, Thomas Santen, and Dominik Zimmermann. A generic system architecture of strategy-based software development. Technical Report 95-8, Technical University of Berlin, 1995.

[HSZ95b]  Maritta Heisel, Thomas Santen, and Dominik Zimmermann. Tool support for formal software development: A generic architecture. In *Proceedings 5-th European Software Engineering Conference*, Springer LNCS, 1995.

[Jon90]   Cliff B. Jones. *Systematic Software Development using VDM*. Prentice Hall, 1990.

[Lev86]   Nancy Leveson. Software safety: Why,what, and how. *Computing Surveys*, 18(2):125–163, June 1986.

[Lev91]   Nancy Leveson. Software safety in embedded computer systems. *Communications of the ACM*, 34(2):34–46, February 1991.

[MMS90]   Louise E. Moser and P.M. Melliar-Smith. Formal verification of safety-critical systems. *Software – Practice and Experience*, 20(8):799–821, August 1990.

[MS93]    Paul Mukherjee and Victoria Stavridou. The formal specification of safety requirements for storing explosives. *Formal Aspects of Computing*, 5:299–336, 1993.

[PST91]   Ben Potter, Jane Sinclair, and David Till. *An Introduction to Formal Specification and Z*. Prentice Hall, 1991.

[SM92]    Sally Shlaer and Stephen J. Mellor. *Object Lifecycles – Modeling the World in States*. Yourdon Press, Englewood Cliffs, 1992.

[Spi92a]  J. M. Spivey. The fuzz manual. Computing Science Consultancy, Oxford, 1992.

[Spi92b]  J. M. Spivey. *The Z Notation – A Reference Manual*. Prentice Hall, 2nd edition, 1992.

[Wil94]   Lloyd Williams. Assessment of safety-critical specifications. *IEEE Software*, pages 51–60, January 1994.

[Woo91a]  J.C.P. Woodcock. An introduction to refinement in Z. In S. Prehm and W.J. Toetenel, editors, *Proc. 4-th International Symposium of VDM Europe, Vol. 2*, LNCS 552, pages 96–117. Springer-Verlag, 1991.

[Woo91b]  J.C.P. Woodcock. The refinement calculus. In S. Prehm and W.J. Toetenel, editors, *Proc. 4-th International Symposium of VDM Europe, Vol. 2*, LNCS 552, pages 80–95. Springer-Verlag, 1991.

# Formally Verified Firmware Modules for Industrial Process Automation

Wolfgang A. Halang, Bernd J. Krämer, and Norbert Völker

Faculty of Electrical Engineering, FernUniversität

D-58084 Hagen, Germany

### Abstract

Society increasingly uses computer based systems, which take care of control and automation functions in safety critical applications. With programmable logic controllers replacing traditional hardwired control devices, the problem of software dependability becomes more and more virulent. As a step towards its solution, the foundation for provably correct software to be used in industrial process automation is laid. In the form of function block diagrams, industrial process automation software is constructed from small sets of high level, application oriented modules. It is shown that the latter can be rigorously verified with formal methods. Then, as firmware, they become part of the computing architecture.

## 1  Introduction

Our society increasingly uses computer based systems, i.e., systems in which software and hardware components implement central parts of the overall system functionality. Often computer based systems take care of control and automation functions in safety critical systems used in (air) traffic control, patient monitoring, or process and production line control. A specific class of computer based systems are programmable logic controllers (PLCs) intended to replace traditional hardwired switching networks based on relay or discrete electronic logic. PLCs are typically used for binary and sequence control, process supervision, data acquisition, signal processing, or for communications and other tasks closely related to industrial processes. Their advantages over pure hardware solutions are flexible adaptation to modifications of the controlled processes just by re-programming instead of re-wiring and higher information processing capabilities. These advantages are, however, partly out-weight by the increasing complexity of the control software and a lack of sound methods to thoroughly understand, specify, design, implement, maintain, and assess properties of such systems. In difference to hardware, software does not wear out and environmental circumstances cannot cause software faults. Rather, all software errors are design or programming errors, which cannot be detected solely by program testing, peer reviews or other, mostly informal methods prevailing in software development practice. As a consequence, licensing authorities are extremely reluctant in approving safety critical systems whose behavior is exclusively program controlled.

Erroneous software or hardware may have disastrous effects in industrial applications, and the situation is aggravated by stronger legal regulations concerning product liability. Thus, the proper functioning and application possibilities of computer based systems primarily depend on the question to what extent safety and dependability requirements can be guaranteed. As a consequence, system developers are forced to use production methods and tools that enable sound proofs of quality aspects such as reliability, safety, security, or functional correctness. In this context, software is of special concern for reasons that were extensively discussed, e.g., in [Parnas,1990].

As a step towards a remedy for this unsatisfactory situation, the work reported here lays the foundation for provably correct software to be used in industrial process automation. It is based on the observation that Function Block Diagrams (FBDs) [IEC,1992], the graphical design method commonly used to represent the functionality of industrial control systems, can be constructed from sets of basic function modules containing rather limited numbers of elements — such as the one introduced in [VDI,1993]. This analysis of FBDs leads to a specific programming paradigm, viz., to compose software out of high level application oriented building blocks instead out of low level (machine oriented) ones. Whereas a single (machine) instruction taken out of a program context does not reveal its purpose, the occurrence of a certain function module instance usually gives already a clue about the problem, its solution, and the module's rôle in it. The interna of basic function blocks are described in Structured Text (ST), a Pascal-like procedural language defined in [IEC,1992]. The behavior of more complex applications is defined by "wiring" basic function modules together to diagrams giving a graphical representation of structured function blocks.

First, in the next section, we shall shortly present this graphical software engineering paradigm. The purpose of this paper is to demonstrate that basic function blocks as commonly used in industrial applications can be rigorously verified with formal methods. To this end, we consider two cases, a timer and a polygonal interpolation function block. In either case, we employ a different formal technique, each considered to be most appropriate. After formal verification, the code of these basic function modules is to be transformed to firmware in order to become an unalterable part of the computing architecture.

# 2 Graphical Programming Paradigm Oriented at Industrial Process Automation

As illustrated in Figure 1, this — actually specification level — programming method consists of graphically interconnecting instances of the above mentioned basic function modules with each other by lines, i.e., single basic functions are invoked one after the other and, in the course of this, they pass parameters. The interconnections between function blocks have to meet just one restriction: each input must be connected to exactly one output. Besides the provision of constants as external input parameters, the basic functions' instances and the parameter flows between them are the only language elements required by this programming paradigm.

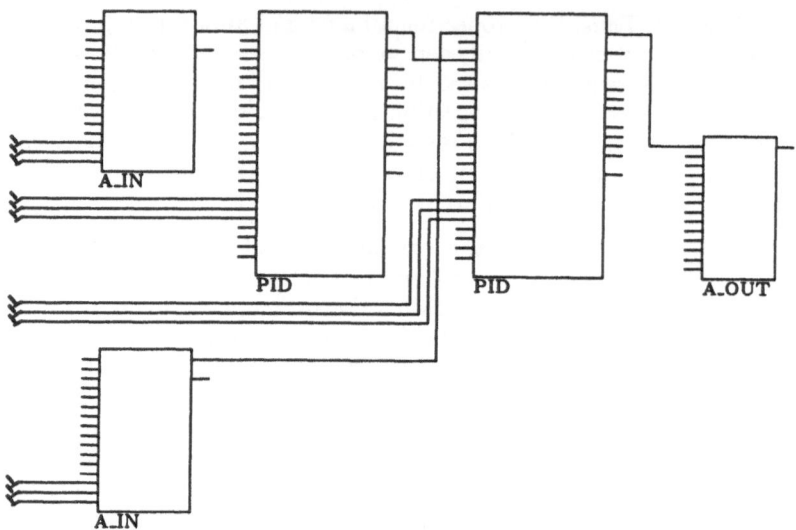

Figure 1. A typical function block diagram

A compiler transforms the graphically represented program logic into object code. Owing to the simple structure, this logic is only able to assume, the generated programs contain no other features than procedure calls and some internal moves of data. The rigorous verification of a compiler transforming graphical software representation into object code is still impossible — but also not necessary, because for FBD software only the module interconnections need to be verified. For this purpose loaded object code implementing an interconnection pattern can be subjected to diverse back translation [Krebs,1984]. This is necessary, because prevailing legal requirements demand that object code must be considered for the correctness proofs of software.

Back translation is a safety licensing method which was developed in the course of the Halden nuclear power plant project. The method is — although rigorous — essentially informal, easily conceivable, and immediately applicable without any training. Thus, it is extremely well suited to be used on the application programming level by people with the most heterogeneous educational backgrounds. The ease of understanding and use inherently fosters error free application of the method. It consists of reading machine programs out of computer memory and giving them to a number of teams working without any mutual contact. All by hand, these teams disassemble and decompile the code, from which they finally try to regain the specification. The software is granted a safety license if the original specification agrees with the inversely obtained re-specifications. In [Halang,1994] a specialized, simple computer system in the form of a programmable logic controller was presented. Closing the semantic gap between software requirements and hardware capabilities, it relinquishes the need for not safety licensable compilers and operating systems.

Instead, the safety licensing of application software developed according to the here considered paradigm is enabled by supporting back translation architecturally. Thus, the effort required to employ this software verification method reduces by several orders of magnitude as compared to verifying code running on conventional von Neumann machines.

# 3 A First Case: Emergency Shutdown Systems

Many technical systems have the potential of disastrous effects on, for instance, the environment, equipment, employees, or the general public in case of malfunctions. An important objective of the design, construction, and commissioning of such systems is, therefore, to minimize the chances that hazards occur. One possibility to achieve this goal is the installation of a system whose only function is to supervise a process and to take appropriate action if anything in the process turns dangerous. So, to prevent hazards, many processes are guarded by these so called safeguarding systems. A special kind of such systems are emergency shutdown systems (ESDs) which are defined as:

*A system that monitors a process, and only acts — i.e., guides the process to a static safe state (generally, a process shut-down) — if the safety of either human beings, the environment, or investments is at stake.*

The mentioned monitoring consists of observing whether certain physical quantities such as temperatures or pressures stay within given bounds and to supervise Boolean quantities for value changes. Typical ESD actions are opening or closing valves, operating switches etc. Structurally, ESDs are functions composed of Boolean operators and delays. The latter are required, because in start-up and shut-down sequences often some monitoring or actions need to be delayed. Originally, safeguarding systems were constructed pneumatically and later, e.g., in railway signaling, with electromagnetic relays. Nowadays, most systems installed are based on integrated electronics and there is a tendency to use microcomputers.

All emergency shutdown systems can be constructed from a set of function modules containing just four elements, viz., the three Boolean operators *And*, *Or*, *Not* and a *timer*. It is sufficient to use only one type of timer. All other forms of timers used in hardwired logic can be implemented by adding inverters. The timer's functionality can be informally described (A formal description of the timer function was given in [Halang,1995] in form of a procedure formulated in the programming language Structured Text.) as follows:

A non-retriggerable, mono-stable element with selectable delay, i.e., an element with the two internal states *triggered* and *non-triggered* and with a Boolean input, such that when the element is in the non-triggered state and it detects a rising edge, i.e., a transition from False to True, at its input, then it switches into the triggered state for a duration as specified by the delay.

The correct implementation of an ESD basically relies on the correct implementation of the individual function blocks composing a corresponding FBD. The correctness proof is trivial for the three Boolean operators we need in ESD designs. But the correctness proof of the basic timer element is already rather difficult; it involves the formalization of

- the syntax and semantics of a small imperative programming language, suitably selected for our purposes as a subset of ST,

- the timer program, and

- the requirements of the timer function block.

This formalization is performed in higher order logic (HOL) [Gordon,1989], which is a typed variant of Church's higher order predicate logic. Since HOL is very expressive, it has become a popular logic to reason about the correctness of hardware and, increasingly, also firmware and software systems. As mechanical proof assistant we use the generic theorem prover Isabelle [Paulson,1994], which supports the interactive construction of proofs in a variety of logics including HOL. The verification task then amounts to proving a certain theorem using that theorem prover. In Isabelle, theorems belong to a special data type and objects of this data type can only be constructed by a rigorous deduction process from axioms or other theorems, respectively. Since the set of in-built axioms and inference rules is quite small, we can be very confident that Isabelle will only prove theorems which follow logically from the definition of the theories. On the other hand, the correctness of the formalization or the validity of the specification cannot be proven formally correct. Hence, in practice, formal correctness proofs should be seen as a way to increase the trustworthiness of a system, but not as a guarantee for 100 % error free software.

Isabelle has a polymorphic type system and allows overloading of operators using a class concept. Theory development in Isabelle is hierarchical, i.e., new theories are defined as logical extensions of already existing ones by introducing new types, constants, or axioms. As our base theory, we choose theory "List" from the Isabelle/HOL library. This theory contains definitions of the quantifiers and logical operations of higher order logic. In addition, it provides natural numbers and polymorphic types of sets, products, sums, and lists.

To enable the automated verification of a program in HOL, the syntax and semantics of the programming language — in our case a subset of Structured Text — have to be modelled in HOL. Representing the syntax is straightforward due to the automated support for the definition of recursive types available in the HOL system. This allows to define a type containing all syntactically correct programs. The HOL system automatically proves certain theorems which characterize this type and provides tools for defining primitive recursive functions on it. These functions and the operations on the built-in types Boolean and numbers are sufficient to define an operational semantics for a simple, imperative programming language such as the employed subset of Structured Text.

The verification of the timer program is performed in three steps:

1. representation of the program as a term in HOL,

2. specification of expected timer properties in HOL, and

3. proof that the program satisfies these properties.

The first step is trivial as it represents just a one-to-one mapping from ST statements into an abstract syntax tree. The formal specification of the timer properties as predicate logic formulae is also fairly straightforward. The main techniques applied for the actual proof are induction over the intermediate states, implicational reasoning in the form of bottom-up and goal directed proofs, re-writing, case introductions, and arithmetic simplifications. Although somewhat tedious and requiring a certain experience with the HOL system, the mechanization of the proof in HOL does have the salient feature that it is basically impossible to prove a non-theorem. Unfortunately, due to space limitations the proof cannot be elaborated here. In a sketchy way it is described in [Halang,1995], whereas the complete code of the proof together with documentation is obtainable via *ftp*[1].

# 4 A Second Case: Engineering Function Modules

Our second function block example is taken from the draft guideline [VDI,1993] defining standard function blocks to be used in vendor independent descriptions of process control programs. This function block, called nonlin, approximates a non-linear static characteristic $f$ by a polygon $f'$, which is given by $n$ base points defining the edges of the polygon. Function block nonlin, once instantiated with an arbitrary finite positive number $n$, receives $2n + 1$ more inputs to set the $n$ base point coordinates and one abscissa, and provides the corresponding ordinate as output. The functionality of nonlin is shown below.

Table 1: Parameters of function block nonlin

| Inputs | Meaning | Type |
|--------|---------|------|
| U | Input Value | Num |
| N | Number of Base Points (some positive integer n) | Num |
| U1 | U-Value of Base Point 1 | Num |
| V1 | V-Value of Base Point 1 | Num |
| U2 | U-Value of Base Point 2 | Num |
| V2 | V-Value of Base Point 2 | Num |
| ... | | |
| Un | U-Value of Base Point n | Num |
| Vn | V-Value of Base Point n | Num |
| Outputs | | |
| V | Result Value | Num |

---

[1] Ftp-server ftp.fernuni-hagen.de, directory /pub/fachb/et/dvt/projects/verification

For U < U1 it is required that V becomes V1 and for U > Un the output value is Vn.

In the following iterative solution to function approximation we assume that

- U- and V-coordinates of base points are kept in two arrays UI[1.. N] and VI[1.. N] of Num,
- the elements of UI are stored in ascending order, i.e., UI[I] < UI[J] for I < J, and
- UI[I] and VI[I] correspond to each other, i.e., $VI[I] = f'(UI[I])$.

The following ST program describes our algorithm in a way adapted from the draft guideline [VDI,1993]:

```
FUNCTION_BLOCK NONLIN
...
VAR_INPUT
    U : NUM;        (* input value *)
    N : INT;        (* number of base points *)
    ...
END_VAR;

VAR
    I : INT         (* local variable used to compute V *)
END_VAR;

VAR_OUT
    V : NUM;        (* result value *)
END_VAR;
        (*** FB body ***)
IF U < UI[1]
    THEN V := VI[1]
    ELSE
        IF UI[N] < U
            THEN V := VI[N]
            ELSE
                I := 1;
                WHILE UI[I+1] < U AND U
                    DO I := I+1
                END;
                V := fp(I,U);
        FI
FI

END_FUNCTION_BLOCK
```

where $fp(i,U)$ is used as an abbreviation for the following arithmetic expressions:

$$((VI[I + 1] - VI[I])/(UI[I + 1] - UI[I])) * (U - UI[I]) + VI[I].$$

The aim of program verification is to increase one's confidence in the correct functioning of a piece of software. One fundamental program verification technique is Dijkstra's predicate transformer approach [Dijkstra,1984]. It relies on logic assertions about values of program variables upon entry and exit of an individual statement or a whole program and is concerned with verifying the (partial) correctness of a program $S$ with respect to a given specification $(P, Q)$. $P$ is called the precondition; it describes relevant properties of program variables on entry of $S$, while $Q$ denotes the postcondition of $S$ and describes the expected effect of computation on the program state assuming that the precondition $P$ ensures termination of $S$. The notation

$$\{P\}S\{Q\}$$

is used to claim that the program $S$ is partially correct, i.e., if $P$ holds before the execution of $S$ and $S$ terminates, then the final state of $S$ will satisfy $Q$. For example, if $S$ denotes a program implementing the behavior of function block nonlin, $Q$ might include the assertion that the value of output V satisfies its requirements. The predicate transformer approach allows us to derive assertions for individual statements from global assertions relating pre- and postconditions of an entire program such as nonlin and construct a correctness proof for it from correctness proofs of its constituent statements. The core method provides a calculus to compute the weakest precondition of $S$ with respect to $Q$, written $wp(S, Q)$, defining states from which program $S$ could execute and terminate in states meeting $Q$.

Our objective is now to prove that the approximation algorithm encoded in the body of function block nonlin is correct with respect to the requirements defined above. For convenience of presentation we use a mathematical notation of our algorithm, which is shown in Figure 2, and simply call it $A$. As before we set

$$fp(i, u) = \frac{v_{i+1} - v_i}{u_{i+1} - u_i} = \frac{v - v_i}{u - u_i}.$$

```
IF u < u_1
    THEN   v := v_1
    ELSE
           IF u_n < u
               THEN v := v_n
               ELSE
                      i := 1;
                      WHILE u_{i+1} < u
                            DO  i := i + 1
                      END;
                      v := fp(i, u)
           FI
FI
```

Figure 2: Approximation algorithm $A$

The correctness proof consists in the calculation of $wp(A, Q)$, where $Q$ is the postcondition:

$$
\begin{aligned}
Q \;=\; & (u < u_1 \text{ and } v = v_1) \text{ or} \\
& (u > u_n \text{ and } v = v_n) \text{ or} \\
& (u_1 \leq u \text{ and } u \leq u_n \text{ and } v = fp(i, u))
\end{aligned}
$$

The calculation of weakest preconditions of statements often depends on weakest preconditions of inner statements, as in the case of conditional and while statements, or it relies on weakest preconditions of subsequent statements as in statement compositions. Figure 3 illustrates the structure of the calculation process shown below by decorating the statements of $A$ with their preconditions. This figure also introduces some abbreviations for constituent statements of $A$, to which we refer in the recursive correctness proof.

```
{wp(A, Q)}
IF u < u₁
     THEN   v := v₁
     ELSE
          {wp(S₁, Q)}
          IF uₙ < u
              THEN   v := vₙ
              ELSE
                    {wp(S₂, Q)}
                    {wp(S₃, wp(S₄, wp(S₅, Q)))}
                    S₃ { i := 1;
                         {wp(S₄, wp(S₅, Q))}
                         WHILE u_{i+1} < u
                    S₄ {   DO   i := i + 1
                         END;
                         {wp(S₅, Q)}
                    S₅ { v := fp(i, u)
          FI
FI
{Q}
```

Figure3: Algorithm $A$ decorated with weakest preconditions

$$
\begin{aligned}
wp(A, Q) \;=\; & wp(\text{IF } u < u_1 \text{ THEN } v := v_1 \text{ ELSE } S_1 \text{ END}, Q) \\
=\; & (u < u_1 \text{ and } Q[v/v_1]) \text{ or} \\
& (u_1 \leq u \text{ and } wp(S_1, Q)) \\
=\; & (u < u_1 \text{ and } v_1 = v_1) \text{ or} \\
& (u_1 \leq u \text{ and } wp(S_1, Q)) \\
=\; & (u < u_1 \text{ and } \mathbf{true}) \text{ or} \\
& (u_1 \leq u \text{ and } wp(S_1, Q)) \\
=\; & u < u_1 \text{ or} \\
& (u_1 \leq u \text{ and } wp(S_1, Q))
\end{aligned}
$$

The simplifications in the first clause of the disjunction are due to the laws of propositional calculus and inconsistencies between conditions in $Q[v/v_1]$ and the condition $u < u_1$. Using again the rule of conditional statements and performing similar simplifications, we obtain:

$$
\begin{aligned}
wp(A, Q) \;=\; & u < u_1 \text{ or} \\
& (u_1 \leq u \text{ and} \\
& \quad wp(\text{IF } u_n < u \text{ THEN } v := v_n \text{ ELSE } S_2 \text{ END}, Q)) \\
\;=\; & u < u_1 \text{ or} \\
& (u_1 \leq u \text{ and} \\
& \quad ((u_n < u \text{ and } Q[v/v_n]) \text{ or} \\
& \quad (u \leq u_n \text{ and } wp(S_2, Q))))) \\
\;=\; & u < u_1 \text{ or} \\
& (u_1 \leq u \text{ and } u_n < u) \text{ or} \\
& (u_1 \leq u \text{ and } u \leq u_n \text{ and } wp(S_2, Q)) \qquad (1)
\end{aligned}
$$

We observe that:

$$
wp(S_2, Q) \;=\; wp(S_3; S_4; S_5, Q) = wp(S_3, wp(S_4, wp(S_5, Q))) \qquad (2)
$$

Hence, to complete the calculation of Condition 1 (which is equivalent to Condition 2), we have to determine the weakest preconditions of $S_3, S_4,$ and $S_5$ first. We apply the assignment rule to calculate the postcondition of the while statement:

$$
\begin{aligned}
wp(S_5, Q) \;=\; & wp(v := fp(i, u), Q) \\
\;=\; & Q[v/fp(i, u)] \\
\;=\; & (u < u_1 \text{ and } fp(i, u) = v_1) \text{ or} \\
& (u > u_n \text{ and } fp(i, u) = v_n) \text{ or} \\
& (u_1 \leq u \text{ and } u \leq u_n \text{ and } fp(i, u) = fp(i, u)) \\
\;=\; & (u < u_1 \text{ and } fp(i, u) = v_1) \text{ or} \\
& (u > u_n \text{ and } fp(i, u) = v_n) \text{ or} \\
& (u_1 \leq u \text{ and } u \leq u_n) \qquad (3)
\end{aligned}
$$

We refer to Condition 3 by $W$. Using $W$ and the rule of statement composition, we can compute the weakest precondition of the while statement:

$$
wp(S_4, wp(S_5, Q)) \;=\; wp(\text{WHILE } u_{i+1} < u \text{ DO } i := i + 1 \text{ END}, W)
$$

We recall the procedure to calculate the precondition of while statements inductively:

$$
\begin{aligned}
P_0 \;=\; & \text{not } (u_{i+1} < u) \text{ and } W \\
\;=\; & u \leq u_{i+1} \text{ and } W \\
\;=\; & u \leq u_{i+1} \text{ and} \\
& ((u < u_1 \text{ and } fp(i, u) = v_1) \text{ or} \\
& (u > u_n \text{ and } fp(i, u) = v_n) \text{ or}
\end{aligned}
$$

$$(u_1 \leq u \text{ and } u \leq u_n))$$
$$= \quad (u \leq u_{i+1} \text{ and } u < u_1 \text{ and } fp(i,u) = v_1) \text{ or}$$
$$(u \leq u_{i+1} \text{ and } u > u_n \text{ and } fp(i,u) = v_n) \text{ or}$$
$$(u \leq u_{i+1} \text{ and } u_1 \leq u \text{ and } u \leq u_n)$$

$$
\begin{aligned}
P_1 \quad &= \quad u_{i+1} < u \text{ and } wp(i := i+1, P_0) \\
&= \quad u_{i+1} < u \text{ and } P_0[i/i+1] \\
&= \quad u_{i+1} < u \text{ and} \\
&\qquad ((u \leq u_{i+2} \text{ and } u < u_1 \text{ and } fp(i+1,u) = v_1) \text{ or} \\
&\qquad (u \leq u_{i+2} \text{ and } u > u_n \text{ and } fp(i+1,u) = v_n) \text{ or} \\
&\qquad (u \leq u_{i+2} \text{ and } u_1 \leq u \text{ and } u \leq u_n)) \\
&= \quad (u_{i+1} < u \text{ and } u \leq u_{i+2} \text{ and } u < u_1 \text{ and } fp(i+1,u) = v_1) \text{ or} \\
&\qquad (u_{i+1} < u \text{ and } u \leq u_{i+2} \text{ and } u > u_n \text{ and } fp(i+1,u) = v_n) \text{ or} \\
&\qquad (u_{i+1} < u \text{ and } u \leq u_{i+2} \text{ and } u_1 \leq u \text{ and } u \leq u_n)
\end{aligned}
$$

$$
\begin{aligned}
P_2 \quad &= \quad u_{i+1} < u \text{ and } wp(i := i+1, P_1) \\
&= \quad (u_{i+2} < u \text{ and } u \leq u_{i+3} \text{ and } u < u_1 \text{ and } fp(i+2,u) = v_1) \text{ or} \\
&\qquad (u_{i+2} < u \text{ and } u \leq u_{i+3} \text{ and } u > u_n \text{ and } fp(i+2,u) = v_n) \text{ or} \\
&\qquad (u_{i+2} < u \text{ and } u \leq u_{i+3} \text{ and } u_1 \leq u \text{ and } u \leq u_n)
\end{aligned}
$$

Note that $u_{i+1} < u$ is implied by condition $u_{i+2} < u$, the requirement $u_{i+1} < u_{i+2}$, and the transitivity of the "$<$" operator.

$$
\begin{aligned}
P_k \quad &= \quad (u_{i+k} < u \text{ and } u \leq u_{i+k+1} \text{ and } u < u_1 \text{ and } fp(i+k,u) = v_1) \text{ or} \\
&\qquad (u_{i+k} < u \text{ and } u \leq u_{i+k+1} \text{ and } u > u_n \text{ and } fp(i+k,u) = v_n) \text{ or} \\
&\qquad (u_{i+k} < u \text{ and } u \leq u_{i+k+1} \text{ and } u_1 \leq u \text{ and } u \leq u_n)
\end{aligned}
$$

From this we conclude that the weakest precondition of $S_5$ is:

$$
\begin{aligned}
wp(S_4, W) \quad = \quad &\exists k \bullet k \geq 0 \text{ and } u_{i+k} < u \text{ and } u \leq u_{i+k+1} \text{ and} \\
&((u < u_1 \text{ and } fp(i,u) = v_1) \text{ or} \\
&(u > u_n \text{ and } fp(i,u) = v_n) \text{ or} \\
&(u_1 \leq u \text{ and } u \leq u_n))
\end{aligned}
$$

which simplifies to:

$$wp(S_4, W) \quad = \quad 1 \leq i \text{ and } i < n \text{ and } u_1 \leq u \text{ and } u \leq u_n \tag{4}$$

Now we return to Condition 2:

$$
\begin{aligned}
wp(S_2, W) \quad &= \quad wp(S_3, wp(S_4, W)) \\
&= \quad wp(i := 1, 1 \leq i \text{ and } i < n \text{ and } u_1 \leq u \text{ and } u \leq u_n) \\
&= \quad 1 \leq 1 \text{ and } 1 < n \text{ and } u_1 \leq u \text{ and } u \leq u_n \\
&= \quad \text{true and } 1 < n \text{ and } u_1 \leq u \text{ and } u \leq u_n \\
&= \quad 1 < n \text{ and } u_1 \leq u \text{ and } u \leq u_n
\end{aligned}
$$

This yields the postcondition we need to complete the computation of the weakest precondition of algorithm $A$ from the intermediate result achieved in Condition 1.

$$
\begin{aligned}
wp(A,Q) \;=\; & u < u_1 \text{ or} \\
& (u_1 \le u \text{ and } u_n < u) \text{ or} \\
& (u_1 \le u \text{ and } u \le u_n \text{ and } wp(S_2,Q)) \\
\;=\; & u < u_1 \text{ or} \\
& (u_1 \le u \text{ and } u_n < u) \text{ or} \\
& (u_1 \le u \text{ and } u \le u_n \text{ and } 1 < n)
\end{aligned}
\tag{5}
$$

Remembering that a general requirement for the design of function block `nonlin` is that the number of base points is larger than one, because otherwise no useful approximation algorithm can be defined, the weakest precondition of $A$ guarantees that the algorithm will terminate and satisfy $Q$, independent of the initial state the computation started from.

# 5  Conclusion

Hitherto, in industry, verification activities consist of program testing based on compiled code only. We have presented here approaches for the rigorous verification of simple programs which specify single components of industrial reactive system. Our emphasis was on the functional correctness of elementary function blocks, such as the ones defined in [VDI,1993] for the chemical process industries. The extra effort necessary to develop formal specifications and to employ program verification techniques is justified by a number of reasons. Standard process control software will be used in hundreds or thousands of installations and its correct operation is often crucial to satisfy hard safety and dependability requirements of entire systems. Formal specification and verification provides certainty about the consistency between specification and program for all possible input data. Moreover, these additional activities provide deeper insight into a specification and program, and they reduce the costs for error detection and maintenance. Formal specifications further enhance the systematic re-use of function blocks, and verified interface specifications support modular and, thus, more efficient proof techniques, because proofs can be re-used, too.

It should be noted that we have achieved a separation of concerns: the correctness of the FBD building blocks' implementation was established once and for all and does not have to be performed over and over again. We are convinced that such abstractions are indispensable for the verification of more complex systems. Although the complete verification of more complex systems is still out of reach, our work demonstrates that existing technology already allows for machine checked correctness proofs of certain software components.

# References

[Dijkstra,1984 ] Dijkstra EW, Feijen WHJ. Een methode van programmeren. Academic service cop., 1984

[Gordon,1989 ] Gordon MJC. Mechanizing Programming Logics in Higher Order Logic. In: Current Trends in Hardware Verification and Automated Theorem Proving, Birtwistle G and Subrahmanyam PA (eds.), pp 387 – 439. New York-Berlin-Heidelberg-London-Paris-Tokyo: Springer-Verlag 1989

[Halang,1994 ] Halang WA, Jung S-K. A Programmable Logic Controller for Safety Critical Systems. High Integrity Systems 1994; 1, 2: 179 – 193

[Halang,1995 ] Halang WA, Krämer BJ, Völker N. Formally Verified Building Blocks in Functional Logic Diagrams for Emergency Shutdown System Design. To appear in High Integrity Systems 1995

[IEC,1992 ] IEC International Standard 1131-3. Programmable Controllers, Part 3: Programming Languages. Geneva: International Electrotechnical Commission 1992

[Krebs,1984 ] Krebs H, Haspel U. Ein Verfahren zur Software-Verifikation. Regelungstechnische Praxis 1984; 28: 73 — 78

[Parnas,1990 ] Parnas DL, van Schouwen J, Kwan SP. Evaluation of safety-critical software. Communications of the ACM 1990; 33(6): 636 – 648

[Paulson,1994 ] Paulson LC. Isabelle: A Generic Theorem Prover. Lecture Notes in Computer Science, Vol. 828. New York-Berlin-Heidelberg-London-Paris-Tokyo: Springer-Verlag 1994

[VDI,1993 ] VDI/VDE-Richtlinie 3696 (Entwurf). Herstellerneutrale Konfigurierung von Prozeßleitsystemen. Berlin: Beuth-Verlag 1993

# Session 6
## Assessment

# Programmable Electronic Controllers (PEC) Performance Assessment – an Approach for Reliability Quantification

Authors

G. Picciolo – Research & Technology
Basic Chemicals, EniChem
Milan (Italy)

P. Gianninò– Innotec Engineering
Florence (Italy)

## 1 Introduction

It is worldwide recognised the relevant role of computer–based system safety–related for industrial applications. Electronic command and control systems are extensilvely used for military, air, ground transport, space and ship control. Similar systems are applied in nuclear power plants and in the process industries. In the last years advanced technologies both hardware and software allow high level performance of such systems being attained and great effort is spent by Manufacturers and approval Laboratories to guarantee their *quality* into the framework of national and international Norms and Codes (ISO/IEC). Computer–based systems operations typically involve [TAY94]:

– control capabilities for electromechanical hardware,
– general information processing and management,
– communications,
– man/machine interface

Despite Product reliability growth continous process, industry realizes that careful attention should be paied for the *Safety Aspects* of the computer–based systems in hazardous applications, where human limb[1] and economics might be heavily involved. Past lessons due to accidents occurred in large chemical processes and refinery Plants,
learned Customers for identifying causes and measures to be undertaken to avoid either dangerous conditions to happen, or mitigate consequences. It is very rare, nowadays, for a major accident to occur as a result of a combination of simple component faults. Most .system reliability and safety analyses focus on the hardware (general). The reason for this is that the failure consequences are in most cases mediated by the hardware itself. It is here
[1] Safety definition relates generally to a *state* of a system where either dangerous or undesired events are not likely to occur

that most hazards exist, and it is here that consequences are felt. However, it is not in the hardware that most failures and errors arise. For large systems and particularly one off systems, hardware failures have historically been the smallest contributors. The biggest contributors are generally operator error (management error), software error and design error. When major accidents do occur, they typically involve a combination of these. However, as specified from various international Norms and Codes and User's significant experiences, safety assessment of hazardous installations should be analyzed as an interpretation of the compromize between the probability of potential dangerous events and the severity of consequences. Due to the severities of such events, depending on the specified scenarios and the inherent operational context, great efforts have been spent accordingly, for safety–related systems and equipment performance reliability (performability) design Valuation and Validation (V & V). On the other hand, system performance requirements reveal significant for an installation being safe, which should be properly identified and verified according to international Norms and Standards. Manufacturers and final Users did perceive the need to attain a common reference framework for verifying and applying [VDE89, IEC94] safety–related systems in own Facilities, where equipment turnover and technical development require new technologies. Programmable Electronic Controllers (PEC) play an important role in safety–related systems.

## 2 Need of quantification – Dependability

Assessment of Reliability metrics of safety systems computer–based is evolving in the last years according to a great number of theoretical and applicative procedures, often depending on the context and domain of application. Furthermore, future commercial systems for air, ground and space applications will require innovative solutions to dependability problems. However, functional performances should be verified against potential failures leading to hazards. Emerging Standards and Norms not yet fully assessed [VDE89, IEC94], justify theoretical and practical effort to face the entire topics into an "Overall Safety Integrity Assessment" in order to achieve safety targets. Product *Life–Cycle* and *dependability*[2] concepts, sinergically integrated together, attempt to achieve this objective[3].

The classic approach of designing for performance and then addressing dependability issues is no longer suitable or practical.

Only advanced design techniques that combine performance and dependability aspects can ensure the production – in a timely and cost effective fashion – of computer system meeting stringent real–time and safety–critical requirements.

[2]The official scope of IEC/TC 56 "Dependability" is as follows:
Standardization in the field of availability, reliability, maintenability andmaintenance support, in any such technological area as may be considered
appropriate, including those not normally dealt with by IEC/TC's.

[3] Particular interest is now invoked by the IEC/TC56 action on the Proposal for harmonization of work programmes of IEC/TC56/WG10 and ISO/IEC/JTC1/SC7/WG9 relating to *software*.

The aims of dependability programmes (IEC–300–1) are to ensure that adequate and effective effort is brought to bear on dependability as a principal *quality measure* during all phases of the Life–Cycle of a Product. A dependability programme should provide continuous study of both qualitative and quantitative requirements throughout all phases of the project. Dependability assessment should be updated, specified requirements should be verified and activities should be integrated with other elements of the developments, production and operation programme. The extent and contents of the dependability programme should be governed by the particular characteristics of the project and any specific constraints and the importance of dependability of actual product.

Dependability assessment, complying with emerging Standards and Norms, is, therefore, of paramount importance as well as validation and certification into an international (EC) Product Market. Noteworthing, the main Standards and Guidelines in force or under development for assessing Dependability (reliability) metrics, use a *qualitative* approach.

They first define a set of *levels*, so that the safety requirements for the system of interest can be classified as belonging to one of these levels. For each level, they specify a set of practices (design and methods for production and test, management of the development process) which are considered useful and effective for attaining that level of safety. A certification that the specified performance system reliability metrics belongs to the given *level*, thus consists in certifying (where possible) that all these practices were (properly, to the extent that the Certifier can assess) used.

However, these standards and Guidelines do not require the Developer (or any other party seeking certification of the system performance reliability metrics) to demonstrate that the identified metrics actually reaches the required level of *Integrity*. This last step would usually coincide with what is called *Quantitative assessment* of *Safety Integrity Level* [IEC94]. Users of PECs, especially in safety–related applications, have a common need for quantitative evaluation of dependability; particularly, for performance reliability: hardware and software (Operating System).

For many potential Users of PECs safety–related, safety requirements are stated in quantitative terms.

They may take forms like, for instance: *"the probability of failure on demand must be less than $10^{-3}$"*, or *"this PEC–based system must be at least as safe as the relay–based system it will substitute"*. To these Users, a qualitative certification of the functional performances reliability metrics says only *"this system has been produced according to specified procedures and techniques and with a given level of care"*, not *"this system satisfies specified functional requirements within a well defined level of confidence (the requirement being that system's potential failures will occur more or less likely: within a specified range of probability)"*. The Standards or Guidelines do say, of course, that a certain set of practices is required when developing the system for applications with, say, a required probability of failure on demand lower than $10^{-6}$; they do not (and of course, cannot) say that a certificate that those practices were used guarantees that probability. If, in some sense, the certificate supports a belief in that being the maximum probability of failure on demand, it supports a very low *level of confidence* for the estimate of that bound.

Quantitative reliability analysis might be, thus, mandatory, as pointed out by national and international guidelines and Norms (e.g. ISA, IEC Drafts). Moreover, complexity of hardware and the embedded software shows that major difficulties encoutered in performance reliability metrics consist in the estimation of the *coverage* of the implemented procedures, in terms of qualitative modelling and subsequent quantitative figure estimation for a credible assessment.

Infact, the major sources of errors and uncertainties in the procedure coverage might be summarized as:

– system modelling for a specified Integrity Level class (hardware and software: Operating System),

– faults and failures causes deriving from physical factor, as electrical, thermal mechanical, etc.. and their common modes, which cannot be modelled exhaustively with single simulators and single reliabilistic procedures or tools; loss or missing of informations concerning the design data and system functional behaviour, leading to an erroneous or uncomplete reliabilistic system model.

Furthermore quantitative certification of the safety of software is uncommon, and demonstrating very high levels of safety is currently impossible [LAP91, STR92].

This is due to two main difficulties: the need for a prohibitively expensive amount of testing, and the (application–caused) complexity of the designs employed.

The last point is certainly critical and is one of the major causes of coverage uncertainty in the metrics estimation.

There are, however, many simple applications, with comparatively modest requirements of safety (e.g. many "safety" or "interlock" applications), where those two obstacles are absent.

A pragmatic approach to carry out Reliability evaluation and validation for computer based–based safety systems for critical application has then been undertaken under the Italian Ministery of Healt: "Istituto Superiore per la Prevenzione e Sicurezza del Lavoro" (ISPESL) on late 1993. It has been set up a specific and multidisciplinary Working Group inside the "General Committee for Equipment and Electric Plants specification, validation and testing".

## 3 ISPESL Working Group Organization and Activities

ISPESL Working Group (Sub–Commitee of the General Committee), in connection with national (CEI: Comitato Elettrotecnico Italiano) and international Organizations (e.g. EWICS), aims to provide guidance (Guidelines) for assessing objectively and pragmatically (from an operative point of view) PEC performability quantitative (*) indices.

Reliability metrics quantification methodologies and procedures are developing for PECs Safety– related, which apply concepts and criteria based on the Life–cycle and Safety

Integrity Assessment; both necessary for identifying and quantifying likelihood (probabilities) of failure occurrences and associated errors (ranges) of specified functional system performances indices.

Numerical targets (ranges) shall be achieved in order to fulfil the SIL (Safety Integrity Levels) requirements (IEC/TC65A, WG10, ISA–ds0.1 SP84 Drafts).

Procedures and methodologies, therefore, shall apply only for the specific integrity Class. The selected Class is 2 as Class 2 complies well with experienced electromechanical

(*) where reasonably achievable

protective systems (interlock) reliability (FdT: Fractional Dead Time: $10^{-2} - 10^{-3}$, as failure probability on demand). Work started with the above assumption on SIL in order to assess methodology and lay down basic operative keystones for further guidelines developments for higher SIL's. Furthermore the choice supports the feasibility to verify practical installation for validating the methodology.

Particular attention is paied for the Human factor[4], as the main cause of errors affecting the whole process of quantification. Working Group organization covers:

– Manufacturers (national and international),
– Final Users (Chemical, Petrochemical, Refinery, National Electric Generating
  Board, National RailRoad)
– Certification Laboratories,
– Universities,
– National Research Council.

All Parties provide each own contribution to set out and verify the operative Documents released. Work development – shared in subgroups, thus identifies several Steps being satisfied. At present:

I°– Full review of relevant national and international Norms, Codes and Practices for PECs safety–related in order to assess key points (system identification and performances, appropriate dependability metrics as well, etc.),

II°– vocabulary and glossary review for identifying and lay down uniquely reference definitions and still missing,

III°– guideline framework for hardware metrics (operating system software is taken as perfect: apriori assumption for further development of the full procedure) assessment, according (possibly) with Manufacturer's internal procedures and the state of the art in reliability analysis techniques and methodologies,

IV°– benchmark exercitation for a test case system (reference), based on the procedure being assessed and for further adjustment.

Even if procedures are presently restricted to the Low Integrity Levels and only for the hardware, a position is highlighted for the

operating system embedded reliability quantification, work which is still in progress (IInd Step of Work).

## 4. Quantitative Hardware Reliability Analysis

Quantitative reliability analysis plays the relevant task to be performed on the above assumptions.

Reliability analysis shall apply to specified system performances in order to define proper functional reliability indices (Performability).

All phases require:

[4] Human aspects of reliability (IEC/TC56, WG12)

- a suitable procedure for the spefied SIL,
- information correctness
- modeling system exaustively
- coverage quantification

to be carried out according to the Life Cycle approach.

## 4.1 Procedure

The Life Cycle approach deserves a logical path toward a specified target and it is so conceived to riduce, or to control within well identified ranges, the potential errors associated to each phase of the quantification process.

Essential definition relates with: System Failure

According to the Definition [IEC94] "System Failure" is defined as such an event, if and only if a specified Manufactuer's functional characteristics shall not comply with own Reliability Design figure (range).

A System *derated* might be, therefore, a system well complying with Manufacturer's specifications, if its performances are within specified accepted limits.

PEC Reliability is thus strictly related to the Manufacturer's Specifications and not with the User Requisitions.

We shall, thus, refer only to *intrinsic system* probabilistic behaviour to failure occurrencies; therefore it will not be taken into account Operator intervention on the System for either substituting failed components (subsystems) or testing (periodically), or external automatical reconfiguration for failure recovery.Errors must comply with the selected SIL (Class 2); errors ranges shall be justified.

The appropriate System Performance Reliability index (or parameter) might be identified by the **MTTF** (Mean Time To Failure).

**MTTF** shall be associated to a specific performance functions (e.g. total transfer time).

## 4.2 Battery Limits

Definite PEC Battery Limits – hardware and software – will be stated, in order to completely and uniquely, associate reliability figures.

Referring to Hardware reliability Valuation and Verification, it will be identified Hardware and Software Battery Limits as reported in IEC–1069.1 Norm and specifically:

– **Hardware**: maximum hardware configuration input and output modules number, full documentation according IEC–1131, part 1 & 2 and Manufacturer's full design data – **Software**: full characterization of Operating system version.

## 4.3 Failure Data

Components (or subsystems) failure rates will be taken as **constant.**

Data Bank will be mentioned if available (e.g. MIL–HNDK –217/F); otherwise Manufacturer will apply either own internal Failure Data or other else documented.

In absence of such Failure Data, Manufacturer will apply Data extracted from an accepted (national or International) Data Bank; Data Source will be mentioned.

In case of components with unknown or impredictable reliability data (e.g. MIL and/or other references) accelerated life tests will be carried out, or equivalent where certifiable.
Failure Data Parameter will be calculated according to the prescribed Manufacturers operative conditions (*).
Assumptions shall be made on components (and/or subsystems) Failure Rate behaviour.

## 4.4 Mission Time: T

Mission time – T –, defines the period of time through the reliability analysis should be carried out (without any Operator intervention) under specified system boundary limits and environment (**)
Mission time will be continuous; the system (PEC) will be taken as continuously in operation. An appropriate Applicative Software Profile will be identified, in order to excitate extensively the Operating Software and Hardware. Reliability Analytical and Simulations Models should be certified and execution errors be demonstrated (range).Failure Rate should be certified and errors be demonstrated (range).

## 4.5 Assumptions

Operating system will be considered as **perfect**; thus no Faults (either systematic or random) and their evolution, as well as induced from hardware Faults, will be taken into account.

# 5 System performances evaluation through suitable tool and classification range.

## 5.1 Functional representation for reliability modeling

System representation will be set up on the basis of the performance and target required.
For each defined SIL, an *Equivalence Class* of procedures and models will be identified.
A defined procedure will be followed in order to achieve the stated Safety Integrity Level Class.
Functional reliability analysis and structural reliability analysis will be carried out.
The former, in order to identify the functional relationships leading to system performance parameter; the latter, to recognize the failure mechanisms of each hardware blocks organizing the functions previously identified.
The increasing complexity of programmable systems, allow on one hand realization of complex functionalities with a large level of integration and low production costs, brought on the other side some difficulties in the management of designs evolutions and assessment, one of the major problematics beeing the loss of observability of subsystems behaviour, resumed commonly in the literature as the problematic of embedded systems.
From the design point of view, a great enhancement to correct functionalities
implementation was introduced with thw methodologies of **Structured Design and Analysis** [YOU79, ABB, ESA].
(*) Refer to IEC–1131, part 1 and 2.
(**) Refer to IEC–1131, part 2 and IEC–1069.1.

The benefits of the use of such techniques during implementation can be resumed as:

– Optimization of the number and of the complexity of subfunctions

– Ease of modularization during architectural design with optimal allocation of functionalities to physical subcomponents

– Good intrinsecal observability with optimal coverage of testing procedures versus expense (number of parameters to be tested)

– Possibility of programmed maintenance, and optimal failure localization for corrective maintenance.

## 5.2 Hardware and software modelling compatibility

Beeing the reliability of the system strictly correlated to the quality of the project management, basically resumed through the previously described objectives, significative enhancements with a structured approach can be reached:

– Real improvement of system reliability requires optimal reliability data allocation to subfunctions when designing system architecture, and control during development of partial reliable objectives thank functional hierarchisation

– Minimization of uncertainity in the estimation, permitting a previsional coverage of system behaviour in degraded mode

– Minimization of the number of components life test (Black Box Models) or Integration Reliability tests (White Box Models).

The two last points are particular efficient when dependability is estimated using a real "Life Cycle" approach; i.s. defining a dependability analisys strategy from the design to the industrialization phase with an interactive process evolving with the development, based conceptually on:

– A Top–Down / Bottom–Up approach and inherent methodologies, the qualitative and quantitative results of the dependability analysis leading to a subsequent revision of the functional description

–A convergence of reliability data prevision to measured and estimated data within the specified uncertainity class, resuming the validity of the procedure

Accordingly with Quality Management glossary, the Life Cycle approach integrated to Structured Functional Analysis and Design should represent a real "Total Dependability Management" [LIG89].

## 5.3 Applicability for quantitative reliability indices estimation [ASA]

Several efficient and standardized methods are available, as Data flow Diagrams or Structured Design analysis Technique (SADT); the latter beeing more efficient for hardware electronics analysis.

The explosion of the system main function into a sufficient detail of subfunctions, mantains the integration coherency throughout the main informative parameters:

– Input functional parameters $I_i$,
– Constraints and bounds $C_i$,
– Hardware physical components $M_i$ affecting subfunction $F_i$,
– Output functional parameters $O_i = O_i (I_i, C_i, M_i)$

The exhaustivity of the functional relationships can be directly correlated to the analysis of uncertainty reduction:

– accurate functional path identification between blocks,
– correct hardware allocation to blocks.

Enhancement in the coverage are to be foreseen as in the simple example of two I/O redundant cards (Figure 1 and 2), where masked links during the design phase are put into evidence:

– unexpected common mode mechanical failure mechanism between A/D and D/A converters,

– undetected conditional failure mechanism from DC/DC 1 converters, to the DC/DC 2 due to overheating when DC/DC 1 fails,

– electrical failure propagation from buffer to bus in degraded mode.

### 5.4 Parameters errors and reliability quantification

Once defined a functional representation by the inherent hardware mapping, reliability models and metrics can be formally estimated from the overall system parameters, directly affecting loss of performance (Battery Limits). At this step, the following assumptions and information modelling have to been set up:

– Operative conditions and hardware mission profile
– Reliability targets in terms of ferformances failure
– correlation between functional parameters deviations and system fault events.
Detailed analysis at the required subfunctional level might be improved:

– a univoque relationship between system parameters deviation and the parameters of the subfunctions.
– formal subdivision of erroneous subfunctional output as distinct contributor of single

entities:
- deviation of input parameters
- contribution of derated constraints and bounds
- hardware derating affecting specific subfunction

If

$O_{ik}$ represents the k outputs of subfunction i

$I_{il}$ represents the l inputs

$M_{im}$ represents the m hardware allocated components

$C_{in}$ represents the n constraints

and $O'_{ik}$ the erroneous output, corresponding to system fialure, the estimated failure (performance) probability for each subfunction, can bet set as:

$P(O'_{ik}) = P(O'_{ik} [I'_{il}, C_{in}, M'_{im}])$, where k =1, $k_{max}$, number of output parameters and $I'_{il}$, $C_{in}$, and $M'_{im}$ are the deviations from rated state.

The system full failure mode functional modelling can be exploded using well reliability assessed techniques (Fault–Tree, Petri Net, complementary with the design analysis).

Particular care should be taken when using simulators combined with Fault injection methodologies, as the physical state of faulty hardware has changed and the electronic models cannot be valid anymore.

Additionally, fault injection can mask some internal failure mechanisms, as failure propagation or common modes.

In some cases, single direct relationships between effects and causes are not so evident and deeper analysis might be carried out analytically or simulative.

Corruption of information due to the translation procedure from System Design documents (e.g.: Block Diagram, etc.) to Data Flow Diagram might be of concern.

- coverage

Coverage is essential prerequisite for achieving the specified Safety Integrity Level.

A Coverage figure must be identified for each procedure phase being developed and associated with to assess the target figure error range [DUG89].

In Figure 3 is reported the Main procedure for hardware performance reliability evaluation.

# 6 Procedure Validation & Verification

Validation is *essential prerequisite* of a "reliable" procedure.

Procedure reliability *must be demonstrated* for validation.

A twofold approach may be applied in order to tackle the problem, attempting to achieve the task.

A Benchmark exercitation shoud be carried out by reliability Practitioners which is based on the *procedure (and methodologies)* application on test case.

Procedure validation should be based on practical comparison with field data (failure statistics) derived from Manufacturers and Endusers proprietary records.

The task may be feasible, as the PEC selected woukd provide safety integrity level is 2, equivalent to electromechanical equipment reliability performances (relay–based).

The task may be feasible, as the PEC selected woukd provide safety integrity level is 2, equivalent to electromechanical equipment reliability performances (relay–based).

In appendix A are reported some Tools applied in the industry for design, valuation and validation of computer–based systems dependability.

# 7 Conclusion

Quantitative computer based safety–critical systems performability Valuation and Validation is an important objective to achieve for such products in an international open market.

National and international activities are running in this field whose industry feels unambigous assessment according to emergencies Norms and Standards.

Practical and theoretical difficulties still exist, particularly for ultrahigh operating systems reliability and efforts are spent also in national and international technical communities in order to define operative documents, which, even if in progress, attempt to approach and attain the final goal of PECs safety in an harmonized context. Under Ministery of Health, guidelines are in Progress (ISPESL) to assess dependability for PECs reliable performances, as final Users and Manufacturers may find effective support for safety critical applications.

# References

[ABB]        Abbaneo, Biondi, Ferrando, Mongardi "Testing of a computer based interlocking software: methodology and environment", SAFE COMP 92; 28–30–Oct–Zurigo

[ASA]        ASA "Functional modeling design and analysis"; VERI LOG, Paris, 1991

[DUG89]        Dugan J. B., Trivedi K. S.,"Coverage modeling for dependability analysis of fault–tolerant systems" IEEE Transaction on Computers, vol. 38, no. 6, pp 775–787, 1989

[ESA]        European Space Agency, Nordwijk, PSS–05 Norms

[IEC94]        IEC/TC65A: System aspects – Draft 1508 (Oct. 1994)– Functional Safety: Safety–related Systems

[LAP91]        Laprie J. C., Littlewood B."Quantitative assessment of safety critical software". University of New Castle upon Tyne, 1991

[LIG89]        Ligeron J., C:, "Le management des grands contrats" Ed. Lavoisier, 1989, Paris

[STR92]        Strigini L."Considerations on current research issues in software safety"– 9th Conference on software safety – Luxemburg, 1992

232

[TAY94]    Taylor J. R. (1994b) *Developing Safety Cases for Command and Control Systems*. Technology and Assessment of Safety Critical Systems, F. Redmill and T. Anderson, eds. Springer Verlag, 1994.

[VDE89]    VDE V DIN 19250:"Grundlegende Sicherheitbetrachtungen fuer MSR-Schutzeinrichtungen", Beuth Verlag, Berlin (Januar 1989)

[YOU79]    Yourdon E., Constantine E. Structured design, Prentice Hall, 1979

# Appendix A

Some available Tools for dependability V & V.

**- SHARPE** (Symbolic Hierarchical Automated Reliability and Performance Evaluator) is a tool for analyzing hybrid, hierarchical models for a class of performance, dependability and combined performance/dependability (performability) models. It has been installed at over 140 sites. It provides a specification language and solution methods for thr following model types: series–parallel reliability block diagrams, fault trees, reliability graphs. Markov chains (acyclic, irreducible cyclic and cyclic with absorbing states), semi–Markov chains (acyclic and irreducible cyclic), series–parallel directed (acyclic) graphs, product-form(closed) queueing networks and generalized stochastic Petri nets.

**SURF–2** is a tool for dependability evaluation of hardware–and–software fault–tolerant systems. A graphical and interactive interface allows to describe the system behavior in the form of a generalized stochastic Petri net (GSPN) or in the form of a Markov chain Structural verifications of the GSPN model are learned out before deriving the associated Markov chain; the Markov chain is then solved analytically. The measures of interest include instantaneous and asymptotic measures, mean sojourn times, and reward measures(performance, cost,etc.).
Model management facilities enable the one–step comparison of the measures corresponding to different architectures. SURF–2 has been widely distributed both in industry and academia.

**UltraSAN** is a software package for model–based performance, dependability and performability evaluation of hierarchically represented systems. At the highest level, a system is a composed model and individual components are given as stochastic activity networks (SANs), a stochastic extension to Petri nets. Measures of interest are derived by defining reward structures at the SAN level. If one chooses to analytically solve the model, automatic state space lumping, using so–called reduced base model construction techniques makes it possible to solve very large models. Besides many transient and steady–state analytical methods, the UltraSAN includes terminating and steady–state simulators as well as importance sampling tool UltraSAN has found wide–spread use in industry and academia.

**TOMSPIN** (Tool for Modeling with Stochastic Petri Nets) uses generalized stochastic Petri nets (GSPNs) to get both performance and dependability results for evaluated systems. The evaluation of those GSPNs is done by transforming them into Markov chains and computing them analytically (more than three million–state models). TOMSPIN supports the steady–state analysis as well as the transient analysis of the model describing the evaluated systems.

**DEPEND** provides an object–oriented framework that allows the evaluation of highly dependable systems. The tool provides facilities to rapidly model fault–tolerant systems. It provides fault injection primitives to simulate realistic fault scenarios. New methods for time acceleration and hybrid simulation have been developed to simulate extended periods of time and the impact of faults on software Time acceleration consists of simulatintg time regions around faults in great detail and leaping forward to the next region. Hybrid simulation, a marriage of detailed functional simulation with analytical or simple Monte–Carlo models, permits evaluation of complex fault scenarios.

**MEFISTO** supports several techniques for injecting faults into VHDL models. Either specific components (called saboteurs, can be inserted in the VHDL model, or some of its components or processes can be mutated. The tool can also use the command language of the underlying simulation engine to modify the values of variables and signals of the target model. MEFISTO is useful to investigate different mechanisms for mapping errors from one level of abstraction to another, and to validate fault and error models applied during fault injection experiments carried out on the implementation of a fault–tolerant system (e.g., software–implemented or pin–level fault injection).

**SOFIT** (Software Object–oriented Fault injection Tool) is intending for validating software under hardware faults by way of fault injection by establishing a link between the behavior of the program and the fault occurrence process. Faults can be injected in memory, registers, data bus and address bus. Fault–types include set–bit, reser–bit, toggle–bit, set–byte, reset–byte and toggle–byte. The duration of faults can be either one–cycle or N–cycle. The object–oriented approach facilitates the test of any other system component, or the incorporation of any other fault–type. In the analyses, emphasis is put on derivation of experimental measures of the coverage provided by software fault tolerance mechanisms.

**D_RAMP** is an integrated environment and an underlying design methodology which supports the design and quantitative (rather than purely qualitative) analysis of dependable, real–time systems. The design process is supported from initial concept to final implementation. Initial concepts are in the form of blocks diagrams and flow diagrams, which are easily developed by the designer. Final implementation consists of printed circuit boards, integrated circuits, workstations, and other hardware elements, as well as the actual software. The environment supports the concurrent design and analysis of hardware and software throughout the design process. A key component of D_RAMP is the integration of the performance and dependability aspects of a system design into a single model representation. Here dependability encompasses the metrics of reliability, availability, maintainability. Performability, refers to the system performance or throughput. Thus, the D_RAMP tool is capable of performing analysis in two general important areas: dependability and performability.

234

DyQNTool is a Tool for performability modelling based on the dynamic queuing network concept; an approach in which queuing network analysis and generalized stochastic Petri nets are combined. With this Tool dynamic queuing network models can be easily constructed. The Tool takes care of the automatic translation of the dynamic queuing network model to a Markov reward model.

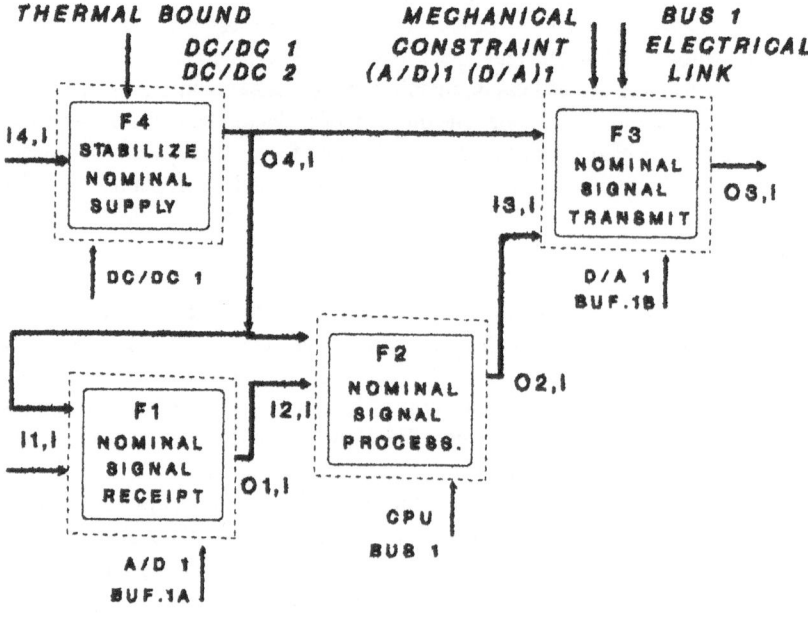

**Figure 1**

235

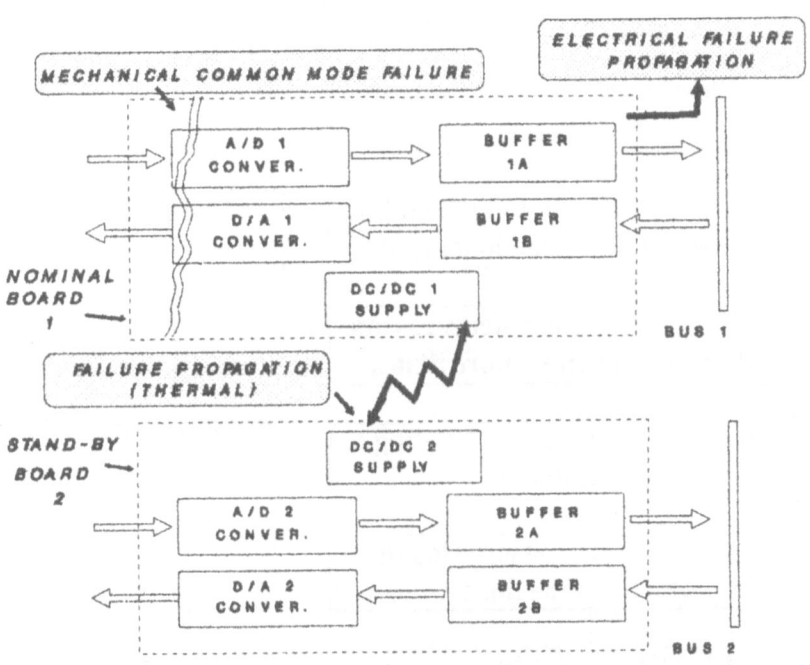

**Figure 2**

## PROCEDURE FLOW CHART HARDWARE PERFORMABILITY ANALYSIS

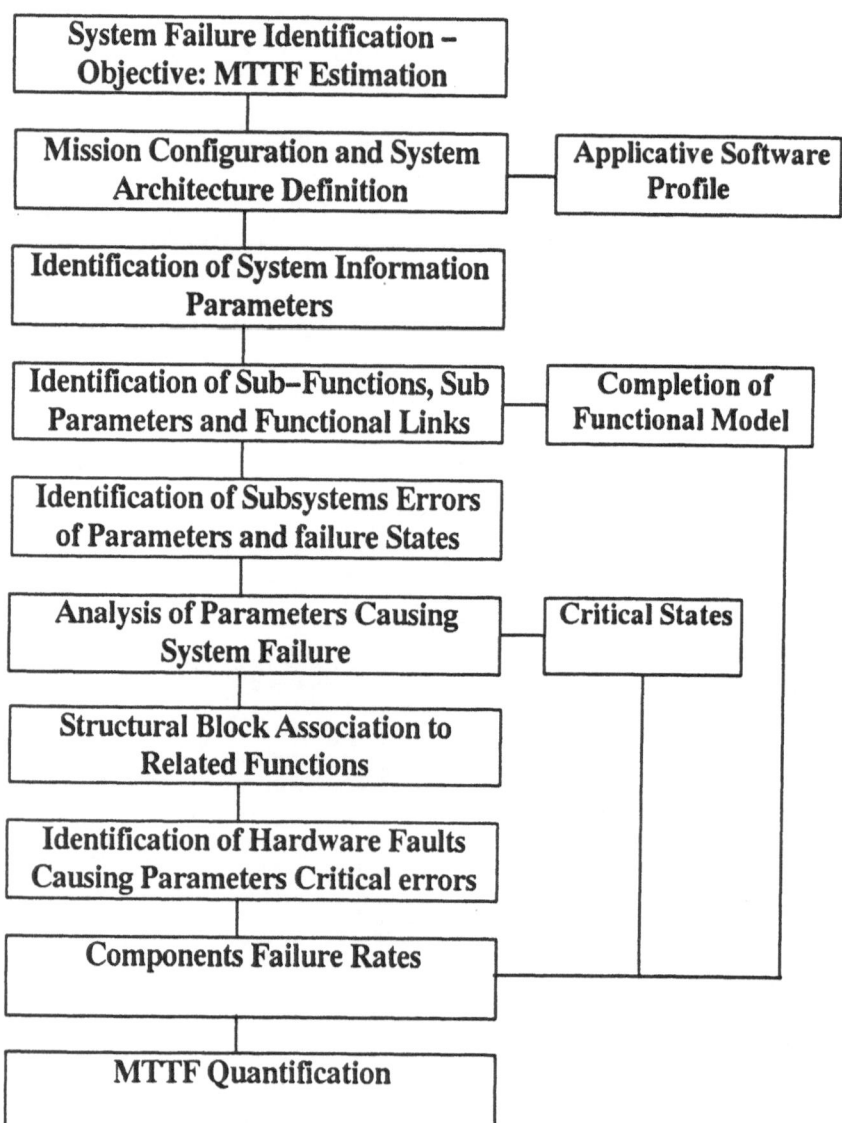

Figure 3

# BOOTSTRAP: SOFTWARE PROCESS ASSESSMENT EXPERIENCES AND FURTHER DEVELOPMENTS

F. Engelmann [‡] , W. Schynoll [§], H. Stienen [‡]

[‡] SYNSPACE AG, Binningen, CH, [§] SWE, Bietigheim, D

### Abstract

*After a brief description of the BOOTSTRAP-method, its main experiences are reported. BOOTSTRAP proved to be a very efficient and effective means not only to assess a current status of software process quality but also to initiate appropriate improvement actions.*

*The second part deals with future developments of the method. In 1993, a joint ISO/IEC working group established a project called SPICE, Software Process Improvement and Capability Evaluation. The BOOTSTRAP Institute, which brings together all parties interested in the dissemination and strengthening of the BOOTSTRAP method, is currently working on a ISO/SPICE compliant version.*

*The last part is devoted to the development of a variant of the BOOTSTRAP process assessment method that is especially suited for safety-critical software systems. Specific requirements for these kinds of highly dependable software are outlined. Also, an approach to process verification and validation is proposed.*

## 1 INTRODUCTION

Between Oct. '90 and Feb. '93, a software process assessment and improvement method - BOOTSTRAP - has been developed in an ESPRIT project by a multi-national consortium of industrial companies and research institutes across Europe [Kuva93]. BOOTSTRAP evolved on the basis of the CMM, the ISO 9000 standard and the European Space Agency's Software Engineering Standards [BSSC93].

The resulting assessment and improvement method is characterized by the following points :

- ♦ It provides detailed maturity levels and capability profiles of organizations and projects.
- ♦ It allows optionally to determine the degree of fulfillment of the 20 elements of the ISO 9001 Standard.
- ♦ It also allows as an addition to calculate the maturity level according to the SEI method.
- ♦ It allows benchmarking, i.e. the comparison with specific industrial sectors, countries and other assessment data stored in the BOOTSTRAP database.
- ♦ It includes immediate feedback from the organization assessed.

# 2 SOFTWARE PROCESS ASSESSMENT METHODS

A software *product* is always the result of a process. It is therefore essential to consider the software development *process* carefully in order to prevent errors in the product and to *produce quality* instead of attempting to achieve it by (late) testing and verification activities [Stev91, BSSC93].

With BOOTSTRAP, an organization and its process quality is assessed with respect to organization (O), methodology (M) and technology (T), as shown in Figure 1. The underlying concept of BOOTSTRAP is that an organization has to satisfy basic organizational requirements first, and then to take care of engineering methods before making investments in technology as third priority [Koch93].

Rather than to perform a plain assessment of the current status, the main objective of a BOOTSTRAP assessment remains to generate a improvement action plan.

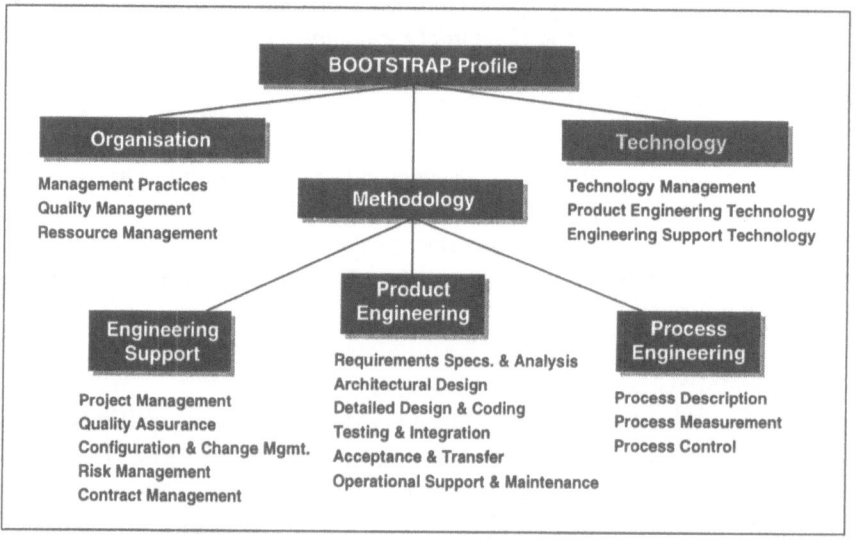

**Figure 1: Overview of the BOOTSTRAP Attributes**

*Improving software process quality requires to measure it.* As most approaches that aim at improving the capabilities of an organization, BOOTSTRAP uses the concept of stepwise improvement. BOOTSTRAP recognizes that within an organization, there may exist several obstacles to overcome. Transitions should take place step by step.

Typical steps that require organizational or cultural changes, are:

◆ Software development must be subdivided in distinct phases.

◆ Quality assurance and testing are essential parts of software development.

◆ Standard procedures and methods are required for managing large projects.

◆ Activity based accounting and technical reviews are the basis for reliable estimation, planning and tracking.

◆ Quality is not the same as perfect; quality can and must be defined.

In 1986, the Software Engineering Institute (SEI) at Carnegie Mellon University developed under the direction of W. Humphrey with substantial financial support of the U.S. Department of Defense (DoD) a process quality framework [Hump89]. This framework, later referred to as the *Capability Maturity Model* (CMM), has become the basis for software process assessment and improvement. Since the first publication of CMM, process assessments have found a major interest in software industry.

SEI has identified 5 major levels. According to CMM a software development organization evolves from an "initial" quality level to an "optimizing" quality level. Figure 2 shows the five CMM levels.

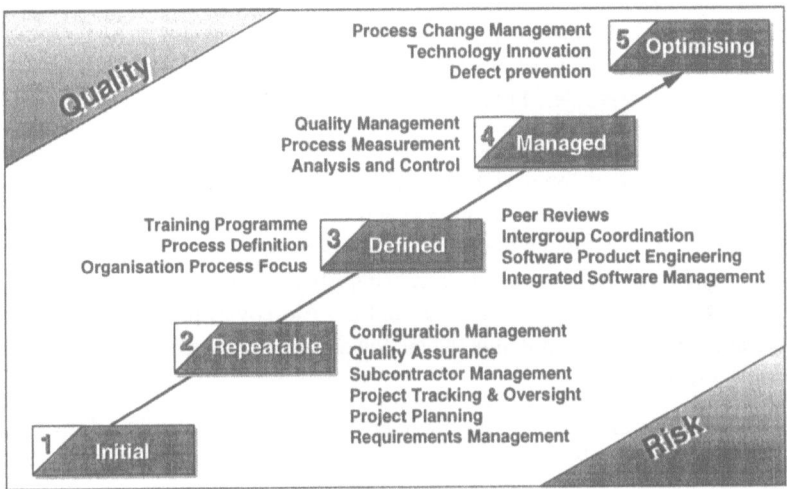

**Figure 2: The 5 CMM Levels.**

CMM provides an organization a roadmap to a mature, stable and disciplined way of software development. For each level a set of key process areas is defined that an organization must be able to perform, before the next level can be approached. Consequently, to each of these process areas a level on a scale ranging from 1 to 5 is assigned. Also, criteria are defined to assess the capability of an organization to perform each key practice. The SEI assessment method which is based on CMM encompasses currently 18 key process areas.

The SEI method is primarily focused on large organizations that develop software for the US government or affiliated agencies. Its main goal is to provide an instrument for selecting the most appropriate contractor for a given task and to encourage a uniform way of developing software within the software supply industry. On the other side, BOOTSTRAP acknowledges the size and the cultural differences across organizations and countries, which are typical for Europe.

The main goals of BOOTSTRAP are:

♦ To demonstrate how management of process quality can be used as a means to produce superior product quality and user satisfaction.

♦ To introduce the principles of Total Quality Management in software developing organizations.

The central objective of SEI is the alignment of individual projects within a common framework, irrespective of their technical content. Therefore, the establishment of an organizational structure and responsibility for defining, measuring, controlling and refining this common framework for individual software development projects is a major concern. All product engineering activities form together only one of 18 key process areas within the SEI scheme. Furthermore, almost all activities on level 3 concentrate on the establishment of an overall software development framework. Moreover, quality management is addressed explicitly on level 4 where systematic measurement is introduced.

SEI uses a kind of benchmarking approach. For each of the identified key process areas SEI is probing the main characteristics of what is supposed to be best practice in that area. Practices are verified by collecting formal evidence from those process features that are observable. From these facts the conclusion is drawn whether the associated process exists or not.

A major problem encountered with the SEI approach is that outside of the culture that provided the yard sticks, deviations may be so large that none of the characteristics may be observed, although the organization performs evidently good in that process area. CMM may be often a too strong simplification, as this model assumes that only one way to perform a given process exists. Moreover, probing a key practice on a certain level is only possible when lower level processes are performed in accordance with CMM.

At the time CMM was proposed not many empirical data were available to validate the underlying assumptions. Meanwhile, several slightly different assessments methods emerged, many assessments are performed now, and there exists ample literature from organizations reporting assessment results.

## 3 MAIN FEATURES OF BOOTSTRAP

Already as part of the ESPRIT project, BOOTSTRAP has been used intensively in order to collect feedback from a large variety of organizations, and to make the method robust and reliable. The assessment experiences made at Robert Bosch GmbH, a large manufacturer of automotive and telecommunications equipment, were very valuable and have been reported elsewhere [Schy93]. Also many small and medium-sized organizations in various industrial sectors across Europe were investigated. It turns out that BOOTSTRAP is a *suitable assessment method for all sizes of software development organizations*. Moreover, the fact that the method does not assume a strict adherence to a distinct key practice model like CMM, but allows for alternative development approaches, is a key factor to the success of BOOTSTRAP.

Mean features of BOOTSTRAP are:

♦ Uniform procedure, mandatory assessor qualification and training scheme
♦ Carefully designed questionnaires for both the site and the project assessments
♦ Analytical, not normative approach
♦ Open questions, multivalued scores
♦ Immediate feedback and action planning

A typical schedule of a full assessment encompasses several briefings and debriefing activities. Main objective of the briefings is to inform about the BOOTSTRAP procedure, to agree on the objectives of the assessment, to select representative projects, and to obtain commitment for improvement actions afterwards from both senior and line management as well as engineering staff. During assessment preparation a pre-assessment questionnaire is distributed in order to collect that kind of mainly statistical information concerning the organization that is usually not readily available.

Major phases of the BOOTSTRAP process are:

- General management briefing
- Assessment preparation
- Line management briefing
- On-site assessment week
- Report preparation and delivery
- Final presentation

The core activity of a BOOTSTRAP assessment is the on-site week. At the beginning, all participating staff members are informed about the method in detail. After that, the interviews are conducted. The interviews concentrate on the key aspects of the software development process. During the interviews as much level of detail is addressed as is needed to determine those key areas that may be candidates for further improvement.

Using computer-supported tools, a preliminary evaluation is also being done on-site, and its results are presented and discussed at the end of the week. Confidentiality is an important aspect here. Results of a single project are presented only to the project representatives during the individual project debriefings. Beyond this, only average figures are presented and discussed.

Action planning starts at the end of the assessment week with the presentation of the preliminary results and includes a feedback loop with the organization in order to ensure acceptance of the results. A BOOTSTRAP assessment is formally finished with the delivery of the assessment report and a final presentation, typically twelve weeks after the kick-off meeting.

Organizations experience BOOTSTRAP as very motivating and as a real support in order *to make improvements really happen* in contrast to other approaches which result very often only in an increased amount of paperwork. BOOTSTRAP stimulates an organization to evaluate and subsequently to improve its own working practices.

Furthermore, most organizations appreciate that BOOTSTRAP reduces *initial costs to a minimum* required to establish a solid baseline for necessary improvement actions and corresponding investments.

For some organizations, the BOOTSTRAP assessment was the starting point of an ISO 9000 certification process. But also organizations already certified since several years were visited. Their motivation was to better define appropriate actions for further improvements.

# 4 THE BOOTSTRAP V2.3 HIGHLIGHTS

The fist version of the BOOTSTRAP Assessment Instruments (Questionnaires, Data Capture and Evaluation Tools) was delivered by the ESPRIT Project BOOTSTRAP in the beginning of 1993. Since then, SYNSPACE has conducted several dozen assessments using this version. In the mean time, the BOOTSTRAP Institute has been established. All founding members of the BOOTSTRAP Institute participated in the original ESPRIT Project. One of the major objectives of the BOOTSTRAP Institute is to promote and strengthen the use of the method.

Based on the experiences of the past 2 or 3 years an improved version of the Assessment Instruments has been prepared. This Version 2.3 is ready for official release now. With the advent of the new release of the BOOTSTRAP method and its related products, it is time now to present and to evaluate typical strengths and weaknesses of the approach. Also a first survey of the assessment results obtained in the last 3 years is presented here.

Future enhancements and extensions are discussed, with special emphasis on the ongoing work within SPICE. Attention is also paid to the development of reliable, safety-critical software and systems.

## 4.1 The BOOTSTRAP Procedure

BOOTSTRAP uses a cooperative approach in order to characterize and analyze an organization or project. The perspectives of management and engineering staff are equally important. For this purpose two questionnaires are used that are nearly symmetrical and cover all major processes which may be observed within a software development organization. BOOTSTRAP uses open questions to assess the actual situation and to solicit opinions of the participants in those areas that are assumed to be key factors for the improvement of software development practices. Depending on the scope of the assessment between 110 and 150 questions are discussed. An interview takes between 3 and 6 hours of time.

With each question a number is assigned, that indicates the capability level on which this question is particularly important, and also a single attribute or list of attributes, that provides reference to one or more process areas. Evaluation is done along these two axes. The first one provides the capability level. The result is one single number. This figure is useful to track progress over time, but it is not very helpful for the identification of areas that are likely candidates for improvement. For that purpose, several bar charts are provided allowing an in-depth analysis of the results.

The first chart, which is shown here, gives the overall scores for the individual levels. The results of one particular organization are compared here with the overall scores of all organizations with similar profiles.

It is frequently observed that within an organization attention is paid to all processes, irrespective of the level. As a matter of fact, better results are usually found on the lower levels, but to expect that all effort is concentrated on level 2 practices only is too rigid. Typically, the overall rating for the next higher level is about half as high as for the previous one.

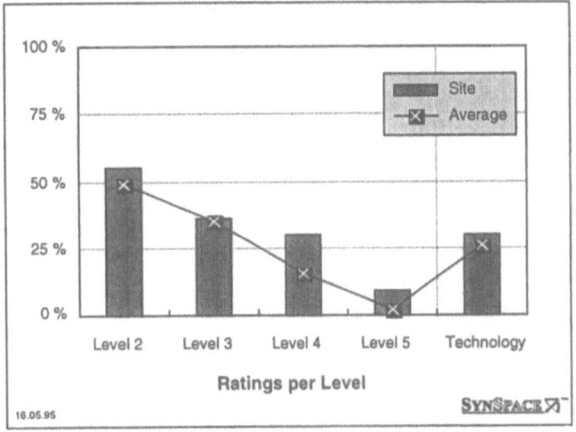

Figure 3: Ratings per capability level for a typical site compared with average results.

In several cases it happened that level 2 ratings were significantly below average, or level 4 ratings were far above average, not following the general rule. This may occur, when an organization has already identified those processes that are assumed to be most important, deliberately neglecting some of the basic processes.

Experience has shown that there may exist good reasons for an organization to emphasize some processes that belong to the higher levels or to skip one of the lower level processes. Typical examples are on one hand processes that address quality, i.e. quality management and quality assurance, and on the other hand some of the product design processes. This is often the case in software development units that produce systems of different types. In this case, the responsibility to define the individual engineering processes is delegated to each of software engineering groups within such a unit. If during the assessment it was found that this approach was valid, questions belonging to these aspects were set to 'not applicable'. By doing so, the criteria associated with these questions were excluded from the computation of the overall BOOTSTRAP level.

## 4.2 Rating and Scoring

BOOTSTRAP associates with each question a four point scale. Values usually range from basic to excellent, but values can vary independently for each question. By using this feature, for specific questions the range may be specified from defined, partially compliant to fully compliant. This is particularly useful in areas where adherence of an organization or project to specific standards is a requirement. Notice that for each question, there is always the possibility, to decide that a particular process is 'completely absent, but required' or 'not-applicable', although this decision may be sometimes very difficult.

The use of computer based tools allows specific analyses, evaluations and presentations of the results. An typical example of such an analysis is given below. About

40% of all relevant software development processes are completely absent in an average organization. On the other hand, only 10 % of all existing processes is performed in an optimal way.

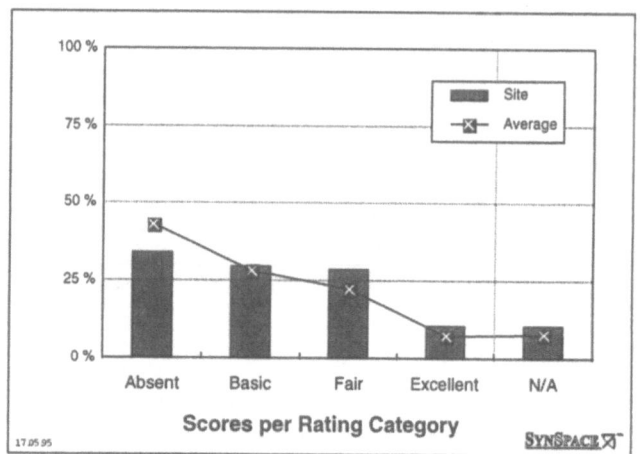

**Figure 4: Distribution of the Scores per Rating Category.**
In this example, about 30% of the processes are completely absent. Only 10% satisfies all criteria. Not-applicability (N/A) is an indication of how well the BOOTSTRAP model fits for a given organization.

A unique feature of BOOTSTRAP is that in each individual case it must be determined whether a particular question is relevant in that specific context. When it is decided that the process or the activity that is addressed by a given question is not relevant, that question is excluded by setting the score to 'not-applicable'. As illustrated by Figure 4, it is not unusual to find that about 10 % of the generic processes that are identified by BOOTSTRAP, may be not relevant for a particular organization.

An example of an even more detailed analysis is given below in Figure 5.

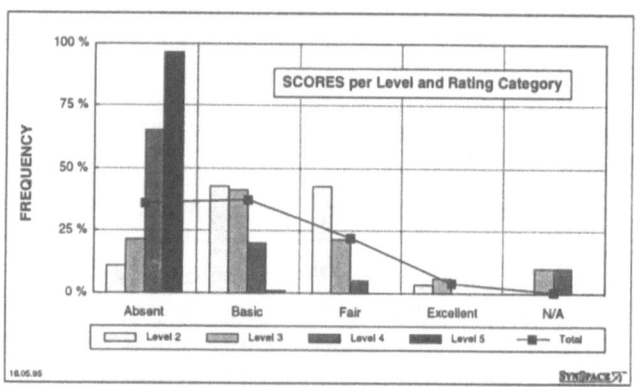

**Figure 5: Frequency of the various possible scores per level.**
On Level 2, processes that are completely absent, are rarely observed.
No Level 5 processes are found in this case.

This chart presents a typical distribution of the scores over the various levels and rating categories. As it may be expected, questions on the lower levels were answered more often positively than questions on the higher levels.

In the case presented here, almost none of the level 5 criteria got a positive score. A bit more surprising is here the fact that excellent scores were hardly found. This implies that for each level the total rating can not be higher than 75%. However, this does not mean that the overall rating of an organization can not be higher than 1.75, as BOOTSTRAP takes also into account the individual ratings on the other levels.

## 4.3 Tool Support

Although printed questionnaires are used during the interviews, an assessment relies heavily on tool support.

BOOTSTRAP uses an integrated set of computer based tools for:

- Data capture
- Evaluation and presentation
- Comparative analysis and benchmarking

Questionnaires are tailored in advance to each individual case in order to reflect the scope of the particular assessment. For that purpose, questions are grouped into several classes. For example, there exist specific questions addressing ISO 9000 compliance. However, to ensure comparability between various software development organizations overall rating and scoring is based on a common set of core questions which are always considered. Analysis and presentation of the results require a fine tuning of the procedures to the needs of the client organization, which is virtually impossible without proper tool support.

## 5 CHARACTERIZATION OF THE ASSESSMENTS

During the last two and a half year SYNSPACE has visited more than 30 sites in the US and Europe. About half of them is located in Germany and Switzerland.

**Figure 6: Size distribution of the software development organizations visited.**

A short survey of the first 25 Software Development Units, which belong to 13 different organizations, is presented here. The number of staff members ranged from 5 to 850 with an average value of 140 per site. Development units in the industrial sector tend to be smaller than in the financial, administrative and logistics area, i.e. about 80 viz. 220 staff members. The total number of software development projects assessed amounts to 120. An overview of the software development unit size distribution is given in Figure 6.

## 5.1 Motivation for Performing a BOOTSTRAP Assessment

The population of those organizations that has been assessed so far, may be not representative for the software industry in general. For example, our sample only includes one independent software house. All other software producing units are Software Engineering or IT departments of mostly larger organizations.

Nearly in all cases, the driving force behind the assessment came either from staff members or executives responsible for software development. Major reason was the feeling of leading persons that software development processes could be improved, but that it was unclear what kind of activities would be the most promising ones. For a decision, it was felt, an objective picture of the current state was needed.

Another important reason for performing an assessment was ISO 9000. BOOTSTRAP is able to large extent to decide whether a software development unit satisfies the criteria for an ISO 9001 certificate. Some of the assessed organizations had already an ISO 9000 certificate for some time, but the software development responsibles were not very happy about that, as in most cases the ISO 9000 audits did address only a limited number of issues that are relevant in software engineering practices.

Looking in detail at the reasons during the assessments, we observed that most organizations had serious problems in consistently delivering software of the expected quality. Users, software engineers and management as well were more or less frustrated. It was expected that a BOOTSTRAP assessment would demonstrate either that the current situation was not as good or bad as senior management or user departments may think, or that as a result of the BOOTSTRAP assessment resources and funding would be made available by management to improve the situation.

## 5.2 State of Practice in Software development

To get a first impression of the actual situation, the overall BOOTSTRAP level of an organization provides a useful measure. Not surprisingly, most organizations end up in the range between level 2 and 3. Only few organizations perform worse or better.

No significant tendency towards higher values over time can be observed. At least organizations visited for the first time fall frequently between level 2 and 3. This seems to be the maturity level on which an organization typically stays when they become convinced that an investment in a BOOTSTRAP assessment will pay off.

Not shown here, as the data are quite sparse at this moment, is progress made by organizations after the initial assessment. A delta assessment makes sense in a distance from one to one and a half year of the first assessment, so it is too early to expect that comparative data already exist. In those cases where data were available, it is estimated that ongoing improvement activities will lead to a higher rating of about 0.5 on the BOOTSTRAP scale within one year after the first assessment.

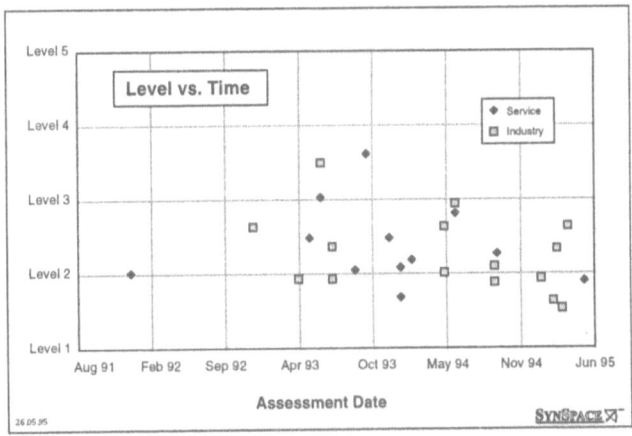

Figure 7: BOOTSTRAP Levels over Time.

At the time the BOOTSTRAP model was developed, it was expected that the level could be estimated with an accuracy of at most 0.25. The calculated uncertainty seems to be a very pessimistic estimate. As may be seen from Figure 7, the conclusion that one should use a quartile scale is incorrect. It is to prefer to report the computed value instead of rounding to the nearest quartile. As there are more than 4 major key practices per level, the method should be able to discriminate at least at that level of information.

Another interesting question is whether the results are dependent from the software development unit size. In Figure 8, the BOOTSTRAP level of each organization is plotted against the number of staff members involved in software development.

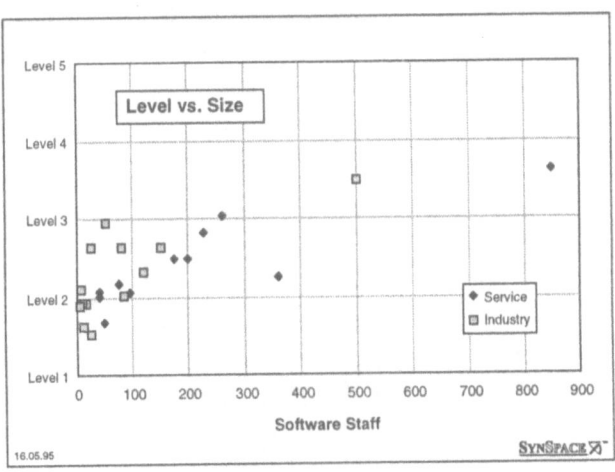

Figure 8: BOOTSTRAP Levels vs. Unit Size.
(Only sites of known size are represented)

A trend towards higher levels for larger organizations can be observed. One of the outliers at the larger sizes may be due to the fact that shortly before the assessment took place staff members from 3 different sites were merged into one single unit.

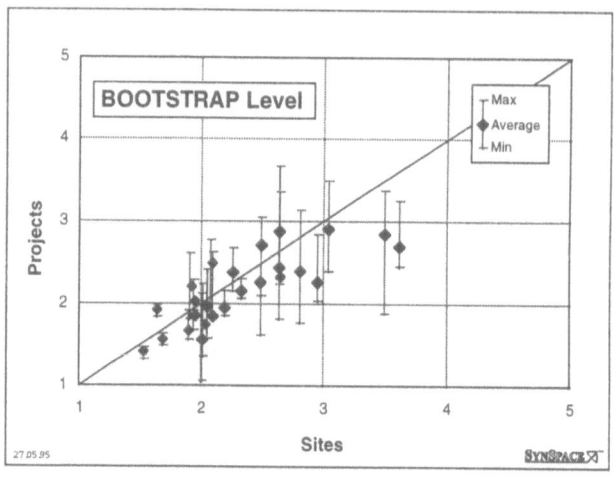

**Figure 9: Average Project Performance vs. Management View.**

However, it is too simplistic to conclude that it is easier for large organizations to reach higher BOOTSTRAP levels. Software engineers in the larger organizations were higher educated, had more years of practical experience and were working in the same area over a longer time. From an even more detailed analysis, it can be shown, that in general organizational stability, customer orientation and staff training seem to be the dominant factors for attaining the higher BOOTSTRAP levels.

Looking at individual projects within the various organizational units, as illustrated in Figure 9, several examples of excellent performing small engineering teams can be observed.

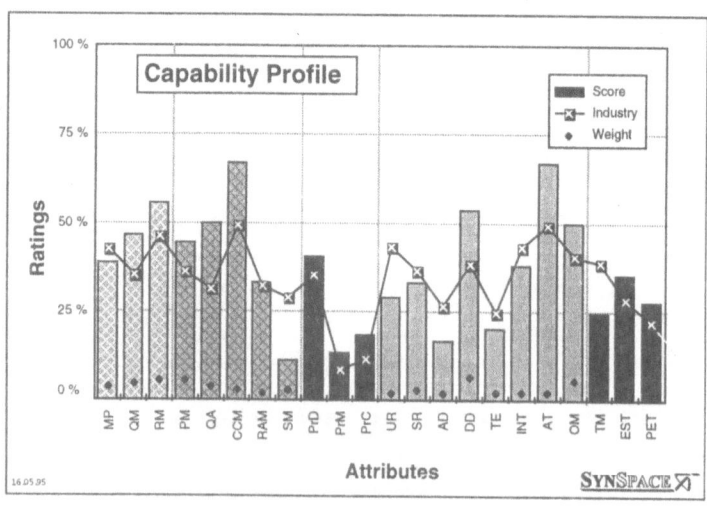

**Figure 10: Example of a BOOTSTRAP Attribute Profile.**

A closer analysis based on individual profiles like the one shown in Figure 10 revealed that most organizations had difficulties in areas, like process description, measurement and control as well as architectural design and unit testing.

Organizations developing safety-critical and highly reliable software systems were not an exception to this general observation. Weaknesses in these areas are compensated by both extensive integration and acceptance testing as well as rigorous quality assurance procedures.

# 6 FURTHER DEVELOPMENTS

## 6.1 SPICE

In 1993, a joint working group of ISO and IEC established a project called SPICE, Software Process Improvement and Capability Evaluation [Dorl93]. SPICE's Baseline Practices Guide defines a very detailed process framework. Software development activities are organized in 5 process categories, encompassing 35 base practices, each subdivided in 3 to 10 separate activities. For each of the practices 6 levels of generic practices are defined. Over 900 items are currently identified. Under these circumstances, it is not a straightforward task to develop an assessment method, as simple as BOOTSTRAP, taking into account all the details incorporated in the SPICE framework [Paul95]. One of the major more practical constraints of an assessment is that it should only consume a reasonable amount of time and effort.

A deeper analysis of the ISO/SPICE working documents reveals that, although the framework is much more detailed, the BOOTSTRAP approach can be made compliant. This will require a few additional attributes, some more questions, and an extension of the scoring mechanism and the evaluation scheme.

Until now, BOOTSTRAP uses a two-dimensional scoring mechanism. First, for each specific process it is determined whether the process exist, and if it exists, what the quality attributes are. In each process area, several processes on different maturity levels may exist.

## 6.2 Self-Assessments

A more generic evaluation approach is used in SYNQUEST, a self-assessment toolkit, developed by SYNSPACE in cooperation with the Technical University of Graz [Müll94].

SYNQUEST is based on the assumption that each process will exist in variants depending on the maturity level of that process within an organization. On the lowest level a process will merely exist, on the higher levels the same process will be defined formally, planned and tracked, the outcome will be documented and reviewed, the process itself and its result will be measured and controlled, and finally supported by appropriate tools. Each process, and also each process area, is therefore measured in several dimensions depending on what kind of evidence on a specific level might be expected. The result of such an evaluation is shown in Figure 11.

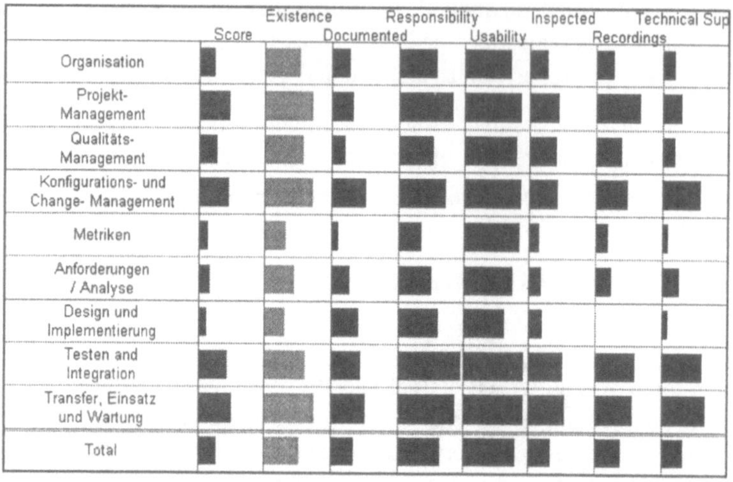

| | Score | Existence | | Responsibility Documented | | Inspected Usability | | Technical Sup Recordings | |
|---|---|---|---|---|---|---|---|---|---|
| Organisation | | | | | | | | | |
| Projekt-Management | | | | | | | | | |
| Qualitäts-Management | | | | | | | | | |
| Konfigurations- und Change- Management | | | | | | | | | |
| Metriken | | | | | | | | | |
| Anforderungen / Analyse | | | | | | | | | |
| Design und Implementierung | | | | | | | | | |
| Testen and Integration | | | | | | | | | |
| Transfer, Einsatz und Wartung | | | | | | | | | |
| Total | | | | | | | | | |

Figure 11: The SYNQUEST method of evaluating software process capability.

## 6.3 Assessment Methods for Safety-Critical Software Processes

BOOTSTRAP, and also CMM, claim to be suitable for any type of software development. As assessment methods become more and more detailed, this general objective are hardly to achieve. BOOTSTRAP allows for that reason to decide that a small number of processes are not applicable within an organization. ISO/SPICE envisages to provide a procedure for scoping the assessment to reflect the kind of application and the needs of a particular customer. As there are no experiences available at this moment, it is not clear whether this tailoring will work in practice or not.

Software development for the European Space Agency (ESA) is currently ruled by two sets of standards. The ESA PSS-05 Series defines in general terms the process requirements for the development of any kind of software product [Stev93]. PSS-05 describes the tasks performed by software development staff. Specific safety and reliability related procedures and methods are not covered. The ESA PSS-01 Series defines product assurance requirements. Product Assurance within ESA covers all reliability, availability, maintainability and security aspects of space systems as well as quality and configuration management. Within the PSS-01 Series there exist a high level standard for the application of software product assurance (ESA PSS-01-21) and also as draft two more detailed guidelines, one for software reliability and safety and one for software quality assurance. These documents do describe the tasks that must be performed by product assurance responsibles.

Using these both sets of standards it is extremely difficult for ESA to evaluate tenders, to select potential contractors and to control software development [Krie94]. This is a major risk, as within the European Space Agency most of the responsibilities are delegated to the individual projects. Product and Quality Assurance is under the responsibility of the contractor. For example, in contrast of what ISO 9000 requires, there is no independent reporting line with respect to quality issue. ESA is only allowed to audit organizations and to participate in reviews and progress meetings.

It is obvious, that in this situation there is an urgent need, firstly to device a uniform set of rules, that will be applicable for all processes that are relevant for the development of safety-critical software, and secondly to assure that the procedures defined will be known and applied by all software contractors. As it is extremely difficult to harmonize the existing sets of standards within a short period of time, a different approach is chosen. ESA recognized that an assessment is a perfect instrument to introduce a uniform, rigorous approach to the development of safety-critical software, that is not a burden and also cost effective.

Three companies, Det Norske Veritas, Intecs Sistemi and SYNSPACE is given a contract to examine all existing ESA standards which may influence the development of safety-critical software, to compile a technical note which describes how each project can establish a consistent framework for related software development practices, and to define a procedure for assessing the capabilities of contracting organization for developing safety-critical software. A first release of the assessment method is expected to be ready for use in 1996.

The assessment method will focus not only on the product assurance tasks, but also on the tasks that must be under control of product assurance.

The new method will incorporate many of the features of Bootstrap, but to be applicable for the in-depth assessment of a larger variety of software systems, additional requirements must be defined and further classified in order to reflect the different capabilities that are required for the various types of systems to be developed.

# 7 LITERATURE

[Kuva93]  P. Kuvaja, et. al., BOOTSTRAP: Europe's assessment method, IEEE Software, Vol. 10, Nr. 3 (1993) pp. 93-95

[BSSC93]  BSSC, ESA's Software Engineering Standards, The Foundation for Reliable Software, ESA Bulletin Nr. 96, 1993, pp. 97 - 104

[Stev91]  R. Stevens, Creating Software the Right Way, Byte, Aug. 1991, pp. 31 - 38

[Koch93]  G.R. Koch, Process Assessment: the BOOTSTRAP Approach, Information and Software Technology, Vol. 30, Nr. 6/7, 1993, pp. 397-403,

[Hump89]  W. Humphrey, Managing the Software Process, Addison-Wesley, 1989

[Schy93]  W. Schynoll, Experiences in the use of BOOTSTRAP, Unicom Seminar 9, Nov. 1993

[Dorl93]  A. Dorling, Software Process Improvement and Capability Determination, Software Quality Journal, Vol. 2, Nr. 3, 1993, pp. 209-224

[Paul95]  M. Paulk, M. Conrad, S. Garcia, CMM versus Spice architectures, IEEE Software Process Newsletter, Nr. 3, 1995, pp. 7 - 11

[Müll94]  G. Mülleitner, C. Steinmann, Minimum Metrics for Software Production, IIG Report No.395, Technical University Graz, Nov. 1994

[Stev93]  R. Stevens, A. Scheffer, Developing a Structured Set of Standards, 'IEEE Software Engineering Standards, 1993, pp. 19 - 24

[Krie94]  W. Kriedte, Y. El Gammal, A New Approach to European Space Standards, ESA Bulletin Nr. 81, 1994, pp. 38 - 43

# Analysis and Assessment of Advanced Road Transport Telematic Systems

Keith M Hobley and Peter H Jesty
Safety Critical Computing Group
School of Computer Studies
University of Leeds
LEEDS, UK

## Abstract

The hazard analysis of an Advanced road Transport Telematic system is an essential part of the safety life-cycle. A systematic methodology for performing this task has been produced by the project PASSPORT. For each of the two phases, preliminary safety analysis and detailed safety analysis, novel modelling techniques have been devised upon which to perform the hazard analysis. The assurance that these analyses give, can be used as part of a certification process which covers security, reliability and environmental issues, as well as safety.

# 1 Introduction

The objective of DRIVE is to improve road safety, transport efficiency and environmental quality. Such benefits are being promised through the use of telematic systems. It is, however, clear that systems sufficient to have the power to advise on, or to control, road safety, transport efficiency or environmental quality might exercise this power for either good or ill, and the latter condition could be quite disastrous in the worst case. Experience shows that, unless positive action is taken, then at some stage in the life of a system a failure to perform as expected will indeed occur. Random faults may occur for many reasons (e.g. component wear and communication breakdown), and systematic faults may be made in the software or in the overall design of the system. The undesirable effects of component wear, communication breakdown and software faults can be overcome by using a suitable design, but this is only effective if the design is correct. No responsible developer will deliberately produce an incorrect design, nevertheless design faults in both hardware and software do occur, and the reason for them can usually be traced to an incomplete understanding of the system due to its complexity. Integrated road transport telematic systems are usually complex.

In order to assure the functional system safety of an Advanced road Transport Telematic (ATT) system the development process must include a safety life-cycle that runs in parallel with the normal development life-cycle. The contents of this life-cycle were laid down during the DRIVE I project DRIVE Safety[DRIVE Safely 92], and the DRIVE project PASSPORT (V2057/8) has continued this work by producing frameworks for the Prospective System Safety Analysis (PSSA) [PASSPORT 95a] as well as for the Certification process[PASSPORT 95b].

# 2 Prospective System Safety Analysis

The framework for PSSA divides the safety analysis task into two separate phases:

1. A Preliminary Safety Analysis (PSA) is performed as part of the feasibility study when the system concept is being proposed. The objective is to discover whether there are any safety hazards associated with the system, and if so, to identify the top-level safety requirements and the safety integrity levels associated with them.

2. A Detailed Safety Analysis (DSA) is performed in parallel with the system design. The objective is to analyse the design for confirmation that all the safety requirements have been implemented to the required integrity level, including any new ones introduced as a result of the architecture chosen for the system.

Figure 1 shows the relationships of PSA and DSA with the main system life-cycle.

The methodology for both the PSA and the DSA follows the same basic pattern:

a. Produce a model of the system.

b. Check the model for completeness and consistency.

c. Undertake hazard analysis.

Whilst the techniques for performing task (c), e.g. Failure Mode and Effects Analysis (FMEA) and Fault Tree Analysis (FTA) are well known, we discovered that there were no suitable modelling techniques available upon which one could perform such a hazard analysis that could be easily validated by a third party. The modelling techniques that we devised are novel and are therefore given prominence in this paper.

# 3 Preliminary Safety Analysis

The aim of a PSA is to identify how the proposed system, or Target of Evaluation (TOE), interacts with its environment, and then to discover whether any of these interactions could result in a hazardous situation in the case of a failure of one or more parts of the TOE. The first task is therefore to produce a model that clearly shows the relationship between the TOE and its environment.

## 3.1 The Passport Diagram

A safety hazard will be caused by *what* the TOE might do to its environment, it is not necessary to know during this phase *how* it might happen. Indeed, since many design decisions will not have been taken, the 'how' may still be unknown. For

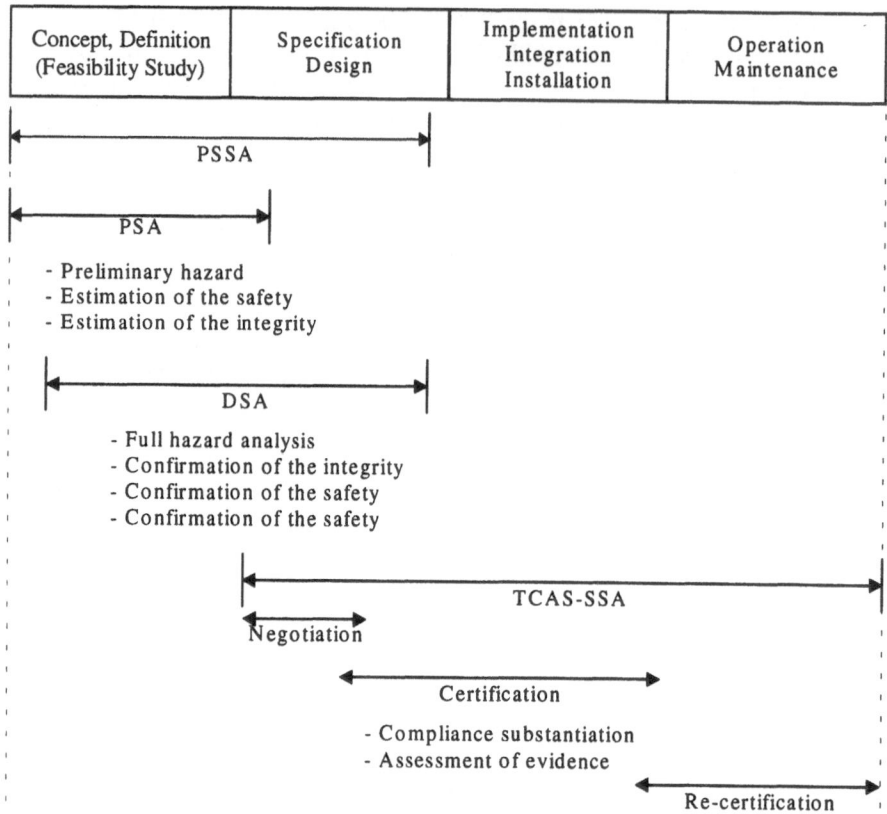

Figure 1 - Relationship between Safety Analysis Tasks and Assessment Tasks

this reason the model that we devised, the PASSPORT Diagram, is a functional model.

The basic building blocks of a PASSPORT Diagram (see Figure 2) are the *Nucleus of the TOE*; a set of *terminators* which either take input from, or provide output to, the operating environment of the TOE; the *information sets* that pass between the terminators and the nucleus of the TOE; and the *flow* of these information sets. Whilst the diagram is based on the data flow diagrams used to model informatic systems, the types of terminator and information sets include anything and everything that might effect the safety of the TOE (e.g. movement of items, algorithms used and the development process itself).

### 3.1.1 Completeness and Consistency Checks

The PASSPORT Diagram model can be checked for completeness by confirming that each of the top-level system requirements can be achieved. Since a PSA should be performed with a team of people with a wide range of expertise, we have found that the act of creating the PASSPORT Diagram model can help to highlight misunderstandings between engineers at a very early stage in the life-cycle.

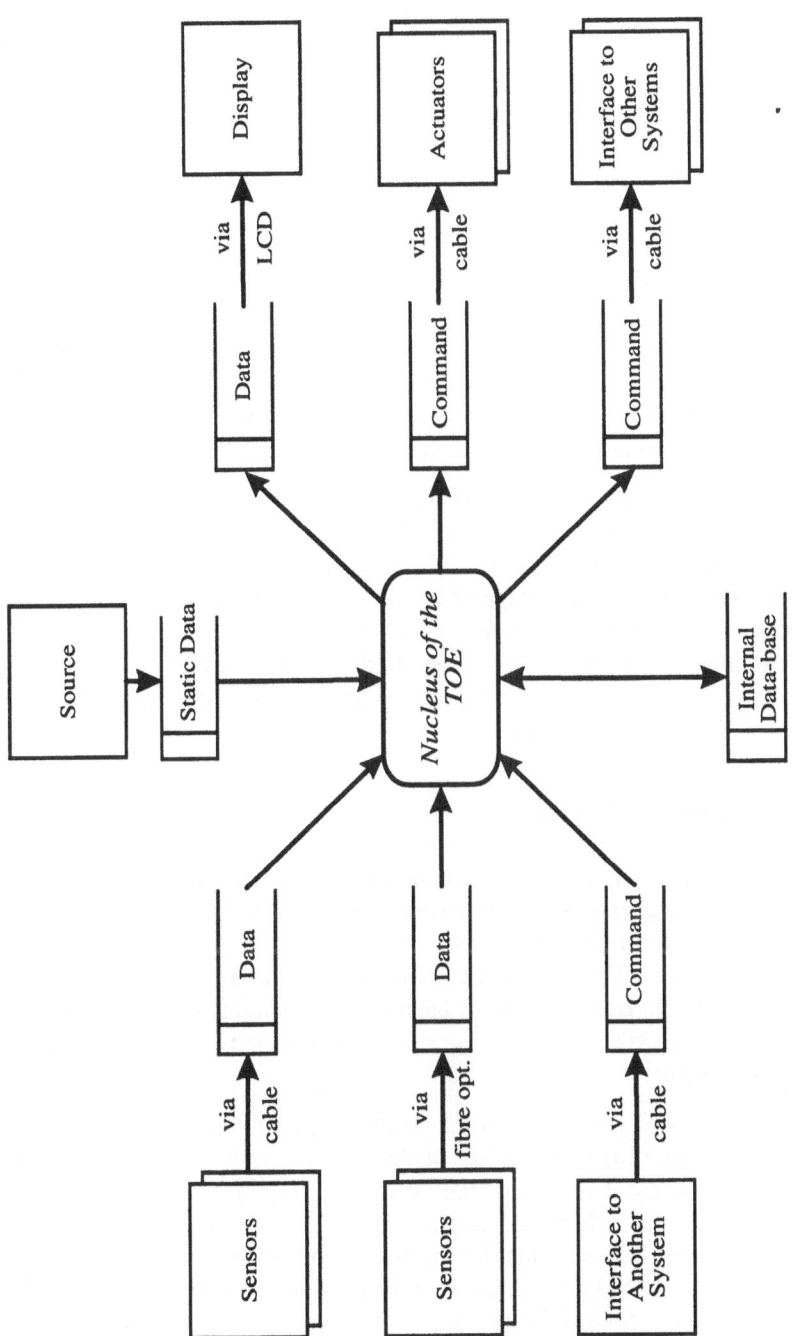

Figure 2 - An Example PASSPORT Diagram

The (self-)consistency of the PASSPORT Diagram model can be checked by applying the maxim "what goes in must come out, and what comes out must have gone in". The checks ensure that there are no unnecessary inputs to the TOE, and that all the data is present for the system to function as intended.

## 3.2 Preliminary Hazard Analysis

Once the PASSPORT Diagram has been shown to be complete and consistent then we know that the only way that the TOE can effect its environment, for good or ill, is contained within it. Each element of the PASSPORT Diagram is then systematically analysed and the question "what if ....?" asked (e.g. what if the actuator failed to operate; operated with no command, etc?, or what if the information was corrupted; failed to arrive, etc?). The effectiveness of this task can be increased by the use of checklists and guidewords to remind the analysis team of all the modes of operation that the system may undertake. By this means a preliminary hazard list can be built up for the system.

Very few, if any, real ATT systems start with a blank sheet of paper. They are usually built from a number of existing sub-systems with some new items to provide the additional functionality. There is therefore normally sufficient information, even at the concept stage of the life-cycle, to perform an analysis, to discover how each of the hazards in the hazard list might occur, by building up a tree of preliminary events that could lead to the final undesirable event. By analysing the leaves of each tree it is then possible to identify the top-level safety requirements necessary to reduce the risk of the hazard.

## 3.3 Safety Integrity Levels

The degree of care that will be taken to implement each of the safety requirements will depend on the importance of each hazard. The concept of a safety integrity level arises naturally from the fact that some activities are perceived as being more hazardous than others. Their use is desirable because the costs associated with high integrity levels can be very great. A balance must therefore be struck between using too low a level, which will increase risk, and using too high a level, which will result in unnecessary costs.

DRIVE Safely proposed that each hazard should be categorised in terms of the degree of loss of control over the safety of the situation after the failure, which might lead to that hazard, has occurred[DRIVE Safely 92]. This concept has now been adopted by the UK Motor Industry[MISRA 94]. The assessment process considers both the type of control that has been lost, and whether any other control features remain that might help to alleviate the situation in time. Each hazard is placed into one of five controllability classes (uncontrollable, difficult to control, debilitating, distracting and nuisance only). These five controllability categories are then mapped directly onto five safety integrity levels (see also Section 6.4).

# 4 Detailed Safety Analysis

The second phase of the PSSA is an analysis of the detailed design of the system. The aim is to confirm the findings of the PSA, check whether the design chosen needs some additional safety requirements and to perform a hazard analysis on all the critical items. However, before these tasks can begin it is necessary to ensure that we have a complete and consistent model of the system.

By this stage in the life-cycle there are normally two models of the system, though their distinction may not be clear. One model is the functional model, normally used by the software engineers, which describes what the system is to do. The other model is the physical architectural model, normally used by the electronic engineers, which describes how the system is to be implemented. We need to ensure that these two separate models are self-consistent, but we found no technique to do this. We therefore devised the PASSPORT Cross model.

## 4.1 The Passport Cross

The PASSPORT Cross model is based upon the Business Systems Planning system, developed by IBM[IBM] to provide a means of performing consistency checks between two different representations of a system, but it has been extended to cover all aspects of DSA.

The main PASSPORT Cross consists of four matrices sharing common axes (see Figure 3). There are two *connection* matrices and two *projection* matrices.

The first connection matrix defines the functional model in terms of Functional *Elements* (FE) and *Information Sets* (IS) upon which they operate. The matrix, FE-IS, is completed by indicating which Information Sets are being used by each Functional Element. The second connection matrix defines the physical architectural model in terms of *Architectural Elements* (AE) and *Communication Facilities* (CF) that join them. The matrix, CF-AE, is completed by indicating which Communication Facilities permit the Architectural Elements to communicate with each other.

The matrices CF-IS and FE-AE are referred to as projection matrices. They relate the elements of the functional model to the elements of the physical architectural model. The matrix CF-IS identifies which Communication Facilities are being used to transmit each Information Set, and the matrix FE-AE identifies which Architectural Elements are being used to implement each Functional Element.

### 4.1.1 *Consistency Checks*

There are a number of different consistency checks that may be performed on the PASSPORT Cross model. They may be grouped together into two basic types: *Intra-matrix* (those checks which are performed upon each matrix independently to confirm that it is well-formed) and *Inter-matrix* (those checks which are performed across the set of matrices as a whole to confirm that they fully relate to each other).

258

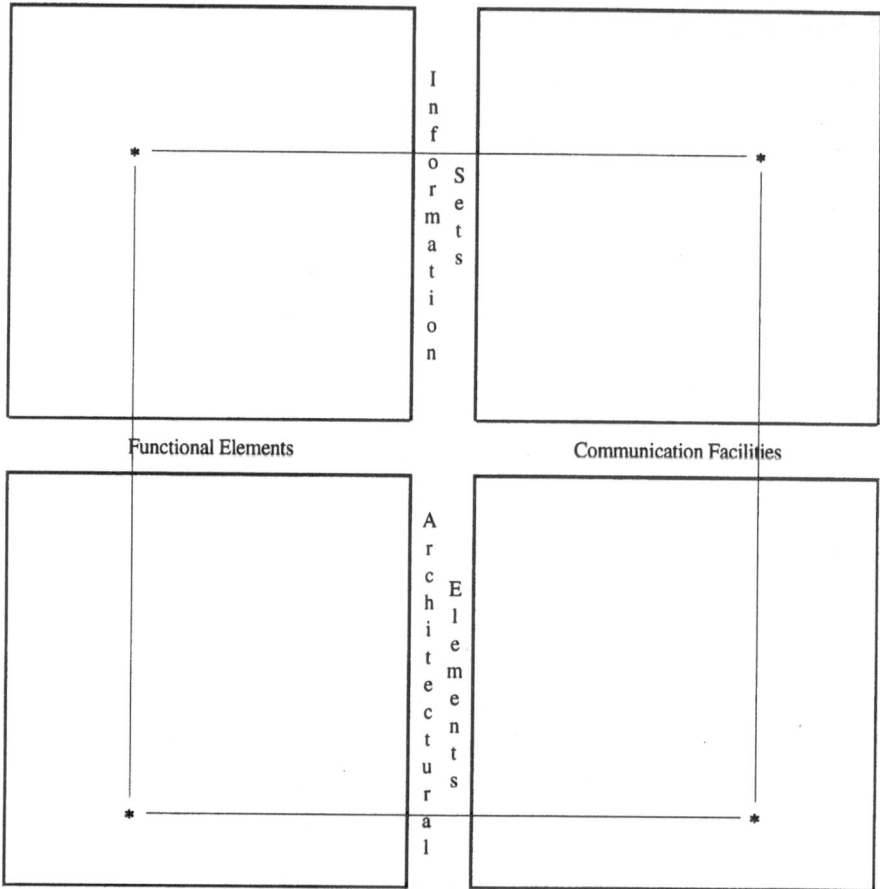

Figure 3 - The PASSPORT Cross

The basic property of the inter-matrix consistency checks (for a simple system) is that for each entry in FE-IS there should be an entry in CF-IS on the horizontal, this should have an entry in CF-AE on the vertical, which itself should have an entry in FE-AE on the horizontal. This final entry should be vertically below the original entry in FE-IS (see Figure 3). For complex systems which might use a particular Architectural Element to perform more than one function (to support reuse), or in which a single function might be performed by more than one Architectural Element (to provide redundancy), the above simple rule needs to be supplemented by additional rules. The full set of consistency rules can be found in [PASSPORT 95a].

### 4.1.2 Levels of Decomposition

During system development both the functional model and the physical architectural model will be refined into greater detail. The PASSPORT Cross model enables the traceability of information throughout this hierarchy of models.

This is particularly useful when assessing the results of the hazard analysis when it is necessary to trace a safety requirement down to its detailed implementation in both hardware and software. There are a number of *Hierarchical* checks which confirm the consistency between the hierarchical levels of decomposition.

It should be noted that it is not necessary to decompose the functional model to the same level as the physical architectural model. However, when the two models are decomposed to different levels it is then necessary to introduce two new matrices, CF-FE and IS-AE, to correctly model the system. Whenever these two matrices, known as *flow* matrices, are not empty, this indicates a logical difference between the level of decomposition of the functional model to the physical architectural model.

### 4.1.3 Passport Cross Selection

When analysing a system using a PASSPORT Cross model, the number of elements and hence the number of cells in each matrix can be very large. We have therefore devised rules that permit an engineer to focus in on an area of interest. These selection rules guarantee that the sub-matrices will pass the intra-matrix and inter-matrix consistency checks if and only if the full set of matrices have passed them.

## 4.2 Hazard Analysis

There are two forms of hazard analysis that must be performed during a DSA, FMEA and FTA.

Ideally every failure mode of every element should be subjected to an FMEA. However, this is impractical and it is therefore necessary to create a list of *sensitive elements* for which it is prudent to perform an FMEA. The PASSPORT Cross model enables such sensitive elements to be identified.

One type of sensitive element is an Architectural Element that implements a safety requirement; this can be readily identified using the traceability facilities of the PASSPORT Cross model. Another type of sensitive element would be an Architectural Element that implements a number of functions; this can be identified by looking at the rows of the FE-AE matrix.

After the FMEA has been performed a check should be made to ensure that the hazards identified are the same as those found during the PSA. Any additional hazards found may require additional safety requirements.

An FTA should be performed upon each hazard identified during the FMEA. Each FTA will identify a list of *common events* and *weak links* in the physical architectural model which have potential safety consequences, and for which care should be taken during the remainder of the development life-cycle. The PASSPORT Cross model can assist the FTA process as it naturally provides information on the inter-connections between various elements.

# 5  Evaluation of the Techniques

The PASSPORT methodology for PSA has been used on ten DRIVE II projects. These projects were selected to form a representative cross-section of the different forms of ATT systems being proposed within DRIVE II. In most cases the PSA process identified hazards that had previously been overlooked by the project being aided. There is no reason to believe that these hazards would have remained undetected, but due to their early identification it is expected that costs will have been saved.

The PASSPORT methodology for DSA has been used on two DRIVE II projects, one being an urban application and the other being an inter-urban application. In both cases the projects concerned expressed their gratitude for the help that was provided by PASSPORT in identifying safety issues.

# 6  Certification

The framework for PSSA was produced to provide developers with a sound methodology for performing safety analysis. However, safety-related or safety-critical systems usually require some form of third party assessment before they may be put into public use. Certification is the process of obtaining regulatory agency approval for a function, equipment or system by establishing that it complies with all the applicable statutory regulations. There are normally two classes of regulations against which a certification process may be performed. The first ensures that the target of the certification process conforms to one or more Standards so that it will be able to function correctly. The second ensures that the target will be able to function safely. Whilst this paper is primarily concerned with the latter class of regulations, much of the discussion below will also be applicable to the former.

## 6.1  The Need for Certification

The general public has become used to being protected from potentially dangerous equipment; this protection is currently achieved by design standards (e.g. an inability to touch live electric circuits) or by a testing process (e.g. vehicle Type Approval and annual safety checks). Since they will expect this protection to continue, there will have to be some sort of assessment of ATT systems before they are released for public use. There are reasons why the Certification of ATT systems will be necessary. Most of the failures to implement a fully functioning system can, in part, be traced to an inability to manage the actual, as opposed to the perceived, complexity. It is very difficult to demonstrate that a telematic system will function correctly in all situations, and it is usually impossible to show that it will never function incorrectly. Indeed only 25% of all large systems function as was originally intended[Gibbs 94], and yet there is still strong pressure to create ever more complex large systems, such as those for ATT. Most ATT systems will be safety-related to some degree, and thus their failures may cause death or injury to the road user either directly or indirectly.

Does this mean that all ATT systems must be assessed? One problem with a formal assessment mechanism, especially one that has legal status, is that it is first necessary to define what is, and what is not, an ATT system. Unlike an aircraft or a nuclear power station, whose boundaries are well defined, many ATT systems, in particular Urban and Inter-Urban Integrated Road Transport Environments (IRTE), will be unbounded and will change continuously over time. Some of the proposed systems are likely to be the most complex systems yet considered with many and varied sub-systems being integrated together. It will therefore be necessary to define both the scope of the system that is to be certified, and the scope of the certificate.

The primary reason for Certification is that of safety, but there are other, commercial, ones as well. There are in fact four basic issues that need to be covered in any assessment process, which together form the pillars of a unified assessment framework. These are *Safety, Security, Reliability* and the *Environment*[Jesty 95].

Traditionally, regulatory agency approval for road transport equipment has been performed by a Type Approval process. This takes the form of a series of tests on a (number of) prototype(s) to ensure conformance to a limited set of regulations. Whilst this is satisfactory for mechanical and electrical equipment, the assessors are only too aware of the limitations when the product contains programmable electronic devices. In order to gain the confidence required that such a product does conform properly a knowledge of the development process is also needed.

## 6.2 ATT System Development

This development may consist of many sub-system developments, each of them consisting themselves of multiple equipment developments. Furthermore, nested within each equipment development, there can be multiple hardware development cycles as well as multiple software development cycles.

Because of the iterative nature of all but the simplest design programmes, the development of ATT systems should be considered to be a cycle rather than sequential. Furthermore, the entry point for any given function may occur at any point in the cycle. For a new ATT function, the process begins with the top level definition of the requirements. For functional additions to an existing ATT system, the entry point may occur in the context of changes to a particular piece of equipment.

The development of an ATT system can be considered as three connected processes:

- the core development process which produces the ATT system itself,

- the supporting processes which act across all levels of the development processes. These processes include:

  - Prospective System Safety Analysis,

- – requirements validation,

- – design verification,

- – configuration management,

- – process assurance.

The level of rigour necessary for each of these supporting processes depends on the complexity of the functional implementation, and the safety integrity level,

- • the certification process.

## 6.3 The Certification of ATT Systems

The PASSPORT project has produced a framework for the Certification of ATT Systems - System Safety Aspects (TCAS-SSA)[PASSPORT 95b], whose relationship with the main system life-cycle is shown in Figure 1. The objective of certification planning and co-ordination is to establish communication between the applicant and the certification authority, and to reach agreement on the means intended to show that the ATT system complies with the regulatory requirements. Figure 4 describes the certification process, and shows the relationships between[PASSPORT 95b]:

- • the applicant and the certification authority,

- • the ATT system development and supporting processes, and the certification process.

Because of the high degree of variance in complexity and integration from one ATT system to another, the certification process needs to be flexible. Certification planning will provide applicants with that flexibility, while at the same time providing both the applicant and the certification authority with a high level of confidence in the approach used. Certification planning separates the regulatory aspects of the overall design process into manageable tasks that can be accomplished in a logical and sequential manner.

The applicant proposes a means of compliance that defines how the development of the ATT system will satisfy the basis of the certification. It may be necessary to resolve issues identified by the certification authority before agreement is reached. The certification data is the evidence that the system satisfies the regulatory requirements. The certification authority will determine the adequacy of the data showing regulatory compliance.

### 6.3.1 Re-Certification

Many of the processes described in the framework above rely on the product and process information obtained during ATT system development. Such data may not always be available for existing ATT systems or subsequent modifications. In such cases an alternative means of compliance may be necessary.

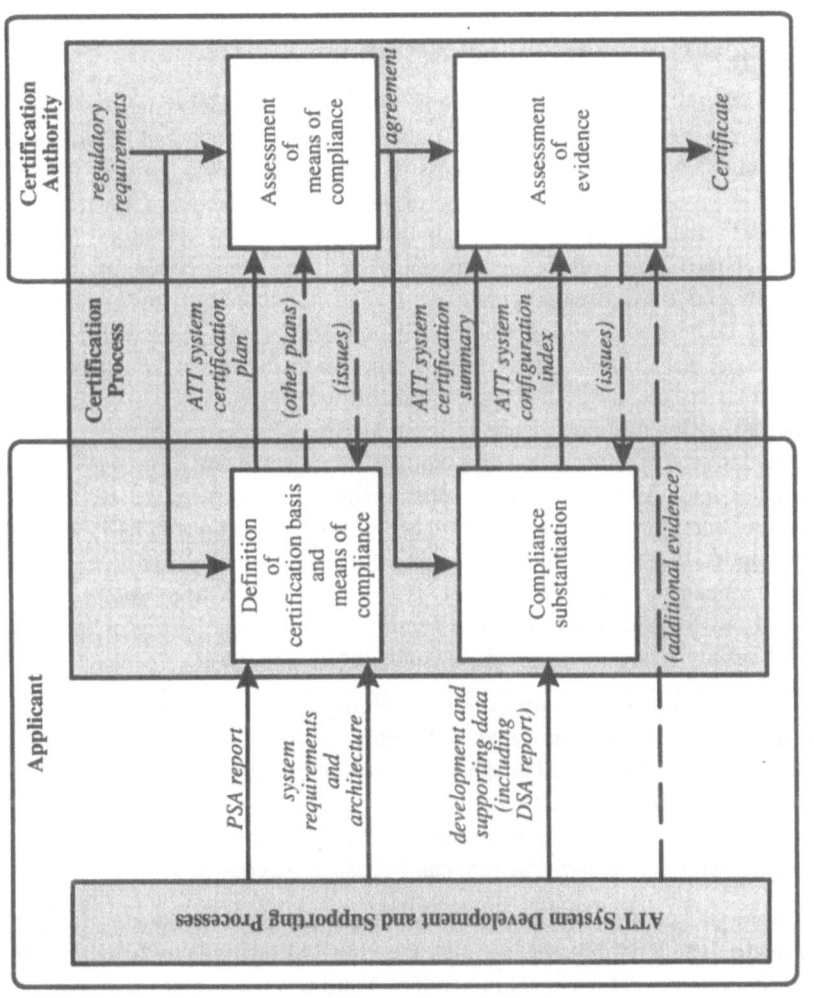

Figure 4 - Description of the Certification Process

When introducing a sub-system or ATT system modification, the certification authority should consider the impact that the modification has on the existing ATT system certification. In some cases a supplement to the existing certification basis may need to be added. The applicant proposes a means of compliance that defines how the revised ATT system will satisfy the basis of the certification. Whether or not this basis has changed, it will be necessary to assess the anticipated means of showing compliance to ensure compatibility with the agreed basis of certification.

## 6.4 Technical Issues for Unified Assessment

Sections 3.3 and 6.2 have made reference to the concept of safety integrity levels. This concept has been formulated in [IEC 1508] so that a system can be produced which is as safe as necessary, without excessive cost. The DRIVE I project DRIVE Safely (V1051) produced an interpretation of an early draft of IEC 1508 for ATT systems[DRIVE Safely 92]. This document covers hardware, software and architectural issues, and makes recommendations for their development and assessment to one of five safety integrity levels. The underlying philosophy requires that a developer should produce evidence for an assessor that will build up a level of confidence in the system that corresponds to the target safety integrity level. Particular aspects of the life-cycle are identified and guidance is given as to how an increasing level of confidence in a product may be demonstrated. Safety-related systems (safety integrity levels 1 and 2) require the effective use of good engineering techniques under a quality management system. For safety-critical systems (safety integrity levels 3 and 4) there is an increasing emphasis on the developers' understanding of the system, and the environment in which it will operate, and ensuring that the former is fit for use in the latter. The recommendations for software have since been revised for the UK Motor Industry Software Reliability Association (MISRA) Guidelines[MISRA 94].

A number of ATT functions require the recognition of the existence of a specific vehicle or road user. When this cannot take place anonymously, such as when a driver makes a request for a parking space, or after an infringement of the law, the security of the data must be maintained. Security issues have already been considered for generic software systems and the ITSEC/ITSEM documents [ITSEC 91 and ITSEM 92] describe how software can be assessed at one of six levels.

It is necessary to ensure that ATT systems will operate at all the locations in which they are placed. The environment in which they have to operate can be extremely hostile (e.g. the engine compartment of a vehicle) and can vary widely (e.g. a road-side controller in direct sun in summer, and in a freezing wind in winter). Most environmental situations are known and are described in the Standards. One particularly important aspect of the environment for ATT systems is their electromagnetic compatibility (EMC) with other electrical and electronic systems. The general requirements, as expressed in the EC Directive [89/336/EEC], are that no equipment should affect other apparatus, and that the equipment should have an adequate level of intrinsic immunity from electromagnetic interference to enable it to operate as intended. The DRIVE II project EMCATT (V2064) has produced recommendations for the assessment of EMC to one of the five safety integrity levels that conform to the DRIVE Safety/MISRA philosophy[EMCATT 95].

Reliability is, in general, concerned with the degree of confidence which one can have in a system, in particular that it will operate as intended whenever required. There are a number of attributes that a system must possess in order to achieve a given level of reliability. These include availability, dependability, maintainability (preventative, corrective, perfective and adaptive) and quality, or fitness for purpose. Conformance to Standards is also essential, especially when there is a need for compatibility or inter-operability with one or more other systems.

## 6.5 Certification or Self Assessment

Sections 6.2 and 6.3 describe a framework for the third party assessment of ATT systems. It is likely that the industry will consider that such a complex process would impose unacceptable costs for all but, possibly, the most safety-critical of systems. The EC Directive [90/683/EEC] provides a mechanism for the assessment process to be under the control of the manufacturer. Under this procedure the entire responsibility for the ATT system will rest with the manufacturer, who will either perform the assessment in-house or call in a third party assessor if necessary. The mechanism, however, breaks down when a developer is unaware that the product is an ATT system which needs a special life-cycle (e.g. the ill fated London Ambulance Computer Aided Dispatch System[SWT_RHA 93]). This is a serious problem because, even with regulatory third party assessment, it is unclear how the authorities can, in general, become aware of a new system before it is too late.

# 7 Conclusions

The project PASSPORT has produced a systematic methodology for the safety analysis of ATT systems, and in order to do this two new modelling techniques, the PASSPORT Diagram and the PASSPORT Cross, have been devised. The methodology has been successfully applied to a number of other DRIVE II projects, thus demonstrating its potential benefit for application to all new ATT systems.

The public will expect any ATT system to be safe to use and fit for its purpose. A framework for certification has also been proposed by the project PASSPORT which incorporates the use of safety integrity levels devised by the DRIVE Safely and EMCATT projects.

# 8 Acknowledgement

The work described in this paper was funded by the European Commission DG XIII Telematics Programmes (DRIVE Safely, PASSPORT and EMCATT), and by the UK Department of Trade and Industry SafeIT programme (MISRA).

# References

[89/336/EEC]    89/336/EEC Council Directive, *On the Approximation of the Laws of the Member States to Electromagnetic Compatibility*, Official Journal of the European Communities (139), 25 May 1989.

[90/683/EEC]    90/683/EEC Council Decision, *Concerning the Modules for the Various Phases of Conformity Assessment Procedures which are Intended to be Used in the Technical Harmonisation Directives*, Official Journal of the European Communities (380), 13 December 1990.

[DRIVE Safely 92]  DRIVE Safely, *Towards a European Standard : The Development of Safe Road Transport Informatic Systems (Draft 2)*, DRIVE Project V1051, 1992.

[EMCATT 95]    EMCATT, *Functional System Safety and EMC*, DRIVE II Project EMCATT (V2064), 1995.

[GIBBS 94]    W W Gibbs, *Software's Chronic Crisis*, Scientific American, pp. 72-81, September 1994.

[IBM]      IBM, *Business System Planning - Information System Guidelines*, IBM Document No. GE/20/02572.

[IEC 1508]    IEC Draft Standard, *Functional Safety of Electrical/Electronic/Programmable Electronic Systems; Generic Aspects*, IEC reference 65A (Secretariat) 123, 1995.

[ITSEC 91]    ITSEC, *Information Technology Security Evaluation Criteria (ITSEC); Provisional Harmonised Criteria*, Version 1.2, CEC DG XIII, 1991.

[ITSEM 92]    ITSEM, *Information Technology Security Evaluation Manual (ITSEM)*, (Draft) Version 0.2, CEC DG XIII, 1992.

[Jesty 95]    P H Jesty and K Wolf, *Should You Trust New Technology*, Traffic Technology International, Spring 1995.

[MISRA 94]    MISRA, *Development Guidelines for Vehicle Based Software*, The Motor Industry Research Association (MIRA), ISBN 0 9524156 0 7, 1994.

[PASSPORT 95a]    PASSPORT, *Framework for Prospective System Safety Analysis*, Deliverable N° 9, DRIVE II Project PASSPORT (V2058), 1995.

[PASSPORT 95b]  PASSPORT, *Towards the Certification of ATT Systems - System Safety Aspects*, Deliverable N° 8, DRIVE II Project PASSPORT (V2058), 1995.

[SWT_RHA 93]    SWT_RHA, *Report of the Inquiry into the London Ambulance Service*, South West Thames Regional Health Authority, 1993, ISBN 0 90513370 6.

# Session 7
## Safe Software

# Loops for Safety Critical Applications*

Johann Blieberger

Department of Automation, Technical University Vienna
Vienna, Austria

## Abstract

In this paper so-called *discrete loops* are described which narrow the gap
between general loops (e.g. while- or repeat-loops) and for-loops. Although discrete loops can be used for applications that would otherwise
require general loops, discrete loops are known to complete in any case.
Furthermore it is possible to determine the number of iterations of a
discrete loop, while this is trivial to do for for-loops and extremely difficult for general loops. Thus discrete loops form an ideal frame-work
for determining the worst case timing behavior of a program and they
are especially useful in implementing real-time and safety related systems
and proving such systems correct.

## 1 Introduction

Ordinary programming languages support two different forms of loop-state-
ments:

**for-loops** A loop variable assumes all values of a given integer range. Starting
with the smallest value of the range, the loop-body is iterated until the
value of the loop variable is outside the given range.

Some programming languages allow for starting with the largest value
and decrementing the loop variable, others allow for defining a fixed step
by which the loop variable is incremented or decremented.

**general loops** The other loop-statement is of a very general form and is considered for implementing those loops that can not be handled by for-
loops. These loops include while-loops, repeat-loops, and loops with exit-
statements (cf. e.g. [Ada95]).

If a general loop does not complete, the corresponding program usually
is incorrect. It is possible to use some logical devices, such as *Hoare logic* (cf. [Hoa69]), to prove that a certain loop (and thus the corresponding
program) completes. This however implies that the programmer has to be an
expert in formal logics. In most cases, however, programmers convince themselves by testing that their loops complete.

Concerning real-time systems the program behavior must not only be correct but the result of a computation must be available within a predefined
deadline. It has turned out that a major progress in order to guarantee the
timeliness of real-time systems can only be achieved if the *scheduling problem*
is solved accordingly. Most scheduling algorithms assume that the runtime of
a task is known a priori (cf. e.g. [LL73, HS91, Mok84]). Thus the *worst case
performance* of a task plays a crucial role.

*Supported by the Austrian Science Foundation (FWF) under grant P10188-MAT.

The most difficult task in estimating the timing behavior of a program is to determine the number of iterations of a certain loop.

Determining the number of iterations of a for-loop is trivial. For example the loop-body of the loop

```
for i in 1..N loop
    -- loop body
end loop;
```

is performed exactly $N$ times.

General loops, however, represent a very difficult task. In order to estimate the worst case performance of general loops many methods and tools have been developed, e.g. [HS91, PK89, NP93, Par93]. Most researchers, however, try to ease the task of estimating the number of general loop iterations by *forbidding* general loops, i.e., by forcing the user to supply constant upper bounds for the number of iterations. Another approach is to let the user specify a time bound within the loop has to complete (cf. e.g. [ITM90]). In any case the user, i.e., the programmer, has to react to such exceptional cases.

In this paper we will narrow the gap between general loops and for-loops by defining *discrete loops*. These loops are known to complete and are easy to analyze (especially their number of iterations) and capture a large part of applications which otherwise would have been implemented by use of general loops.

Clearly, discrete loops form an ideal frame-work for determining the worst case timing behavior of a program and they are especially useful in implementing real-time and safety related systems and proving such systems correct.

# 2  Discrete Loops

In this section we give an informal introduction to discrete loops, before we perform a theoretical treatment, i.e., an exact definition and some mathematical results. Further results can be found in [Bli94].

## 2.1  Introduction to Discrete Loops

In contrast to for-loops, discrete loops allow for more complex dependency between two successive values of the loop-variable. In fact an arbitrary functional dependency between two successive values of the loop-variable is admissible, but this dependency must be constrained in order to ensure that the loop completes and to determine the number of iterations of the loop. Details of this constraints will follow below.

Like for-loops discrete loops have a loop-variable and an integer range associated with them.The fact that the loop is allowed to range over discrete values, coined the name *discrete loop*. The major difference to for-loops is that the loop-variable is not assigned each of the values of the range. Which values are assigned to the loop-variable, is completely governed by the loop-body. The loop-header, however, contains a list of all those values that can possibly be assigned to the loop-variable during the next iteration. In fact each item of this list of values is a function of the loop-variable.

A simple example is shown in Figure 1. In this example the loop-variable k

```
discrete k := 1 in 1..N new k := 2*k loop
  -- loop body
end loop;
```

Figure 1: A simple example of a discrete loop

will assume the values $1, 2, 4, 8, 16, 32, 64, \ldots$ until finally a value greater than N would be reached. Of course the effect of this example can also be achieved by a simple for-loop, where the powers of two are computed within the loop body.

A more complex example is depicted in Figure 2. In this example the loop-

```
discrete k := 1 in 1..N new k := 2*k | 2*k+1 loop
  -- loop body
end loop;
```

Figure 2: A more complex example of a discrete loop

variable k can assume the values $1, 2, 4, 9, 18, 37, 75, \ldots$ until finally a value greater than N would be reached. But it is also possible that k follows the sequence $1, 3, 6, 13, 26, 52, 105, \ldots$. Here the same effect can not be achieved by a for-loop, because the value of the loop variable can not be determined exactly before the loop body has been completely elaborated. The reason for this is the *indeterminism* involved in discrete loops.

The term "indeterminism" requires some explanation: Clearly the loop body *determines* exactly which of the given alternatives is chosen, thus one can say that there is definitely no indeterminism involved. On the other hand, from an outside-view of the loop one can not determine which of the alternatives will be chosen, without having a closer look at the loop body or without exactly knowing which data are processed by the loop. It is this "outside-view" indeterminism we mean here. Furthermore this indeterminism enables us to estimate the number of loop iterations quite accurately without having to know all details of the loop body.

By the way, a loop like that in Figure 2 occurs in a not-recursive implementation of *Heapsort* (cf. [Knu73] or [SS93] for a more readable form in a high-order programming language).

There are two main reasons for stating this functional dependency between successive values of the loop-variable in the loop-header:

1. The compiler or, if it can not be done statically at compile-time, the runtime system should check if the loop-variable does in fact obtain one of the possible values stated in the loop-header. This will evidently ease debugging and shift some runtime errors to compile-time errors. In fact, if the information given in the loop-header is incorrect, this results in a *programming error*, not in a *timing error*. Of course this programming error could cause a timing error.

2. Under some circumstances, the information in the loop-header will make determining the number of loop iterations feasible.

## 2.2  Theoretical Treatment

Discrete loops can be defined using a range of any discrete type, e.g. an enumeration. In our theoretical treatment, however, we will assume that the range is $1..N$ and that the loop-variable starts with $k_1 = s$, where $s$ is the starting value of the loop. This restriction, however, does not inhibit transferring our results to the cases mentioned above. If $s$ is not in the range $1..N$, the loop-body is not executed, rather the control-flow of the program is transferred to the first statement after the loop.

**Definition 2.1.** A *discrete loop* is characterized by $N \in N$ and a finite number of functions $f_i : N \to N$, $1 \leq i \leq e$.

**Definition 2.2.** An *iteration sequence* $(k_\nu)$ is defined by the recurrence relation

$$k_1 := s, \quad s \in [1, N]$$

$$k_{\nu+1} := f_i(k_\nu)$$

for some $i$. The set of all possible iteration sequences is denoted by $\mathcal{K} = \{(k_\nu)\}$.

*Remark 2.1.* Note that $k_\nu \in N$ for all $\nu \in N$.

**Definition 2.3.** An iteration sequence $(k_\nu)$ is said to *complete* if $1 \leq k_\nu \leq N$ for all $\nu \leq \omega$ but $k_{\omega+1} < 1$ or $k_{\omega+1} > N$ for some $\omega \in N$. The number $\omega$ is denoted by len $k_\nu$ and called the length of $(k_\nu)$. It corresponds to the number of iterations of the discrete loop if the loop variable iterates through $(k_\nu)$.

**Definition 2.4.** A discrete loop is called a *completing discrete loop* if all $(k_\nu) \in \mathcal{K}$ are completing sequences for all $N$ and for all $s \in [1, N]$.

## 3  Monotonical Discrete Loops

**Definition 3.1.** A sequence $(k_\nu)$ is called *strictly monotonically increasing* if $k_{\nu+1} > k_\nu$ for all $\nu \geq 1$. It is called *strictly monotonically decreasing* if $k_{\nu+1} < k_\nu$ for all $\nu \geq 1$.

**Definition 3.2.** A discrete loop is called a *monotonically increasing discrete loop* if all $(k_\nu) \in \mathcal{K}$ are strictly monotonically increasing sequences. It is called a *monotonically decreasing discrete loop* if all $(k_\nu) \in \mathcal{K}$ are strictly monotonically decreasing sequences. A discrete loop is called a *monotonical discrete loop* if it is either monotonically increasing or monotonically decreasing.

**Lemma 3.1.** *A monotonical discrete loop is completing.*

*Proof.* If all $(k_\nu)$ are strictly monotonically increasing, there certainly must exist some $\omega \geq 1$ such that $k_\omega \leq N < k_{\omega+1}$. Thus the loop completes.

On the other hand, if all $(k_\nu)$ are strictly monotonically decreasing, there certainly must exist some $\omega \geq 1$ such that $k_\omega \geq 1 > k_{\omega+1}$. Thus the loop completes in this case too.

**Lemma 3.2.** *Let a monotonically increasing discrete loop be characterized by $N$ and the functions $f_i$. Then all functions $f_i$ fulfill*

$$f_i(x) > x$$

*for all $x \in [1, N]$.*

*Proof.* If there would exist some $f_d$ such that $f_d(x) \leq x$, there would exist an iteration sequence $(k_\nu)$ such that $k_{\nu+1} = f_d(k_\nu) \leq k_\nu$ which contradicts Definition 3.2.

**Lemma 3.3.** *Let a monotonically decreasing discrete loop be characterized by $N$ and the functions $f_i$. Then all functions $f_i$ fulfill*

$$f_i(x) < x$$

*for all $x \in [1, N]$.*

*Proof.* If there would exist some $f_j$ such that $f_j(x) \geq x$, there would exist an iteration sequence $(k_\nu)$ such that $k_{\nu+1} = f_j(k_\nu) \geq k_\nu$ which contradicts Definition 3.2.

## 3.1  Syntactical and Semantical Issues of Monotonical Discrete Loops

Although the syntax of discrete loops is certainly important, we consider the semantical issues more important. In order to be able to demonstrate the advantages of discrete loops over conventional loops, however, we define an Ada-like syntax which will be used in the following examples. But it is important to note that an appropriate syntax can be defined for other languages too.

The syntax of a monotonical discrete loop is given by a notation similar to that in [Ada95].

```
loop_statement ::=
    [loop_simple_name:]
        [iteration_scheme] loop
            sequence_of_statements
        end loop [loop_simple_name];

iteration_scheme ::= while condition
    | for for_loop_parameter_specification
    | discrete discrete_loop_parameter_specification

for_loop_parameter_specification ::=
    identifier in [reverse] discrete_range

discrete_loop_parameter_specification ::=
    identifier := initial_value in [reverse] discrete_range
        new identifier := list_of_iteration_functions

list_of_iteration_functions ::=
    iteration_function { | iteration_function }

iteration_function ::= expression
```

For a loop with a **discrete** iteration scheme, the loop parameter specification is the declaration of the *loop variable* with the given identifier. The loop variable is an object whose type is the base type of the discrete range. The initial value of the loop variable is given by initial_value. The optional keyword **reverse** defines the loop to be monotonically decreasing; if it is missing the loop is considered to be monotonically increasing. Within the sequence of statements the loop variable behaves like any other variable, i.e., it can be used on both sides of an assignment statement for example.

Before the sequence of statements is executed, the list of iteration functions is evaluated. This results in a list of *possible successive values*. It is also checked whether all of these values are greater than the value of the loop variable if the keyword **reverse** is missing, or whether they are smaller than the value of the loop variable if **reverse** is present. If one of these checks fails, the exception **monotonic_error** is raised.

After the sequence of statements has been executed, it is checked whether the value of the loop variable is contained in the list of possible successive values. If this check fails, the exception **successor_error** is raised.

If the value of the loop variable is still within the discrete range stated in the loop header, the loop is iterated (at least) once more. If it is not within the range, the loop completes.

*Remark 3.1.* The semantics of monotonical discrete loops ensure that such a loop will always complete, either because the value of the loop variable is outside the given discrete range or because one of the above checks fail, i.e., one of the exceptions **monotonic_error** or **successor_error** is raised.

*Remark 3.2.* A corresponding compiler is free to perform as many checks as it likes in order to inhibit one of the runtime exceptions **monotonic_error** and **successor_error**. This can be done by ensuring that the iteration functions are monotonical functions and by performing data-flow analysis to make sure that **successor_error** will never be raised. Thus a lot of runtime checks can be avoided.

Moreover the compiler might even detect the number of iterations of the loop, which is a valuable result for real-time applications. Clearly the number of iterations depends on the initial value of the loop variable, on the discrete range (especially the number of elements in the range), and on the iteration functions.

# 4 The Number of Iterations of a Monotonical Discrete Loop

Because of the indeterminism involved in the definition of discrete loops, the number of iterations of such a loop cannot be determined exactly. We can, however, find lower and upper bounds for the number of iterations. Corresponding theoretical results are given in the following subsection.

## 4.1 Lower and Upper Bounds

**Definition 4.1.** Let $\omega(\mathcal{K})$ denote the multi-set of the length of all sequences $(k_\nu) \in \mathcal{K}$ of a monotonical discrete loop and let

$$\text{L} = \min \omega(\mathcal{K}) \quad \text{and} \quad \text{U} = \max \omega(\mathcal{K})$$

denote the lower and upper bound of the length of the sequences. These represent lower and upper bounds for the number of iterations of the discrete loop too.

In the rest of this section we will only be concerned with montonically increasing discrete loops. Of course the following treatment can easily be modified in order to deal with monotonically decreasing discrete loops.

In order to calculate U and L we can use algorithms given in [Meh84a]. The following Theorem 4.1, however, will show that under certain conditions U and L can be determined much easier. Before that we need one further definition.

**Definition 4.2.** Let a monotonically increasing discrete loop be given by the number $N$ and the iteration functions $f_i(x)$. Then we denote by

$$k_{\nu+1}^{(\min)} = \min_i f_i(k_\nu^{(\min)}) \quad \text{and by} \quad k_{\nu+1}^{(\max)} = \max_i f_i(k_\nu^{(\max)})$$

the sequences that always assume the smallest and largest possible values, respectively.

**Theorem 4.1.** If for all $1 \leq i \leq e$ $f_i(1) > 1$ and $f_i(x+1) - f_i(x) \geq 1$ for all $x \in N$, then

1. the corresponding discrete loop completes,

2. the length of $(k_\nu^{(\max)})$ is equal to L, and

3. the length of $(k_\nu^{(\min)})$ is equal to U.

A proof of Theorem 4.1 can be found in [Bli94].

If $f_{\min}(x) = \min_i\{f_i(x)\}$ and $f_{\max}(x) = \max_i\{f_i(x)\}$ can be determined independently of $x$, Theorem 4.1 enables us to restrict our interest to two single functions in estimating lower and upper bounds of the number of iterations of a discrete loop.

## 4.2 Some Results on Special Iteration Functions

In this subsection we prove some theorems which cover many important cases. We study monotonically increasing discrete loops which are characterized by $N \in N$ and the iteration functions $f_i(x)$ and we assume that $f(x) = f_{\min}(x)$ can be determined independently of $x$. The initial value of the loop variable is assumed to be $k_1 = 1$, but our results can easily be generalized.

**Theorem 4.2.** If $f(x) = \lceil \alpha x + \beta \rceil$, $\alpha > 1$, $\beta \geq 0$, then the length of the corresponding loop sequence is bounded above by

$$\left\lfloor \log_\alpha \left( \frac{N(\alpha - 1) + \beta}{\alpha + \beta - 1} \right) + 1 \right\rfloor.$$

*Proof.* We clearly have

$$f(x) = \lceil \alpha x + \beta \rceil \geq \alpha x + \beta.$$

Thus

$$k_\nu \geq \alpha^{\nu-1} + \frac{\alpha^{\nu-1} - 1}{\alpha - 1} \beta = \alpha^{\nu-1} \left( \frac{\alpha + \beta - 1}{\alpha - 1} \right) - \frac{\beta}{\alpha - 1}.$$

To estimate len $k_\nu$ we must have

$$\alpha^{\nu-1} \left( \frac{\alpha + \beta - 1}{\alpha - 1} \right) - \frac{\beta}{\alpha - 1} > N$$

which is equivalent to

$$\alpha^{\nu-1} > \frac{N(\alpha-1)}{\alpha+\beta-1} + \frac{\beta}{\alpha+\beta-1}.$$

Taking logarithms we have proved the theorem.

By similar methods lower bounds for the number of iterations of monotonically increasing discrete loops can be derived.

Integrating the results of Theorem 4.2 and similar theorems into a compiler, the number of iterations of discrete loops can often be estimated at compile time.

# 5   Discrete Loops with a Remainder Function

**Definition 5.1.** In contrast to the previous sections we now define a *loop sequence of remaining items* to be the sequence of *the number of data items* that remain to be processed during the remaining iterations of the loop. Such a loop sequence is denoted by $(r_\nu)$ and the set of all loop sequences by $\mathcal{R} = \{(r_\nu)\}$. A corresponding discrete loop is called a *discrete loop with a remainder function*.

*Remark 5.1.* Definition 5.1 is justified by the fact that normally each iteration of a loop excludes a certain number of data items from future processing (within the same loop statement). Thus the sequence of the number of the remaining items is responsible for the overall number of loop iterations. This situation is typical for *divide and conquer* algorithms. For example in *binary search* the number of the remaining items is equal to the length of the remaining interval.

**Definition 5.2.** A loop sequence of remaining items is called *monotonical* if $r_{\nu+1} < r_\nu$.

**Definition 5.3.** A discrete loop with a remainder function is called *monotonical* if all its loop sequences $(r_\nu) \in \mathcal{R}$ are monotonical.

**Lemma 5.1.** *A monotonical discrete loop with a remainder function is completing.*

*Proof.* Since a monotonically decreasing discrete function will become smaller than 1 in finitely many steps, the corresponding loop will complete.

## 5.1   Syntactical and Semantical Issues of Discrete Loops with Remainder Functions

The syntax of a discrete loop with a remainder function is again given by a notation similar to that in [Ada95]. In fact we add to the syntax definition of Section 3.1.

```
loop_statement ::=
    [loop_simple_name:]
        [iteration_scheme] loop
            sequence_of_statements
        end loop [loop_simple_name];

iteration_scheme ::= while condition
```

| **for** for_loop_parameter_specification
| **discrete** discrete_loop_parameter_specification

for_loop_parameter_specification ::=
    identifier **in** [**reverse**] discrete_range

discrete_loop_parameter_specification ::=
    monotonical_discrete_loop_parameter_specification |
    discrete_loop_with_remainder_function_parameter_specification

monotonical_discrete_loop_parameter_specification ::=
    identifier := initial_value **in** [**reverse**] discrete_range
        **new** identifier := list_of_iteration_functions

discrete_loop_with_remainder_function_parameter_specification ::=
    [identifier := initial_value
        **new** identifier := list_of_iteration_functions]
        **with** *rem*_identifier := initial_value **new** remainder_function

list_of_iteration_functions ::=
    iteration_function { | iteration_function }

iteration_function ::= expression

remainder_function ::=
    *rem*_identifier =   expression |
    *rem*_identifier <= expression [ **and** *rem*_identifier >= expression ]

For a discrete loop with a remainder function, the corresponding loop parameter specification is the optional declaration of the *loop variable* with the given identifier. The loop variable is an object whose type is the base type of result type of the iteration functions, which must be the same for all iteration functions. The initial value of the loop variable is given by initial_value. Within the sequence of statements the loop variable behaves like any other variable, i.e., it can be used on both sides of an assignment statement for example.

After the keyword **with** the *remainder loop variable* is declared by the given identifier (*rem*_identifier). Its type must be a subtype of **natural** in the cases (1) and (2) below or an interval between two **natural** numbers in the case (3). Its initial value is given by initial_value. The remainder function itself may have three different forms:

1. If the remainder function can be determined exactly, it is given by an equation.

2. If only an upper bound of the remainder function is available, it is given by an inequality ($<=$).

3. If in addition to (2) a lower bound of the remainder function is known, it can be given by an optional inequality ($>=$). The second inequality must be separated from the first one by the keyword **and**.

The base type of the expressions defining the remainder function or its bounds must be **natural**.

In case (1) the remainder loop variable behaves like a constant within the sequence of statements. In cases (2) and (3) the remainder loop variable behaves like any other variable within the sequence of statements. If the value of the remainder loop variable is changed during the execution of the statements, we call the original value *previous value* and the new value *current value*.

Before the sequence of statements is executed, the list of iteration functions is evaluated if a loop variable is given. This results in a list of *possible successive values*.

After the sequence of statements has been executed, it is checked whether the value of the loop variable is contained in the list of possible successive values. If this check fails, the exception **successor_error** is raised.

After the sequence of statements has been executed, the remainder function or its bounds (depending on which are given by the programmer) are evaluated. In case (1) the new value of the remainder loop variable is set to the value calculated by the remainder function if it is smaller than the previous value, otherwise the exception **monotonic_error** is raised.

In case (2) the new value of the remainder loop variable is set to the value calculated by the remainder function if the previous value of the remainder loop variable is equal to its current value and if the calculated value is smaller than the current value, otherwise the exception **monotonic_error** is raised. If the previous and the current value differ, the remainder loop variable is set to the current value if it is smaller than or equal to the calculated value, which in turn must be smaller than the previous value. If this is not true, the exception **monotonic_error** is raised.

In case (3), at the beginning both the lower and upper bound of the remainder loop variable are set to the initial value provided by the programmer. After the loop body has been executed the new upper and lower bounds of the remainder loop variable are set to the values calculated by the appropriate remainder functions if the current value of the remainder variable is equal to the previous value and if the calculated upper bound is smaller than the current value of the upper bound and if the calculated lower bound is smaller or equal to the current value of the lower bound. If the current value and the previous value differ, both the upper and lower bound are set to the current value if the current value is smaller than the calculated upper bound, which in turn must be smaller than the previous upper bound, and if the current value is greater than the calculated lower bound, which in turn must be smaller or equal than the previous lower bound. Otherwise the exception **monotonic_error** is raised. This exception is raised too if the interval does not contain at least one element.

If in cases (1) and (2) the value of the remainder loop variable is zero or if in case (3) the upper bound is zero, the exception **loop_error** is raised, otherwise the loop is continued.

The regular way to complete a discrete loop with a remainder function is to use an *exit* statement, before the remainder loop variable is equal to zero.

*Remark 5.2.* The semantics of discrete loops with remainder functions ensure that such a loop will always complete, either if the loop is terminated by an *exit* statement or because one of the above check fails, i.e., one of the exceptions **monotonic_error**, **successor_error**, or **loop_error** is raised.

*Remark 5.3.* A corresponding compiler is free to perform as many checks as it likes in order to inhibit one of the runtime exceptions **monotonic_error**, **successor_error**, and **loop_error**. This can be done by ensuring that the remainder function or its bounds are monotonical, by performing data-flow analysis to make sure that **successor_error** will never be raised, or by ensuring that the loop will complete before the remainder loop variable is equal to zero. Thus a lot of runtime checks can be avoided.

Moreover the compiler might even detect bounds of the number of iterations of the loop, which is a valuable result for real-time applications.

## 5.2 Some Examples of Monotonical Discrete Loops with Remainder Functions

### Traversing Binary Trees

Discrete loops with remainder functions are especially well-suited for algorithms designed to traverse binary trees. A template showing such applications is given in Figure 3. In this figure **root** denotes a pointer to the root of the tree, **height**

```
1   discrete node_pointer := root
2      new node_pointer := node_pointer.left | node_pointer.right
3    with h := height
4      new h := h-1 loop
5
6    -- loop body:
7    --   Here the node pointed at by node_pointer is processed
8    --   and node_pointer is either set to the left or right
9    --   successor.
10   --   The loop is completed if node_pointer = null.
11
12  end loop;
```

Figure 3: Template for Traversing Binary Trees

denotes the maximum height of the tree, and **node_pointer** is a pointer to a node of the tree. The actual value of **height** depends on which kind of tree is used, e.g. standard binary trees or AVL-trees.

### Weight-Balanced Trees

So-called *weight-balanced trees* have been introduced in [NR73] and are treated in detail in [Meh84b].

**Definition 5.4.** We define:

1. Let $T$ be a binary tree with left subtree $T_\ell$ and right subtree $T_r$. Then

$$\rho(T) = |T_\ell|/|T| = 1 - |T_r|/|T|$$

is called the root balance of $T$. Here $|T|$ denotes the number of leaves of tree $T$.

2. Tree $T$ is of bounded balance $\alpha$ if for every subtree $T'$ of $T$:

$$\alpha \le \rho(T') \le 1 - \alpha$$

3. BB[$\alpha$] is the set of all trees of bounded balance $\alpha$.

If the parameter $\alpha$ satisfies $1/4 < \alpha \le 1 - \sqrt{2}/2$, the operations *Access*, *Insert*, *Delete*, *Min*, and *Deletemin* take time $O(\log N)$ in BB[$\alpha$]-trees. Here $N$ is the number of leaves in the BB[$\alpha$]-tree. Some of the above operations can move the root balance of some nodes on the path of search outside the permissible range $[\alpha, 1 - \alpha]$. This can be "repaired" by *single* and *double rotations* (for details see [Meh84b]).

BB[$\alpha$]-trees are binary trees with bounded height. In fact it is proved in [Meh84b] that

$$\text{height}(T) \le \frac{\text{ld } N - 1}{-\text{ld } (1 - \alpha)} + 1,$$

where $N$ is the number of leaves in the BB[$\alpha$]-tree $T$.

A template for the above operations is shown in Figure 4, where floor(x) is supposed to implement $\lfloor x \rfloor$. The remainder function of Figure 4 has the

```
1  discrete node_pointer := root
2     new node_pointer := node_pointer.left | node_pointer.right
3     with r := N -- N = number of leaves of tree
4     new r := floor((1-alpha)*r) loop
5
6     -- loop body
7
8  end loop;
```

Figure 4: Another Template for Operations on BB[$\alpha$]-trees

advantage that it does not need logarithms since it works with the number of leaves instead of the height of the tree.

## 5.3 The Number of Iterations of a Monotonical Discrete Loop with a Remainder Function

**Theorem 5.1.** *If a loop sequence of remaining items fulfills*

$$r_1 = N,$$

$$r_{\nu+1} = \lfloor r_\nu/\mu \rfloor,$$

*where $\mu > 1$, then len $r_\nu$ is bounded above by*

$$\lfloor \log_\mu N + 2 \rfloor.$$

*Proof.* We clearly have

$$\lfloor r_\nu/\mu \rfloor \le r_\nu/\mu.$$

Thus

$$r_\nu \leq \frac{N}{\mu^{\nu-1}}$$

and to estimate the length of $(r_\nu)$ we must have

$$N < \mu^{\nu-1}.$$

Taking logarithms the theorem is proved.

# 6 Discrete Loops and Safety

There are several reasons why safety related systems can profit from discrete loops:

- The syntax and semantics of discrete loops are easy enough to permit validation or even verification of a suitable compiler.

  This is especially true if only runtime checks are to be performed, i.e., no compile time checks such as solving recurrences or data-flow analysis.

- Since discrete loops are known to complete in any case, *no* endless loop can occur.

  Thus during verification or validation no effort has to be spent in order to prove that the application will complete. (We assume that tasks can be scheduled periodically.) Note that this can be done without having to rely on formal logical devices such as *Hoare Logic* (cf. [Hoa69, LS87]).

- Since the number of iterations of a discrete loop is bounded from below and from above, it is easy to derive lower and upper bounds for the timing behavior of an application. Even if no automated tool can be used for that purpose, information on the timing behavior can be derived by hand.

  Thus the validation process can provide exact bounds for the timing behavior of the application, again without use of formal logics such as in [Sch92]. This is a valuable basis to start schedulability analysis.

- Since all important steps can be done by automated tools, validating or even verifying these tools can save validation effort for applications. Such tools include compilers and schedulability analyzers (cf. e.g. [HS91]).

Of course in the discussion above we have assumed that no while loops are present in the application to be validated or verified.

# 7 Conclusion

In this paper we have described discrete loops which narrow the gap between general loops and for-loops. Since they are well-suited for determining the number of iterations, they form an ideal frame-work for estimating the worst case timing behavior of real-time programs and safety related applications.

Thus we conclude that only for-loops and discrete loops should be allowed for implementing safety related and real-time systems.

Development of a precompiler implementing discrete loops is part of Project WOOP which is carried out at the *Department of Automation* at the *Technical University of Vienna*.

# References

[Ada95]    ISO/IEC 8652. *Ada Reference manual*, 1995.

[Bli94]    J. Blieberger. Discrete loops and worst case performance. *Computer Languages*, 20(3):193–212, 1994.

[Hoa69]    C. A. R. Hoare. An axiomatic basis for computer programming. *Communications of ACM*, 12:576–580, 1969.

[HS91]     W. A. Halang and A. D. Stoyenko. *Constructing predictable real time systems*. Kluwer Academic Publishers, Boston, 1991.

[ITM90]    Y. Ishikawa, H. Tokuda, and C. W. Mercer. Object-oriented real-time language design: Constructs for timing constraints. In *ECOOP/OOPSLA '90 Proceedings*, pages 289–298, October 1990.

[Knu73]    D. E. Knuth. *Sorting and Searching*, volume 3 of *The Art of Computer Programming*. Addison-Wesley, Reading, Mass., 1973.

[LL73]     C. Liu and J. Layland. Scheduling algorithms for multiprogramming in a hard real-time environment. *Journal of the ACM*, 20(1):46–61, 1973.

[LS87]     J. Loeckx and K. Sieber. *The foundations of program verification*. Wiley-Teubner Series in Computer Science. John Wiley & Sons and B.G. Teubner, New York and Stuttgart, second edition, 1987.

[Meh84a]   K. Mehlhorn. *Graph Algorithms and NP-Completeness*, volume 2 of *Data Structures and Algorithms*. Springer-Verlag, Berlin, 1984.

[Meh84b]   K. Mehlhorn. *Sorting and Searching*, volume 1 of *Data Structures and Algorithms*. Springer-Verlag, Berlin, 1984.

[Mok84]    A. K. Mok. The design of real-time programming systems based on process models. In *Proceedings of the IEEE Real Time Systems Symposium*, pages 5–16, Austin, Texas, 1984. IEEE Press.

[NP93]     V. Nirkhe and W. Pugh. A partial evaluator for the Maruti hard real-time system. *The Journal of Real-Time Systems*, 5:13–30, 1993.

[NR73]     I. Nievergelt and E. Reingold. Binary search trees of bounded balance. *SIAM Journal of Computing*, 2(1):33–43, 1973.

[Par93]    C. Y. Park. Predicting program execution times by analyzing static and dynamic program paths. *The Journal of Real-Time Systems*, 5:31–62, 1993.

[PK89]     P. Puschner and C. Koza. Calculating the maximum execution time of real-time programs. *The Journal of Real-Time Systems*, 1:159–176, 1989.

[Sch92]    D. J. Scholefield. *A Refinement Calculus for Real-Time Systems*. PhD thesis, University of York, 1992.

[SS93]     R. Schaffer and R. Sedgewick. The analysis of heapsort. *Journal of Algorithms*, 15:76–100, 1993.

# Ontario Hydro's Experience with New Methods for Engineering Safety Critical Software

M. Viola
Ontario Hydro
Toronto, Ontario, Canada

## Abstract

Ontario Hydro has had experience in designing and qualifying safety critical software used in the reactor shutdown systems of its nuclear generating stations. To govern this work, a high level *Standard for Software Engineering of Safety Critical Software* has been jointly developed by Ontario Hydro and Atomic Energy of Canada Limited (AECL). Detailed sub-tier standards and procedures have also been developed which define the specific detailed methodology to be used for the specification and implementation of each software engineering process.

The first use of the high level standard and sub-tier standards and procedures at Ontario Hydro was for a microprocessor based Digital Trip Meter which will be used as part of the safety shutdown system at our Pickering B Nuclear Generating Station.

An earlier presentation made at the SAFECOMP'94 conference, held in Anaheim, California USA, during October 24-26, 1994 [OH94], covered the methodology adopted under the standard which was used to formally specify and verify the software for the Digital Trip Meter application. This paper reports on our experience with the application of the high level standard and sub-tier standards and procedures on this project. The following issues will be discussed in the paper:

   (a)   Identification of the strengths and weaknesses of the methods used for key software engineering processes.

   (b)   Assessment of the effectiveness of the development processes at minimizing the introduction of errors and the effectiveness of the verification and validation activities at detecting errors.

   (c)   Assessment of the overall costs for the various software engineering processes.

   (d)   Identification of areas for improvement both in the high level standard and in the individual methods applied and in the tool support for these processes.

## 1.0   Introduction

In 1990, the Darlington Nuclear Generating Station was added to Ontario Hydro's generating capacity. This station represented the first time that Ontario Hydro had utilized computers to implement the trip decision logic in each of two independent shutdown systems for each reactor. Due to the lack of a recognized standard governing safety critical software development and verification at that time, considerable effort was expended in gaining acceptance of this software with the federal regulator (Atomic Energy Control Board - AECB).

Based on that licensing experience and substantial experience in applying digital systems, the high level *Standard for Software Engineering of Safety Critical Software* [Std90] was jointly developed by Ontario Hydro and AECL. This standard defines requirements on the software engineering process, the outputs from the process, and requirements that must be met by each output. The requirements have been defined to not unnecessarily constrain the methodology which can be used to achieve the requirements of each process output. This approach facilitates gaining concurrence with the regulator on the acceptance criteria for engineering safety critical software and provides flexibility for the utility to revise or adopt new methodologies as appropriate for a specific application.

The first use of the high level standard and methodology by Ontario Hydro was for a microprocessor based Digital Trip Meter that has been designed for the Pickering B Nuclear Generating Station, for use in the Heat Transport High Temperature (HTHT) trip in Shutdown System No. 1. Detailed sub-tier standards and procedures were developed to define the methodology for each activity in the software engineering lifecycle.

The following sections provide an overview of the systems engineering process, software engineering process, background on the Digital Trip Meter application, and our assessment of the strengths, weakness, effectiveness and costs of the methods employed. Identification of the areas of improvement in both the high level standard and in the individual methods are provided together with our assessment of key processes.

# 2.0    Engineering Processes

Software engineering is a distinct set of activities, and does not proceed in isolation from the engineering of the system and hardware. It is an integral part of an overall iterative process, involving the input of system and hardware requirements and design specifications to the software engineering process, as well as ongoing changes to these specifications.

## 2.1    System Engineering Process

The system engineering process for the Digital Trip Meter project utilized the technique of step-wise refinement to simplify the overall system into smaller, more manageable subsystems. Figure 1 illustrates the system level document hierarchy for the project. At each level, additional (and more specific) requirements were included to satisfy more abstract requirements identified in the preceding level.

The collection of specifications that is fed into the software engineering process is collectively referred to as the *Design Input Documentation (DID)*. The Trip Meter Design Requirements, Digital Trip Meter Design Description and the Hardware Reference Manuals collectively are referred to as the DID for the Digital Trip Meter project and form the basis for production and review of subsequent documents for the software project.

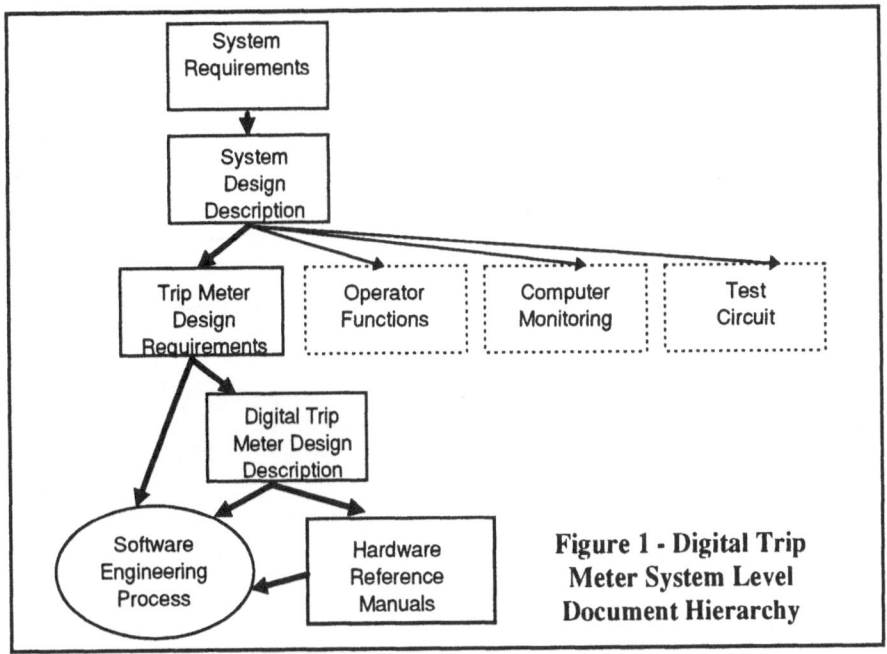

**Figure 1 - Digital Trip Meter System Level Document Hierarchy**

The Trip Meter Design Requirements document contained the requirements for the trip meter portion of the System Design. Other requirements such as the human-machine interface requirements, required performance of the meter at extreme environmental conditions, reliability targets, ease of maintenance, adherence to applicable standards and codes were also added to the document.

The last system level document, Digital Trip Meter Design Description, partitioned the trip meter requirements into the Digital Trip Meter **hardware** and **software** components, documented rationale for the design decisions, defined defensive mechanisms (such as the addition of a fail safe watchdog timer), and partitioned the system so that the safety critical subsystem would be isolated from other subsystems. The design decisions were made with the intent to meet all of the trip meter requirements and to allow sufficient flexibility so that the meter could be configured for other applications in the future with minimal re-qualification effort.

The *Hardware Reference Manuals* are documents developed by the meter hardware manufacturer. These documents provide the necessary information to document the software-hardware interfaces.

## 2.2    Software Engineering Process

The high level standard requires a planned and systematic software engineering process to be followed over the entire lifecycle of the software. A key element of the software engineering process is that it is forward-going, utilizing the concept of stepwise refinement from requirements, to design, to code. Figure 2 illustrates the software engineering lifecycle adopted. The sub-tier standards and procedures define the methodology for the execution of each of the software engineering processes shown in Figure 2.

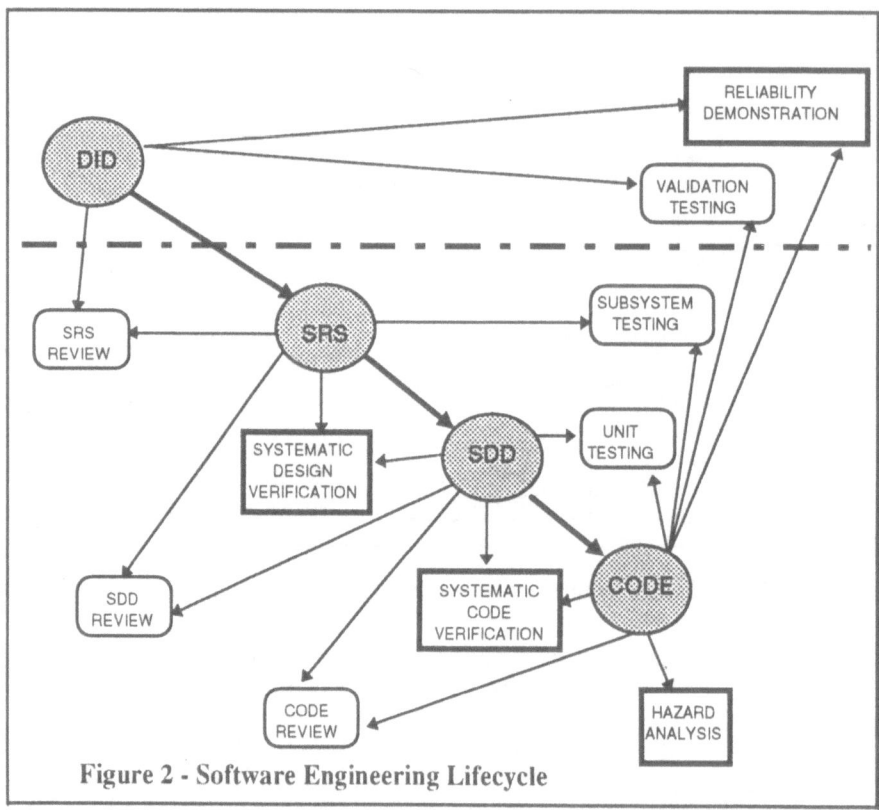

Figure 2 - Software Engineering Lifecycle

Key fundamental principles of the standard that characterize the approached adopted by Ontario Hydro and Atomic Energy of Canada Limited are:

Formal Specification: Documentation must be prepared to describe clearly the required behaviour of the software using mathematical functions written in a notation which has a well defined syntax and semantics.

Review and Verification: The outputs from each development process must be reviewed to identify that they comply with the requirements specified in the inputs to that process. In particular, those outputs written using mathematical functions must be systematically verified against the inputs using mathematical verification techniques.

Reliability Testing: Reliability of the safety critical software must be demonstrated using statistically valid, trajectory-based random testing.

Software Hazard Analysis: Analyses must be performed to identify and evaluate potential unsafe failures in the computer system and in the software component of the system with the aim of either eliminating them or assisting in the reduction of any associated risks to an acceptable level.

# 3.0    Pickering NGS B Digital Trip Meter - Application

The Digital Trip Meter will be used to replace an obsolete analog meter which is used to detect Heat Transport system High Temperature (HTHT) and initiate a protective reactor trip action. The old analog trip meters relied on operator interaction for the safety performance of the HTHT trip system. The operating procedures required the operator to periodically monitor both the process temperature and the trip setpoint from the analog trip meter display, and to maintain the trip setpoint within a pre-determined margin of the process value, by manually adjusting the trip setpoint. The maximum permissible margin is determined by the plant safety analysis, and if exceeded, the HTHT trip will be impaired.

Key features of the meter are support for Resistive Temperature Detectors (RTDs), current or voltage input; 35 segment dual colour bar graph display; dedicated setpoint digital display; process variable/margin to trip digital display; LED status indicators; trip setpoint and margin to trip alarm adjustment pushbuttons; trip, alarm, and signal out-of-range contact outputs.

# 4.0    Results

The following sections provide our assessment of the strengths, weakness and costs of the methods employed.

## 4.1    Software Requirements Specification and Review Processes

A Software Requirements Specification (SRS) was prepared to document the DID requirements formally using a methodology which employs a tabular representation of mathematical functions, an underlying finite-state machine model and is based on the four variable model described in [Parnas91]. The use of a tabular representation of mathematical functions provides for more complete specifications which are uniquely interpretable and facilitates mathematical verification for demonstrating conformance of the software design to specifications.

The objective of the SRS Review process was to verify the SRS met the requirements of the DID and to verify the justification for including requirements and design constraints in the SRS which were not defined in the DID. The requirements review process consisted of two main activities. The first was a review and comment process which did not only include the traditional review mechanism (i.e. issuing the document for general comment to the relevant domain experts), but also took the form of a more systematic and "active" review process [Parnas85]. This active review included the creation of selected questions to be answered by specific participants and the required completion of exercises to cause the SRS to be used and any difficulties to be revealed. The second activity consisted of a technical review meeting, as well as a requirements walkthrough inspection.

**Strengths:**

*A Tabular representation of mathematical functions provides for more complete specifications* - In spite of the amount of rigour applied to the production of the DID documents, the translation of requirements into the mathematically precise SRS revealed a number of areas where the DID specification was inconsistent or incomplete. For the most part, this occurred when simultaneous conditions existed, for which different actions were required. Twenty-two changes to the DID were identified as a direct result of creating the SRS.

*The tabular notation was understandable to domain experts and helped identify areas of undue complexity* - The domain experts responsible for the process system, nuclear safety and human factors were able to confirm functionality for the meter from the SRS. In addition, areas of complexity in the user interface became apparent when the requirements were formalized. In one case, a minor change to the user interface dramatically simplified the logic required, and the load on the operator by reducing the number of operating modes of the meter.

*Errors were eliminated earlier in the lifecycle* - This benefit is a result of having a complete specification. The SRS is complete since input domain coverage can be checked to determine if the required behaviour of the outputs has been specified for the entire valid range of each input and for all combinations of inputs that affect each output. Without a complete specification, errors can be introduced in which the software responds incorrectly when exposed to unanticipated input scenarios.

*Mathematically precise methodology facilitates formal verification between the Software Requirements Specification and the Software Design* - The tabular representation of mathematical functions permitted the use of mathematical techniques to demonstrate conformance of the software design to the requirements specifications.

*The review process provides independent assessment of the requirements* - The review and assessment of the justification for including requirements or design constraints in the SRS which were not derived from the DID, resulted in the identification of 10 discrepancies.

**Weaknesses:**

*The SRS Procedure immaturity caused some difficulty during the first pass through the software lifecycle* - Some difficulties with the SRS Procedure were discovered through its first application. As a result of these difficulties, 138 changes to the SRS were required. Improved wording in the updated SRS Procedure and the availability of a practical example will benefit future applications.

*Some difficulties with expressing SRS timing requirements were observed -* Difficulties were experienced in trying to express the functional timing requirements in a mathematically precise notation and yet ensure that the notation does not imply a particular design or implementation. This has been addressed in the updated SRS Procedure.

*Formalizing requirements too early in the project can lead to costly re-work -* Hardware design was in a state of flux during the initial preparation of the SRS. This resulted in 21 changes being made to the SRS which could have been avoided had some sections of the SRS been completed later in the project.

## 4.2    Software Design and Review Processes

A tabular, mathematically precise notation was used to document the software design. The software design was derived from the SRS and its structure was based on Information-hiding techniques [Parnas72]. The software design is documented in the Software Design Description (SDD).

The objective of the Software Design Review process was to provide assurance that the Software Design met the intent of the DID, was traceable to the SRS and followed good software engineering practice. The Design Review process consisted of an inspection by the domain experts and a review by a software design expert. The review included an assessment of the SDD against the following quality attributes: completeness, correctness, predicatability, robustness, consistency, structuredness, verifability, modifiability, traceability, modularity, and understandability.

Based on our experience with producing the SRS, it was recognized that formalizing the design too early can be costly. This was avoided on the software design process by initially producing a draft version of the document complete with the module decomposition and interface design. Sections of the design were prototyped and benchmarked to confirm key resource estimates such as performance, RAM and ROM capacity estimates. Once this was performed, the formal design document was completed and released for independent verification.

**Strengths:**
*The document formality was found to help minimize errors -* The SDD was produced with relatively few errors being introduced. This is attributed to the rigor in the forward going design process, and the completeness of the specifications.

*Information-hiding design technique was found to lessen the impact of changes -* Functionality changes resulting from the hardware prototype evaluation were integrated with minimal impact on the software design as a result of the information-hiding design approach adopted.

*The Software Design process supports improvement of the software robustness -* The design process encourages the identification of additional self-checks to enhance the robustness of the design with respect to hardware failures or other system level hazards. A hardware FMEA led to the addition of a number of software self-checks.

*The review process provides independent assessment of the design -* The design review process permitted the domain experts to verify that the SDD met the actual intent of the DID requirements, allowed them to assess the justification for inclusion of any additional functionality and confirm that the resulting design remained consistent with the intent of the requirements. The review process also confirmed traceability back to the SRS, assessment of the quality of the design based on the quality attributes in the high level standard and determined if design decisions were made consistent with good software engineering practice based on expert opinion.

**Weaknesses:**
*Tool support for creating the SDD was lacking -* A majority of the errors uncovered by the design review process were categorized as editorial (30) or improvements to wording in the SDD to clarify the design intent or missing information (29). Many of these errors could have been eliminated if proper tool support were available to support the creating and checking of the SDD. The remaining changes (21) pertained to improvements in maintainability, information-hiding aspects of the Software Design and changes to provide closer conformance to the SRS to simplify comparison.

*Expert opinion on good software engineering practices is subjective in nature -* Part of the review process is dependent on the existence of knowledgeable staff in the application of information-hiding design and good software engineering practice. This activity can be subjective in nature and the results will be heavily biased by the individual's experience, background and knowledge.

## 4.3 Systematic Design and Code Verification

The Systematic Design Verification (SDV) process is intended to show compliance between the SRS and SDD by using mathematical verification techniques. Similarly, the Systematic Code Verification (SCV) process is intended to show compliance between the SDD and Code by using mathematical verification techniques.

**Strengths:**
*The tabular representation of mathematical functions facilitated the use of mathematical techniques to show compliance between the software and its*

*specification* - The SDV process was able to successfully show compliance between the SRS and SDD. The results of the process found 2 discrepancies between the SRS and SDD. The remaining discrepancies were attributed to editorial changes, clarifications and suggested changes to the SRS and SDD to facilitate mathematical verification. The SCV process was able to successfully show compliance between the SDD and the Code. The SCV process identified 46 discrepancies. Forty-two of these were deemed to be minor "editorial type" changes to the SDD which could have been avoided if tool support were available. Of the four remaining changes, two required logic changes in the SRS and SDD, one resulted in additional detail in the SDD to simplify the verification and the fourth identified a temporary variable used in the code which was of the wrong type (i.e. integer instead of byte). This code discrepancy did not have an impact on the external behaviour of the software.

The few actual discrepancies observed between the SRS and the SDD, and between the SDD and the Code can be attributed to the rigour of the forward going design process, and the completeness of the specifications. Another factor is the use of small teams of highly qualified staff.

*The mathematical verification technique provides confidence in completeness of the Software Design Description and Code* - The functional discrepancies discovered by the formal verification processes on this particular project would have been immediately noticed during testing. However, the mathematical verification processes provided assurance and confidence in completeness which would not have been possible to achieve with testing.

**Weaknesses:**
*Tool support for creating the development documents was lacking* - A majority of the problems found were of an editorial nature. These types of problems could have been prevented if tool support had been available for the production of the SRS and SDD.

*Some difficulties were noted in expressing the code in a tabular mathematically precise format* - A key element in the SCV process is translating the code into a black box tabular representation that can be mathematically compared against the Software Design Description. This process is performed manually and was difficult to audit. As such, it is subject to error.

*The systematic verification activities are costly in relation to other processes* - Our results have shown the systematic verification activities to be costly processes which represent almost a quarter of the total software engineering costs. The addition of tool support and experience gained on this project will contribute to reducing these costs on future applications.

## 4.4 Code Implementation and Review

The source code is comprised of 31 modules, each with an average of 3 access programs. Approximately 1500 statements of ANSI C code were produced.

The Code Review process provided assurance that the code implementation met the requirements of the Coding Guidelines Procedure and followed good software engineering practice. The Code Review process consisted of a manual code inspection and completion of a comprehensive checklist.

**Strengths:**
*Mathematically precise specifications and strict Coding Guidelines have resulted in the production of code with very few errors* - Very few errors were found in the code implementation. This was attributed to the mathematically precise Software Design Description document which provided a complete decomposition of the software. A total of 20 errors were logged against the Code. Of these, 4 resulted in changes to the code to improve its robustness against system level hazards, 7 changes were made to correct or improve comments, 1 logic coding error was found, and 8 changes were identified to correct Coding Guideline inconsistencies.

*The review process provides independent assessment of the code* - This process provided assurance that the code implementation decisions were made consistent with good software engineering practice based on expert opinion. No changes were made as a result of this aspect of the review.

**Weaknesses:**
*The Code Review process is manually intensive* - Code Reviews consisted of a manually intensive exercise which consisted of performing an inspection of the code and completing a checklist of 154 questions. Ensuring the consistent application of the checklists over all of the software is a demanding task. Automation of some of these checks could reduce this weakness.

*Expert opinion on good code implementation decisions is subjective in nature* - Part of the review process is dependent on access to knowledgeable staff for assessing code implementation decisions. This activity can be subjective in nature and the results will be heavily biased by the individual's experience, background and knowledge.

## 4.5 Hazards Analysis

The objective of the Hazards Analysis Review process was to verify that the software required to handle system failure modes does so effectively. The technique consists of identifying the system level hazards and then reviewing the software from a perspective of attempting to determine all possible causes for the undesired system hazard to occur [Leveson83].

**Strengths:**

*The review of the software from a safety perspective provides greater confidence in the software meeting its safety objectives* - The hazard analysis process was found to complement the other validation efforts in that it undertakes an orthogonal review to the traditional functional perspective. An example of the success of this process was the identification of a possible configuration of the Digital Trip Meter which could potentially delay the trip action during an accident. Safeguards are in place to ensure this failure mode cannot occur in the configuration specific to the Pickering application.

**Weaknesses:**

*Errors are found late in design lifecycle* - The Hazards Analysis is not performed until late in the design cycle when the code is almost completed. Hazards are sometimes found which warrant changes to the SDD and/or SRS. It would be more cost effective to apply this technique on earlier development outputs if impact on upstream documents can be minimized.

## 4.6   Testing

Unit Testing was performed against the SDD, Subsystem Testing was performed against the SRS, Validation Testing was performed against the DID and Reliability Demonstration testing was performed against the DID and SRS.

**Strengths:**

*Ensures the executable code of each program behaves as specified in the SDD, shows that the executable code does not perform unintended functions and that program interfaces behave as specified in the SDD* - Unit testing identified 2 discrepancies between the SDD and the code. The remaining errors reported were identified during the process of creating the test cases which resulted in 13 changes to the SDD.

*Shows that software integrated with the target hardware meets the requirements in the SRS* - Subsystem testing detected 2 errors in the code, 1 of which resulted in a change to the SDD. Four errors were identified during the process of creating the test cases which resulted in changes to the SRS.

*Methodology facilitated definition of the test coverage criteria* - The SRS and SDD documents make use of the methodology employing a tabular representation of mathematical functions. This representation permitted the definition of explicit rules that should be followed in the creation of test cases. This permitted us to achieve a known test coverage for test processes which use the SRS and SDD documents as input.

*Shows that the entire executable code integrated with the target hardware meets the requirements in the DID* - The validation test cases were prepared by an individual who did not participate in the original production and review of the DID documents. Given that these documents are written in English prose, they are subject to ambiguities and incompleteness. In spite of the rigour applied in creating the documents, the validation tester identified 11 problems with the DID documents. Six problems with the DID were identified in the process of creating the test cases and another five after the testing was completed.

One error was found in this process against the configuration data file which was not subject to the formal verification process. The error survived the earlier testing processes since the magnitude of the error did not exceed the accuracy requirements of the meter.

*Demonstrates that the reliability hypothesis is achieved for the executable code.* - Reliability qualification testing was successfully completed and we were able to demonstrate a software unavailability target of $10^{-4}$ years/year with a confidence interval of 50 %.

**Weaknesses:**
*Tool support for creating and verifying the test cases are lacking* - Tools were put in place to assist with the execution of the test cases for Unit Testing, Subsystem Testing and Reliability Demonstration Testing. However, the process of creating the test case scripts and their review was an extensive manual process. Automation of this aspect of the processes would significantly reduce testing costs.

*Testing did not find functionality errors in the SRS or SDD that were not already detected by other verification processes* - This is not as much a weakness of the testing processes as it is a strength of the formal specification and verification methodologies.

# 5.0    Assessment

Figure 3 shows the number of errors detected by each process and which development document (Design Input Documentation (DID), Software Requirements Specification (SRS), Software Design Description (SDD), and Code) was impacted by a corrective change.

## 5.1    Effectiveness of the Software Engineering Processes

Our experience with the methodology developed to meet the high level standard has shown that very few errors were found in the Code on completion of the process. One logic error was found and this was discovered before testing began. Correctness of the software was further shown by three levels of testing and Reliability Demonstration Testing which executed over 7,000 test cases without detecting an error. The following section discusses the effectiveness of the Software Engineering Processes under the headings of Software Engineering

Model, Formal Specifications, Systematic Verification, Review Processes, Hazard Analysis and Testing.

*Software Engineering Model* - The conventional waterfall model has been identified as weak when not everything is known at the start of the project or will not become well known in the same sequence that work phases are scheduled. When all the early documents must be fully completed as entry criteria to the later phases of the project work, there is a risk of extra cost and schedule slippage due to re-work of these documents if changes are required as a result of later project work. This can be significant in projects which employ formal methods. The high level standard has been revised to address this issue.

*Formal Specifications* - The SRS and SDD were both written using a tabular representation of mathematical functions as identified earlier in this paper. With this methodology, we are confident that the SRS specification is unambiguous since the notation has a well defined syntax and semantics. Completeness of the specifications are assured since input domain coverage can be checked to determine if the required behaviour of the outputs has been specified for the entire valid range of each input and for all combinations of inputs that affect each output.

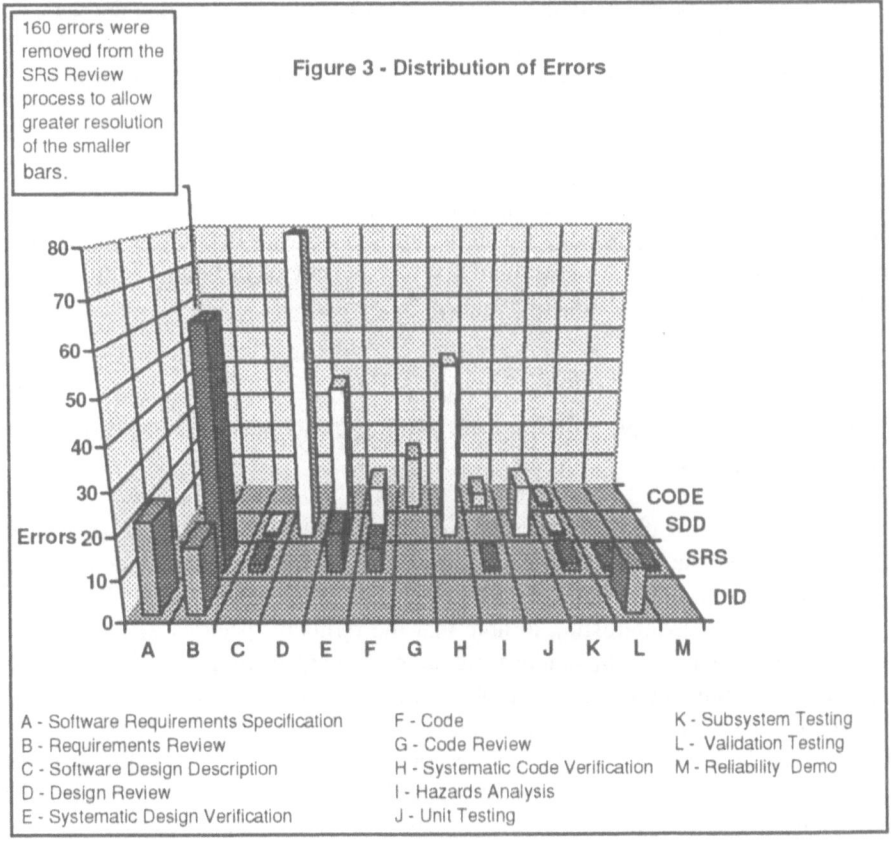

Figure 3 - Distribution of Errors

160 errors were removed from the SRS Review process to allow greater resolution of the smaller bars.

A - Software Requirements Specification
B - Requirements Review
C - Software Design Description
D - Design Review
E - Systematic Design Verification

F - Code
G - Code Review
H - Systematic Code Verification
I - Hazards Analysis
J - Unit Testing

K - Subsystem Testing
L - Validation Testing
M - Reliability Demo

*Systematic Verification* - Our experience with the systematic verification methodologies has shown very few discrepancies being identified between the SRS and SDD and between the SDD and Code. We attribute this to the rigor of the forward going design process, and the completeness of the specifications. This demonstrates that the development processes are effective at minimizing the introduction of errors. Another factor is the use of small teams of highly qualified staff. The teams are for the most part made up of staff which have participated in addressing the licensing issues with the Darlington Shutdown Systems, and the subsequent development of the processes and methodologies which were used on this project. As a result the staff are extremely knowledgeable of the concepts, issues, and processes, enabling them to effectively apply the procedures with great success.

*Reviews Processes* - The informal review processes were found to be effective at assessing qualitative attributes of the development outputs such as meeting design intent and assessing good software engineering practice and coding implementation decisions.

*Hazard Analysis* - This process has proven to be a valuable tool for detecting errors that would not have been detectable by other verification activities. The concern regarding finding issues late in the design process is being addressed in the revision to the high level standard. The Hazard Analysis process will now be required at several points in the software design lifecycle. This decision reflects our experience that this type of analysis is a valuable complementary verification tool which focuses on safety.

*Testing* - Our experience with testing on this project has shown that very few errors were detected by these processes. This is not a negative reflection on the testing processes but rather it is a confirmation of the benefit of utilizing complete and mathematically precise specifications.

## 5.2    Assessment of Costs

Figure 4 shows the relative costs for each of the software development, verification and validation processes.

Key processes which characterize the high level standard are the use of formal specifications for the SRS and SDD, systematic verification between the SRS and SDD and between the SDD and Code, Hazards Analysis, and Reliability Demonstration. It is interesting to note that the formally specified SRS and SDD required 10% of the software engineering budget but are attributed with the greatest impact on minimizing the number of errors detected on the project. The two systematic verification activities combined required 24% and the four levels of testing combined required 27%. From the chart it is clear that the greatest opportunity for improvement would be tool support to facilitate systematic verification and testing which had large components of manual activity.

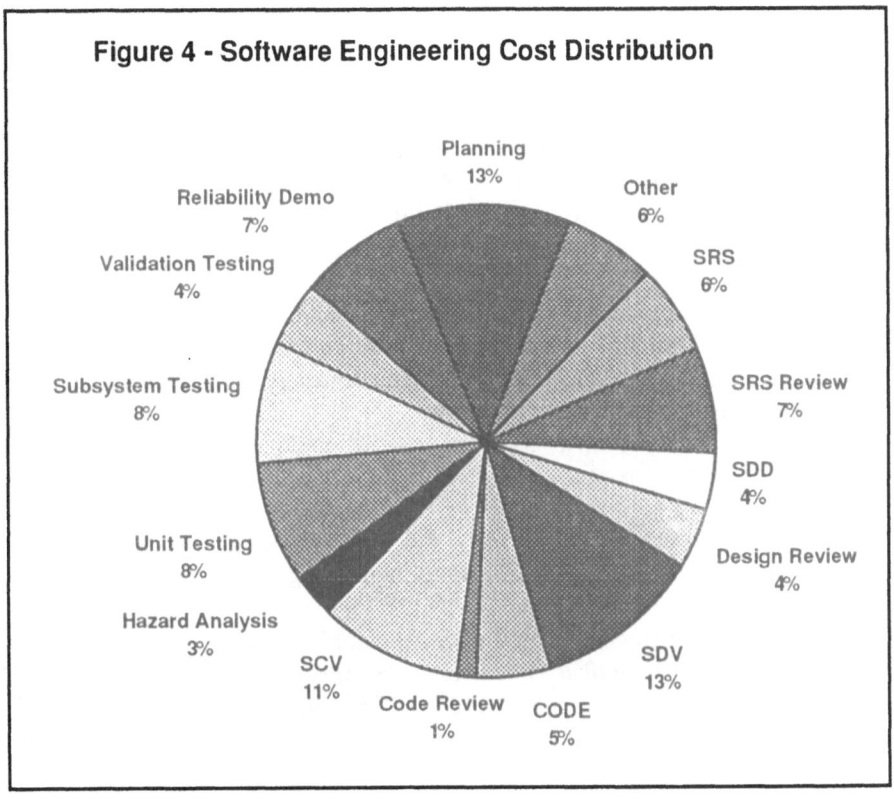

**Figure 4 - Software Engineering Cost Distribution**

Planning 13%
Other 6%
Reliability Demo 7%
SRS 6%
Validation Testing 4%
SRS Review 7%
Subsystem Testing 8%
SDD 4%
Unit Testing 8%
Design Review 4%
Hazard Analysis 3%
SDV 13%
SCV 11%
Code Review 1%
CODE 5%

# 6.0    Conclusions and Future Work

Our experience on this project has shown that the standards and methods developed by Ontario Hydro and AECL have proven to provide a practical and effective approach for the development of safety critical software for nuclear power plant applications. This review has shown the strengths of the formal specification methodology in minimizing the number of errors in the software and the benefit of the complementary verification techniques such as the review processes, Hazards Analysis and testing to provide confidence in the quality of the software.

Weaknesses of the methodologies are being addressed as part of our continuous improvement programs and as part of the effort to establish the necessary procedures to redesign the Darlington Shutdown System Trip Computer software. Work is currently underway to revise the Darlington software to incorporate functional improvements accumulated over the past five years of the plant operating history.

Our experience on the Digital Trip Meter project has helped us to refine our methods and procedures and identify the areas where tool support will be of the most benefit. Many of the errors detected by the verification processes could have been avoided if proper tool support had been available.

Work is currently underway to produce tools that support the creation, use and maintenance of the primary documents for a safety critical application. These include the development documents, Software Requirements Specification and the Software Design Description and the verification documents Systematic Design Verification Report and the Systematic Code Verification Report.

Tools are also being produced to support testing. The intent is to take advantage of the formally specified SRS and SDD documents and create a test oracle that is essentially an executable model of the requirements to support the generation of test cases and the establishment of test coverage.

# References

[Leveson83]    Leveson, N.G., Harvey, P.R. *"Analyzing Software Safety"*, IEEE Transactions On Software Engineering, SE-9, 5, (Sept 1983).

[OH94]    M<sup>c</sup>Dougall, J., Moum, G., Viola, M., *"Tabular Representation of Mathematical Functions for the Specification and Verification of Safety Critical Software"*, SAFECOMP'94 Proceedings of the 13<sup>th</sup> International Conference on Computer Safety, Reliability and Security, Anaheim, California, USA, 24-26 October, 1994.

[Parnas72]    Parnas, D.L., *"On the Criteria to be Used in Decomposing Systems into Modules"*, Communications of the ACM Vol. 15, No. 12, December 1972, pp. 1053-1058.

[Parnas85]    Parnas, D.L., Weiss, D., *"Active Design Reviews: Principles and Practices"*, Proceedings of the 8th International Conference on Software Engineering, London, August 1985.

[Parnas91]    Parnas, D.L., Madey, J., "Functional Documentation for Computer Systems Engineering (Version 2), Faculty of Engineering, McMaster University, Communications Research Laboratory Report No. 237, September 1991.

[Std90]    Joannou, P.K. et al. *Standard for Software Engineering of Safety Critical Software*, Ontario Hydro, Electrical and Controls Engineering Department Standard 982-C-H-69002-0001, Rev 0, December 1990.

# Is Software Safe to Fly?

By Mr. K.N. Narahari, Mrs. Shylaja Prasad & Dr. K. Karunakar,
Aeronautical Development Agency,
Bangalore 560037.
INDIA.

**Abstract**

A systematic approach to the software development process using accelerometer sensor assembly of a flight control system as an example with emphasis on safety is detailed in this paper.

## 1 Introduction

With the advent of high speed computing processors, more and more processing of real time embedded system is transferred to software. Safety tends to become an important attribute of such software system if software implements safety control functions also. Such software must protect the personnel working in its operating environment. This is essential for modern fighter aircraft employing Digital Fly By Wire (DFBW) technology for its Flight Control System (FCS).

Safety is an attribute that is related to the _use_ of software but the scope of _use_ covers the full software life cycle. It is almost impossible to make an error-free software but, by putting good and systematic software engineering practices, a hazard free software can be ensured. A safe software should ensure that nothing catastrophic shall happen during its operation. For certification purposes, we should be able to say with some degree of confidence that software is safe to fly. Presently, safety is relied on qualitative assessment and engineering judgement. This is specially true for aircraft industry.

_Simplicity is the first step towards safety_. Simple systems have fewer failure modes and are easier to inspect, analyze and test for correctness. A complex system can be easily built over the simple components.

The software life cycle starts with the requirement analysis phase. The requirement analysis accumulates knowledge from experienced professionals and puts it as a document, commonly known as Software Requirement Specification (SRS).

The software implementation starts with software design. The design has a major impact on the testability, maintainability and efficiency of the software. A safe software should have a predictable behavior with functionality, timing behavior and memory requirements. The implementations like interrupts, tasking, processor caching, etc., which are difficult to predict should be avoided in safe software. The programming language used for writing safety critical software must be free from language insecurities and the language must have a precise definition. For critical portions of software, the defensive programming could be adopted to ensure the continued safe operation of the software even in the presence of exceptional events.

A step by step systematic approach to the software development and testing process, using flight critical system as an example, with emphasis on safety is detailed.

## 1.1 Accelerometer Sensor Assembly of modern fighter aircraft

The FCS is a safety critical component of the aircraft and in modern DFBW aircraft it happens to be software intensive. The FCS usually consists of redundant channels. The redundancy is built into the system for the reasons of increasing reliability.

The case study assumes four redundant channels in FCS. Each channel has one computer and each channel has one set of sensor sub-system connected to this computer. The output of the computer is connected to aircraft actuation system. Each channel communicates through a high speed link with other channels. The Accelerometer Sensor Assembly (ASA) is a sub-system of FCS. This assembly is responsible for measuring linear accelerations of the aircraft. Each channel receives one set of ASA signals from the ASA assembly directly and redundant set from other three channels. It transmits the signals received directly to other three channels along with failure information, if any. The best signal is selected out of the good signals in each channel and passed on to computation model for further processing. It is assumed that each channel runs similar software. This is shown in Figure 1.

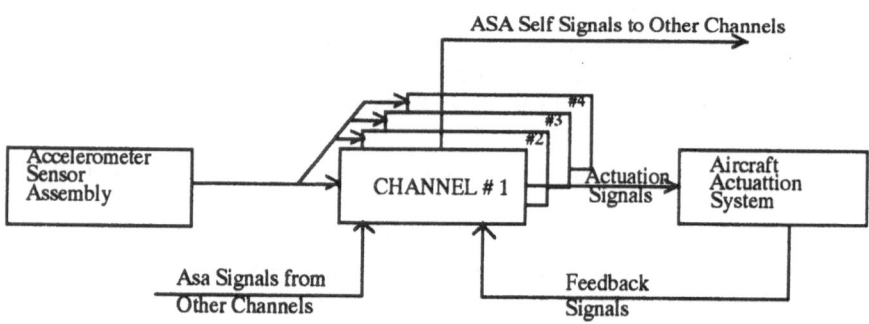

Figure 1 - Typical FCS Architecture of Modern DFBW Aircraft

# 2 Software Development Process

## 2.1 Requirement Analysis of ASA

The sample behavior diagram, generated using RDD 100 requirement analysis tool [RDDrm] for ASA is shown in figure 2. The figure shows the functions in a time

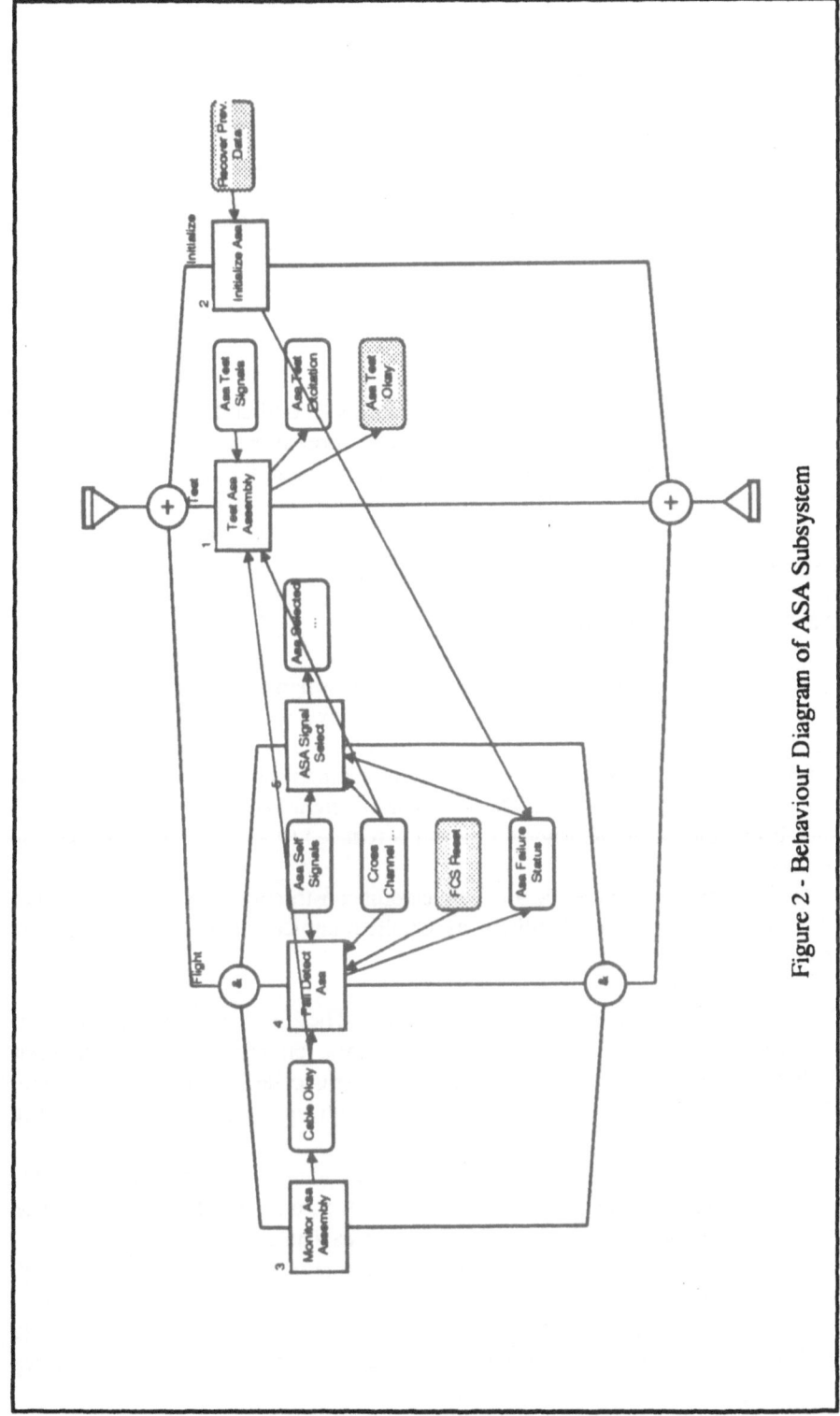

Figure 2 - Behaviour Diagram of ASA Subsystem

sequence beginning at the top of the figure and ending at the bottom of the figure. Each parallel flow is entered by an ampersand (&) with a circle around it. This symbol is also used to show exit from parallel flow. A plus sign (+) indicates a select function that combines two or more branches.

Good requirement analysis depends on the requirements being understandable, unambiguous, complete, verifiable, consistent, modifiable and traceable. As the new requirement(s) are added, the originating requirement(s) are linked backward and forward to the new requirement by using RDD100 tool. This ensures traceability throughout the development phase. SRS for ASA module is generated using RDD100 tool.

## 2.2 Software Design

The design has a major impact on the testability, modifiability and efficiency of the system. The design has often two levels - preliminary design and detailed design. Preliminary design focuses on identifying modules, their specifications and their interconnections. Detailed design expands the preliminary design by specifying the data structures and computational logic in sufficient detail to start coding.

### 2.2.1 Preliminary Design

2.2.1.1  Design by Abstraction: The preliminary design starts with decomposition of problem into sub problems. Decomposition allows us to conceive a program in terms of components that can be combined to solve the original problem. Abstraction helps us to make a good choice of components that are easy to specify, build, and use in combination. Design involves alteration between decomposition and abstraction, until the original problem is reduced to a set of problems that can be solved easily.

The design through abstraction, with careful construction of the specification of its components, greatly simplifies its detailed implementation and subsequent maintenance.

2.2.1.2   Language Support for Abstraction: The most common method of abstraction is the facility to employ subroutines, allowing separation to their definition and invocation. This type of abstraction is called functional abstraction. Functional abstraction is most useful in dealing with a problem that is conveniently decomposed in terms of independent functional units.

Often, however, it is more fruitful to think of adding new kinds of data objects, with operations to create such objects, to obtain information from them, and to modify them in meaningful ways. This is known as data abstraction.

Functional abstraction forms the basis of the structural design methodology, while data abstraction forms the basis of the object-oriented design methodology.

Ada language supports both the design methodologies. Data abstraction in Ada language is implemented by using package. The package specification specifies operation defined on the package with the parameters and their types. The body of a package consists of the implementation of all subroutines given in the specification.

During Preliminary Design, the goal is to generate the package specifications. Preliminary Design of ASA is carried out using TeamWork/Ada Ada Structure Graph (ASG) [TEM 4.0]. ASG is a design tool used to provide a visual representation of Ada programs. Figure 3 shows the sample ASG diagram of ASA package. The packages are shown by rectangles, with their names in the top left corner. The inheritance ("withing" relation) is shown by double arrows. The single arrow shows the calling of subroutines or accessing of data object. The tool provides annotation editor to write comments and explanations. There are 30 modules (functions and procedures) defined for ASA subsystem.

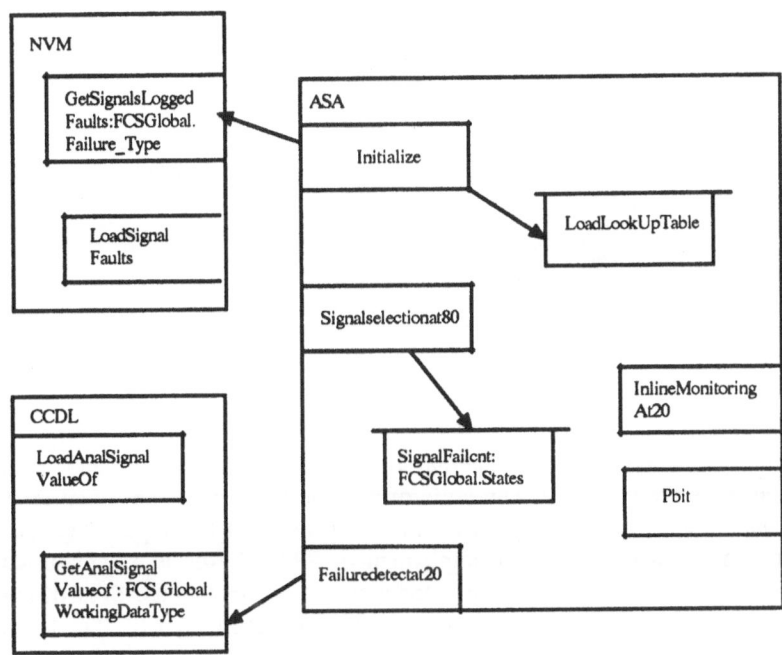

Figure. 3 - ASG Diagram of ASA Package

2.2.1.3 Preliminary design quality metrics: The goal of having metrics for the Preliminary Design is to assess the quality of the design. A high quality Preliminary Design leads to a high quality detailed design.

2.2.1.3.1 Fan in and Fan out of modules: This metric specifies the complexity in maintaining the software, once it is delivered. A high fan-in means that a number of other modules depend on this module. A high fan-out means that a module depends on too many modules. Fan in and Fan out value of 4 to 6 is acceptable from the maintenance point of view. Fan in and Fan out values of ASA modules is shown in figure 4 (a) and (b).

2.2.1.3.2 Number of parameters: This metric specifies the coupling between modules. More the number of parameters, more is the coupling and less is the cohesion as the module is performing too many functions. A maximum of six parameters is good choice, provided most of the parameters are of elementary type and not of structured type like arrays and records. Number of parameters of ASA modules is show in figure 4 (c).

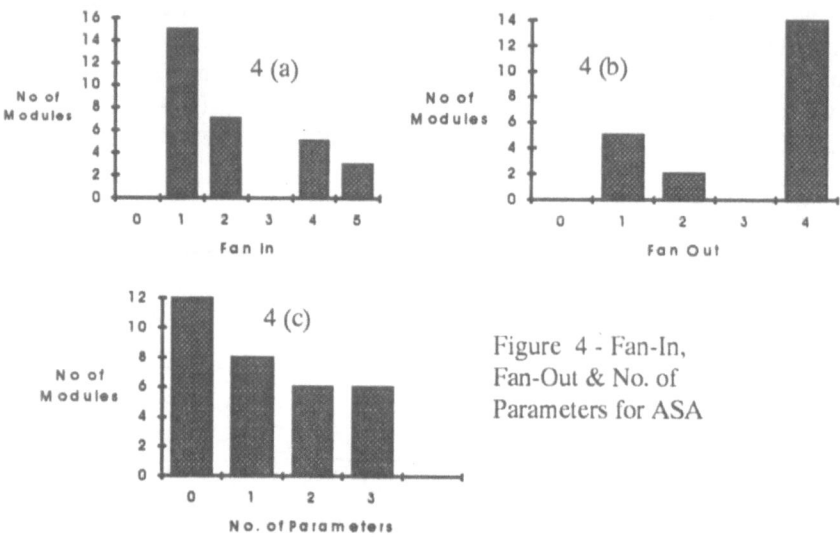

Figure 4 - Fan-In, Fan-Out & No. of Parameters for ASA

## 2.2.2 Detailed Design

During detailed design, the internal logic of each of the modules and details of data structures are specified. The logic of a module is usually specified in high level design description language, which is independent of the target language in which software will eventually be implemented. SPARK Ada is a safe subset of Ada language. SPARK imposes the mandatory use of annotations. SPARK annotations are introduced at the design stage to ensure verification of the design. These annotations start with "--#" and are ordinary comments to Ada compiler, but it is a specification language to SPARK Examiner.

2.2.2.1 Detailing module specifications: The first step in detailed design for a module is to provide specification of the module precisely. Once the module is precisely specified, the internal logic for the module that will implement the given specifications can be decided.

The desirable properties of specifications are that it should be complete, unambiguous and understandable. The specification language should be such that the specifications can be easily written and implementation independent. Structured English is used to describe the specification of all modules of ASA subsystem.

2.2.2.1.1    SPARK Annotations for ASA: The SPARK annotations for ASA package specification and body are given in figures 5 & 6 respectively [SEM 1.3]. The annotations in package specification are based on abstract data like *IncomingSignals*. These variables define the data abstraction. These abstract variables are refined to actual physical variables (consume memory) in package body of ASA.

## 2.2.3 Software Design Verification

The focus of design verification is on showing that detailed design is traceable to preliminary design and preliminary design is traceable to requirement specifications. Methods like design walkthroughs, critical design review and consistency checkers are often employed to verify the design.

```
with FcsGlobal, Nvm, CCDL;
--# inherit       FcsGlobal,
--#         Nvm,
--#         Ccdl;

package Asa
--# own          IncomingSignals ,
    .....    Contains list of variables declared in
    .....    Specification of ASA and Abstract objects
    .....
is
      procedure Initialization ;
      --# global        Nvm.SignalsFaultLog,
      .....    contains  list of variables used by
      .....    procedure Initialization
      --#          LookUpData;
      --# derives       Failures     from
      Nvm.SignalsFaultLog,
      .....    contains I/O relationship between Global
      .....    variables, parameters
      .....    and   abstract objects

      --* Specifications of all subroutines are not
listed *--
      ..........
end Asa ;
```

Figure 5 - Package Specification of ASA with SPARK Annotations

```
package body Asa
--# own            IncomingSignals    is
     AsaNormAccelIn,
..... contains relationship between abstract objects
..... and actual variables.
is
     ..... contains local type definitions, etc.

     Procedure  Initialization
     --# global       AsaTable,
     .....   contains actual global variables used by
     .....   the module and I/O relationship between
     .....   Actual variables

     is separate;
     .................

End Asa
```

Figure 6 - Package Body of ASA with SPARK Annotations

2.2.3.1 SPARK Examiner Tool:    SPARK examiner [SEM 1.3] checks
conformance to the rules of SPARK Ada. It also checks consistency between
SPARK Ada code (implementation logic) and annotations by control flow, data
flow and information flow analysis. All modules of ASA subsystem were passed
through SPARK examiner and the software development is progressed to coding
phase only after confirming that the design is complete and consistent.

## 2.3 Software Coding

### 2.3.1 Language Issues and Requirements

The integrity of software is vital in case of safety critical applications. So the
choice of programming language should be made carefully. At one end, a language
like C is aimed at convenience of use and efficiency and on the other extreme, a
language like Ada is aimed at more expressive power and generality.

In safety related applications, the language requirements are quite limited, in
terms of applicability, but very strict. The first requirement of a language is that
its definition be precise and logically coherent. The language itself must not
contain any ambiguities that allow the construction of programs of uncertain
meaning. The definition must also be complete. For high integrity work, a formal
definition of the language must be established as the essential basis of its use.

The language should be strongly typed. The language should be a deterministic
language that means the code should have a predictable behavior in terms of space
and time. Memory usage for features like dynamic allocation, recursion and
unconstrained arrays and timing for interrupts, microprocessor instruction cache,

tasking and garbage collections are difficult to predict and should be avoided in safety critical applications.

The language should not have any insecurities in its definition. A language is insecure if a program can violate the definition of the language in any way that is impossible, or even very difficult, to detect prior to program's execution. All language violations must be detected prior to program execution. The SPARK Examiner detects the insecurity.

Finally, language should make the verification process easy. It should support separate compilation of program units and program units should not be allowed to have side-affects.

2.3.1.1 SPARK Ada - The Ada language in its complete form is not suitable for safety critical applications, for two closely related reasons - Inadequacy of its definition and excessive complexity [ARM 83].

SPARK Ada defines a subset of Ada language by removing dangerous features or features that could lead to intractable validation problems - such as variant records, use of functions and procedures as parameters, tasks, exceptions, generics, etc. It also introduces an "annotation" language to solve other Ada language insecurities.

All modules of ASA subsystem were implemented in SPARK Ada. Each subunit is a separate compilable unit that is specified through "separate" clause. The modules along with ASA package specification and body were passed through SPARK Examiner successfully.

*2.3.2 Programming Issues*

The first requirement to program safety critical system is to have well defined and detailed coding standards containing:

1. *Program Layout*: How the program is organized and presented can have great effect on readability and verification of programs.

2. *Naming Conventions*: The names of process and variables should reflect its purpose. It aids in verification and understanding.

3. *User-defined types*: The typing facility of modern language should be exploited, wherever applicable. This provides more clarity to the code.

4. *Module Size*: Modules should not be very big in size. A module should be around 50 lines of executable code (LOC). LOC is used as one of quality metric during coding phase.

5. *Module Interface*: More complex the interface, more will be the coupling. As a rule of thumb, any module having more than six parameters should be broken into multiple modules, if possible.

6. *Control Constructs*: The single entry and single exit constructs should be used. loop constructs with multiple exits should not be allowed. Gotos should be strictly prohibited in case of safety critical applications.

7. *Side Effects*: It should be clearly documented whenever sub units are having side effects. As a rule of thumb, functions should not be allowed to have side effects.

8. *Robustness*: A program is robust if it does something planned even for exceptional condition. This could be achieved by employing consistent *defensive programming* like validation of input data, checking of state data and protecting control flow. In the next section, few defensive techniques will be depicted through examples.

2.3.2.1  Programming Problem Areas.

1. *CASE Statements*: At no point, the corruption or inaccuracy of DATA should cause a change in control flow. The most prone to this is CASE statements as most compilers implement CASE statements using jump tables. Hence every CASE statement shall be protected by an 'IF' statement to guard the  CASE variable to be in the nominal range.

2. *Loops*: To protect control flow from data corruption, loops should have a defined static iteration count.

3. *Invariant conditions in loops*: Invariant condition in loop should be avoided.

4. *Constants and Variables*: It should be ensured that constants are statically determined and mapped to ROM area. Also, variables should not hold constant values and should be mapped to RAM area.

5. *Interfacing with Hardware:* A care must be taken to ensure that hardware mappings are computed at compile time. The hardware read routines should be capable of predictably handling both good and bad data.

6. *Fixed and Floating Point representation*: Many languages represent real in two forms - fixed and floating. Fixed representation has absolute error bounds and has a single zero whereas floating representation has relative error bounds and has multiple zeros. Fixed representation results are easily predictable and repeatable whereas floating representation results are less predictable and usually employ complex algorithms.

7. *Protection from Overflow*: The problem of overflow is specially encountered with trigonometric and inverse functions. It should be properly tackled by analyzing the boundary conditions and using defensive techniques.

### 2.3.3  Code Analysis and Code Inspection

Code analysis and Code Inspection are used to verify that code has adhered to specified standards, and is simple and easy to maintain. The code verification can be done by code walkthrough and independent audits. Language subset checker like SPARK Examiner can be employed to find any deviations from the subset. These deviations must be justified at code walkthrough. The static analysis on code should be performed at this stage. This should be ideally tool supported. Tools like LDRA testbed or Logiscope could be employed at this stage.

2.3.3.1 Static Analysis: The aim of static analysis is to detect errors, potential errors, and generate information about the structure of program that can be useful for documentation or understanding of program. Extensive static analysis reduces the effort during testing.

2.3.3.1.1 Data Flow Anomalies: Data flow anomalies are suspicious uses of data in a program. Data flow anomalies are technically not errors and can go undetected by the compiler. Usually, three types of data flow anomalies are found in a program, namely, (i) DD anomaly: Variable is not used between two definitions, (ii) DU anomaly: Variable is given a value which is never used and (iii) UR anomaly: Variable is undefined and then referenced.

Since ASA modules are passed through SPARK examiner successfully, all known insecurities in the Ada programming languages have been avoided in the development of ASA, most of the programming problems presented is avoided automatically and also the software is free from data flow and control flow anomalies.

2.3.3.1.2 Complexity Analysis: Complexity is an important factor affecting the productivity, cost estimation and time schedule. The complexity analysis should essentially be tool supported. The following complexity metrics are of interest in safety critical applications:

1. *LOC*: The number of executable or deliverable lines of code (LOC) is language dependent but traditionally trusted metric. For high level language like Ada, a LOC of 50 is considered good whereas for assembly, it may be 150 to 200.

2. *McCabe Complexity*: This metric represents the complexity of algorithm. It represents the number of distinct regions in flow graph of a program. McCabe should be kept low. A McCabe of 10 for 50 lines of high-level language code is good.

3. *Knot Count*: This metric represents the added complexity of implementation of algorithm. A knot is essentially the intersection of two or more control flow jumps. Depending on the directions of the control flow jumps, we can have three types of knots - up-up knot, down-down knot and up-down knot A Knot count of six to eight is acceptable for 50 line Ada program.

4. *Essential McCabe and Essential Knot Count*: The McCabe and Knot count are usually computed from flow graph of the program. For most of the cases, it is possible to perform reduction on this flow graphs by replacing structured programming primitives by a single node. The McCabe and Knot count of this resultant graph are called essential McCabe and essential Knot count. These are used to reflect the lack of structure in the program. For a structured program, essential McCabe should be 1 and essential Knot count should be 0.

5. *LCSAJ Density*: An LCSAJ (Linear Code Sequence And Jump) is characterized by three quantities: a Start point, an End point, and a point to which control flow jump is made. The start point is either the start of program or any point to which control may be jump during program execution. Then the

body of code through which the flow of control may proceed sequentially constitutes the linear code sequence. The end of the sequence is a point where control flow may perform a jump.

The LCSAJ density is a maintainability metric. If a line of code is to be changed then density informs the user how many LCSAJs will be affected by that change. If density is high then confidence that changes are correct for all LCSAJs will be reduced, and therefore an increased amount of regression testing will be required.

2.3.4 Static and complexity analysis results: The detailed static and complexity analysis results of ASA sub units are show in figure 7. The results are generated using LDRA Testbed analysis tool. Essential McCabe of all modules of ASA is 1 and essential knots is 0.

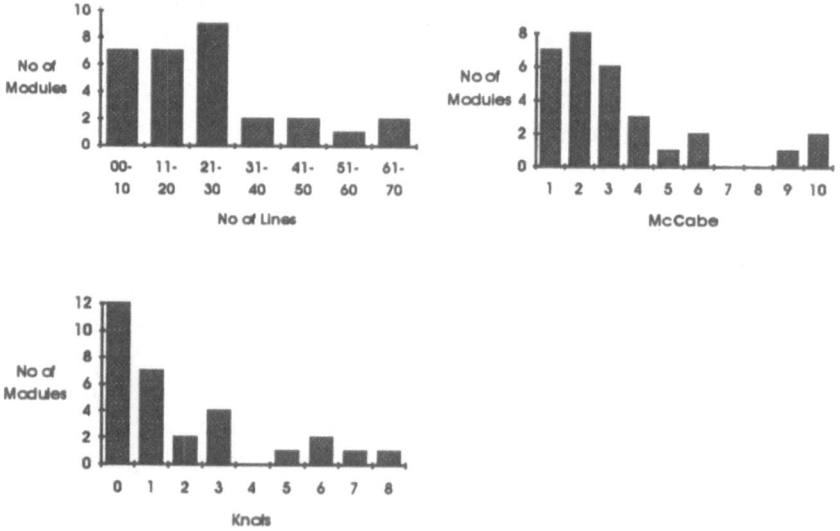

Figure 7 - Static and Complexity Analysis Results of ASA Subsystem

## 2.4 Software Testing

Testing is a dynamic method for verification and validation and plays an important role in quality assurance for software. During testing phase behavior of the system is observed.

### 2.4.1 Levels of Testing

The basic levels of testing are unit testing, integration testing, system testing and acceptance testing. These different levels of testing detect different types of faults.

Unit testing is used for testing a module or a small collection of modules. Its goal is to test internal logic of the modules. During integration testing, modules are combined into sub-systems, which are then tested. The goal here is to test the interfaces between modules. In system testing and acceptance testing, the entire system is tested. The goal here is to test the system against the requirements and test the requirements themselves. This is essentially a validation exercise.

### 2.4.2 Test Coverage

Test effectiveness is dependent on achieving a suitable level of test coverage. Test coverage can be split into two types - structural coverage and value coverage. The structural coverage ensures that all parts of code under test are exercised. The value coverage ensures that sufficient range of values is used to demonstrate the entire functionality.

2.4.2.1 Structural Coverage: The structural coverage criteria are based on the number of statements, branches and paths that are exercised by test cases.

2.4.2.2 Value Coverage: Since it is difficult to test every combination of inputs, input values should be divided into partitions within which the behavior of the modules is essentially same. The discontinuities of behavior should be tested by selecting test cases on and close to the partition boundaries covering (i) Just below minimum value, (ii) Minimum value, (iii) Intermediate value (iv) Maximum Value and (v) Just above maximum value.

2.4.2.3 LDRA Testbed: LDRA testbed is one of the testing tools. [LDRA 8.1] LDRA performs both Static and Dynamic analysis on the code and gives results in graphical format (flow graph, histogram, Kiviat diagram) also.

For dynamic analysis, the code is to be instrumented. The instrumentation inserts probes at the strategic points. Instrumented code is then compiled, linked and executed with test data. The execution trace is stored in a history file that is used to perform dynamic analysis.

All the modules of ASA were instrumented using LDRA Testbed and nearly 1500 test data vectors were generated to test the modules. The tests are repeated without instrumentation to ensure that the tool has not introduced any errors.

### 2.4.3 Test Results

Test results of ASA modules are as follows. 100% statement coverage, 100% branch coverage and 84% LCSAJ coverage have been achieved during testing. Uncovered LCSAJs have been reviewed and found to be not feasible to test.

## 3 Other Issues

### 3.1 Object Code Verification

Object Code Verification is essential since it is possible that the compilation process could introduce faults or unsafe features into the object code that could not

312

be detected by any of the above mentioned means has an effect on the final product.

## 3.2 Compiler Verification

Compiler verification is needed as it is difficult to perform 100% object code verification for the code produced for safety critical application. This paper cannot justify the full scope of compiler verification.

## 3.3 Run-Time System tailoring

The compiler run-time system (RTS) inherits the same safety criticality as the remainder of the application but unfortunately its design is fixed by the compiler supplier and hence it is necessary to tailor RTS.

All tailored and not tailored components remaining in the RTS must be independently audited to ensure correctness and freedom from unexpected behavior that might prejudice safety of the system.

## 3.4 Software Build Verification

The purpose of software build verification is to ensure
* The correct library units have been linked into the load (software to run on target)
* Units have been mapped to correct physical locations
* Each segment (code, constant and data) has been mapped onto the correct memory areas for e.g. Constants should be mapped to ROM area and variables should be mapped to RAM area
* The different memory regions don't overlap
* Sufficient Stack space has been allocated to satisfy the requirements of the code for all input conditions

## 3.5 Software Down Loading

Target loading process should be controlled by high integrity software and target image should be read back to ensure that load is correct to ensure that no computer can startup without correct software loaded on the aircraft.

## 4 Conclusion

A practical approach to develop software for safety critical application has been explained citing the example of Accelerometer Sensor Assembly sub-system of modern Digital Fly By Wire aircraft. Emphasis has been given in explaining the usage and need of safe subset of a language, quality metrics and software verification process. Programming issues with programming problem areas have been highlighted. Issues like compiler verification, object code verification, run-time system tailoring, and software build verification are also outlined briefly.

*Finally, there is only one certain fact that there is no such thing as certainty. In spite of performing all the above mentioned activities, we can not say with certainty that software is safe to use but we can say with higher level of confidence that software is safe to use. REMEMBER IN THE CONTEXT OF REAL TIME EMBEDDED SOFTWARE, SAFE MEANS LESS DANGEROUS.*

# 5 Acknowledgment

Authors acknowledge the contribution of Mr. Navin Srivastava for preparing this paper. Also authors thank the management of Aeronautical Development Agency, Bangalore, India for providing the necessary support for preparing this paper.

### References

[ARM83]    The Annotated Ada Reference Manual ANSI/MIL-STD-1815A-1983 (Annotated) 2nd edition, Kavr.A.Nyberg (editor).

[LDRA8.1]  LDRA Testbed Ada 9x (UNIX/VMS) Manual, Revision 8.1, September 1994.

[RDDRm]    RDD-100 Reference Manual, Release 4.0.3.

[SEM1.3]   SPARK Examiner Manual Release 1.3, Program Validation Ltd., May 1994.

[TEM 4.0]  TeamWork/Ada User's Guide Release 4.0

1. Ada is a registered trademark of the US Government, Ada Joint Program Office.
2. LCSAJ is a trademark of LDRA Ltd. and Subsidiaries.
3. LDRA testbed is a registered trademark of Program Analysers Ltd.
4. RDD-100 is a registered trademark of Ascent Logic Corporation.
5. TeamWork is a registered trademark of Cadres Technologies Inc.
6. VAX/VMS is a registered trademark of Digital Equipment Corporation.

# Session 8
## Applications I

# A Software Development Approach for Robotics Control Systems

Emilio RUIZ MORALES

Institute for Systems Engineering and Informatics

Joint Research Center, Ispra (Varese) - Italy

## Abstract

This paper presents the tailoring strategies used for applying the ESA Software Engineering Standards to the development of an embedded real-time robotics control system. The paper details also the software design approach.

## 1    Introduction

This paper describes the adopted strategies and methodologies for the design, development and testing of a safety-critical real-time embedded software devoted to control the *Robertino* robot at JRC Ispra.

In particular, the ESA Software Engineering Standards (PSS-05) [ESA] has been used for guiding the development process of this software project.

The aims of this paper are:

- to show how PSS-05 has to be tailored for this type of software projects, explaining the reasons for such a tailoring and the difficulties applying PSS-05;
- to detail the issues applying the Ward & Mellor [W&M] method for the software design phase and to suggest a more suitable software design approach.

The section 2 describes *Robertino* and its tasks to introduce the main functional features and quality the requirements of the *Robertino* Control System (RCS).

Section 3 details the development strategies for each project's development phase.

## 2    The *ROBERTINO* Robot and its Control System

### 2.1    Description of *Robertino* Tasks

The *Robertino* robot, located at JRC-Ispra site, is a Cartesian gantry robot with 4 degrees of freedom (see figure 1) able to carry a payload up to 6.5 tons over the whole workspace (2.2m x 3m x 6.5m) with an accuracy of 0.1mm.

*Robertino* is a 1/3 scaled version of the Blanket Handling Device (BHD) that will be used for blankets maintenance of the International Thermo-nuclear Experimental Reactor (ITER). *Robertino* is used to carry out the following tasks:

- validate the blankets design of the ITER fusion reactor by checking the ability of the blanket segments to be removed from a 1/3 scaled mock-up of an ITER fusion reactor section. As part of this activity, a 3D kinematics simulator,

containing a collision detection module and calibrated models, will be used to generate trajectories to be experimentally validated with *Robertino*;
- study remote handling operations for the substitution of these blanket segments: this involves the study of adequate gripping systems and attachments locks, dynamic aspects during the blankets substitution, in-vessel inspection systems and devices, and appropriate systems for reliable and safe control of remote handling operations.

In addition, the use of *Robertino* could be extended to industrial heavy robotics applications requiring the precise positioning, assembly and joining of large heavy components.

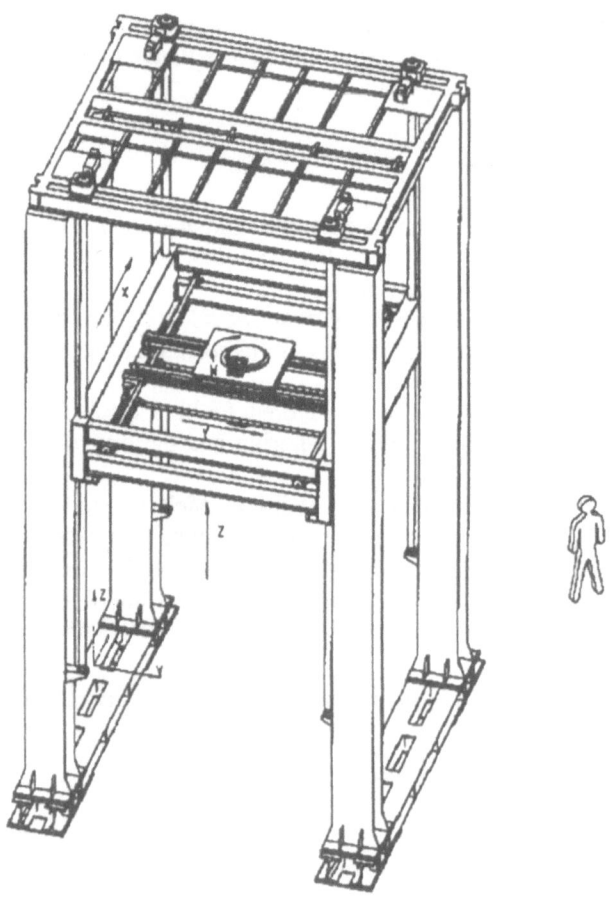

Figure 1: The *Robertino* Robot

## 2.2 Environment of the *Robertino* Control System (RCS)

The *Robertino* Control System (RCS) will be connected to (see figure 2):
- the Remote Handling Workstation (RHW) containing a kinematics simulator (TELEGRIP from Deneb Inc.) used for tasks planning and execution;
- the PC-based Platform for *Robertino* Instrumentation (PPRI) in order to synchronise the *Robertino* motion with its on-board data acquisition processes.

The RHW, the PPRI will also provide a user interface to remotely control the *Robertino* motion. A Hand Box and a Local User Interface Console are also available to locally control *Robertino*.

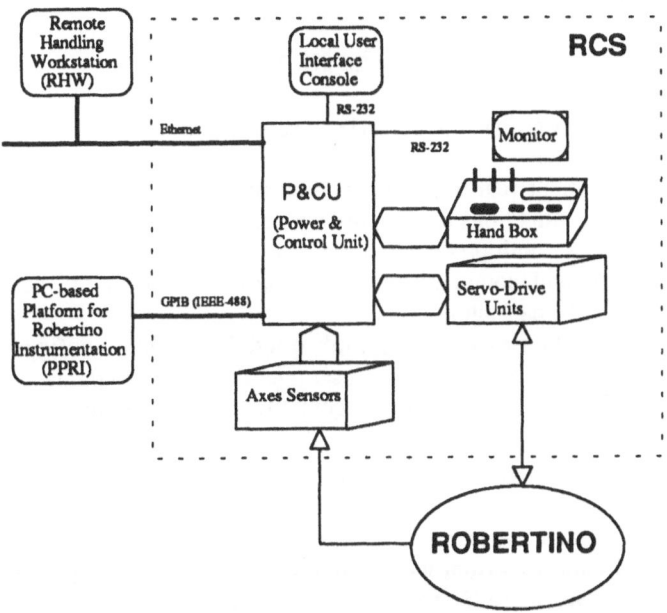

Figure 2: The RCS Environment

## 2.3 Quality Requirements of the *Robertino* Control System

The *Robertino* Control System (RCS) is an embedded real-time control system. Due to the nature of the *Robertino* tasks, RCS is a **safety-critical system** that has been designed taking into account a set of quality characteristics [ISO] in order to be:
- **safe** and **reliable** to minimise the risks that RCS hardware or software failures endanger humans, damage the reactor vessel, the transported items or *Robertino* itself;
- **adaptable** in order to allow RCS software functional expansions and integration of modular sensors-based systems enabled to interact with the motion controller;
- **configurable** to integrate additional end-effectors degrees of freedom to the axes motion control.

These quality requirements have been considered throughout the hardware design and the software development process. Regarding the software aspects, the software **maintainability** is another main quality requirement.

## 2.4 Hardware Platform for the RCS Software

The embedded hardware platform of the RCS software uses a VXI bus backplane [VXI] in order to facilitate the hardware integration of additional sensors-based modules required for experimental purposes. This backplane contains one MC68030 CPU for the Communication Interface (CI) with external systems and another MC68040 CPU for the Numerical Axes Control (NAC) functions.

## 2.5 Software Platform for the RCS Software

The VxWorks 5.1 operating system has been selected because it offers the following features:
- it is a performant real-time and multitasking operating system: adequate to run concurrent functions that require a precise and guaranteed timing;
- it supports a multiprocessor architecture: adequate for several CPUs synchronised and communicating with each other over the bus backplane (with the VxMP package);
- the VxWorks libraries are very complete and the kernel is configurable;
- a UNIX workstation is used as user interface for programs development, compilation, downloading to the target CPU, and debugging.

## 2.6 RCS software characteristics

The RCS software is a **real-time software** characterised by the fact that it has to manage several interfaces from which unexpected asynchronous events can arrive requiring a deterministic RCS response. For instance, the user interface commands have to be handled in 500 msec while emergency conditions must be processed in 50 msec. Deterministic behaviour is also required for the motors servo-loops execution (run every 5msec) and the micro-interpolation (run every 40 msec and synchronised with the motors servo-loops).

On the other hand, the transitions between RCS software states depend essentially on external and internal events. Furthermore, for each state, combinations of events can cause transitions to other states; this implies that the software has a high number of states and transitions, and that the **RCS software complexity results more from the behavioural features rather than from the functional aspects**. Therefore, the software design phase has to be performed carefully with an appropriate methodology (see section 3.6).

In addition, the RCS software is also an **embedded system** requiring a certain degree of autonomy to ensure the system safety.

The RCS software is about 25.000 LOC and the required resource for its development is about 4 man/year spread over a period of 2 years.

As PSS-05 defines a small projects as requiring less than 2 man/year but a large projects as requiring more than 20 man years of effort, we will consider the RCS software project more as a **small project**.

As far as functionality is concerned, the RCS software is divided into two functional modules:

1.  the Communication Interface (CI) module, installed on the CI-CPU
2.  the Numerical Axes Control (NAC) module, installed on the NAC-CPU.

### 2.6.1   NAC Module Capabilities

The NAC module provides the following capabilities:
*   motors synchronisation: for those axes composed by several motors;
*   trajectory generation with smooth acceleration profiles in order to avoid the induction of vibrations on the transported load;
*   execution of the generated trajectory consisting of a position profile;
*   axes motion through joysticks commands (manual mode);
*   axes motion in point to point mode with linear trajectories and synchronised axes;
*   automatic reset procedure for axes position;
*   single motors motion mode;
*   axes status, axes position and error messages monitoring on Monitor;
*   interface with the Hand Box, the PLC, the motors speed variators and the *Robertino* motion sensors;
*   errors reporting to CI;
*   connection monitoring with the CI module;
*   software configuration from ASCII files for the number and type of axes to control, etc. .

### 2.6.2   CI Module Capabilities

The CI module provides the following capabilities
*   communication interfaces management with the following external devices: the Remote Handling Workstation (RHW), the PC-based Processing Platform for *Robertino* Instrumentation (PPRI) and the Local User Interface Console;
*   checks on the communication links with external devices;
*   motion control sessions management for the external devices;
*   motion monitoring sessions management for the external devices;
*   error logging on files;
*   software configuration from ASCII files for the number and type of devices to be connected to RCS;
*   connection monitoring with the NAC module.

# 3 Strategies for the RCS Software Project

## 3.1 Selection of a Standard for the Software Development

Due to the safety-critical nature of the RCS software, the product quality has particularly been emphasised. The strategy consisted of adopting the "PSS-05 European Space Agency's Standard for Software Development" [ESA] in order to:
- obtain a maintainable software by a formalised documentation;
- follow a consistent development process;
- ensure the software product quality in terms of safety and reliability.

The selection of PSS-05 has been done after a survey of the publicly available software development standards. PSS-05 is based on IEEE standards but presents an organised set of recommendations that make this standard the most complete among the publicly available standards. Indeed, the standard is general enough to cover all types of softwares (i.e. databases, user interfaces, real-time softwares, etc.).

PSS-05 is a typical example of a traditional waterfall type of software development in which every development phase is rigorously defined. For instance, at each development phase, PSS-05 settles the required inputs and outputs documents, a set of recommendations on how to carry out the phase activities and the management and Quality Assurance procedures to ensure the final product quality. The PSS-05 practises must be applied in all the ESA's projects but are also increasingly used in other industrial fields.

## 3.2 Tailoring of the PSS-05 Standard

The application of the PSS-05 standard has given some difficulties in our specific project context.

Firstly, the standard requires many documents to be produced and an organisational structure which is too costly for small to medium sized projects. Therefore a specific tailoring of the standard had to be done for the particular needs of the RCS software project. Unfortunately, PSS-05 does not advise the way to perform such a tailoring.

Secondly, the standard covers many types of softwares but it has been noticed that the concept of each life-cycle phase can differ according to the type of software. This is particularly true for the real-time embedded systems and in particular during the user requirements phase and the software requirement phase. As a consequence, the standard should clarify these points by practical examples for each software type, and also advise the way to tailor the standard for these specific applications.

The PSS-05 tailoring for the RCS software project consisted of the following adaptations:

1. reduce as far as possible the number of documents to be produced guaranteeing the development and product quality: this means combining project plans, combining the requirements documents and combining the design documents;
2. reduce duplication between the source code and the detailed design documents;
3. combine the system test and the acceptance test specification;
4. insert a software safety analysis at the life cycle phases;
5. organise project reviews at the end of each life cycle phase.

Next sections detail how these guide lines have been practically implemented in the RCS software project.

## 3.3 Project Plans

The PSS-05 requires the definition of the following plans at the start of a project:
- the Software Project Management Plan;
- the Software Configuration Management Plan;
- the Software Verification and Validation Plan;
- the Quality Assurance Plan.

At each development phase, these plans have to be updated in order to control the project development.

However, In small projects as in the RCS software project, it is preferable to combine these plans in a general software project management plan. This plan should be detailed when starting the project and updated at each phase of the life cycle.

## 3.4 Requirements Phase and Document

PSS-05 divides the specification phase into the user requirements phase and the software requirements phase, the user being considered as the customer of a software product.

This section shows why and how these two phases can be combined into one single requirements phase in the case of embedded systems. This principle could also be applied to small projects for economical reasons.

### 3.4.1 The User Concept in Embedded Systems

In the case of our real-time embedded system, the user in the User Requirements Document (URD) is considered as the hardware using the software. In particular, the RCS software user is the *Robertino* robot itself.

Hence the user requirements are functional requirements describing in detail the system behaviour and capabilities with respect to the hardware interface signals and to the tasks to be performed.

### 3.4.2   Assumption on Hardware Design

As the functional requirements depend strongly on interface signals, the project stability would be compromised if the specifications of the hardware interface signals suffer changes during the software development. As a consequence, the hardware design of the embedded system should be complete before starting the software project. Actually, PSS-05 assumes that the hardware interfaces design have been done before starting the software project.

### 3.4.3   Constructing a Logical Model with the Ward and Mellor method

PSS-05 requires to build the logical model during the software requirements phase to allow reasoning about the software in order to find software requirements; in addition, the logical model is also used for identifying the software components.
In our case, the Ward and Mellor method for real-time systems [W&M] has been applied to build the essential model, i.e. the logical model. The method consists of:
- building a RCS context diagram that details the data exchanges with the external systems;
- listing all the external events and the related actions by the software system;
- constructing the essential model, i.e. the logical model, identifying the main functional transforms and the data flows between them: many iterations have been required to obtain a consistent and complete model.

Afterwards, the model is used for requirements identification.

### 3.4.4   Rules Adopted to State Requirements

In the RCS software project, the following rules have been followed to state the requirements:
1. Identify main requirements as main functional areas: operation modes, abnormal conditions handling, interfaces capabilities, etc.
2. Decompose main requirements in sub-requirements.
3. Avoid duplication of sub-requirements in different main requirements.
4. Define sub-requirements as functions that manage events (commands, signals, etc.) specifying clearly the execution conditions and the related actions; the execution conditions and the actions should be considered as requirements of sub-requirements as well.
5. Formalise sub-requirements in structured English type, if possible.

These rules will facilitate the internal states identification in the software design phase, and the design of acceptance tests.

### 3.4.5   Merging the PSS-05 User and Software Requirements Documents

As the user and software requirements have the same meaning in our case (see section 3.4.1), this view is particularly strengthened in the case of embedded systems, the URD can be combined with the Software Requirements Document (SRD) in a single 'Requirements Document' (RD) to avoid useless duplication.

This requirements document is to be used by the customer as base-line for the acceptance tests and integration tests specification but also by the developer as input to the software design phase.

Merging of requirements documents should be considered by PSS-05 in the case of small software projects for embedded systems.

### 3.4.6    Structure of Requirements Document

As the only differences between the SRD and the URD are the requirements classification and the logical model description of the SRD, the RD structure should consist of the sections common to the URD or the SRD. However, a descriptive section on the system capabilities and the hardware interfaces should be added to improve the readability and clarity of the requirements.

If the requirements are stated in terms of capability and constraint requirements (as in the URD) then a table identifying the type of each requirement (i.e. functional requirements, performance requirements, interface requirements, etc.) should be added in appendix (see [URD]).

The Logical Model should be added in appendix or in a separate document.

## 3.5    Software Design Phases and Documents

PSS-05 divides the software design activities in two phases:
1.    the Architectural Design (AD) phase: the aim is to identify the software components, their interfaces and their behaviour; afterwards, the project can be divided into sub-projects and assigned to different teams of developers;
2.    the Detailed Design (DD) phase: each single software component is detailed in terms of single functions by the developer.
The results are reported in the Architectural Design Document and the Detailed Design Document following the PSS-05 structure.

## 3.6    The Architectural Design Phase

### 3.6.1    Issues of the Logical Model Built Using Ward & Mellor

In our project, the software components identification has been obtained refining the logical model of the requirements phase. Although the logical model was complete, it was not easily readable due to:
• the high coupling between data transforms in terms of data flows;
• the complex behavioural representation by control transforms: for a system with such complex behaviour, the diagrams containing control transforms were overloaded by functions activated during the transitions;
• the shared data: accessed by too many processes and suspected to cause dead-locks problems.

Therefore, this model was not adequate to obtain an implementation model. The design issues were mainly due to the Ward an Mellor method which is not quite suitable for systems having such a complex behaviour and so many input-outputs.

### 3.6.2 Refinements of the Logical Model

The Logical Model had to be rearranged to provide a clear definition of the software components. The RCS software designers decided to apply a components design and description inspired on the Hierarchical Object Oriented Design concepts (HOOD) (see [HOOD] and [ERXA]). Namely, the key points in the model rearrangements were:

- minimise the communication between main processes by a message exchange protocol;
- minimise access to global data;
- define processes as entities behaving autonomously providing and requesting services through their interfaces.

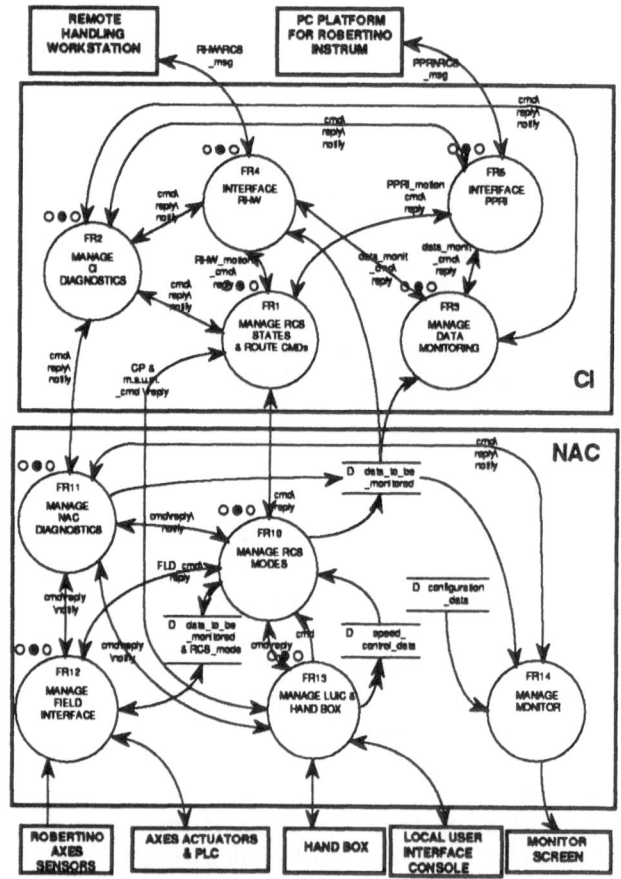

Figure 3: RCS Software Objects (using System Architect case tool).

In addition, the components were classified in static or dynamic objects and the HOOD hierarchy principles have been applied to objects organisation. As a result, the expandability, modularity and maintainability (including namely the ease for installation and debugging) requirements have been ensured by the resulting architecture. Figure 3 shows a view of the RCS software objects.

During this activity, the designer's experience was essential to obtain a model providing a clear definition of the software components. In fact, the philosophy of the error handling, the configuration modalities and the process intercommunication have been clearly defined as well.

### 3.6.3   Sub-projects Definition

This RCS architectural design has enabled the project division into two sub-projects: one concerning the Numerical Axes Control (NAC) module and another regarding the Communication Interface module. The NAC module has been outsourced to a software house specialising in robotics and automation systems (ERXA s.r.l. from Torino-Italy), and the CI module has been developed in-house at JRC-Ispra.
This clear component specification has facilitated the test cases specification for integration testing of both and single CI and NAC modules.

## 3.7   The Detailed Design Phase

After the architectural design definition, the detailed design of the objects has to be performed. The detailed design consists of defining the implementation aspects of the objects, i.e. the internal functions, the functions organisation, the interfaces specification, the internal data structures and values.
However, the functional specifications had to remain at a general level to avoid duplication between code and detailed design. A set of coding rules were specified at this stage to provide a readable commented source code.
The detailed design document structure could be the same as the one of the architectural design document. In the RCS project, one detailed design document has been written for each object.

## 3.8   Project Reviews

### 3.8.1   Requirements Review

During the RCS requirements definition phase, a major review has been organised involving users, developers and external consultants from industry and university in order to guaranty the quality of the document in terms of clarity, understandability, completeness, consistency and verifiability.
As the project stability depends on the requirements phase and as this document is taken as base-line for acceptance tests, small projects should consider this a useful practice.

### 3.8.2   Architectural Design Review

Before passing to the Detailed Design phase, the user and the developer should reviewed the Architectural Design Document in order to ensure that all the requirements were met by the software architecture.

### 3.8.3   Detailed Design Review

Before coding, the detailed design was reviewed in detail to check possible implementation conflicts with the requirements.

### 3.8.4   Acceptance Tests Specification Review

This review is important to check the tests coverage versus requirements.

## 3.9   Software Safety Study

In case of safety critical softwares, the project management should consider a safety analysis at the requirements phase and the architectural design phase.
The safety analysis can be divided into two steps:
1.    the Failure Mode, Effects and Criticality Analysis (FMECA): its purpose is to understand how the software can fail and to recommend how failures can be prevented, thereby improving software reliability;
2.    the Fault Tree Analysis (FTA): its objective is to analyse the effects of faults and measure the fault tolerance of each capability.

A preliminary safety analysis has been applied to the requirements phase of the RCS project. The results are:
*   FMECA: about 40 failure modes that could result in damage to people, property or *Robertino* itself have been identified in the RCS software; 239 failure modes were identified in total. As RCS has little redundancy, it has been recommended to inspect the code's critical functions and to achieve a 100% branch coverage during testing.
*   FTA: the result of the FTA is that the RCS contains very little redundancy at the functional level; this means that every function is a single point failure and that more fault tolerance should be given into the design; this has led to a design review to improve fault tolerance.
This study has been performed with the support of external consultancy form CRAY systems.

## 3.10   System Tests and the Acceptance Tests

### 3.10.1   Combining System Tests and the Acceptance Tests Phases

As the user requirements and the software requirements phases have been combined, the acceptance tests and system tests phases should be combined as well.

In the RCS software project, the way the requirements have been settled has facilitated the test cases identification and design (see section 3.4.4. and [ATS]).
This phase should be done at the end of the software design to avoid to re-writing documents due to requirements changes or refinements during the project development.

### 3.10.2  Tests Coverage

The acceptance tests document shall provide test design and test cases definition addressing the functional requirement of the software. The functional requirements coverage by test cases must be of 100%.
In the case of safety critical software, the testing effort should be concentrated on statement coverage that should be of at least 90%. The statement not covered by the testing should be reviewed. This can be quite expensive in effort for small projects, therefore each small project should set a statement testing target.

### 3.10.3  Tests Results

Tests on *Robertino* have been reported in the Test Report Document. The testing strategy consisted of dividing the tests campaign in two steps:
1.  perform all test cases on a test rig simulating *Robertino* axes and sensor signals: this would allow to test all RCS functionalities and abnormal situations that could damage *Robertino* itself;
2.  perform functional test cases on *Robertino*.
At the current phase of the project, the NAC sub-project has been tested on the test rig and the CI is still under unit testing.
During NAC tests on test rig, 30 failures over the 600 acceptance test cases have been identified until now. Therefore, such a positivie result has to be granted to:
*  the design method;
*  the effort devoted to the requirements and design phases;
*  the safety analysis;
*  the developers experience.

However, although the author is confident on the results, the test campaign must still to be completed with the CI component and on *Robertino*.

## 4    Conclusions & Future Works

The RCS project is an example of tailoring PSS-05 for a small safety-critical software project of a real-time embedded robotics control system.
Although each software project is a different world according to its requirements, the presented RCS project can be considered as a useful experience to other software robotics control systems applications.
Next issues of PSS-05 should consider suggesting a PSS-05 tailoring for small projects according to the software type. Practical examples on how the standard has been applied in practical cases should also be added to clarify usefully PSS-05.

Finally, the RCS software development will be very used as a test bed for the implementation of a dependability measurement framework [DEPEND].

# 5 Acknowledgements

The author wishes to thank particularly Marc Wilikens for his helpful personal involvement to the safety and reliability aspects of the *Robertino* project.
A special thank also to Flaviano Farfaletti-Casali, David Maisonnier and the ERXA people for their support and contribution to the RCS project.

# 6 References

[ATS]          Acceptance Test Specification for the Robertino Control System
               Project, JRCs Technical Note N°I.94.162 ISEI/IE/2825/94 Issue
               1.1, E. Ruiz Morales, JRC-Ispra, July 1995.

[DEPEND]       Dependability Measurement of Safety Critical Systems Feasibility
               Study, JRC Technical Note N°I.94.116 ISEI/IE/2776/94, T.O.
               Jackson, J.A. McDermid, I.C. Wand, M. Wilikens, August 1994.

[ERXA]         Metodologia di Progetto, Reference CFMD/MDP Issue 1.1, M.
               Ricci, Erxa s.r.l., 1994.

[ESA]          ESA Software Engineering Sandards, ESA PSS-05-0 Issue2,
               Board of Software Standardisation and Control, ESTEC
               publications, 1991.

[HOOD]         Hood Reference Manual, Issue 3.1, Reference HRM/91-07/V3.1,
               HOOD Working Group, ESTEC publications, 1991.

[ISO]          ISO/IEC 9126, Information Technology - Software Product
               Evaluation - Quality characteristics and guidelines for their use,
               1991.

[URD]          User Requirements Document for the Robertino Control System
               Software, JRC Technical Note N°I.94.159 ISEI/IE/2828/94 Issue
               1.2, E. Ruiz Morales, JRC-Ispra, July 1995.

[VXI]          VXIbus System Specification, Revision 1.4, VXI bus Consortium,
               National Instruments, 1992.

[W&M]          Structured Development for Real-Time Systems, P.T. Ward &
               S.J. Mellor, Yourdon Press, 1985. (Three Volumes),

# An Attempt to Evaluate Functional Diversity Employed in a Reactor Protection System

Jörgen Christmansson, Zbigniew Kalbarczyk, Jan Torin
Laboratory for Dependable Computing
Department of Computer Engineering
Chalmers University of Technology
S-412 96 Göteborg, Sweden

## 1. Introduction

Computers are currently employed to control safety critical applications such as nuclear power plants and aircraft. A failure in a computer system that controls a safety critical application can lead to significant economic losses or even the loss of human lives. The requirements for computers used in safety critical applications is therefore in the order of $10^{-10}$ - $10^{-8}$ failures per hour. Software faults are believed to account for many operational failures in safety critical systems, whereas hardware is known to be quite dependable owing to the use of extra hardware. The software dependability can be improved by:

(i) Methods and techniques (e.g. reliable design, testing and verification) aimed at producing a fault-free software, i.e. fault avoidance; and

(ii) Methods and techniques (e.g. error processing) aimed at providing a service complying with the specification in spite of faults, i.e. fault tolerance.

The presently obtainable level of verifiable dependability (i.e. with the use of fault avoidance) in a single software unit before operational use is about $10^{-4}$ failures per hour [Laprie 91]. However, there are cases in which actual field data for software failures have shown operational dependability to be better than pre-operational forecasts (e.g. $10^{-6}$ failures per hour [Kanoun 87]). Nevertheless, there is still a gap between operational dependability and the requirements. Fault tolerance has been proposed as a means to fill the gap.

The basic principle for fault tolerance is redundancy, and systems using this principle can adequately tolerate independent faults (i.e. faults that are attributable to different causes). However, related faults (i.e. faults that are attributable to a common cause) are a serious problem as they can provoke common cause failures (CCF). Design diversity, data diversity and functional diversity have been proposed as means to deal with related faults.

In design diversity [Avizienis 85], [Randell 75], programs must exist in several versions and be produced using the same specifications by different programming teams. There are doubts, however, about the benefits of software design diversity with respect to gains in fault tolerance [Knight 86], [Bishop 86] as a result of the probability that the same bugs are produced in several of the program versions and the extra cost for redundant software. Design diversity has been the subject of a number of evaluations and will not be considered in this paper.

Data diversity [Ammann 87], [Christmansson 94] requires an algorithm for the generation of logically equivalent data sets (data re-expression) and requires no design diversity. The fault tolerance of a system employing data diversity depends upon the ability of the re-expression algorithm to produce data points that lie outside the failure domain (the set of input points that cause program failures), given that an initial data point caused a failure. Data diversity will not either be investigated in this paper.

Functional diversity exploits the fact that some systems have multiple ways of achieving the same end result. This paper will try to evaluate the ability of functional diversity to reduce the probability of a system failure (unavailability). A sub-system, an emergency core-cooling system (ECCS) in a reactor protection system (RPS) of a boiling water reactor, was used as an example for the analysis. The dependability of a system can be reduced by CCFs, and the impact of a CCF on the dependability of a system employing functional diversity is therefore estimated. The evaluation and estimation are done with a probabilistic risk assessment technique called fault tree analysis. The quantification of the fault trees indicates that functional diversity with functional separation can contribute to a great improvement in dependability as compared with a system without functional diversity.

The description is organized according to the following: the Emergency Core Cooling system with three candidate architectures is presented in section 2. The failure logic of the architectures considered is presented as fault trees in section 3. Section 4 describes the approach used to model common cause failures. The results of the quantification of the fault trees (the system unavailability as a function of time) are presented in section 5. A summary and concluding remarks are given in section 6.

# 2. An Emergency Core Cooling System

A subsystem, an emergency core cooling system (ECCS) in a reactor protection system (RPS) of a boiling water reactor, was used as an example for the analysis. A channel in the ECCS consists of: sensors (e.g. reactor pressure vessel instrumentation), components for communication, adjudicators and actuator chains. The sensors monitor specific plant parameters, e.g. temperature, pressure and level. Whether an actuator chain shall or shall not take protection actions is decided by an adjudicator on the basis of two-out-of-four logic. That is, at least two channels must signal that plant parameters have exceeded predefined limits.

The ECCS under consideration has four identical channels that communicate via four (see figure 1) or eight asynchronous communication links. Each channel works as an independent cyclic (time-triggered) system. An operator can monitor the ECCS via the safety section overview (SSO), which is continuously updated by each channel.

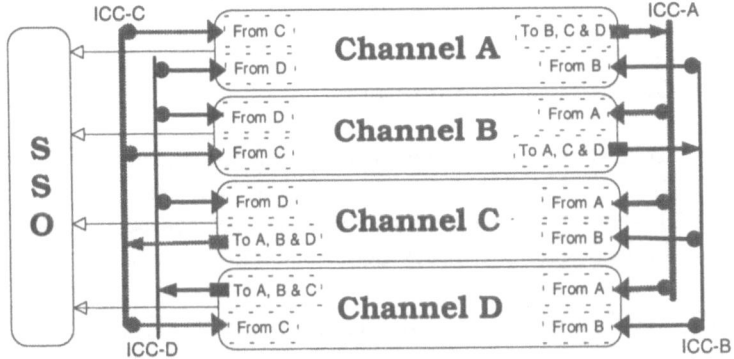

Figure 1. Communication between channels in the ECCS.

In this study, three different architectures of the system are considered:

- Configuration 0, ECCS without functional diversity.

- Configuration 1, ECCS with functional diversity, although without functional separation.

- Configuration 2, ECCS with functional diversity and functional separation.

These configurations will be described by block diagrams in the chapters below.

## 2.1. Abbreviations Used in the Block Diagrams

| | |
|---|---|
| CI | Containment Instrumentation |
| CI-HP | Containment Instrumentation for High Pressure Emergency Core Cooling |
| CI-LP | Containment Instrumentation for Low Pressure Emergency Core Cooling |
| ECC | Emergency Core Cooling |
| ECCS | Emergency Core Cooling System |
| HP-ECC | High Pressure Emergency Core Cooling |
| ICC | Inter Channel Communication |
| ICC-HP | Inter Channel Communication between High Pressure Emergency Core Cooling function |
| ICC-LP | Inter Channel Communication between Low Pressure Emergency Core Cooling function |
| LP-ECC | Low Pressure Emergency Core Cooling |
| MMI | Man Machine Interface |
| PL-HP | Protection Logic for High Pressure Emergency Core Cooling |

| PL-LP | Protection Logic for Low Pressure Emergency Core Cooling |
|---|---|
| PRS | Pressure Relief System |
| RPVI | Reactor Pressure Vessel Instrumentation |
| RPVI-HP | Reactor Pressure Vessel Instrumentation for High Pressure Emergency Core Cooling |
| RPVI-LP | Reactor Pressure Vessel Instrumentation for Low Pressure Emergency Core Cooling |
| SSO | Safety Section Overview |
| SSSF | Safety System Support Features |
| TB | Token bus |
| TB-HP | Token bus communication between the components that form the High Pressure Emergency Core Cooling function |
| TB-LP | Token bus communication between the components that form the Low Pressure Emergency Core Cooling function |

## 2.2. Configuration 0, ECCS without Functional Diversity

The implementation of one of four identical channels of the ECCS for configuration 0 is shown in figure 2. The components in one channel communicate via the token bus (TB). An important characteristic of this configuration is that the computer components in one channel have: different application software, identical system software and the same computer hardware. The ECCS shall monitor the reactor and provide emergency cooling when required by the process.

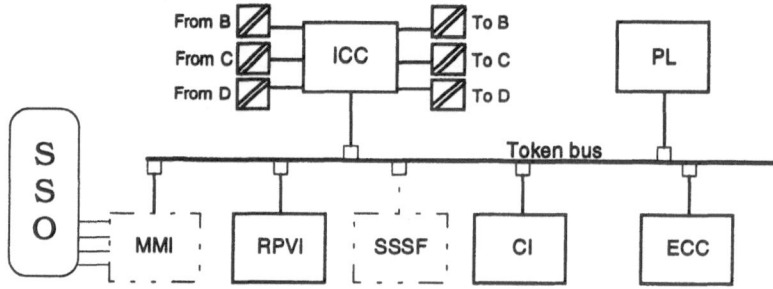

Figure 2. Block diagram of channel A in configuration 0.

The process is monitored by means of the instrumentation (RPVI & CI), and two plant parameters are observed: the level of water in the vessel and the temperature in the reactor containment. The protection logic (PL) determines whether any of the two sampled parameters has exceeded predefined limits. The result of that judgement is sent to the other channels via the inter channel communication (ICC) unit. The PL furthermore decides whether the channels' emergency core cooling (ECC) shall or shall not be started on the basis of two-out-of-four logic. That is, at least two channels must signal that plant parameters have exceeded predefined limits. The ECC operates as a high pressure coolant injection system; to cope with the loss of coolant, it is sufficient for two channels to start their ECC. The ECC takes its water from special storage tanks.

When the ECC pump has been started and the power supply to the pump fails, a standby power diesel generator must be started within 30 seconds. That generator can be started by the safety system support feature (SSSF) or by an operator. The human operators can monitor the ECCS via the SSO, which is continuously updated from each channel's man machine interface (MMI). Note that the rectangles that represent MMI and SSSF in figures 2, 3 and 4 have dashed lines. This is meant to indicate that those components are not considered in the fault trees.

## 2.3. Configuration 1, ECCS with Functional Diversity

The implementation of one channel of the ECCS for configuration 1 is shown in figure 3. This configuration exploits the fact that ECC can be achieved in multiple ways: high pressure emergency core cooling (PL-HP and HP-ECC) and low pressure emergency core cooling (PL-LP, PRS and LP-ECC). The diversified functions are not functionally separated, however, as they use the same instrumentation, communicate within the channel via the same bus and communicate with other channels via the same component (ICC). The computer components in one channel of this configuration have: different application software, identical system software and the same computer hardware.

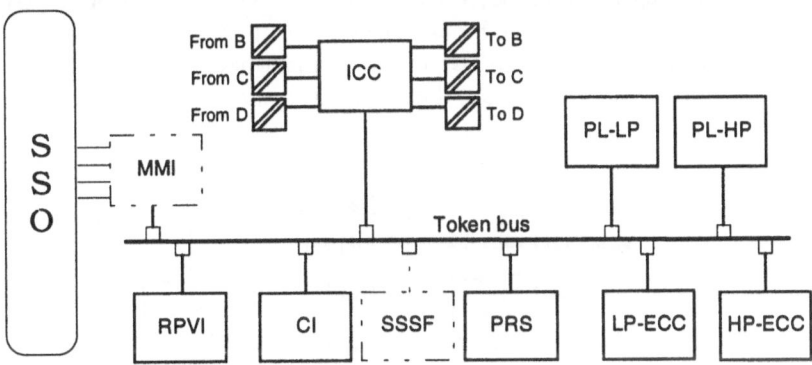

Figure 3. Block diagram of channel A in configuration 1.

The HP-ECC function is the same as that described in the previous section. The LP-ECC function monitoring the process uses three plant parameters: the level of water in the reactor pressure vessel, the pressure in the reactor containment and the derivatives of the pressure in the containment. The protection logic for low pressure emergency core cooling (PL-LP) determines whether any of those sampled parameters have exceeded predefined limits. The result of that judgement is sent to the PL-LP in the other channels via the ICC unit. The PL-LP furthermore decides whether the channels' pressure relief systems (PRS) and LP-ECC shall or shall not be started on the basis of two-out-of-four logic. The LP-ECC operates at low pressure, feeding spray water into the reactor when the pressure drops as a result of the PRS actions.

When the ECC-HP or ECC-LP pumps have been started and the power supply to a pump fails, a standby power diesel generator should be started within 30 seconds. The ECC-LP draws water from a condensation pool inside the containment. The water in this pool is cooled by a containment vessel spray system, which in turn is cooled by sea water via a heat exchanger. The condensation pool can accept decay heat for about eight hours without any cooling by sea water. Thus, the containment vessel spray system must be started within four to six hours. The generator and the spray system can be started by the SSSF or by an operator. The human operators can monitor the ECCS via the SSO, which is continuously updated from each channel's MMI.

## 2.4. Configuration 2, Functional Diversity and Separation

The implementation of one channel of the ECCS for configuration 2 is shown in figure 4. This configuration (as configuration 1) exploits the fact that ECC can be achieved in multiple ways: high pressure emergency core cooling (RPVI-HP, CI-HP, PL-HP and HP-ECC) and low pressure emergency core cooling (RPVI-LP, CI-LP, PL-LP, PRS and LP-ECC). This means that components with the same function (e.g. PL-HP and PL-LP) have different application software, identical system software and the same computer hardware. The diversified functions are also functionally separated, as they use different buses within the channel (TB-HP and TB-LP) and communicate with other channels via different components (ICC-HP and ICC-LP). The ICCs and TBs in one channel have: identical application software, identical system software and the same computer hardware. However, the ICCs and TBs have different workloads, e.g. the software in TB-HP and TB-LP will be exercised in different ways.

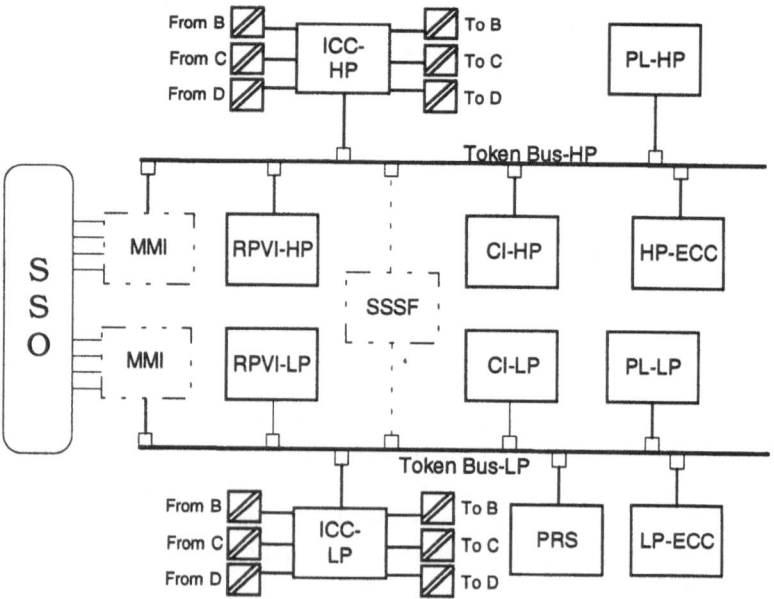

Figure 4. Block diagram of channel A in configuration 2.

The functions of HP-ECC, LP-ECC, SSSF and MMI are the same as those described in the two previous sections.

# 3. Fault Trees for the Different Configurations

Prior to the construction of fault trees, a preliminary Failure Mode, Effect and Criticality Analysis (FMECA) was carried out on the basis of the block diagrams of the three possible configurations for a ECCS. Only the failure mode crash failure (i.e. a component loses its internal state or halts for more than one cycle) was used in the FMECA. The FMECA was a useful basis for the development of fault trees, as it forces a systematic and structured walk-through of the system with respect to component failures and their consequences.

The component level fault trees for configurations 0, 1 and 2 are shown in figures 5, 8 and 11, respectively. In those fault trees, the component failures were divided into independent and related failures. The related failure events were split into: events that affect three components (CCF 3ch.); and events that affect four or more components (CCF 4ch.). Note that failures in MMI and SSSF, as well as related failures between two components, are not considered in the fault trees. Observe also that the basic events (e.g. independent failure ICC channels A, B, C and D) are not shown in the presentation of the trees.

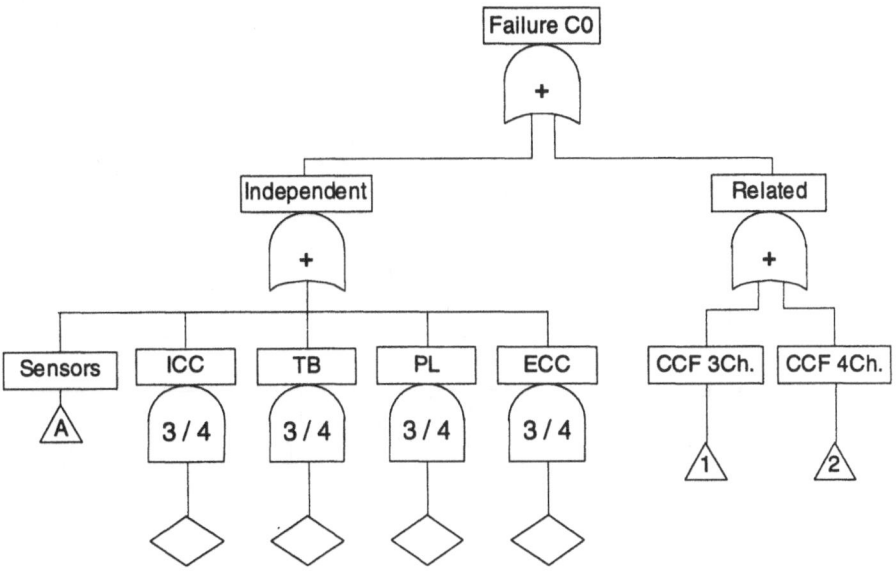

Figure 5. Component level fault tree for configuration 0.

## 3.1. Fault Trees for Configuration 0

Configuration 0 (see figure 5) fails if at least three out of four redundant components fail either independently or in a related way. Related failures between three and four redundant components are shown in subtrees one and two (see figure

6), respectively. The sensors are an exception to this, however, as both the CI and the RPVI must fail (independently or owing to a common cause) in at least three of four channels. Independent failures of the sensors are modelled in subtree A, which is shown in figure 7. This figure also shows related failures of six or more sensor systems in subtree 3.

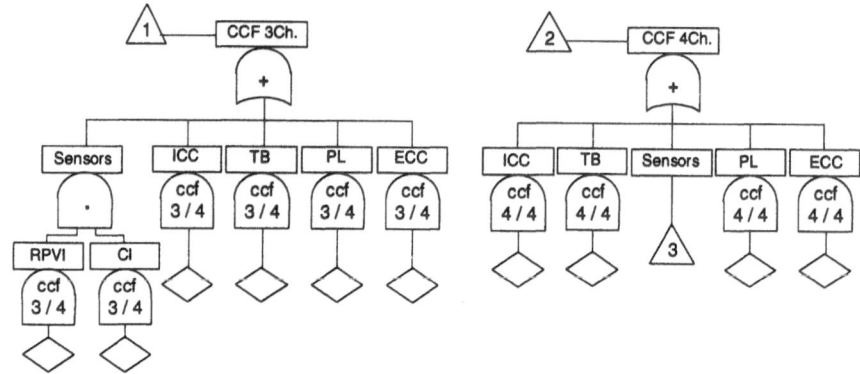

Figure 6. Subtrees 1 and 2, related failures three or four redundant components.

The sensor systems in one channel (RPVI and CI) have the same functionality, different application software, identical system software and the same computer hardware. Related failures can consequently occur within the group of eight sensor components. That is, six, seven or eight of eight instrumentation components in the four channels can fail as a result of a common cause. This is modelled by the branch "CCF All" in subtree 3, which is shown in figure 7.

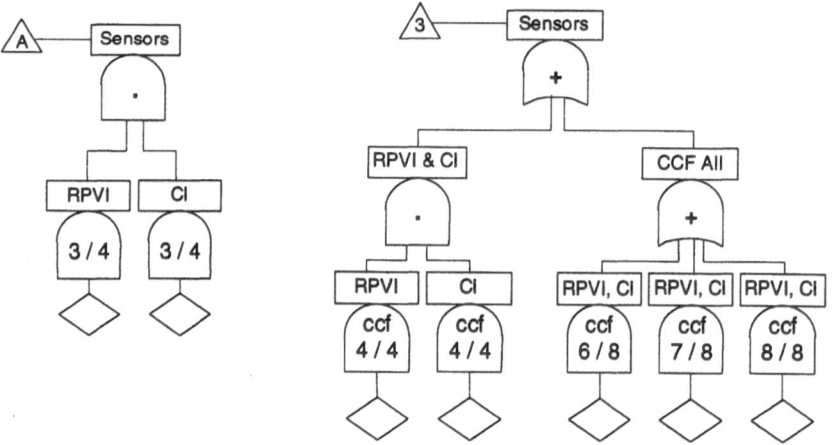

Figure 7. Subtrees A and 3, independent or related failures for the sensors.

## 3.2. Fault Trees for Configuration 1

There are two main groups of redundant components in configuration 1, those that handle communication (TB and ICC) and those that have similar functions ({RPVI, CI}; {PL-HP, PL-LP}; {HP-ECC, (PRS, LP-ECC)}). Configuration 1 (see figure 8) fails if at least three of four redundant communication components within one group fail, or if at least three of four components within one group fail for both the high pressure train and the low pressure train. Related failures can occur between six or more components of the groups with similar functions (i.e. instrumentation, protection logic and actuators). That is, six, seven or eight of eight components in a group of redundant components can fail as a result of a common cause. This is modelled by the branch "CCF All" in subtrees 3, 5 and 6 (see figures 7 and 10).

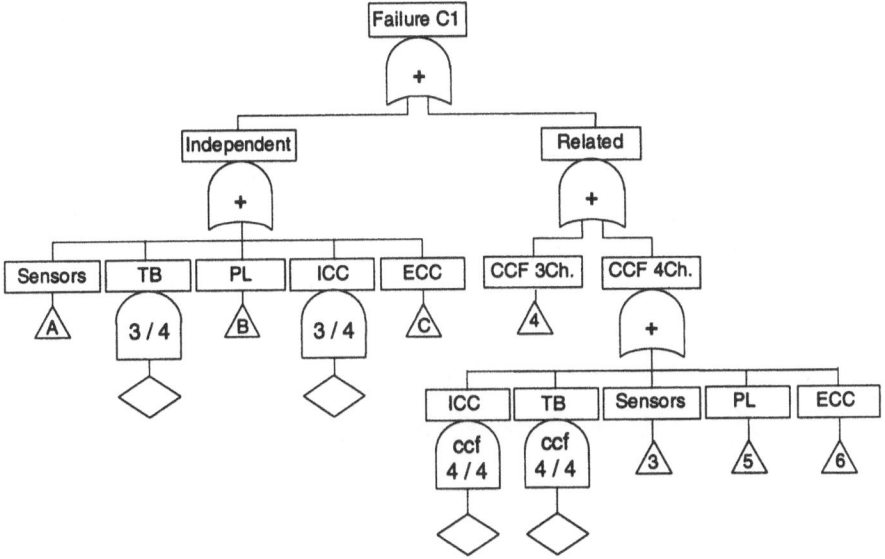

Figure 8. Component level fault tree for configuration 1.

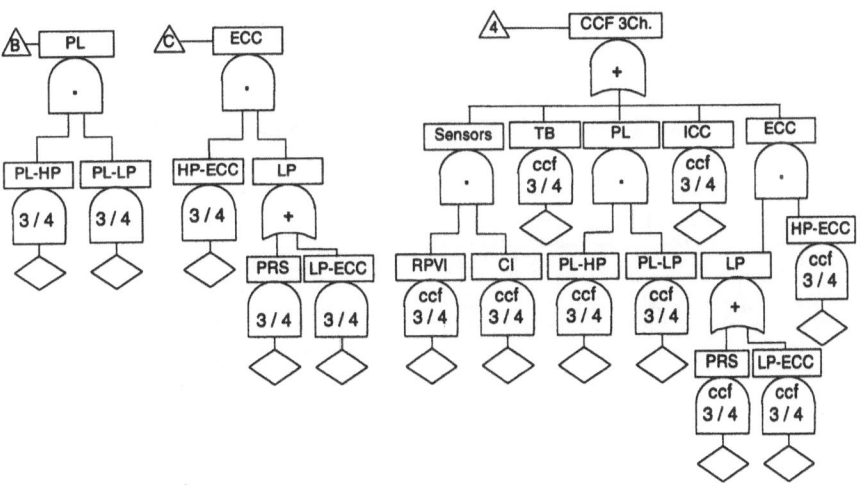

Figure 9. Subtree B, subtree C and subtree 4.

## 3.2. Fault Trees for Configuration 2

There are two main groups of redundant components in configuration 2, those that handle communication ({TB-HP, TB-LP}; {ICC-HP, ICC-LP}) and those that have similar functions ({RPVI, CI}; {PL-HP, PL-LP}; {HP-ECC, (PRS, LP-ECC)}). Each group of redundant components (e.g. TB-HP and TB-LP) that handle communication have identical application software, identical system software and the same computer hardware.

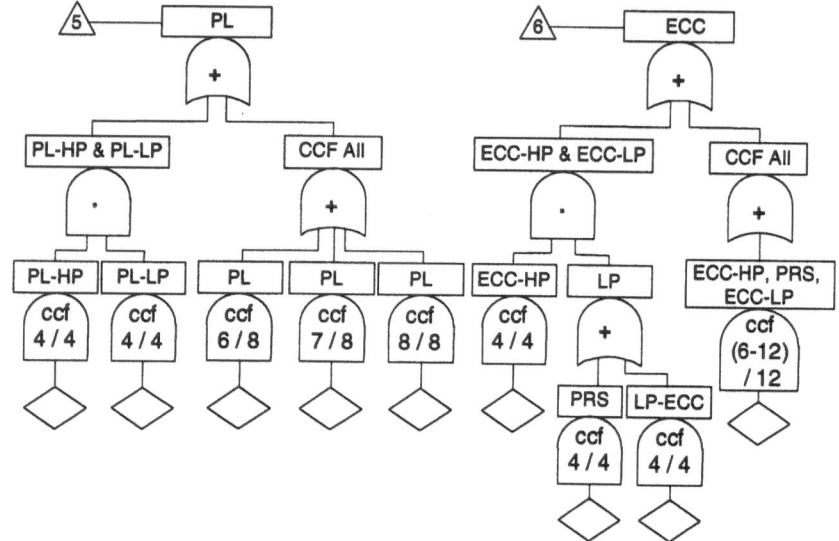

Figure 10. Subtree 5 and subtree 6.

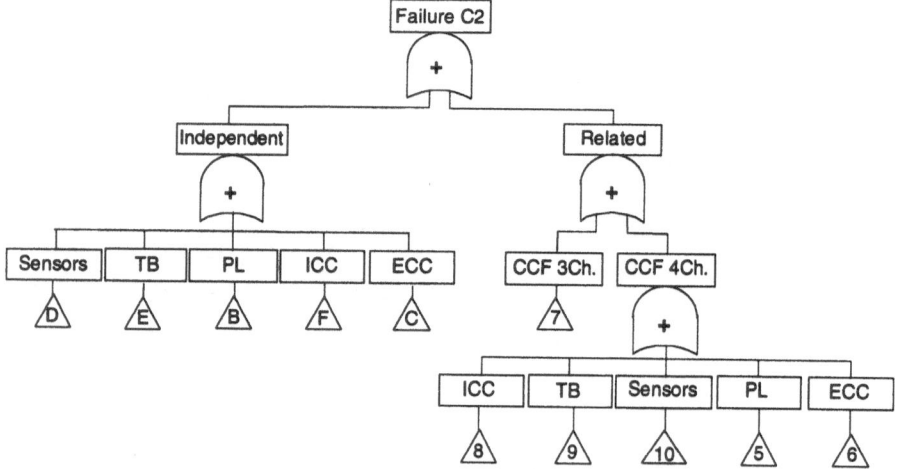

Figure 11. Component level fault tree for configuration 2.

Configuration 2 (see figure 11) fails if at least three of four redundant components fail in both the high pressure train and the low pressure train. Related failures can occur between six or more redundant components. That is, six, seven or eight of eight components in a group of redundant components can fail as a result of a common cause. This is modelled by the branch "CCF All" in subtrees 5, 6, 8, 9 and 10 (see figure 10 and 14).

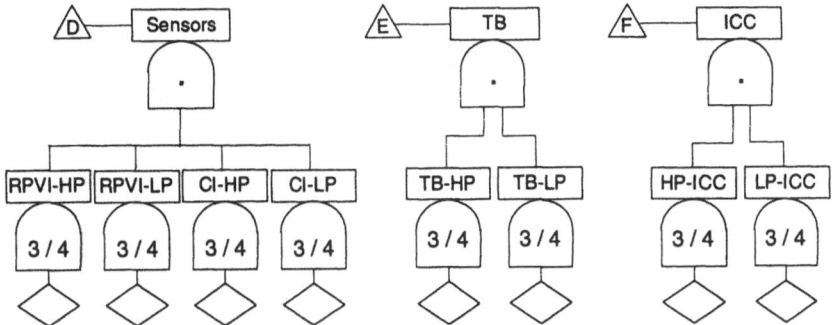

Figure 12. Subtree D, subtree E and subtree F.

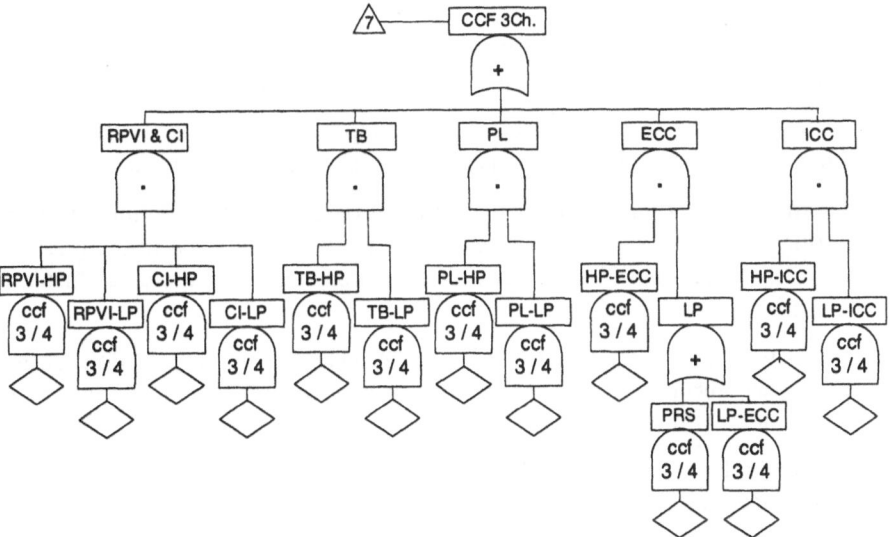

Figure 13. Subtree 7.

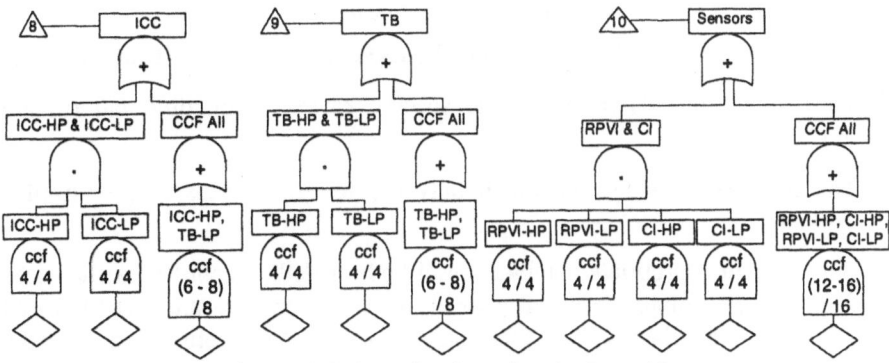

Figure 14. Subtree 8, subtree 9 and subtree 10.

# 4. Model of Common Cause Failures

Many definitions of CCFs have been proposed, although, as emphasized in [Dhillon 94], there is no single and widely accepted definition. The difficulties in identifying CCFs stem from the fact that three types of statistical dependence are involved: dependence among failure events themselves, dependence of a failure event on the conditions under which the failure occurs and dependence among these conditions. Nevertheless, for the needs of the current paper, the following definition of CCFs is accepted:

*A common cause failure is defined as the failure or unavailability of more than one component at the same time and due to the same shared cause [Fleming 86].*

Common cause failures and their sources are extremely diverse. Lala, in [Lala 93], classified the CCF using the same taxonomy as presented in [Laprie 92]. Consequently, five major classes of CCF are identified:

1. (External) physical faults, which are the result of permanent system interference caused by its operational environment, such as heat, sand, salt water, dust etc.

2. Transient faults, which are the result of temporary interference to the system from its physical environment, such as lightning, heat, radiation etc.

3. Intermittent (design) faults, which are introduced by human imperfections during the development of a system. These faults are activated in the presence of rarely occurring combinations of conditions. One example is "pattern-sensitive" faults in semiconductors.

4. (Permanent) design faults, which are introduced by human imperfections during the development of a system. The presence of these faults is not related to combinations of conditions.

5. Interaction faults, which arise from the interaction between the computer system and an operator.

Of these five classes of CCFs, only design faults are considered in the current study.

## 4.1. Existing Common Cause Failure Models

Two basic categories of CCF models are discussed in the literature, [Dhillon 94]:

- The explicit models in which the specific causes of multiple failures are modelled explicitly

- The parametric models in which only the impact of causes are modelled explicitly

It seems that an appropriate combination of both modelling approaches should be used to achieve completeness. The available useful models are usually based directly upon the collected data and analysis of the causal interactions and the effects of events.

In [Watson 86], two simple models are pointed out:

- Limit or cut-off model - this type of model is based on the availability data and engineering judgement. A probability of common-cause failures is selected, usually in the range $10^{-3}$ to $10^{-5}$, and is used as a parameter in the system failure model (e.g. fault tree). This means that the occurrence of all CCFs is incorporated in the fault tree as a basic event with the CCF limit probability.

- Conditional or $\beta$ factor model - the basis for this model is that components share some common cause of failure X. Thus, given a failure of one component, there is some conditional probability that cause X will affect more than one component in a system. In this model, the failure rate of a component is divided into two parts: (i) the component's independent failure rate $\lambda_I$ and (ii) the component's common cause (related) failure rate $\lambda_{cc}$. This is characterized by a factor $\beta = \lambda_{cc} / \lambda_I$. The major problem here is an estimation of the value of the $\beta$ factor. The range of $\beta$ factor from the measured data is 0.03 to 0.41, although the majority of values are in the range of 0.03 to 0.24 [Watson 86]. It should be emphasized that data (for estimating $\beta$ factor) were derived from systems that do not apply diversity. Consequently, it is difficult to judge the extent to which they are representative for the system considered in the current study.

To complete this discussion, it should be mentioned that more sophisticated CCF models have been developed, e.g. the Basic Parameter Model, Multiple Greek Letter Model and Binomial Failure Rate Model. A comparison of these models is found in [Fleming 86]. These three parametric models are derived on the basis of the simple models. Two major differences between simple and more advanced models are that the more sophisticated models: (i) include both time-based and demand-based failure rates, while simple models usually consider only time-based failure rates; and (ii) assume that each component under shock does not have a zero probability of survival while, in the simple models, it is assumed that each component under shock will fail with the conditional probability of 1.

In the current paper, the $\beta$ factor parametric model is used. A characterization of the model is presented in further sections.

## 4.2. Model Development

This section defines parameters used in the model of the system.

As described in the previous sections, the failure logic of the system is presented in the form of the fault tree. It is necessary for further analysis to transform Boolean variables that represent success and failure states of the system and its components to the parameters (failure rates, common cause failure rates) needed to calculate the probability of their states. The parameters must be quantified, that is the probabilities and the probability distributions should be provided.

For the purpose of the current study, the parameters used for the comparison of various system architectures will not be evaluated on the basis of actual data, as such data are not available for the system under consideration. Furthermore, the objective of the study is to compare various system configurations and not to derive the actual probabilities of a system failure. The exact values of the particular parameters are therefore not actually critical.

### 4.2.1. Assumptions and Definitions

The major assumption in the consideration is that the failure rate of a given component can be divided into the following parts:

$$\lambda_{comp} = \lambda_I + \lambda_{ccf}$$

where:

> $\lambda_{comp}$ is the total failure rate of a component; it should be emphasized that an underlying cause of the component's failure is not considered in this study.

> $\lambda_I$ component's independent failure rate

> $\lambda_{ccf}$ component's related failure rate

To incorporate common cause events into the fault tree, the components in the system are divided into common cause groups. Members of a given group share the same cause, i.e. each common cause event can affect k components, $2 \le k \le N$ (N is the number of components in the group). An example of a common cause group is an ICC group, which comprises four ICCs (ICCs from channels A, B, C and D). Common cause failures of a given component generally comprise common cause impacts two components, common cause impacts three components, ..., common cause impacts all N components. When the goal of the analysis is to obtain an exact (as far as possible) estimate of the system unavailability, it is important to distinguish all these common mode failure classes.

However, the major objective of this study is a comparison between various architectures; the exact estimates of system unavailability are not critical. Thus the analysis and fault trees are simplified by neglecting failures provoked by a common cause that affects two components from a given group. Only the common cause that affects three or more components from a given group is taken into account in the further discussion. The component's related failure rate, $\lambda_{ccf}$, can be expressed as:

$$\lambda_{ccf} = \lambda_{ccf3} + \lambda_{ccf\ge4}$$

where:

> $\lambda_{ccf3}$ depicts a component's related failure rate, which comprises failures provoked by common causes that affect three components from a given group; and

> $\lambda_{ccf\ge4}$ depicts a component's related failure rate, which comprises failures provoked by common causes that affect four or more components from a given group.

It is assumed that the total failure rate of a component ($\lambda_{comp}$) is known. A contribution of each type of failure rate (i.e. independent failure rate, related failure rate that impacts three components etc.) to the total failure rate of the component is defined as follows:

$$\lambda_I = \lambda_{comp} * \alpha$$

$$\lambda_{3ch} = \lambda_{comp} * \beta$$

$$\lambda_{ccf\geq4} = \lambda_{comp} * \Gamma$$

where[1] : $\alpha + \beta + \Gamma = 1$.

Furthermore, factor $\Gamma$ is split into two parts and defined as follows for the different configurations:

Configuration 0:
$$\Gamma = \begin{cases} \gamma_1 + \varepsilon & \text{for the group of instrumentation} \\ \gamma_2 & \text{for other component groups} \end{cases}$$

Configuration 1:
$$\Gamma = \begin{cases} \gamma_1 + \varepsilon & \text{for the groups: instrumentation, protection logics and actuators} \\ \gamma_2 & \text{for the groups: token buses and interchannel communications} \end{cases}$$

Configuration 2:
$$\Gamma = \begin{cases} \gamma_1 + \varepsilon & \text{for the groups: instrumentation, protection logics and actuators} \\ \gamma_2 + \delta & \text{for the groups: token buses and interchannel communications} \end{cases}$$

where:

> factors $\gamma_1$, $\gamma_2$ reflect failures provoked by a common cause that affects four components with the same functionality and the same implementation; generally, this factor is different than zero for all three configurations.

> factor $\delta$ reflects failures provoked by a common cause that affects more than four components with the same functionality and the same implementation; this factor is equal to zero for configuration 0 and configuration 1 and different than zero for configuration 2.

> factor $\varepsilon$ reflects failures provoked by a common cause that affects more than four components with the same functionality and different implementation; generally, this factor is different than zero for all configurations.

---

[1] This assumption is similar to the one made in [Arlat 88] where the probability of the activation of a fault in any component (in this particular case, a component is a software program version in an N-version system) is equal to the sum of probabilities of the activation of: (i) an independent fault, (ii) a related fault between two versions, (iii) a related fault between three versions and (iv) a related fault between three versions and the decider.

In the analysis of common cause failures, it is necessary to decide which groups of components have a significant probability for two or more components within the group being affected by a common cause. As emphasized in [Fleming 86], experience of common cause events indicates that most such events involve identical active components that are initially in the same operational mode, i.e. operating or standby. Thus, in the present paper, common cause failures are assumed to affect components with the same functionality and the same implementation (reflected by factors $\beta$, $\gamma$ and $\delta$) or the same functionality and different implementation (reflected by factor $\epsilon$). Components with different functionalities are assumed to fail independently.

The defined failure rates will be used to derive the probabilities of the primary events in the fault trees. The probabilities of primary events are calculated assuming the exponential distribution of the probabilities. The objective of the estimation is to derive the system unavailability (probability of a top event in a given fault tree) as a function of mission time. It is generally assumed that failure rates for all components in the system are the same and have a fixed value of $\lambda_{comp}$ = $10^{-4}$ 1/h (explicit mention will be made if this assumption is not valid).

The actual system quantification is presented in the next section.

# 5. Quantification of the Models

In this section, the system is quantified using the parameters introduced in the previous section. The analysis should result in an estimate of the probability of a Top Event (unavailability) in the fault tree. The fault trees that define the failure logic of the system are solved using the software tool SHARPE, [Sahner 93].

The objective of this analysis is to compare three system architectures: Configuration 0, ECCS without functional diversity; Configuration 1, ECCS with functional diversity, although without functional separation; and Configuration 2, ECCS with functional diversity and functional separation. The system unavailability as a function of the mission time is used to compare the configurations.

First, the system configurations are compared for the "best case" and the "worse case". The impact of common cause failures for more realistic conditions is then evaluated. Finally, the influence of changes in the failure rates of the critical components in configuration 1 is considered.

## 5.1. "Best Case" and "Worst Case"

Three system configurations are compared as the system unavailability is computed for the "best case" and for the "worst case".

It is assumed for the "best case" that all components in the system fail independently, i.e. there are no related failures. This is equivalent to the assumption that factor $\alpha$ (which characterizes the proportion of independent failures of a given component) is equal to 1 and that all other factors are equal to zero.

It is assumed for the "worst case" that both independent and common cause (related) failures are present in the system. The value of factor $\alpha$ is fixed in relation to independent failures and is assigned the value of 0.4. Consequently, the proportion of common cause failures is 0.6, i.e. $\beta + \Gamma = 0.6$. Furthermore, it is decided that all common cause failures are comprised of the factor $\Gamma$, i.e. $\beta = 0$. This is a rather pessimistic assumption, as: (i) the proportion of independent failures of a given component is likely to be greater than 0.4; and (ii) not all CCFs are of the most severe type, that is they do not affect all redundant components in a system, i.e. $\beta$ is usually not equal to zero.

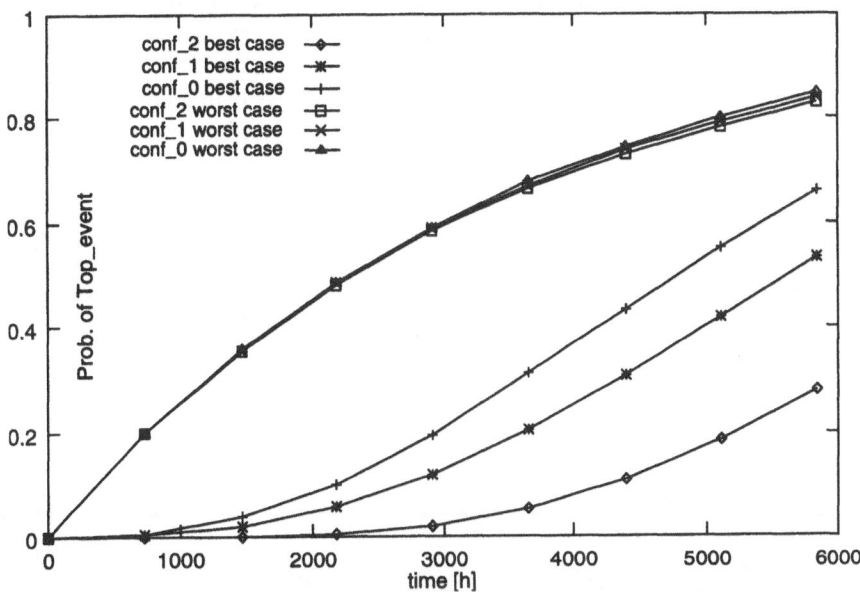

Figure 15. "Best Case" ($\alpha = 1.0$; $\beta = \Gamma = 0$) and "Worst Case" ($\alpha = 0.4$; $\beta + \Gamma = 0.6$) for all configurations.

Results of the comparison of various configurations are presented in figure 15. The figure gives the unavailability of the system as a function of the mission time. Rather obviously, results of the "best case" are that configuration 2 shows a significant advantage over configurations 0 and 1. Also, configuration 1 is better than configuration 0. A different situation is observed for the "worst case", in which all three system configurations are almost equivalent, i.e. there is no significant difference in the system unavailability for any of the configurations considered. This means that, when there is a high percentage of common cause failures, then there is a high probability that all redundant components (with identical or different implementations) might be affected by a failure. The redundancy becomes ineffective (can not mask a failure).

## 5.2. Impact of common cause failures

A further analysis was conducted to compare the considered architectures for more realistic conditions. It is assumed that:

- the proportion ($k_{ccf}$) of common cause failures for all components is the same

- the value of $k_{ccf}$ is varied in an interval [0.1, 0.6], i.e. $\beta + \Gamma = [0.1, 0.6]$

- the distribution of common cause failures between different types is as follows: $\beta = \Gamma$; $\gamma_1 = 2*\Gamma/3$; $\varepsilon = \Gamma/3$; and $\delta = \gamma_2 = \Gamma/2$

The results of the analysis are shown in figure 16, which gives the system unavailability as a function of the mission time[2]. It can be seen that configuration 2 is always the best one, provided that the value of $k_{ccf}$ is the same for all configurations. However, the situation might be different if values of $k_{ccf}$ for different configurations are distinct (the curves can cross each other in figure 2). It was observed, for instance, that configuration 1 was better than configuration 2 for $k_{ccf} = 0.3$, for configuration 2 and for $k_{ccf} = 0.2$ for configuration 1 for small probabilities of a component failure.

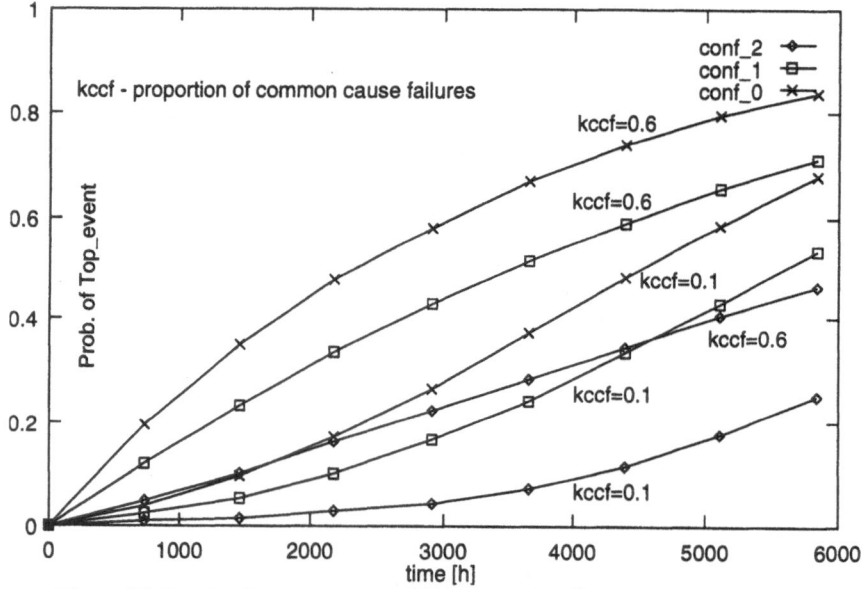

Figure 16. Impact of common cause failures: $\alpha = 0.4$; $\beta + \Gamma = k_{ccf} = \{0.1, 0.6\}$.

## 5.3. Impact of changes in the failure rates of a component

It was concluded from the previous analysis that architecture 2 has an advantage over the two other architectures as long as the proportion of common cause failures

---

[2] Observe that, due to readability, the figure presents only results for $k_{ccf} = \{0.1, 0.6\}$, although the analysis was conducted for the entire interval [0.1, 0.6].

($k_{ccf}$) in all configurations is the same (or close to one another). Thus, using configuration 2, it is possible to improve the system availability as compared with the other two configurations. A question was raised as to whether it is possible to improve the availability of the system with configuration 1 by only improving the reliability of critical components (token bus, ICC).

The results of such a study are presented in figure 17, which gives the system unavailability as a function of the mission time. Two cases are considered:

- configuration 1 and configuration 2, under the assumption that each component in the system has the same failure rate, namely $10^{-4}$ 1/h. This is depicted as conf_1 and conf_2 in figure 17.

- configuration 1 and configuration 2, under the assumption that the failure rates for token bus and for ICC are reduced to the value of $0.2 * 10^{-4}$ 1/h. This is depicted as "conf_1 improved" and "conf_2 improved" in figure 17.

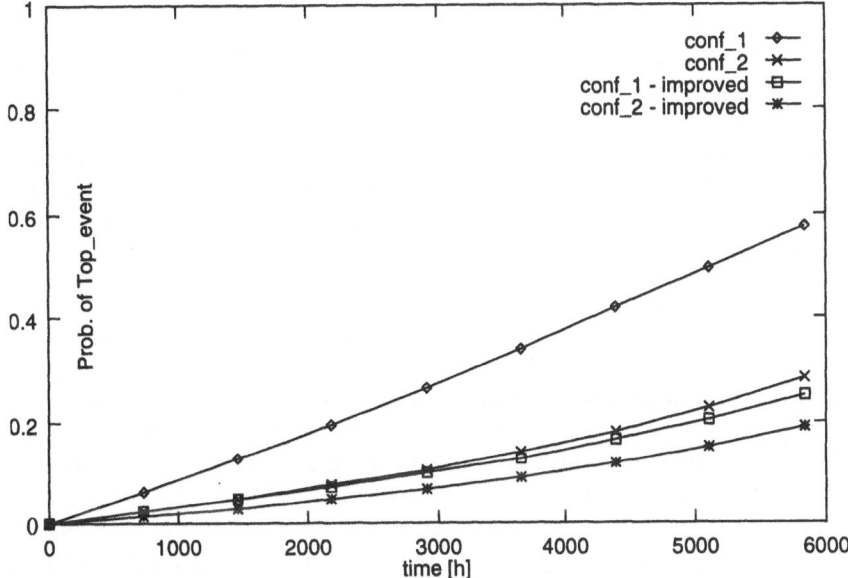

Figure 17. Impact of improvement in the reliability of critical components: $\alpha = 0.7$; $\beta + \Gamma = k_{ccf} = 0.3$; $\lambda_1 = 10^{-4}$ for all components in conf_1 and conf_2; $\lambda_2 = 0.2 * 10^{-4}$ for token bus and ICC; and $\lambda_1 = 10^{-4}$ for other components in conf_1 improved and conf_2 improved.

One obvious result is that, in both cases, configuration 2 is better than configuration 1. However, it can also be seen that configuration 1, with an improved reliability of the token bus and ICC ("conf_1 improved"), offers better system availability as compared with configuration 2 without an improvement in the reliability of token bus and ICC ("conf_2").

This suggests that a designer of the system has two alternatives to improve the system availability (assuming that the proportion of common cause failures, $k_{ccf}$, is the same for all cases considered):

- add components to the system ensuring functional diversity with functional separation (configuration 2)

- increase the reliability of critical components (configuration 1 improved)

In the second case, the failure rate must be reduced (by a factor of about 5 in this particular study). However, developing more reliable components is associated with extra cost, which involves, for instance, better testing. Thus a decision as to which alternative is most suitable should be taken on the basis of a careful analysis of system properties.

# 6. Summary and Concluding Remarks

This study presents an attempt to evaluate the ability of functional diversity to reduce the probability of common cause failures. An emergency core cooling system (ECCS) in a reactor protection system is used as a case for the study.

The evaluation is conducted by the comparison of three different system architectures: configuration 0, ECCS without functional diversity; configuration 1, ECCS with functional diversity, although without functional separation; and configuration 2, ECCS with functional diversity and functional separation. The failure logic of each configuration is presented as a fault tree. The fault trees incorporate both independent and common cause failures. On the basis of the fault tree, the system unavailability as a function of the mission time is computed and used to compare the configurations.

The results of the analysis indicate that the functional diversity has a potential to reduce the probability of a system failure, provided that the proportion of common cause failures is the same for all configurations. It can be seen that configuration 1 is better than configuration 0. Furthermore, it seems that an addition of the functional separation to the system with functional diversity contributes to the ability of the system to cope with common cause failures. It is noticed that configuration 2 has a significant advantage over configuration 1. However, it is also observed that the results are strongly dependent on the proportion of common cause failures in the system and, in particular, those common cause events that cause all channels in the system to fail. Thus, when there is a high percentage of such common cause failures, there is also a high probability that all redundant components might be affected by a failure. The redundancy becomes ineffective (can not mask a failure). As a result, all three configurations become nearly equivalent (the "worse case").

This study should be considered as a starting point for a more detailed analysis of the actual system. It seems worthwhile, for instance, to:

- include failures provoked by common causes that affect two components from a given group (these failures are neglected in this study);

- develop subtrees for the critical components in the system; subtrees, together with the actual data on component's failures, should allow a more accurate estimate of the probability of a component's failure;

- investigate a system that uses fault tolerant components.

## Acknowledgement

We would like to acknowledge the valuable comments and insight of Anders Ericsson, ABB Atom. This work was supported by grants from: SKI (the Swedish Nuclear Power Inspectorate) and NUTEK (the Swedish National Board for Industrial and Technical Development).

# References

[Ammann 87]    Ammann P.E., Knight J.C., "Data Diversity: an Approach to Software Fault Tolerance", in Proc. IEEE Fault Tolerant Computing Symposium , 1987, pp. 122-126.

[Arlat 88]    Arlat J., Kanoun K., Laprie J.C., "Dependability Evaluation of Software Fault-Tolerance", in Proc. IEEE Fault Tolerant Computing Symposium, FTCS-18, 1988, pp.142-147.

[Avizienis 85]    Avizienis A., "The N-Version Approach to Fault-Tolerant Software", IEEE Trans. on Soft. Eng., Vol. SE-11, No. 12, 1985, pp. 1491-1501.

[Bishop 86]    Bishop P.G., Esp D.G., Barnes M., Humphreys P., Dahll G., Lahti J., "PODS - A Project on Diverse Software", IEEE Trans. on Soft. Eng., Vol. SE-12, No.9, 1986, pp. 929-940.

[Christmansson 94] Christmansson J., Kalbarczyk Z., Torin J., "Dependable Flight Control System Using Data Diversity with Error Recovery", International Journal of Computer Systems Science & Engineering, vol.9, no.2, pp.98-106, 1994.

[Dhillon 94]    Dhillon B.S., Anude O.C., "Common-Cause Failures in Engineering Systems: A Review", International Journal of Reliability, Quality and Safety Engineering, Vol.1, No.1, 1994, pp.103-129.

[Fleming 86]    Fleming K.N., Mosleh A., Deremer R.K., "A Systematic Procedure for Incorporation of Common Cause Events Into Risk and Reliability Models", Nuclear Engineering and Design 93, 1986, pp.245-273.

[Kanoun 87]    Kanoun K., Sabourin T., "Software Dependability of a Telephone Switching System", in Proc. IEEE Fault Tolerant Computing Symposium , 1987, pp. 236-241.

[Knight 86]    Knight J.C., Leveson N.G., "An Experimental Evaluation of the Assumption of Independence in Multiversion Programming", IEEE Trans. on Soft. Eng., Vol. SE-12, 1986, pp. 96-109.

[Lala 93]      Lala J.H., Harper R.E., "Reducing the Probability of Common-Mode Failures in the Fault Tolerant Parallel Processor", 12th IEEE/AIAA Digital Avionics Systems Conference, 1993, pp. 221-230.

[Laprie 91]      Laprie J.C., Littlewood B., "Quantitative Assessment of Safety-Critical Software: Why and How?", ESPRIT BRA Project 3092 Predicting Dependable Computing Systems, Second Year Report, Volume 3.

[Laprie 92]      Laprie J.C. (ed.), "Dependability: Basic Concepts and Terminology", Dependable Computing and Fault-Tolerant Systems series, Vol.5, Spring-Verlag, 1992.

[Randell 75]      Randell B., "System Structure for Software Fault -Tolerance", IEEE Trans. on Soft. Eng., Vol. SE-1, 1975, pp. 220-232.

[Sahner 93]      Sahner R.A., Trivedi K.S., "A Software Tool for Learning About Stochastic Models", IEEE Trans. on Education, Vol.36, No.1, 1993.

[Watson 86]      Watson I.A., "Analysis of Dependent Events and Multiple Unavailabilities with Particular Reference to Common-Cause Failures", Nuclear Engineering and Design 93, 1986, pp.227-244.

# Requirements Analysis and Safety: A Case Study (using GRASP)

A. Coombes[*], J. McDermid, J. Moffett

*High Integrity Systems Engineering Group and*
*BAe Dependable Computing Systems Centre,*
*Department of Computer Science,*
*University of York*
*United Kingdom*

P. Morris

*Institute for Systems Engineering and Informatics,*
*Commission of the European Communities Joint Research Centre,*
*21020 Ispra (VA)*
*Italy*

## Abstract

Modifications to requirements take place under many circumstances. In the case of safety critical systems, the most tragic is following an accident. Although this may ensure that the particular accident will be avoided by the modified system in the future, there is no guarantee that the modifications will not introduce further hazards (or re-introduce previously eliminated ones), particularly as the design of complex systems often involves compromises between different failure modes.

GRASP (Goal based Requirements Analysis Specification and Proof) is an evolving goal-driven requirements specification method, incorporating a causal modelling language, intended for the development of requirements for safety critical systems. In this paper, we show how GRASP can be used to model an accident and use it to shed light on the causes of that accident. We then demonstrate the use of GRASP in redefining the requirements to prevent future occurrences of this accident, while remaining mindful of the need to meet previously established safety requirements.

## 1 Introduction

The development and modification of requirement specifications has long been identified as an important and difficult part of software development. These inherent problems are exacerbated when dealing with safety-critical systems, as the developer must ensure that old problems are not re-introduced, nor new hazards added to the system.

---

[*] Primary Author. Email: andyc@minster.york.ac.uk, Fax +44 1904 432708,
Telephone: +44 1904 432787

It is the belief of the authors that these problems occur because certain information generated in the original design, such as the system rationale and the design philosophy is not recorded. As a consequence of this, the maintainer needs to (try to) infer such information from the structure of the code, and such design documents as exist, a task which does not always succeed. The authors suggest that one way in which such information can be recorded, without the developer being overwhelmed is to record the objectives of the system as a hierarchy of goals. This permits the system developer to add new requirements and modify existing ones in the context of the existing system. In order to illustrate this, the paper demonstrates the analysis of an accident against an existing specification.

The structure of this paper is as follows. Firstly relationships between GRASP and other, related work is considered. In Section 3, the GRASP method is outlined. Section 4 describes the case study itself. Finally conclusions are drawn in Section 5.

## 2 Related Work

GRASP has been evolving over a number of years, and the ideas are still being refined [Coombes92, Morris94, Coombes94]. Some of the ideas on goal structuring originated in work on safety cases in the late 1980's (although only recently published [McDermid94a]). We have been influenced by KAOS [Dardenne93], and the work of Mylopoulos[Mylopoulos92]. Whilst there are strong similarities with these approaches, our use of models and strategies, plus the inclusion of justification/rationale [Conklin91] are distinguishing features.

Our ideas of causal modelling go back several years [Abowd91,Lister90] and have been made more rigorous over time (e.g. [Coombes94, Moffett95]). One of the most similar approaches is FOREST's MAL [Maibaum87], but we chose a different approach due to the technical problems (specifically the difficulty of understanding specifications arising from several higher-order logics being used in the process). We have deliberately avoided *direct* use of classical temporal logics (e.g. [Kröger87]) because causality is more fundamental than temporality, in the sense that causal relations induce temporal ones, but not vice-versa, however they offer one way of axiomatising our causal logic. We have been influenced by work in qualitative physics, and believe that value (e.g. in specification animation) would arise from building links to qualitative physics modelling environments [Kuipers93]. The three-valued qualitative physics used in this paper has been previously used for specification, but in a rather different way (specifically in a discrete domain) [Downing90].

## 3 Overview of GRASP

GRASP is a requirements analysis method intended for development and analysis of requirements for safety-critical systems. Since many such systems encompass both hardware and software sub-systems, it is important for the requirements to cover both (and not to distinguish between them at the highest levels of abstraction).

GRASP consists of two main components: one structural (goal structuring) and one notational (combining causal logic and qualitative physics). The goal structures are used to represent conflicts and trade-offs between requirements, and to achieve traceability. The combination of qualitative physics with causal logic gives an abstract and compact representation of key properties of embedded systems independent of implementation technology. In practice, we would expect to use an eclectic set of notations incorporating causal approaches, classical software engineering methods and natural language text. The components of GRASP are now considered.

## 3.1 Goal Structures

In developing a system, a great deal of useful information which is produced during the development process is subsequently discarded. Such information might include: alternative designs considered and rejected, the rationale for picking one particular implementation, and implicit tensions within the requirements (e.g. between safety and security features). Indeed removing the last of these is an explicit intention of many requirements techniques (i.e. by ensuring that requirements are consistent). The purpose of a goal structure is to capture some of the information which might otherwise be discarded, our belief being that recording this information will assist in change management.

A goal structure is a hierarchy which represents aims and objectives of the system at different levels of detail. The topmost levels of the hierarchy provide the most abstract statements of the objectives of the system. The bottom levels represent low-level (consistent) design statements. Intermediate levels represent the results of analysis, decomposition and trade-offs arising from the system design process.

The goal structuring concepts upon which GRASP relies are shown in Figure 1.

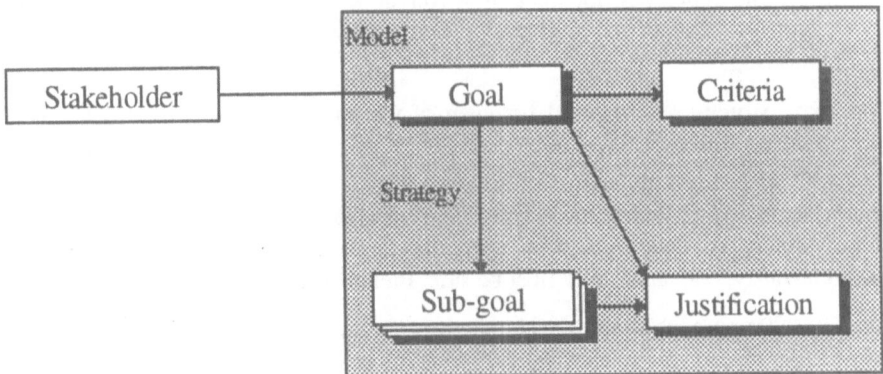

**Figure 1: Goal Structure**

The goal structure consists of six main elements:

1. *Goal.* This represents a single "requirement" that the system is intended to achieve. This goal may be functional, non-functional or both. The term "goal" is preferred to requirement, since a requirements document is typically considered to be consistent (that is, no two requirements are mutually incompatible), whereas it is conceivable that any two goals (mainly at the earlier levels of the analysis) are inconsistent, i.e. they can't both be satisfied by a single system.

2. *Stakeholder.* This represents a legal individual (e.g. an enterprise or a group) who is concerned with the satisfaction of a particular goal. Every top-level goal of the system should be "owned by" a stakeholder. Lower level goals may exist without stakeholders (but in every case, it will be possible to derive a set of "interested parties" by looking at the stakeholders of ancestor goals).

3. *Criteria.* Each criterion describes an objective measurement of whether or not a particular goal has been satisfied. In many cases, the goal will serve as the criteria (e.g. the goal "to provide electronic communication between employees" is either satisfied or not). However, in some cases, the goal will be less deterministic (e.g. "maximize productivity"), in which case, a separate criterion needs to be established which details the evaluation procedure, which is needed to show goal satisfaction.

4. *Strategy.* This denotes the set of sub-goals used to satisfy a particular goal. In some cases, there may be a choice between possible strategies, in which case, a *meta-strategy* is used.

5. *Justification.* Whenever a goal is decomposed into a set of sub-goals, a justification is used to show why it is believed that these sub-goals satisfy the goal. Where meta-strategies exist, the justification indicates why one strategy was chosen and why others were rejected.

6. *Model.* The model is the basis against which goals are articulated. At the initial stages in the requirements analysis, the model will describe the environment of the system under development, however at later stages, the model will also encompass design commitments and decomposition made in system engineering.

The above structure is repeated for each of the sub-goals that arise from a goal (possibly with different stakeholders, and with a richer model). Although Figure 1 seems to imply a tree-like structure for goal decomposition (i.e. with only one root goal), the reality is that there will normally be many root goals, many of which share descendant nodes, as it is normally necessary to satisfy multiple goals simultaneously, e.g. an aircraft must be safe, fuel efficient, maintainable, and so on.

The GRASP approach to the development of a requirements specification consists of:

1. initially identifying the primary goals of the system and the model against which these goals are to be expressed

2. performing various validation and verification tasks upon the combination of model and goals. These may include safety analyses.

3. refining the model and goals to progressively more detailed descriptions.

The purpose of a model is (initially) to describe all pertinent aspects of the environment into which the system being specified will be placed. This encompasses all components with which the system will communicate, and such aspects of the physical environment as are causally relevant. As the development progresses and design decisions are made, the model is progressively enriched. This allows the requirements of lower-level components to be expressed against a model which incorporates the anticipated behaviour of the remainder of the system (thus at any point in the development, the model combined with the requirements define the "total system").

## 3.2 Representation of Goals

In order to represent the elements of the goal structure, a notation must possess a number of properties. These properties include the following:

- The ability to represent behaviour at a variety of levels of abstraction.

- Permitting reasoning between several components represented at different levels of abstraction.

- Independence from a particular implementation technology (e.g. software).

- The ability to reason about continuous and discrete aspects of the system within the same mechanism.

To achieve these objectives, the GRASP method uses two related notations: qualitative physics and causality (in the form of a causal logic).

Causality (i.e. the relationship between cause and effect) is intrinsic to any technical system. Indeed where sub-systems of different technologies (e.g. electronic, mechanical, software) interact, causality is one of the few concepts common to all of them. This is a primary motivation for adopting causality as a basis for GRASP.

The use of qualitative physics (a technique originating from AI) provides a means of representing the characteristic behaviour of entities from some physical environment without needing to provide an overly-descriptive model of these entities. Instead of working with numerical values for properties, it uses a coarse-grained scale (the granularity of which can be adjusted as necessary).

### 3.2.1 Causal Logic

The causal logic which is described here relies upon one primary concept: the *condition*. A condition is intended to capture both events and states, but whereas a state typically implies a discrete nature, the condition is intended to encompass restrictions on values of continuous variables (these restrictions are typically expressed in terms of qualitative expressions, see Section 3.2.2). The logic is deliberately very simple, to facilitate compact and intelligible representations of complex situations.

The logic provides three causal relationships:

**causes ::= resultsIn | leadsTo | sustains**

The relation "**resultsIn**" represents an "immediate" cause, similar to event-response pairs in Ward-Mellor [Ward89]. "**leadsTo**" describes a causal relationship with some delay between cause and effect. The "**sustains**" relationship implies that the effect remains so long as the cause holds.

In evaluating causal logic statements, a distinction is made between external and internal conditions. An external condition is one which occurs only as the cause of causal expressions. Internal conditions occur as the effect and, possibly, as the cause. Internal conditions may change value only as a direct consequence of a causal relationship, whereas external conditions may change arbitrarily.

All of the causal relationships may be predicated on a guard. Such a guarded expression is of the form:

> **while** A
>> B **causes** C

(**causes** is used here to denote that the guard may apply to any of the casual relationships.)

Such an expression means that so long as A holds, then whenever B occurs, then C occurs as a consequence.

When we say that A causes B, we mean that A is an INUS condition [Mackie75]†. We refer to the other conditions, which when combined with A make a sufficient condition, as the *causal field* of A. Under the semantic model, we assume that the causal field of any variable normally holds. This is necessary for practical purposes. Often the causal field is very complex. It would be expensive and perhaps unhelpful to set out all the elements of the causal field. Thus we support a process which allows for information that is thought to be irrelevant to be elided (but can be added later if appropriate).

As the notation described here may lead to contradictions in the specification, a form of default logic [Pearl88] is adopted where A causes B *normally* means that

---

† This stands for Insufficient and Necessary condition that is part of a condition that is Unnecessary and Sufficient. A is an INUS condition if $(A \wedge B \wedge C) \vee (D \wedge E \wedge \ldots) \vee (F \wedge \ldots)$ represents the entire list of causes for some effect; thus A is not necessary for the effect (since the second or third disjuncts can bring it about), nor is it sufficient (since B and C are required before the effect can happen).

whenever A holds, B will hold later (i.e. that A is sufficient for B). In the presence of a contradiction (e.g. A causes B and A causes ¬B), some kind of resolution procedure needs to be invoked which may involve elaborating the causal field.

This semantics is also important methodologically. If analysis shows that a hazard will arise we can add a defence mechanism (typically strengthening the **while** clause), thereby explicating the causal field. The causal analysis is then repeated to show that the hazard cannot now arise. (In principle we need to do probabilistic analysis; at this stage our techniques only enable us to do logical specification and analysis).

Further, as the design models are elaborated during development, we can extend the causal structures, effectively elaborating the causal field to show the role of the system in effecting the required cause. The semantics outlined above allow us to treat this as a valid extension of the model, even though it is not a conservative extension, in the classical sense. We shall see illustrations of this below.

### 3.2.2 *Qualitative Physics*

A single causal condition consists either of a logical condition or of a qualitative expression. Qualitative expressions are intended to provide a means of representing the characteristic behaviour of entities from some physical environment without needing to provide an overly-descriptive model of these entities. What is needed for such qualitative expressions are such relationships as:

- Comparisons between values (bigger/smaller/much bigger/much smaller/about the same). Note there is a special case for bigger/smaller than zero (i.e. value is positive/negative).

- Relationships between values (proportional to/inversely proportional to).

- Comparisons between rates of change of values.

We believe that the use of qualitative expressions is warranted in requirements as it is often necessary to describe general relationships between elements of the domain, and between the domain and the proposed system, but it is impractical to quantify these relationships as the values depend upon yet to be taken design decisions, e.g. we may know increasing one value decreases another, but not what the constant of proportionality is. Clearly it is also a form of abstraction leading to compact specifications (furthermore, it provides more explanation than a specification which contains apparently arbitrary values[‡]).

---

[‡] Indeed we believe this could be very beneficial; many requirements contain premature certain premature commitments to detail because the engineers feel they need to say something.

A variety of *quantity spaces* have been suggested ranging from simple three-value spaces (less than, equal to and greater than some value, typically zero), to the use of fuzzy numbers to represent the number line. Leitch [Leitch93] suggests the use of multiple models to ensure that a suitable trade-off between abstraction and detail is achieved. In engineering large scale systems we would expect to progress towards more precise models as the design becomes more detailed. A typical usage of such models is as follows:

- When adding two numbers, if they are both positive, it is only necessary to note that the result is another positive value (three-valued). However, if one number is positive, the other negative, comparison of the magnitude of the values becomes necessary to avoid an undetermined answer (e.g. if the positive number is large, and the negative one small, then the result is positive).

The notations that we use are limited to the three-valued quantity space [Leitch93, de Kleer84].

$[x]_K = +,0,-$ $\qquad$ x is greater than, equal to or less than K

$[x] = [x]_0 +,0,-$ $\qquad$ x is greater than, equal to or less than 0

$\delta x$ $\qquad$ three valued differential of x with respect to time (equivalent to $[dx/dt]$).

The causal logic can be used to establish generic relationships between certain qualitative expressions, for instance, given two qualitative terms $\delta x$ and $[x]$, the following table summarises possible relationships:

| when [x] = | if δx = | cause type | gives [x] = |
|---|---|---|---|
| + | - | **leadsTo** | 0,-[§] |
| + | 0,+ | **sustains** | + |
| 0 | - | **leadsTo** | - |
| 0 | 0 | **sustains** | 0 |
| 0 | + | **leadsTo** | + |
| - | + | **leadsTo** | 0,+ |
| - | 0,- | **sustains** | - |

Thus, from the table it can be seen that:

> **when** [x] = +
> $\delta x$ = - **leadsTo** [x] =

---

[§] In this case, it is valid to say that either outcome ([x] = 0 or [x] = -) is a valid consequence of the causal relationship.

# 4 A Case Study

In order to illustrate the GRASP approach, we will give an analysis of aspects of the Airbus A320 accident in Warsaw in 1993, based on information provided by the literature and estimates (where data was not available). The analysis will proceed by developing the model and requirements of the system prior to the accident as a specification within GRASP. From this point, we devise strategies which would prevent future occurrences of the accident, while meeting the previously established requirements as closely as possible. Further executions of this model demonstrate improved safety properties.

## 4.1 Accident Description

A Lufthansa A320-200 (flight number LH 2904, registration D-AIPN) from Frankfurt was coming in to land at Okecie airport in Warsaw on September 14th 1993. The aircraft was landing in torrential rain and strong winds. The runway had recently been resurfaced, and was not draining well. Air Traffic Control (ATC) had warned the pilots that there was a strong cross wind.

Under normal conditions an A320 will land at a speed of 135 knots. If there are strong cross winds the plane lands at a higher speed; the reported cross winds at Warsaw implied an appropriate landing speed would be 154 knots. However the actual ground speed was 170 knots.

The plane touched down 750 metres along a runway with a total length of 2800 metres. (This is much further down the runway than is usual for the normal landing procedure.)

Normal braking is accomplished via the combined forces of wheel brakes and reverse thrusters. Most of the braking is accomplished via the wheel brakes and the tailbrakes (spoilers), the reverse thrusters have only a 20% effect on the rate of deceleration. A fully loaded A320 should be able to come to a halt within 1000 metres, at normal landing speed on a dry runway.

Neither the ground spoilers nor the reverse thrusters were deployed until 12 seconds after touch-down. The wheel braking system did not come on until 1700 metres down the runway, by which time the plane had only slowed up to 154 knots. Only about 1100 metres of wet runway was left. About 100 metres from the end of the runway is an earth bank, around 7 metres high. The plane was still travelling at speed when it hit the bank. The left wing and engine were torn loose and fuel from the ruptured tanks ignited. The collision resulted in two deaths (including the co-pilot) and 54 people were injured.

Analysis of the accident is based on:

1. publicly available descriptions of the functions of the A320 on-board computer systems [Ladkin94];

2. press accounts [FI93a, FI93b, FI93c, FI93d, FI93e].

Obviously the basis for the example cannot be viewed as completely reliable, but it should serve to illustrate the applicability of the method to a 'real world problem'. In order to describe and ultimately to analyse the accident clearly it is necessary to describe how the landing system should work; it can then be compared with what is believed to have happened.

## 4.2 Modelling the Accident

In order to show the GRASP approach, it is assumed that a goal hierarchy giving the specification of the existing system already exists. This will allow us to produce goals "out of thin air". Furthermore, the example will, for the sake of brevity, concentrate primarily upon the aspect of GRASP concerned with goals and models (i.e. issues of criteria, strategy and justification will not be dealt with explicitly).

### 4.2.1  Basic Model

In applying the GRASP approach, the first step is to identify initial goals and models. Since goals will be expressed in terms of the model, it is necessary to define the model first.

Initially, it was intended that two objects would exist in the model: the aircraft and the runway. However, consideration of the relevant properties quickly showed that the main aspects of the problem could be formulated purely in terms of the aircraft's properties (although certain properties such as altitude and distance are derived from the runway).

Aircraft properties (relevant to the landing phase only) include:

| Property | Type/Units | Description |
|---|---|---|
| Altitude | m | Height above runway |
| Brakes | true,false | Pilot issues "brake" command |
| Speed | $ms^{-1}$ | Speed of aircraft relative to runway |
| Distance | m | Distance from aircraft to end of runway |
| Deceleration | $ms^{-2}$ | Rate of deceleration of aircraft. |

At this point, causal and qualitative relationships are established between these properties.

$$\text{deceleration} = -\delta\text{speed} \tag{1}$$

$$\text{speed} = -\delta\text{distance} \tag{2}$$

These model properties permit us to infer a number of properties of the relationship between deceleration & speed, and speed & distance (using the generic qualitative relationships defined in Section 3.2.2), for instance:

**while** [distance] = +
$\qquad$ δdistance = - **leadsTo** distance = - $\hspace{4cm}$ 3

The accident occurred because the aircraft overran the runway.

$\qquad$ [distance] = -

It is clear that this should be in the requirements as something to be avoided. Therefore, an initial step is to check whether or not the requirements of the system permit this to happen. If the requirements contain some goal which should prevent the accident from occurring, it is necessary to examine the decomposition of this goal to determine why the goal was not satisfied. If the accident is not ruled out by the requirements, the converse of the accident must be added to the requirements as a new goal. We assume that the existing system requirements contains a goal of the form G1** (below) which should have prevented the accident.

**while** [altitude] = 0
$\qquad$ brakes **leadsTo** [distance] = + ∧ [speed] = 0 $\hspace{3cm}$ G1

Decomposition of this goal leads to the addition of the following property to the model.

**while** [speed] = + ∧ [altitude] = 0
$\qquad$ brakes **sustain** [deceleration] = + $\hspace{5cm}$ 4

### 4.2.2 Increasing the detail of the model

Immediately before the accident occurred, other variables in the basic model were as follows:

$\qquad$ [altitude] = 0

$\qquad$ [speed] = +

$\qquad$ [deceleration] = +

$\qquad$ [distance] = +

$\qquad$ brakes = true

These facts are consistent with the accident occurring later (i.e. model properties 2 and 3 give the accident condition), but examining the goal shows that they are also consistent with the goal being satisfied (i.e. no overrun). Furthermore, none of these variables is at odds with expected aircraft behaviour.

The basic model demonstrates potentially unsafe behaviour; if this model had shown that the behaviour was clearly safe or clearly unsafe, it would have been necessary to alter the major details of the model, since the model does not contain

---

** Note that goals are numbered hierarchically to allow the reader to infer relationships between the goals.

clues as to the nature of the accident. However, in view of the potentially unsafe behaviour, we generate a more precise model. This model, which refines the coarser model, more exactly defines the circumstances in which the behaviour is unsafe.

The increase in precision is gained by adding parameters to the model property 4. Therefore, the expressions are rephrased as follows:

$$\textbf{while } [\text{altitude}] = 0$$
$$[\text{speed}]_v = - \wedge [\text{deceleration}]_a = + \wedge [\text{distance}]_s = -$$
$$\textbf{leadsTo } [\text{distance}] = + \wedge [\text{speed}] = 0 \qquad\qquad 5$$

The relationship between v, a and s can be used to define an envelope of safe values for each point in the time between the aircraft touching down and the aircraft coming to a halt. We consider only one point in this interval: the point at which the brakes are first deployed.

From 5, it can be seen that sub-goals of G1 are as follows:

$$\textbf{while } [\text{altitude}] = 0$$
$$\text{brakes } \textbf{sustains } [\text{speed}]_v = - \qquad\qquad \text{G1.1}$$

$$\textbf{while } [\text{altitude}] = 0$$
$$\text{brakes } \textbf{sustains } [\text{distance}]_s = - \qquad\qquad \text{G1.2}$$

$$\textbf{while } [\text{altitude}] = 0$$
$$\text{brakes } \textbf{sustains } [\text{deceleration}]_a = + \qquad\qquad \text{G1.3}$$

(part of the causal field relevant to this expression is that this is during the landing phase only).

Decomposition of G1.1 and G1.2 (not given here for the sake of space) would show that the pilot is responsible for achieving these goals. Examination of the accident data shows that, although close to the limit both of these were within the recommended envelope for the class of aircraft. This leads us to conclude that, for some reason, G1.3 was not achieved.

Therefore, the next refinement of the model looks at the circumstances surrounding the deceleration of the aircraft.

### 4.2.3 Modelling Deceleration Causes

To make a more precise analysis, it is necessary to understand how G1 is realised. Three main braking systems are used: reverse thrusters, ground spoilers and wheel brakes. To model these, new properties are added to the aircraft object:

| Property | Type | Description | D[††] |
|---|---|---|---|

[††] This column is added to appropriately identify which braking system is included in a particular deceleration rate (in the following table).

| WheelBrakes | true, false | Wheel brakes have been applied | D1 |
| ReverseThrust | true, false | Reverse thrusters have been applied | D2 |
| Spoilers | true, false | Spoilers have been applied | D3 |

We also need to introduce causal relationships:

$$\neg\text{reverseThrust} \wedge \neg\text{WheelBrakes} \wedge \neg\text{spoilers} \text{ \textbf{sustains} } [\text{deceleration}] = 0 \qquad 6$$

In the absence of any braking, deceleration is zero.

**while** [speed] = 0
$$\neg\text{reverseThrust} \text{ \textbf{sustains} } [\text{deceleration}] = 0 \qquad 7$$

If reverse thrust is not being applied when the aircraft's speed reaches 0, the aircraft's speed remains at zero.

Rather than list the effect of each of the possible combinations of braking technology on the deceleration, the following table summarises:

| WheelBrakes | reverseThrust | spoilers | [altitude] | deceleration |
|---|---|---|---|---|
| false | false | true | +,0 | D3 |
| false | true | false | +,0 | D2 |
| false | true | true | +,0 | D23 |
| true | false | false | 0 | D1 |
| true | false | false | + | 0 |
| true | false | true | 0 | D13 |
| true | false | true | + | D3 |
| true | true | false | 0 | D12 |
| true | true | false | + | D2 |
| true | true | true | 0 | D123 |
| true | true | true | + | D23 |

Note that the combination of no braking has already been dealt with by the **sustains** relationship in 6.

Thus each combination of the table identifies a causal relationship of the form:

$$\text{reverseThrust} \wedge \neg\text{WheelBrakes} \wedge \text{spoilers}$$
$$\text{\textbf{resultsIn}}^{\#} [\text{deceleration}]_{D23} = - \qquad 8$$

---

[#] This is not strictly immediate, but may be considered so for this analysis.

Whenever the wheel brakes are on, it is also necessary to give the altitude as a guard (i.e. on those rows where altitude is either +, or 0, but not both).

> **while** [altitude] = 0
>> ¬reverseThrust ∧ WheelBrakes ∧ ¬spoilers **resultsIn**
>> [deceleration]$_{D1}$ = -    9

Given that D2 is the smallest value, the braking strengths can be ordered as a lattice:

Most braking

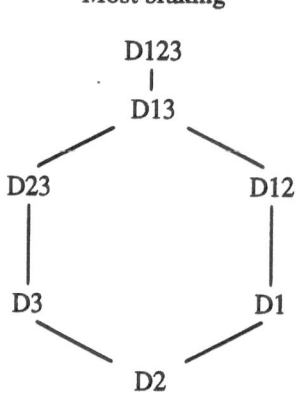

Least braking

We assume that if the plane is to stop within the distance, the value of "a" (from property 5) must be greater than D13

The refinement of G1.3 is, therefore:

> **while** [altitude] = 0
>> brakes **sustains** WheelBrakes ∧ reverseThrust ∧ spoilers    G1.3.1

Other related goals concern the deployment of braking systems during flight (these goals are separated due to differences in criticality and source of the goal):

> [altitude] = + **sustains** ¬reverseThrust    G2

> [altitude] = + **sustains** ¬WheelBrakes    G3

> [altitude] = + **sustains** ¬spoilers    G4

The derivation of G3 indicates that the measurement of altitude (originating from the radio altimeter) is suitable for the wheel brakes. However, due to a higher criticality, the reverse thrusters require a more definite signal that the aircraft is on the ground[§§]. In the case of the A320, the signal that was chosen was "Weight on

---

[§§] The radio altimeter operates with a certain margin of error, and therefore one must either err on the side of caution (in which case the reverse thrusters may not

Wheels" (WoW), which indicates that a weight of 12 tonnes is present on both of the main landing gear. Therefore, the following property is added:

| Property | Type | Description |
|----------|------|-------------|
| WoW | true, false | Weight on both landing gear is greater than 12 tonnes. |

The behaviour of the system is then as follows:

$$\textbf{while } WoW \vee [altitude] = 0$$
$$brakes \wedge [speed]_{antiskid} = + \textbf{ resultsIn } WheelBrakes \qquad 10$$

Note that this behaviour is not derived purely from G1.3.1 and G3, but also from another goal which concerns the prevention of the wheel brakes locking on. Thus, the wheel brakes are only active when the aircraft exceeds some speed, antiskid.

$$\textbf{while } WoW$$
$$brakes \textbf{ resultsIn } reverseThrust \qquad 11$$

$$\textbf{while } WoW \vee [altitude] = 0$$
$$brakes \textbf{ resultsIn } spoilers \qquad 12$$

Therefore, in order to achieve goal G1.3.1, the following property must hold:

$$[altitude] = 0 \Rightarrow WoW \qquad 13$$

If this model was decomposed further, it would be found that this property did not hold in the case of the accident described here: a side wind prevented the aircraft from landing on both wheels, which prevented the reverse thrusters from being deployed until some time after initial touchdown.

## 4.3 Identification of Accident Causes and Redefinition of Requirements

By examining the goal hierarchy, it is possible to identify possible places where modifications to the requirements prevent the same accident from happening again. Traceability relationships to parent goals ensure that existing requirements are not violated.

### Goal G1

This goal is too general to identify any possible alterations to the requirements.

### Goal G1.1

Decrease the recommended landing speed.

---

deploy on every landing), or risk the reverse thrusters deploying as the aircraft comes to land.

**Goal G1.2**

Increase the touchdown distance from the end of the runway.

**Goal G1.3**

Increase the effective deceleration of the aircraft. This leads to:

**Goal G1.3.1**

1. Increase the braking power of the braking technologies, so that (for instance) two out of three of the brake systems alone are sufficient to stop the aircraft within its normal braking envelope.

2. Ensure that all braking systems come on. This involves ensuring that the weight on wheels signal holds whenever the aircraft is at altitude 0.

# 5 Conclusions

One of the difficulties in developing modern aerospace (and other large-scale engineering) systems is their complexity, and their level of integration. The example set out above spans five computer-based systems containing (between them) hundreds of thousands of lines of code. The causal abstractions we advocate enable abstract, but focused, models and specifications to be produced. Our experience in modelling this, and other, aerospace systems leads us to believe that the ideas are of value in systems and software engineering. Similarly, we believe that the goal structuring gives an effective way of managing the interaction (conflicts) between requirements, and making and rationalizing the trade-offs.

However, we should set out some caveats on what we have done. First, nothing in what we have written should be taken as an assertion that the design decisions made by Airbus Industrie are "wrong". We have indicated clearly that the design process involves trade-offs, and these trade-offs involve establishing preferences over failure modes. It is certainly arguable that the Warsaw accident is "preferable" to the Lauda Air crash over Thailand, which involved violating G2.

Second, we have obviously cut some corners in our analysis - and this generates two sub-points. A more complete analysis, at this level of abstraction, is not very much more complex (although too large for this paper). It certainly is tractable, being vastly simpler than the systems that it models. In addition, cutting corners is useful, for example in hazard analysis. The model of "brake" (i.e. issuing a brake command) is simplistic. However, clearly "brake" poses a threat to aircraft safety. In a hazard analysis (e.g. via HAZOP [McDermid94a]) one can do "what if" analyses, considering the effect of "brake" not being issued in time. If this exposes hazards (as it should in this case) then this would suggest a more detailed analysis of causes of "brake". This is not guaranteed to expose the problem, but does increase the chance it will be found. Put another way, the simple model helps in analysis.

Our work on safety case tools [McDermid94b] is providing support for the goal structuring concepts. We hope to be able to extend the tools to support the causal modelling and analysis, and thus to investigate the ability of the techniques to scale to 'real world' problems.

The GRASP specification method, as used in this case study shows two advantages when working with specifications that are subject to modification:

- The use of qualitative physics minimises the amount of detailed analysis needed to show when modifications are necessary.

- When modifications can be incorporated into the goal structure, their potential ramifications can be identified and the work necessary for a thorough re-analysis is minimised.

# 6 Acknowledgments

The authors of this paper gratefully acknowledges the financial assistance of: British Aerospace PLC, the EPSRC and the JRC at Ispra.

# 7 References

[Abowd91]     G. Abowd, "Formal Aspects of Human Computer Interaction", YCS 161, Department of Computer Science, University of York, 1991.

[Conklin91]    E. Conklin, K Yakemovic, "A Process Oriented Approach to Design Rationale", Human-Computer Interactions, 6(3-4), 1991.

[Coombes92]   A. C. Coombes, J. A. McDermid, "High-Level Requirements Modelling - A Causal Approach", BCS-FACS Christmas Workshop, Imperial College, London, 16th-17th December 1992.

[Coombes94]   A. C. Coombes, P. Morris, J. A. McDermid, "Causality as a means for the expression of requirements for safety-critical systems", COMPASS '94, Gaithersburg, MD, 1994.

[Dardenne93]  A. Dardenne, A. van Lamsweerde, S. Fickas, "Goal Directed Requirements Acquisition", Science of Computer Programming, 20(1-2), 1993.

[de Kleer84]  J. de Kleer and J. S. Brown, "A Qualitative Physics Based on Confluences", Artificial Intelligence, 24(1-3), 1984.

[FI93a]       "Rain Factor in Loss of Lufthansa A320" Flight International, 144(4388):14, 22-28 September 1993.

[FI93b]       "Aquaplaning 'An A320 Crash Factor'", Flight International, 144(4389):15, 29 September-5 October 1993.

[FI93c]        "Actuation Delay was Crucial at Warsaw", Flight International, 144 (4391):10, 13-19 October 1993.

[FI93d]        "Early Warsaw Result Provokes Questions", Flight International, 144(4394):14, 3-9 November 1993.

[FI93e]        "Warsaw Overrun was Preventable", Flight International, 144 (4399):8 8-14 December 1993.

[Kröger87]     F. Kröger, "The Temporal Logic of Programs", Springer-Verlag, 1987.

[Kuipers]      B. Kuipers, "Qualitative Simulation: then and now", Artificial Intelligence, 59(1-2), 1993.

[Ladkin94]     P. B. Ladkin, "Analysis of a Technical Description of the Airbus A320 Braking System", CRIN-CNRS & INRIA

[Leitch93]     R. Leitch, "Recent Progress in the Development of Qualitative Reasoning", Herriot-Watt University, 1993

[Lister90]     A. Lister, A. Burns, "An Architectural Framework for Timely and Reliable Distributed Information Systems (TARDIS): Description and Case Study", YCS 140, Department of Computer Science, University of York, 1990.

[Mackie75]     J. Mackie, "Causes and Conditions" in "Causation and Conditionals", Ed. E. Sosa, Oxford University Press, 1975.

[Maibaum87]    T. Maibaum, "A Logic for the Formal Requirements Specification of Real-Time Embedded Systems", Imperial College, 1987.

[McDermid94a]  J. A McDermid, "Support for Safety Cases and Safety Arguments using SAM", Reliability Engineering and System Safety, 43(3), 1994.

[McDermid94b]  J. McDermid, D. Pumfrey, "A Development of Hazard Analysis to Aid Software Design", COMPASS '94, Gaithersburg, MD, 1994.

[Moffett95]    J. Moffett, J. Hall, J. McDermid, "A Model for a Causal Logic", in prep. 1995

[Morris94]     P. Morris, A. Coombes, J. McDermid, "Requirements and Traceability", REFSQ '94, Utrecht, Netherlands, 1994.

[Mylopoulos92] J. Mylopoulos, L. Chung, B. Nixon, "Representing and Using Non-functional Requirements: A Process-Oriented Approach", IEEE Transactions of Software Engineering, 18(6), 1992.

[Pearl88]      J. Pearl, "Embracing causality in default reasoning", Artificial Intelligence, 35(2), June 1988

[Ward89]     P. Ward, S. Mellor, "Structured Analysis for Real-Time Systems", Prentice-Hall, 1989.

# Session 9
## Applications II

# Neural Nets and Diversity

Amanda J.C. Sharkey,

Noel E. Sharkey

and Gopinath O. Chandroth

Department of Computer Science, University of Sheffield

Sheffield, U.K.

### Abstract

Although the issue of reliability is extensively discussed in the software engineering literature, it has received only limited attention in the Neural Computing community. In this paper, the software engineering concept of *diversity* is made use of to improve the performance of a neural net system solution to a problem of fault diagnosis in a marine diesel engine. Essentially the aim here was to find methods of creating a set of solutions which are diverse in the sense that they each fail on different inputs. A truly diverse set of solutions can be combined by means of a majority voter to yield 100% generalisation performance. The issue of identifying the best ways to promote diversity in neural nets was investigated in terms of a fault diagnosis problem in a marine engine. It was concluded that two effective methods for creating diverse solutions were (i) to take data from two different sensors, or (ii) to create new data sets by subjecting the set of inputs to non-linear transformations. These conclusions have far reaching implications for other Neural Net applications.

## 1 Introduction

In this paper an approach is taken which combines the emerging technology of Neural Computing with research in conventional Software Engineering. This approach is developed in the context of training neural nets on the problem of fault diagnosis in an internal combustion engine. An investigation was carried out to identify effective methods of creating neural net solutions to the problem that were *diverse* in the sense that their failures did not overlap and which could be combined by means of a majority voter to produce an improved level of performance. The *diversity* referred to here is a concept taken from the software engineering literature, where it is related to N-version programming. N-version programming is employed with the aim of increasing the fault-tolerance of conventional programs; the traditional approach being the independent development of alternative versions of a piece of software [Avizienes84]. These versions can then be executed in parallel, each receiving identical inputs and producing the required output. The outputs can then be collected by a voter, with the majority vote being chosen in the case of disagreement. N-version programming is made use of in real systems. For example, the Airbus Industry A310 aircraft makes use of dual programming in the slat and flap control system [Martin83]. Similarly, [Taylor81] describes the application of dual programming to point switching, signal control and traffic control in the Gothenburg area by Swedish State Railways.

The original assumption in software engineering was that independently developed N different versions would fail independently, and therefore increase reliability [Dahll80]. However, it has become clear that this is not necessarily the case, and that even when working independently, people tend to make the same mistakes when solving a difficult intellectual problem [Knight86]. Thus, even independently developed programs can show an unacceptably high level of coincident failures. This problem, it is claimed [Littlewood89], can be circumvented through an emphasis on diversity, and the development of programs based on different methodologies. Littlewood and Miller [Littlewood89] define the degree of methodological diversity in terms of the size and the sign of the (product moment) correlation between the probability of failures for two contrasting methodologies.

It has been argued that neural nets are particularly well suited to the promotion of diversity [Sharkey92] [Partridge94]. That is because, in contrast to conventional symbolic programming, there are a number of parameters which can be (almost automatically) altered without implicating the programmer's understanding of the problem. In this paper we make use of the potential of neural nets for diversity in the context of the problem of fault diagnosis of a ship's engine. In the next section, Section 2, the method of training nets of the fault diagnosis problem will be described, together with the resulting degree of generalisation. Section 3 provides an account of the way in which diverse solutions were identified from amongst the training sets, and combined to form a neural net system.

## 2    Training on fault diagnosis problem

Recognition of faulty combustion in a ship's engine usually requires the intervention of a skilled marine engineer, to undertake the time-consuming and fallible process of comparing current indicator diagrams with 'ideal' diagrams; a process which relies on occasional checks rather than constant monitoring. An aim of the current research was to examine the feasibility of replacing this system with a neural net system which could be used to monitor combustion quality during every engine cycle, and to provide immediate warnings of faults. The early detection of combustion condition in a marine engine is crucial since undetected faults can rapidly become compounded and even result in total breakdown.

Data for fault diagnosis of a ship's engine[1] was generated using the MERLIN[2] simulator of diesel performance (a simulator which produces results which are almost identical to data from a real engine, [Banisoleiman93]). During each engine cycle, the piston in a cylinder engages in a reciprocal motion as a result of the combustion of the air/fuel mixture in the cylinder.

In the current study, fault diagnosis was based on two measures taken at fixed intervals during the engine cycle, both of which indicate combustion quality; that is (i) the Pressure in the cylinder and (ii) the Temperature of the gases in the cylinder. The MERLIN simulator was used to produce examples of *Ideal Combustion*, and also two kinds of faulty combustion; *Retarded injection of fuel* and *Advanced injection of fuel*. Normally fuel is injected into the cylinder towards the end of the compression stroke. Retarded fuel injection means

---

[1]The data were generated from the MERLIN simulator by Gopinath O. Chandroth. This particular method of generating faulty data for Neural Net training for the purpose of fault diagnosis of a ship's engine is his idea, and forms the basis of his MSc thesis [Gopinath94].

[2]Engine simulation software developed by Lloyds Register of Shipping, London

that the fuel is injected into the cylinder later than normal. This might be caused by a slipped fuel injection cam. Retarded fuel injection results in 'afterburning', whereby complete combustion of the injected fuel does not occur. The unburned fuel impinges on the turbocharger blades, causing reduced turbocharger efficiency. Reduced turbocharger efficiency implies less air available for combustion, which in turn causes more unburned fuel to leave the cylinders. Accumulated unburned fuel could even ignite in the exhaust uptake and cause a major accident. Advanced fuel injection could also be due to a slipped cam or by a faulty fuel injector. Advanced fuel injection means that the fuel is injected too early, with the result that combustion occurs earlier and high firing pressures develop. The high pressure means that the engine components are under uneven stress, and that more power is developed in the cylinder concerned, resulting in unequal power distribution between the cylinders.

To carry out the condition monitoring, nets were each trained to perform this classification on the basis of a subset of the available data, and then tested for their ability to generalise to a larger, and previously unseen set of data. Training and test sets were generated using the MERLIN simulator, which was used to produce 588 examples of each of the three classes of data, each example corresponding to one engine cycle. On each engine cycle, measures of pressure and temperature were recorded for every 0.5 degrees of crank rotation. The 1440 values taken on each engine cycle were then reduced, using statistical sampling techniques, such that each pressure example consisted of 52 input values and each temperature example consisted of 71 input values. No explicit information about the position of the crank was included in the input; the temporal characteristics of the data were implicitly represented. The two sets of data, based on either pressure or temperature, were treated separately: the training regimes for each are described below. The two data sets of pressure and temperature were divided into a test set of 314 example pairs, a second test set of 100 example pairs, and 9 training sets of 150 example pairs (50 from each class).

**Pressure data:** Nets were trained on nine different training sets, from nine different random initial conditions; resulting in a total of 81 trained nets. Each training set consisted of 150 examples of pressure data, each containing 52 values taken during one engine cycle. Each net had 52 inputs, 2 hidden units, and 3 output units, and was trained with a learning rate of 0.6, a momentum of 0.2 and an error tolerance of 0.1. All nets converged, and the average training time was 343 cycles. The nets were trained to produce an output of 100 if the example was one of Ideal Combustion, 010 if it was an example of Retarded Injection, and 001 if it was an example of Advanced Injection. When the trained nets were tested on a test set of 314 previously unseen examples, they produced the correct classification for, on average, 98.33% of the data.

**Temperature data**: A second set of 81 nets was produced by training on nine different training sets from the starting point of nine different random seeds. Each training set contained 150 examples of temperature data, each of which consisted of 71 values taken during one engine cycle. Individual nets had 71 inputs, 2 hidden units and 3 output units and were trained with learning rates of 0.6 a momentum term of 0.2, and an error tolerance of 0.1. All nets converged (Mean no. of cycles to train, 576). As in the case of nets trained on pressure data, nets were trained to output 100 for an example of Ideal Combustion, 010 for an example of Retarded Injection and 001 for and example of Retarded Injection. When tested on the test set of 314 previously unseen examples, they produced the correct classification for, on average, 98.15% of the data.

**Combined data**: Nine new training sets were produced by combining the pressure and temperature training sets. Each example in the combined training sets consisted of 123 input values (52 pressure and 71 temperature), all taken during the same engine cycle. Nets were trained on these sets, from the starting point of nine different initial conditions, resulting again in a total of 81 trained nets, with architectures of 123 inputs, 2 hidden units and 1 output units. The same learning rate of 0.6 and momentum of 0.2, and error tolerance of 0.1 were used to train the nets; all nets converged and the mean training time was 307 cycles. When tested for generalisation on a combined test set of 314 examples, the mean generalisation percentage was 98.45%.

**Summary**: Training sets were developed based on three different kinds of input data. On each engine cycle, measures of both Pressure and Temperature were taken. Nets were trained on a combination of both 'Pressure and Temperature' inputs, or just on 'Pressure' inputs, or just on 'Temperature' inputs. These three types of input can be seen as equivalent to the different methodologies recommended by Littlewood and Miller [Littlewood89]. Nets belonging to each methodology all exhibited good, but not perfect, generalisation to a test set that had not been used during training; the average generalisation being 98.04% for the Pressure nets; 98.15% for the Temperature nets and 98.46% for the Combined nets.

# 3   Creating diverse solutions

Having trained the nets on the basis of contrasting methodologies, it was then possible to look at the extent to which their failures coincided, since each net was tested on a test set in which each example corresponded to the same engine cycle. Using the Littlewood and Miller [Littlewood89] statistical model, correlations can be computed on the basis of the failures of sets of nets corresponding to different methodologies. When small positive, or even negative, correlations are obtained between nets based on different methodologies, this indicates that nets taken from those methodologies are likely to have a low probability of coincident failures. Thus, a system of such nets, together with a voter, should result in a more reliable system.

Given the aim of creating diverse neural net solutions, the next question is how they may best be created. In software engineering, the manipulations that have been employed in an effort to decrease the number of coincident failures have included the following; (a) working from different specifications [Ramamoorthy81]; (b) using different programmers [Knight and Leveson, 1986]; and (c) using different types of programming language, eg procedural versus logic programming [Adams92]. Although neural nets are not conventionally programmed, their training can be said to involve *Extensional Programming* [Cottrell89][Sharkey in press]. By the term, extensional programming, we refer to the decisions that are made by the researcher about the architecture of the net, its initial conditions, learning parameters, input and output representations, and the composition of the data used to train it. Manipulation of these factors affects what a net will learn, and how long it takes to learn it. Previous investigations [Sharkey95a][Sharkey95b] have suggested that manipulations of either initial conditions, or training sets are less effective than using different sources of data. Since the training methods reported above make use of three different sources of data; Pressure, Temperature and Temperature and Pressure combined, statistical methods derived from Littlewood and Miller were used to examine the pairwise correlations of the failures resulting from nets trained on these three data sources.

## 3.1 Training from different data sources

Pairwise correlations between failures of nets trained on the three data sources, and tested on the first test set are shown in Tables 1,2 and 3. Examination of those tables shows that there are several negative correlations amongst those computed (negating the surmise [Littlewood89] that negative correlations may not be attainable in practice). The rows and columns of these tables are labelled either with a 'T' for temperature, or a 'P' for pressure, or a 'C' for combined pressure and temperature data. In each case the letter is accompanied by a number between 1 and 9; the number indicates which of the nine training sets was used to train the associated nets. Each entry in the tables represents the correlation between two methodologies (data sources), each methodology being represented by nine different versions (nets trained from nine different random seeds). It is possible to use these correlations as a guide to the selection of three nets, one from each methodology, which when linked via a simple Majority Voter always produce a correct response to the test set. Thus, for one set of three nets, each of the nets failed on some of the inputs (Percentage Failure: Pressure 1.27%; Temperature 0.96%; Combined 1.59%), but there were no inputs which failed on more than one of these nets. Thus, when combined with a voter, these three nets exhibited 100% correct generalisation performance. The same result (no coincident failures) was obtained when the nets were tested on the second test set.

On the basis of the data analysis presented so far it seems clear that an effective method for creating diverse neural net solutions to a problem is to base these solutions on different sources of data, as is the case here. It is possible to find nets trained on data from either Pressure or Temperature which represent diverse solutions in the sense that their failures are negatively correlated.

|  | T 1 | T 2 | T 3 | T 4 | T 5 | T 6 | T 7 | T 8 | T 9 |
|---|---|---|---|---|---|---|---|---|---|
| P 1 | -0.02 | 0.01 | 0.01 | 0.01 | 0.01 | 0.01 | 0.01 | 0.02 | 0.01 |
| P 2 | -0.03 | 0.16 | 0.29 | 0.01 | 0.01 | 0.01 | 0.11 | 0.03 | 0.01 |
| P 3 | -0.02 | 0.22 | 0.37 | 0.01 | 0.01 | 0.01 | 0.15 | 0.02 | 0.01 |
| P 4 | -0.03 | 0.02 | 0.01 | 0.02 | 0.01 | 0.01 | 0.03 | 0.03 | 0.02 |
| P 5 | -0.03 | 0.02 | 0.01 | 0.02 | 0.01 | 0.02 | 0.03 | 0.03 | 0.02 |
| P 6 | -0.03 | 0.02 | 0.01 | 0.01 | 0.01 | 0.39 | 0.02 | 0.11 | 0.01 |
| P 7 | -0.02 | 0.23 | 0.40 | 0.01 | 0.01 | 0.01 | 0.16 | 0.02 | 0.01 |
| P 8 | -0.03 | 0.02 | 0.01 | 0.02 | 0.01 | 0.25 | 0.03 | 0.06 | 0.02 |
| P 9 | -0.02 | 0.01 | 0.01 | 0.01 | 0.01 | 0.01 | 0.02 | 0.02 | 0.01 |

Table 1: Correlations between Temperature and Pressure methodologies. Temperature labelled from T1 to T9, corresponding to nine different training sets used. Pressure also labelled from P1 to P9, corresponding to nine different training sets used

|  | P 1 | P 2 | P 3 | P 4 | P 5 | P 6 | P 7 | P 8 | P 9 |
|---|---|---|---|---|---|---|---|---|---|
| C 1 | -0.01 | 0.01 | 0.01 | 0.01 | 0.01 | 0.46 | 0.01 | 0.01 | 0.02 |
| C 2 | -0.03 | 0.20 | 0.23 | 0.02 | 0.60 | 0.25 | 0.18 | 0.05 | 0.03 |
| C 3 | -0.02 | 0.26 | 0.24 | 0.02 | 0.16 | 0.32 | 0.24 | 0.01 | 0.02 |
| C 4 | -0.03 | 0.02 | 0.02 | 0.02 | 0.09 | 0.28 | 0.02 | 0.02 | 0.03 |
| C 5 | 0.11 | 0.02 | 0.01 | 0.00 | 0.41 | 0.31 | 0.02 | 0.03 | 0.03 |
| C 6 | 0.23 | 0.02 | 0.02 | 0.02 | 0.01 | 0.62 | 0.02 | 0.01 | 0.03 |
| C 7 | -0.02 | 0.29 | 0.26 | 0.02 | 0.04 | 0.31 | 0.26 | 0.01 | 0.02 |
| C 8 | -0.03 | 0.02 | 0.03 | 0.01 | 0.56 | 0.45 | 0.02 | 0.04 | 0.03 |
| C 9 | -0.02 | 0.01 | 0.01 | 0.02 | 0.05 | 0.36 | 0.01 | 0.01 | 0.02 |

Table 2: Correlations between Pressure and Combined methodologies.

|      | T 1   | T 2  | T 3  | T 4  | T 5  | T 6  | T 7  | T 8  | T 9  |
|------|-------|------|------|------|------|------|------|------|------|
| C 1  | 0.48  | 0.38 | 0.02 | 0.03 | 0.02 | 0.02 | 0.13 | 0.02 | 0.29 |
| C 2  | 0.06  | 0.30 | 0.30 | 0.02 | 0.01 | 0.01 | 0.30 | 0.01 | 0.02 |
| C 3  | -0.01 | 0.50 | 0.77 | 0.01 | 0.01 | 0.01 | 0.77 | 0.01 | 0.02 |
| C 4  | 0.07  | 0.01 | 0.01 | 0.01 | 0.01 | 0.01 | 0.01 | 0.01 | 0.02 |
| C 5  | 0.06  | 0.01 | 0.01 | 0.01 | 0.01 | 0.01 | 0.01 | 0.01 | 0.02 |
| C 6  | -0.01 | 0.01 | 0.01 | 0.01 | 0.01 | 0.69 | 0.01 | 0.01 | 0.02 |
| C 7  | 0.08  | 0.26 | 0.50 | 0.03 | 0.02 | 0.01 | 0.50 | 0.02 | 0.01 |
| C 8  | 0.03  | 0.02 | 0.02 | 0.03 | 0.02 | 0.22 | 0.03 | 0.02 | 0.04 |
| C 9  | 0.77  | 0.79 | 0.01 | 0.01 | 0.01 | 0.01 | 0.33 | 0.01 | 0.55 |

Table 3: Correlations between Temperature and Combined Temperature and Pressure methodologies.

The data used here was generated by the MERLIN simulator, but the obvious implication is that this effect could be recreated by training nets on data from two different sensor readings. In other words, training nets on either a Pressure sensor, or a Temperature sensor, or a combination of the two inputs.

## 3.2 Transforming the inputs

An effect similar to that of training nets on different sources of data, can be created by transforming the inputs. In this section, research supporting this claim will be described. The idea here is to create two new data sets through the application of non-linear transformations of the original inputs; and that nets trained on these transformed data sets will make different errors from those trained on the original data set, and from each other. Accordingly, in the transformation study described here, the original Pressure data was used to create diverse solutions by transforming the inputs. After initial investigations, two transformation methods were found to yield the required (diverse) results. The first involved transforming the inputs by using them as input to a net where they were trained to produce new outputs, and then treating the resulting hidden unit activations as the transformed inputs. The second method involved transforming the inputs by testing them on an random (untrained) set of weights, and treating the resulting outputs as the transformed inputs.

**Transformation Method 1 (TM 1):** A new data set was created by transforming the original inputs. This transformation was accomplished by feeding the inputs into a transformation net, where they were trained to produce new outputs. The new outputs, in this case, were the same as the inputs; in other words the net was trained on the function of autoassociation, and the original 52 element input vector was trained to reproduce itself as output. The ensuing hidden unit activations were then taken as transformed inputs, and trained on the original classification task (identifying instances of ideal combustion, retarded fuel injection, advanced fuel injection). The resulting set of weights could then be tested for generalisation. To do this, the inputs in the test set were first similarly transformed by passing them across the autoassociation transformation net, and taking the resulting hidden units. When tested on the transformed test set, the net showed a 98.1% level of correct generalisation.

**Transformation Method 2 (TM 2):** A new data set was again created by transforming the original inputs. In this case the transformation was accomplished by passing the original inputs across a set of random weights, in an untrained net of dimensions 52-2-52. The ensuing outputs were treated as the transformed inputs, and trained on the original classification. This net was tested for generalisation (transforming the inputs in the test set by passing them across the random net and collecting the outputs), and showed a 95.5% level of correct generalisation.

**The effect of transformations on diversity:** It was then possible to perform pairwise correlations of the generalisation failures shown by the three types of net (Original pressure net, Orig; Transformation method 1, TM 1; Transformation method 2, TM 2). These are shown in Table 5 (the number of shared failures in each case are shown in brackets). What these results mean is that these nets do not exhibit any coincident failures; and if combined by means of a Majority voter would yield 100% correct generalisation performance. The same results (no coincident failure) were obtained when the nets were tested on the second test set.

It is apparent from these results that the two transformation methods used

|  | Orig | TM 1 | TM 2 |
|---|---|---|---|
| Orig |  | -0.0306 (0) | -0.0173 (0) |
| TM 1 |  |  | -0.0302 (0) |
| TM 2 |  |  |  |

Table 4: Pairwise Correlations between nets from three methodologies; Original, Transformation method 1 and Transformation method 2

(TM 1 and TM 2) provide, in combination with the original pressure data, an effective means of creating diverse solutions. The same idea could be used to create diverse solutions for other problems. Selection of an appropriate transformation method depends on the content of the data sets involved. The challenge is to find a transformation which does permit the original classification to be learnt, but which does not result in any coincident errors. Alternative transformations could be created by training a transformation net on different functions from those described here, or by using different sets of random weights.

# 4    Conclusions

Two diversity-promoting techniques have been examined; training on data from different sensors, and training using transformed inputs. It is possible to provide a visual analysis of what is going on. Both these techniques work because they subtley alter the underlying function inferred by the neural net on the basis of the training data. As discussed by [Denker87] and [Sharkey93], the generalisation performance of a net depends on the extent to which the function inferred on the basis of the input-output pairs in the training set corresponds to that required for the test set. We can represent the function inferred on the basis of the training set as a shaded grey area, as in Figure 1. The generalisation performance of a net can then be shown as the proportion of the test set which is covered by the inferred function. In Figure 1, the net shows good generalisation for Test set B, and poor generalisation for Test set A. It is then possible to show, as in Figure 2, what it means for three nets to fail on different inputs in the test set. In Figure 2, each net (shown as a, b, and c) makes some errors on the test set, but the errors in each case do not overlap. Thus, if the three nets were combined by means of a majority voter, all the examples in the test set would be covered by at least two of the inferred functions.

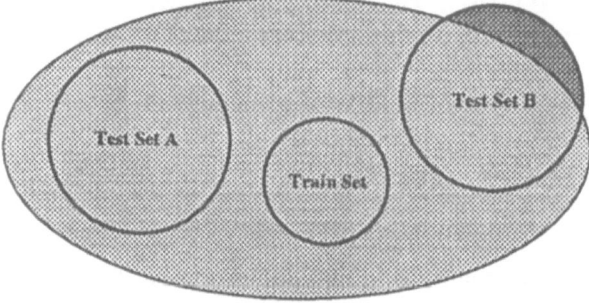

Figure 1: The light grey shaded area represents the set of ordered pairs for the function inferred by a net on the basis of the training set. This net will show 100% generalisation to Test set A, but will make more errors on Test set B (shown by dark grey area).

In conclusion, our findings support the claim that neural nets are good candidates for the promotion of diversity. In a description of a case study application of neural nets to the problem of fault diagnosis of a ship's engine, we provide an example of how the reliability of a neural net system can be increased. This research can therefore be seen as an illustration of the way in which safety issues can benefit from the linking of the disparate areas of neural nets and software engineering.

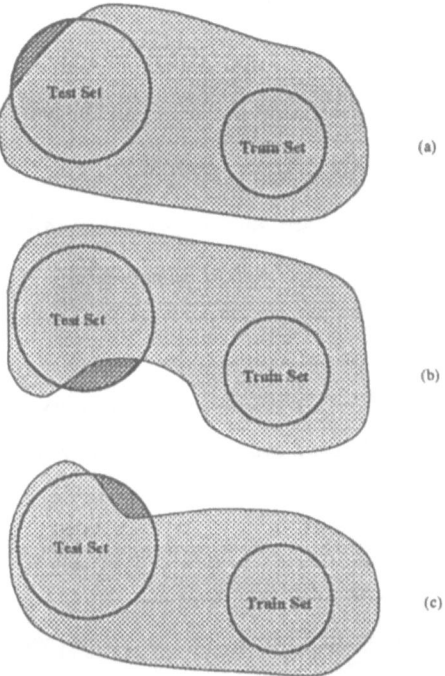

Figure 2: In each of (a), (b) and (c), the light-grey shaded area represents the set of ordered pairs for the function inferred by a net on the basis of the training set. The associated generalisation errors are represented by a dark-grey shaded area. The training sets in (a), (b) and (c) are different, with the result that different functions are inferred, and different errors are made. Since the error sets are disjoint, if the three nets are combined the result will be 100% generalisation to the test set.

In this paper, two alternative techniques for creating diverse neural net solutions are presented. The first is that of taking data from different sensors. In the feasibility study reported here, for each engine cycle, different forms of input data are produced, and the same output classification performed on them. When the generalisation failures of the resulting nets are correlated, it is possible to find amongst them, neural net solutions which are diverse in that they both negatively correlated, and contain no coincident errors. The second diversity-promoting technique is one that is appropriate where the first is not possible. It involves the artificial creation of new data sets; transforming the input data, through the use of transformation nets. Again, the results reported demonstrate that diverse solutions can result from training nets on transformed data sets. In both cases (use of different sensors, or data that has undergone different transformations), the effect of using different sources of data interacts with the actual training sets used. The use of different data sources does not *guarantee* diverse solutions; it is best combined with the use of different training sets with the different data types. However, the resulting diversity results are impressive. They justify the proposed methodology, and imply that similar techniques could be usefully employed in other neural net applications.

### References

[Adams92] Adams, J.M. and Taha, A. (1992) "An Experiment in Software Redundancy with Diverse Methodologies," *Proc of the Twenty-Fifth Hawaii International Conference on Systems Sciences.*

[Avizienis84] Avizienis, A. & Kelly, J.P.J. (1984) Fault diagnosis by design diversity: Concepts and experiments. IEEE Comput. 17, 8, 67-80.

[Banisoleiman93] Banisoleiman, K., Smith, L.A., & Matheieson, N. (1993) Simulation of diesel engine performance, *Trans of the Institute of Marine Engineers*, 105 (3) 117-135.

[Cottrell89] Cottrell, G.W., Munro, P. & Zipser, D. (1989) "Image compression by back propagation: An example of extensional programming", In (Ed.) Noel E. Sharkey, *Models of Cognition: a review of Cognitive Science*, Ablex, New Jersey, 208-241.

[Dahll80] Dahll, G. and Lahti, J. (1980) "An investigation of methods for production and verification of highly reliable software", In L. Lauber (ed) *Safety of Computer Control Systems (Proc. SAFECOMP'79)*, New York: Pergamon.

[Denker87] Denker, J., Schwartz, D., Wittner, B., et al (1987) Large automatic learning , rule extraction and generalisation. *Complex Systems* 1, 877-822.

[Gopinath94] Gopinath, O.C. (1994) "A neural net solution for diesel engine fault diagnosis", MSc thesis, University of Sheffield.

[Knight86] Knight, J.C. and Leveson, N.G. (1986) An experimental evaluation of independence in multiversion programming. *Trans on Software Eng*, Vol SE-12 no 1.

[Littlewood89] Littlewood, B. and Miller, D.R. (1989) Conceptual modeling of coincident failures in multiversion software. *IEEE Trans. on Software Engineering*, 15,(12).

[Martin83] Martin, D.J. (1983) Dissimilar software in high integrity applications in flight controls," In *Software for Avionics* (AGARD Conf. Proc. 330), Jan, 1983, pp36-1-36-9.

[Partridge94] Partridge, D. & Sharkey, N.E. (1994) Neural computing for software reliability. *Expert Systems*, 11, 3, 167-175.

[Ramamoorthy81] Ramamoorthy, C.V., Mok, Y.R., Bastani, E.B., Chin, G.H. and Suzuki, K. (1981) "Application of a methodology for the development and validation of reliable process control software" *IEEE Trans. Software Eng.*, vol SE-7, pp 537-555.

[Sharkey in press] Sharkey, A.J.C. and Sharkey, N.E. (in press) Cognitive Modelling: Psychology and Connectionism. In (Ed.) M.A. Arbib *The Handbook of Brain Theory and Neural Networks*, Bradford Books/MIT Press.

[Sharkey 95a] Sharkey, A.J.C., Sharkey, N.E. and Gopinath, O.C. Diversity, (1995) Neural Nets and Safety Critical Applications. In Proceedings of The Second Swedish National Conference on Connectionism. pp 165-178. Lawrence Erlbaum Associates, Hillsdale: New Jersey.

[Sharkey 95b] Sharkey, N.E., Neary, J. and Sharkey, A.J.C. (1995) Searching weight space for backpropagation solution types. In Proceedings of The Second Swedish National Conference on Connectionism. pp 103-120. Lawrence Erlbaum Associates, Hillsdale: New Jersey.

[Sharkey92] Sharkey, N.E. and Partridge, D.P. (1992) The statistical independence of network generalisation: an application in software engineering. In P.G. Lisboa & M.J. Taylor (Eds) *Neural Networks: Techniques and Applications* Chichester, UK: Ellis Horwood.

[Sharkey 93] Sharkey, N.E. and Sharkey, A.J.C. (1993) Adaptive Generalisation. *Artificial Intelligence Review*, 7, 313-328.

[Taylor81] Taylor, J.R. (1981) "Letters from the editor", *ACM Software Eng Notes*, vol 6, no1, pp 1-2.

We would like to thank the EPSRC/DTI Grant No.GR/H85427 IED4/1/9301 for funding this research.

# On-line Software Error Detection
# by Executable Assertions:
# from Theory to Practice

Christophe Rabéjac
*LIS* - Matra Marconi Space
LAAS-CNRS, 7 Av. du Colonel Roche,
31077 Toulouse Cedex France.
email: rabejac@laas.fr

## Abstract

Software faults are nowadays a major concern for safety-critical systems, particularly because they cannot be entirely removed during testing and validation phases. Executable assertions, inserted in source code and checked throughout operational execution, can detect errors caused by these residual software faults.

This paper describes in a first part a method to design efficient executable assertions for on-line software error detection. The proposed method includes mechanisms to check software execution through the two complementary aspects of data and control-flow.

In a second part, we expose the application of this method on two experimental cases studies, in order to validate the key concepts involved. In the first case, validation involves software fault injection techniques, whereas in the second case the formal method B is used.

## 1 Introduction

Fault-tolerance techniques have been an essential concern in critical systems for last decades, especially for the hardware aspects. But the development of software-based critical systems has lead to an increasing number of failures due to residual *software* faults. In the space industry context, the particularly demanding operational conditions and maintenance difficulties make this problem even worse.

Because of the intensive test procedures usually applied to critical software, most of residual software faults are those which produce only *intermittent* errors: the activation conditions of these faults are seldom satisfied and only during short periods, because they depend on a high number of parameters (context, time, order of occurrence of some events, internal state...).

The well known software fault-tolerance techniques like recovery-blocks [Randell 79] or N-version programming [Avizienis 85; Hourtolle 87] can tolerate this kind of faults, but they are very expensive. In fact, they can tolerate a wider range of faults, but the faults we are interested in can often be tolerated by a simple detection mechanism and re-execution of the same piece of software (from a given recovery point) precisely because they produce only intermittent errors. Anyway, independently from the recovery mechanisms, the detection stage remains of major importance in the tolerance approaches.

Executable assertions provide a cheap and efficient way to perform on-line detection [Leveson and Shimeall 83], provided they are designed and implemented through an accurately defined method[1], and dedicated to specific properties (safety-critical ones in our context). They have already been used either to detect hardware errors [Andrews 79] or to test software [Mahmood 84], but rarely to detect software errors during operational execution.

In this paper, we describe a method to design efficient executable assertions for on-line software error detection. This method is based on a safety specification stating critical properties to be ensured and implemented through data and control-flow assertions.

Two experimental case studies are presented, which demonstrate the ability of assertions to detect intermittent errors.

# 2 Designing executable assertions

In this section, we first define the proposed method for designing assertions able to detect software errors, and then we expose the characteristics of these assertions, from the two complementary data and control-flow viewpoints: errors involving variables, loops or branching local to a functional block can be detected by data assertions; errors regarding functional blocks chaining can be detected by control-flow assertions.

Of course, hardware faults leading to errors visible through software execution can also be detected by these assertions, but with a greater latency than the one obtained with specific mechanisms. For example, a SEU (Single Event Upset) will be far more efficiently handled by an EDAC (Error Detection And Correction) device than by a software assertion.

## 2.1 Method

We define here a functional block as a piece of software with an input, an output and which granularity is coarse enough to be traceable from the specification. It is typically a function or a procedure[2].

---

1. Indeed, [Leveson et al 90] shows that there are great differences in the ability of individual programmers to design effective assertions.
2. An Ada package (or a C++ class) is not a functional block by itself, but it defines a set of functional blocks through the functions and procedures (or the methods) which it encapsulates

The executable assertions technique consists in inserting in source code at this functional block level specific instructions verifying during execution that some conditions hold.

The design of these assertions must follow a few essential rules:
- they must be based on the specifications to be able to reveal design and coding faults,
- they must not disturb nominal execution (neither program data, nor real-time constraints) and thus must be inserted before integration testing (in order for the overhead induced by assertions to be taken into account),
- they should verify not only intermediate data and results but also check the correctness of control-flow (from the point of view of both sequencing and timing),
- they should, in order to keep the overhead low, be used only for errors which may induce catastrophic failures.

The proposed method to design assertions compliant with these rules decomposes in three main successive steps, which relations with phases of a classical life cycle are shown on Fig. 1:
- Pinpointing of catastrophic failures, associated to safety-critical risks. This preliminary step can be carried out by a risk analysis based on the requirements and on an overall knowledge of the system and of its environment.
- Identification of software safety constraints (by a Fault Tree Analysis of the safety risks) and production of a (partially formal) safety specification, complementary to the functional specification, stating the critical safety properties to be ensured. This safety specification not only provides a precise basis to implement the assertions, but can also reveal some ambiguities or mistakes in the functional specification itself. At this stage, a Failure Mode Effect and Criticity Analysis conducted on the software design documents can substantiate on the one hand the appropriateness of the identified constraints and spot on the other hand some new constraints.
- Implementation of assertions (before the testing phase), and validation of them by software fault injection, and by formal proof with respect to the formal parts of the safety specification.

The third step, which consists in coding assertions for data and control-flow checking (see below), is of course partially language-dependent: some high-level languages like Eiffel supply built-in mechanisms to write such assertions [Meyer 88] but are not widely used in industry; on the contrary, some others languages like C have spread all over the world, but are rather exacting to use in safety critical systems. One of the best trade-offs seems to be the Ada language: it is well accepted in aerospace industry and benefits from powerful strong typing and exception mechanisms which are valuable in fault prevention and error handling. This is the language we used for our case studies.

At last, it is worth noting that the existence of executable assertions in source code can be very useful for testing and maintenance purposes: they are particularly helpful during the tests by improving the overall observability and providing a (partial) oracle, and during the maintenance phase by revealing possible regressions.

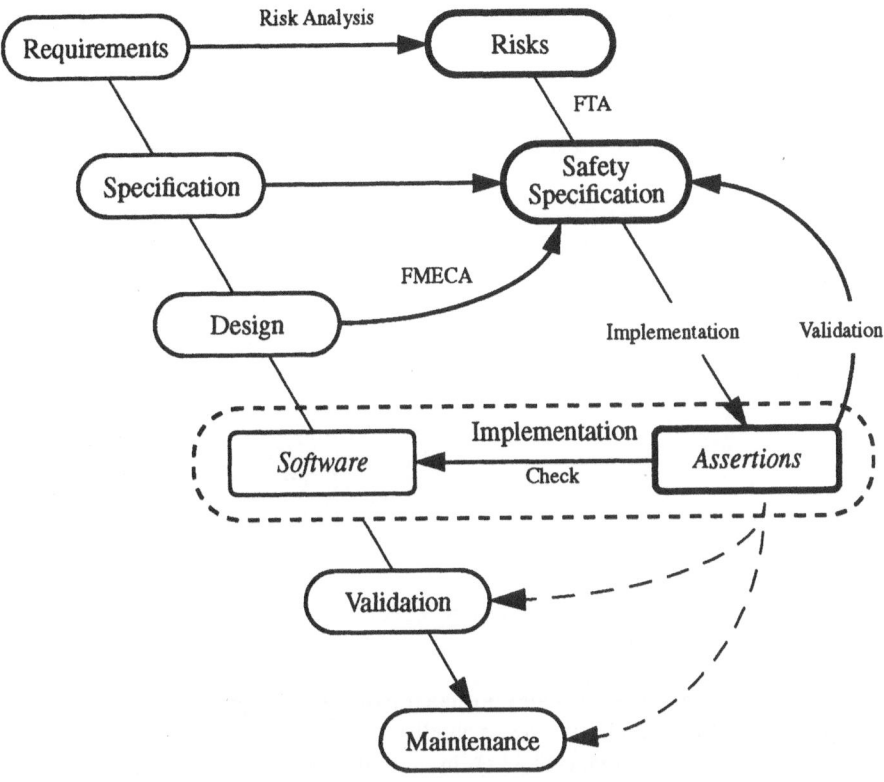

**Fig. 1: Overall view of the method**

## 2.2 Data assertions

Data assertions check data correctness through preconditions and postconditions on functional blocks (see Fig. 2). Preconditions state the properties which have to be verified for the correct execution of the corresponding functional block to be possible. These preconditions consist essentially in checking input data, and can include:

- systematic use of type checking facilities of the programming language (e.g., use two distinct types for a velocity and an altitude, even if the two variables are integers),
- range checks (on data themselves but also on their variations),

- integrity data checks, from semantic (test of possible relations between different data) and structural (verification of the integrity of data structures themselves) viewpoints.

Postconditions state the properties which have to be verified for the execution of the corresponding functional block to be actually correct. The postconditions consist in checking the validity of output data (for functions) or of modified global data (for procedures). They include:

- computation (exact or approximate) of inverse functions,
- verification of underlying implicit laws,
- approximation by simpler functions.

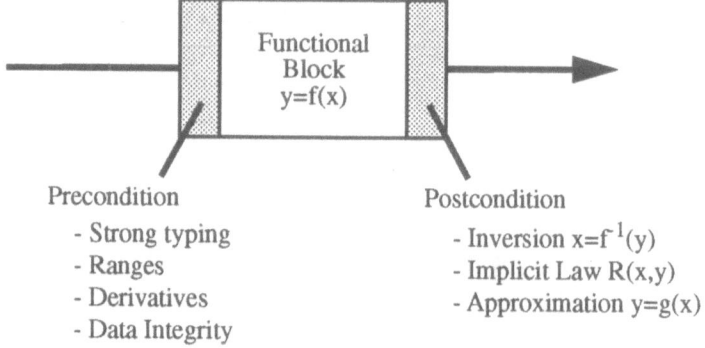

Precondition
- Strong typing
- Ranges
- Derivatives
- Data Integrity

Postcondition
- Inversion $x=f^{-1}(y)$
- Implicit Law $R(x,y)$
- Approximation $y=g(x)$

## Fig. 2: Data Assertions

Systematic use of preconditions and postconditions allows the verification of input, output and global data under the responsibility of the called functional blocks (and not of the calling functional blocks) in a uniform way minimizing the code complexity (a given functional block may be called by several different functional block). But in order to keep the overhead induced by the execution of assertions low, it is important to avoid any redundant checks. For example, when two functional blocks are always executed sequentially, the postcondition of the first and the precondition of the second should not include similar checks (range check of a given parameter for example).

## 2.3 Control-flow assertions

Control-flow assertions are implemented as a set of timed traces. Each trace verifies branching and termination of a (critical) slice of the software by monitoring the execution through a control-flow automaton (see Fig. 3):

- On the one hand, code-points are defined in the source code at the functional block level, and the control-flow automaton of the trace is aware of all possible transitions between these states (code-points constitutes indeed the different possible states of the control-flow automatons). Then, when the execution of a code-point is reported to the automaton, the checking mechanism consists in verifying the validity of the transition from the previous state[3].

- On the other hand, a trace is activated on a given state (its start-state), and may be deactivated on given states (its stop-states): a timer can thus be associated to each trace[4], in order to check the maximum transition time between the start and stop states of the corresponding trace.

  Moreover, on each reached state it is possible to check the minimum elapsed time since the activation of the trace.

  At last, note that a given state may be shared by several traces: the activation of such a state will trigger all the corresponding (active) traces.

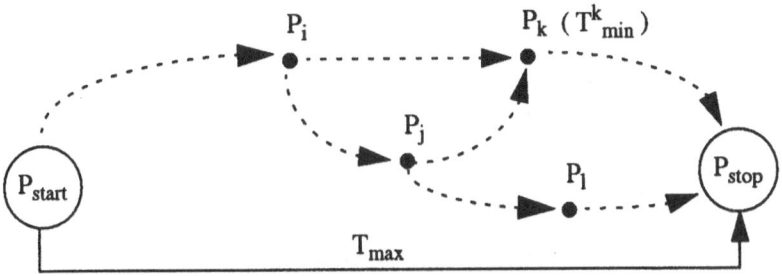

### Fig. 3: Control-flow automaton of a given trace

This mechanism, allowing the detection of sequencing and timing errors is especially valuable in a real-time context, where results delivery constraints (both in terms of relative order and time) must be carefully respected.

From a more practical point of view, it is worth mentioning that traces are described in a standard format, allowing automatic generation of their code by a lex parser. The syntax of this AIF[5] format, which describes in an abstract manner the control-flow automaton associated to each trace, is given in annex.

## 3  Cases Studies

The first case study is based on an Ada software developed during a previous experiment (FTAda) on N-version programming of a part of a spacecraft software [Simon and al. 90] within the CNES[6] R&T program. The second case study (Fault-

---

3. [Ayache and al. 79] describe a similar approach through the use of an "observer" specified by a Petri net. In our context, a plain automaton is sufficient and cheaper to implement, especially in terms of induced overhead (which is essential for real-time applications).

4. This trace admitting at least one stop-point.

5. Automaton Interchange Format

6. Centre National d'Etudes Spatiales.

tolerance validation) aims at validating the software of an on-board monitoring and reconfiguration unit (executable assertions are just one of the multiple techniques used in this CNES R&T project).

The method described above has been applied on these two cases studies, in order to adjust and validate it. In particular, the data assertions developed for the first case study were validated by an intensive software error injection campaign, whereas the control-flow assertions written for the second case study are currently being formally specified and proved using the B-method.

## 3.1 First case study: FTAda

### 3.1.1 Description

The two main functions of the FTAda software are to slow down the satellite's rotation after its separation from launcher (part of the Attitude and Orbit Control System) and to manage the pyrotechnical sequence for solar panel deployment of the satellite's platform.

The software is divided into two cyclic tasks (1 Hz and 8 Hz) and one asynchronous task (Fig. 4): the 1 Hz task reads gyroscope data, computes satellite's rotation speed (MRV Calculation), and identifies adequate thrust nozzles (MRV Piloting); the 8 Hz task activates these thrust nozzles (Nozzles Modulation) and watches their operation (Monitoring); it also computes the orbit position (PSO); lastly, the asynchronous task fires successively the 14 elementary pyrotechnical orders (Pyro) according to the satellite's rotation speed evolution.

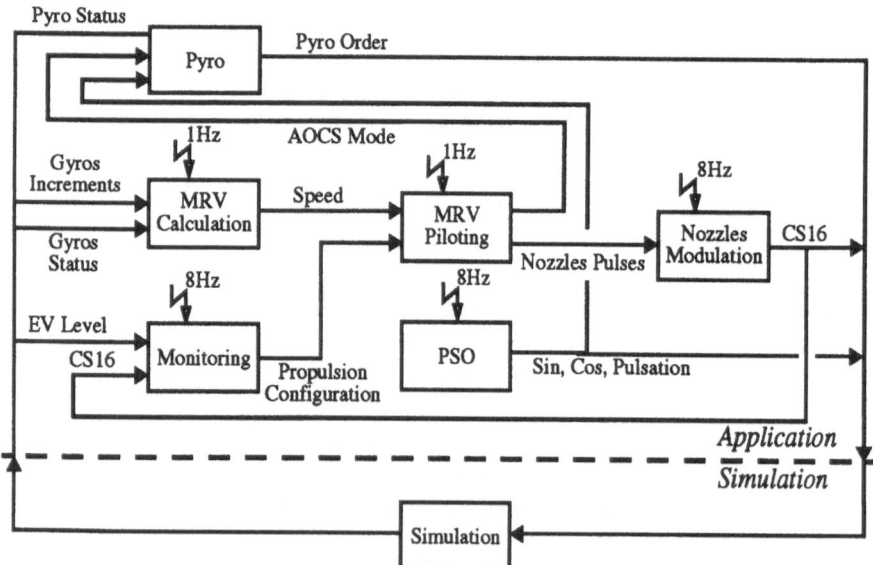

## Fig. 4: Functional overview of the application

The whole software environment of the experiment includes also three other software modules: the simulation (which calculates the satellite's rotation speed according to the activated thrust nozzles and simulates the effects of the pyrotechnical orders), the system manager (launching and managing of other modules), and of course the assertions module. The overall size of the software modules is around 4500 Ada LOC[7].

### 3.1.2 Assertions

Data assertions used cover the critical variables of the software, such as calculated speed, selected gyrometers[8] or nozzles pulses. They have been inserted into code as preconditions and postconditions of functional blocks.

For example, the logic of assertion on calculated speed (functional block "MRV calculating") is as follows:
- wait the gyrometers' desaturation[9],
- as long as speed is not reduced (that is, is greater than a given value), check that speed decreases "normally" (that is, neither too slowly nor too rapidly), and that gyrometers never becomes saturated again,
- as soon as speed becomes reduced, check that it remains reduced.

It represents approximately 50 LOC, to be compared to the size of the corresponding functional block "MRV Calculation" which is about 250 LOC.

In addition of data assertions, Control-flow assertions check the rhythm of cyclic tasks (1 Hz and 8 Hz), and verify the correct sequencing of the pyrotechnical task (including existing delays between some of these orders).

### 3.1.3 Validation

A set of generic software error injection functions was developed and used to validate the assertions. During execution, intermittent errors (characterized by their occurrence probabilities and durations) can be injected on variable values or on condition evaluations: these intermittent errors aim at simulating the effects on the software internal state of a design fault. In addition, specific overhead tasks were designed in order to simulate temporary overloads or marginal synchronisation conditions.

The injection campaign concerning the speed value calculated by the 1 Hz task was conducted on 5000 test cases of 90 seconds each. It consisted in perturbing the "MRV Calculation" functional block output (that is, the calculated speed) with an injection

---

7. Lines Of Code, excluding comments.

8. The piloting algorithm needs 3 gyrometers to be able to determine nozzles pulses, whereas there are 6 gyrometers available on-board (this redundancy allowing the tolerance of 3 gyrometer failures).

9. A gyrometer can measure speeds within some limits: beyond these limits, the gyrometer is said to be saturated.

probability of 0.01 at each cycle, and for an average duration of 10 cycles[10] (1 cycle representing of course 1 second).

A failure was said to occur as soon as the real speed value (computed by the simulation) increased.

A sensitivity study was performed on these 5000 test cases: they were executed several times but with slightly different tolerance parameters. Fig. 6 shows results obtained when the tolerance parameters are reduced[11].

It demonstrates the necessary trade-off between high coverage and low number of false alarms, and shows that for this particular experiment the P point may be a good choice: reducing further the tolerance parameters does not significantly ameliorate coverage factor (logarithm-like curve, due to a saturation phenomenon when approaching the 100% coverage) but does substantially inflate the number of false alarms (exponential-like curve).

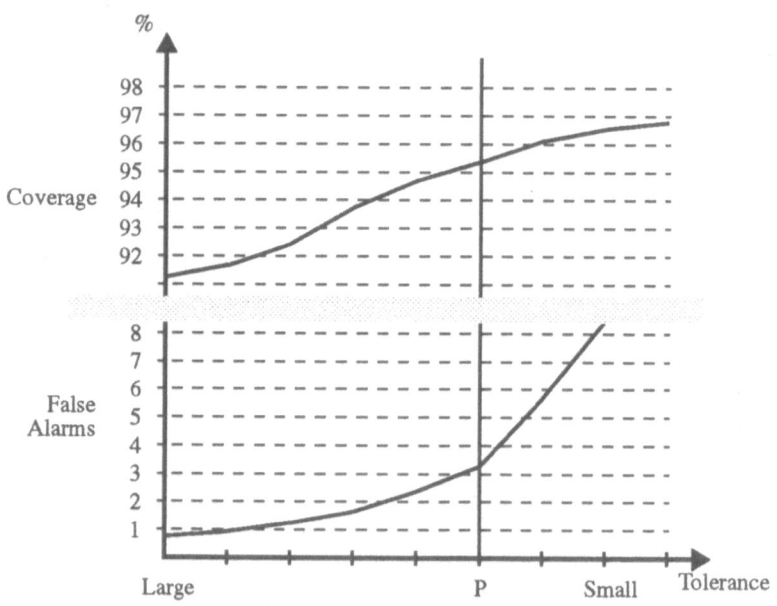

## Fig. 5: Sensitivity study

Table 1 gives numerical results for three distinguished points (large tolerance, point P, small tolerance).

---

10. This relatively high perturbation rate (execution perturbed on average for 10% of the total execution time) was necessary for the piloting algorithm to diverge.
11. In this case, the minimum rate of speed decrease (checked by the assertions) was increased.

| Tolerance | Coverage | False Alarms |
|:---------:|:--------:|:------------:|
| large | 91.2 % | 0.8 % |
| point P | 95.4 % | 3.3 % |
| small | 96.5 % | 8.4 % |

## Table 1: Numerical results

It is worth noting that the "false alarms" cases include the cases where the injected errors modified significantly the behaviour of the software (erroneous internal state) without eventually causing a system failure.

## 3.2 Second case study: Fault-tolerance validation

### 3.2.1 Description

This case study is based on a new spatial equipment, the USR[12], which is in charge of monitoring and automatically reconfiguring the space system in which it is embedded, in order to increase the autonomy of the system with respect to the ground stations. Its architecture (4 identical CPU cards in hot redundancy) allows the tolerance of 2 internal hardware failures not simultaneous and not linked.

In this context the USR software (written in Ada, and based on a specific real-time operating system) is especially critical, and assertions are used (amongst other techniques like N-version programming) to ensure its correct operation.

The main tasks involved in the software monitoring and reconfiguration process are, by decreasing priority order, the clock task, the sequencer task, the reconfiguration task, the monitoring task, and all input-output tasks. The correct sequencing of all these tasks (both in relative order and time) implies sharp real-time constraints.

### 3.2.2 Assertions

Control-flow assertions are used to ensure these constraints: they involve 8 traces totalling 36 transitions over 26 states. For example, the AIF description of the trace "SW_IPN" which checks the synchronisation constraints between software and IPN[13] in nominal mode is as follows[14]:

```
-- Verification of the nominal SW & IPN loops
-- (SW_Mode_Active and Synchro_OK)
```

---

12. Unité de Surveillance et de Reconfiguration, in french.

13. Inter-Processors Network.

14. States "HW_Registers", "Reconf_Requests" and "OBT_SW" correspond to IPN exchanges, whereas states "Acquire_HW_Registers", "USR_Monitoring", and "Reconfiguration" are raised by the corresponding software task.

```
SW_IPN (
  Synchro_Step_6 *          ->      HW_Registers;
  --

  RTC_IT                    ->      Acquire_HW_Registers;
  Acquire_HW_Registers      ->      HW_Registers;
  HW_Registers              ->      USR_Monitoring;
  USR_Monitoring            ->      Reconf_Requests;
  Reconf_Requests           ->      Reconfiguration;
  Reconfiguration           ->      OBT_SW;
  OBT_SW                    ->      RTC_IT )
```

It starts as soon as the software of the local CPU unit is synchronised with other CPU units (state "Synchro_Step_6"), and then loops between cyclic RTC[15] interruptions (generated every 100 milliseconds by the clock task).

Finally, note that the overhead induced by the assertions should be low enough to be supported by the flight version of the software: taken as a whole, control-flow assertions need approximately 15 milliseconds at each RTC cycle to execute, which gives an overhead about 15%.

### 3.2.3 Validation

In this case study, the on-going validation phase uses the formal method B [Abrial 92].

This formal method covers the whole software development cycle, from the specification stage to the generation of code, by successive refinement steps. It belongs to the so-called "model oriented" approaches, by providing a comprehensive set of formal constructions to describe the intended system.

The main algorithms used by the Ada automaton implementing the verification of control-flow (checking of transitions, managing of timers) are presently being specified using this method. Then, this formal specification will be exploited to prove the correctness of these algorithms, by using the B theorem prover.

## 4 Conclusion

The paper proposed a framework for the design of executable assertions able to detect software errors during operational execution. These assertions are described in a "safety specification" of the system, and then implemented by the use of data and control-flow checks.

Furthermore, two cases studies are presented. The first one demonstrates in particular the necessary trade-off between high coverage and low number of false alarms. The

---

15. Real-Time Clock.

second one explains how control-flow assertions can verify the correct execution of several communicating software tasks.

# Annex

The concrete AIF syntax can be described by the following BNF grammar (where all double-quoted strings are of course terminal symbols):

```
AIF                     ::=   /empty/
                        |  AIF Section;
Section                 ::=   "#traces#" Trace_Declaration_List
                        |  "#states#" State_Declaration_List
                        |  "#transitions#" Trace_Definition_List ;
Trace_Declaration_List  ::=   /empty/
                        |  Trace_Declaration_List Trace_Id ;
State_Declaration_List  ::=   /empty/
                        |  State_Declaration_List State_Id ;
Trace_Definition_List   ::=   /empty/
                        |  Trace_Definition_List Trace_Definition ;
Trace_Definition        ::=   Trace_Id "(" Transition_List ")" ;
Transition_List         ::=   /empty/
                        |  Transition_List Transition ;
Transition              ::=   State_Source "->" State_Destionation
                        ";" ;
State_Source            ::=   State_Id
                        |  State_Id "*"
                        |  State_Id "*" Timeout;
State_Destination       ::=   State_Id
                        |  State_Id "*"
                        |  State_Id Earlytest
                        |  State_Id Earlytest "*";

Terminal symbols:       Trace_Id
                        State_Id[16]
                        Timeout
                        Earlytest[17]
```

The two sections "Trace_Declaration_List" and "State_Declaration_List" allows the declaration of traces and states identificators. All declared traces must then be defined (one after another) in the "Trace_Definition_List" section.

Each transition of a given trace has a source state and a destination state: the source state may be the start state of the trace (and is marked by "*" in this case) and a timeout can then be specified (until the trace stops or re-starts); the destination state may be a stop state of the trace (and is marked by "*" in this case) , and may include

16. traces and states identificators (Trace_Id, State_Id) are strings.
17. Timeout and Earlytest are integers.

an early test (which consists in verifying that the state has not been reached too early with respect to the trace's activation time).

# Acknowledgement

The author would like to thank Jean-Paul Blanquart and Jean-Pierre Queille, both for their contribution to this work and for their useful comments on early drafts of this paper.

# Bibliography

[Abrial 92] J.-R. Abrial (1992). "B-Technology Technical Overview", BP International.

[Andrews 79] D. M. Andrews (June 1979), "Using executable assertions for testing and fault tolerance", *FTCS 9, Madison, USA*, pp. 102-105.

[Avizienis 85] A. Avizienis (1985), "The N-version Approach to Fault-Tolerant Systems", *IEEE Transactions on Software Engineering*, 11 (12), pp. 1491-1501.

[Ayache and al. 79] J.M. Ayache, P. Azéma, M. Diaz (1979), "Observer: a concept for on-line detection of control errors in concurrent systems", FTCS-9, pp. 79-86.

[Hourtolle 87] C. Hourtolle (1987), "Conception de Logiciels Sûrs de Fonctionnement : Analyse de la Sécurité des Logiciels, Mécanismes de décision pour la programmation en N-Versions", Doctorate thesis, Institut National Polytechnique de Toulouse, LAAS Report 87.267, 137 p. (in French).

[Leveson et al 90] N.G. Leveson, S.S. Cha, J.C. Knight, T.J. Shimeall (April 1990), "The Use of Self Checks and Voting in Software Error Detection: An Empirical Study.", IEEE Transactions on Software Engineering, 16 (4), pp. 432-443.

[Leveson and Shimeall 83] N.G. Leveson, T.J. Shimeall (1983), "Safety Assertions for Process-Control Systems", *FTCS 13*, pp. 236-240.

[Mahmood 84] A. Mahmood, D. M. Andrews and E. J. McCluskey (Dec. 1984), "Executable assertions and Flight Software", *AIAA/IEEE 6th Digital Avionics Systems Conference*, pp. 346-351.

[Meyer 88] B. Meyer (1988), "Systematic approaches to software construction (chapter 7)", Object-oriented Software Construction, pp. 111-164, Prentice Hall International Series in Computer Science, C.A.R. Hoare.

[Randell 75] B. Randell (1975), "System Structure for Software Fault Tolerance", *IEEE Transactions on Software Engineering*, SE-1 (2), pp. 220-232.

[Simon and al. 90] D. Simon, C. Hourtolle, H.Biondi, J. Bernelas, P. Duverneuil, S. Gallet, P. Vielcanet, S. d. Viguerie, F. Gsell and J. N. Chelotti (June 1990), "A software fault tolerance experiment for space applications", *FTCS 20, Newcastle, UK*, pp. 28-35.

# The Use of Animated Graphical Simulation Techniques to Facilitate Safe Operation, Assembly and Disassembly of Safety Critical Equipment and Systems

David Hughes
University of Plymouth
Plymouth, United Kingdom.

## Abstract

The results of a research project into the problems confronting the Service Engineer in repairing and maintaining complex, technologically advanced equipment and systems are reported. From an analysis of these problems a solution is presented which utilises graphical simulations encapsulated in an Electronic Performance Support System to guide the Engineer through assembly and disassembly operations. The proposed solution would appraise the Engineer of the potential causes of problems, provide information on service histories, offer analytical tools to facilitate problem identification and solution generation and eliminate much of the drudgery and inefficiency associated with the maintenance of service records. The paper concludes with a case study describing the development of a system which could be used by Engineers maintaining safety critical equipment and systems.

## 1 Introduction

The work load of the Service Engineer in many safety critical areas has increased dramatically in the last few years. Clients require rapid response in an environment where the diversity and technological complexity of the equipment, machines and systems are growing rapidly. Consequently, if standards are to be maintained or improved, there is a need to examine the support provided to the Engineer to improve productivity whilst, at the same time, improving safety, [Moubray, 1991].

Unfortunately, whilst the problems in servicing safety critical systems has been recognised, the need to provide appropriate support in the form of computer based tools, has not. In addition the Service Engineer now faces an increasing range of administrative tasks, the completion of service reports and time sheets further restricting the time available. Even though the reporting requirements are onerous, little use is made of the information obtained. Consequently opportunities for sharing information on the types of equipment and systems being maintained and on the problems and solutions encountered is lost. This restricts the ability to solve problems quickly or acquire the necessary spares prior to the service visit. Result poor customer service, additional expense and potentially unsafe operation.

## 2  Fault Diagnosis and Maintenance Tasks

In carrying out fault diagnostics and maintenance tasks on safety critical systems support is usually provided in the form of paper based manuals. Unfortunately, due to the diversity of problems and complexity of the equipment encountered this frequently requires several volumes to be carried. In addition to their sheer bulk manuals provide a relatively static medium being difficult to maintain or update. In addition, such manuals do not show clearly how the equipment or system operates, nor do they show the integration of various sub-assemblies and parts dynamically. In this latter respect that the incorporation of a graphical simulation techniques within a coherent Electronic Performance Support System (EPSS) is particularly useful.

## 3  Electronics Performance Support Systems

Electronic Performance Support Systems, have been described as "an integrated electronic environment that is available to, and easily accessible by, each employee. It is structured to provide immediate individualised on-line access to the full range of information ....." [Geary, 1992].

Clearly, an EPSS is not just a piece of software - it is a concept which explicitly recognises the importance of providing information to those who need it in order for them to carry out their tasks more productively, [Bentley, 1992]. Such a system would alleviate many of the problems of the Service Engineer. If, within an EPSS framework, graphical simulations were provided to assist the Service Engineer to understand how the equipment or systems operated this would offer considerable advantages. If additional help was available in the form of interactive simulations to provide detailed step by step guidance on how equipment and systems should be assembled or disassembled then further benefits would be obtained.

Given recent advances in technology, it is now possible to provide high quality graphical representations of reality via lifelike simulations and demonstrations using text, sophisticated graphics and visual media. This enables the user to formulate new concepts or cognitive schemata with respect to a phenomenon or process. This facility greatly extends the commonly recognised capabilities possessed by the electronic computer for manipulating, storing and presenting information.

The mode of presentation adopted uses a variety of complex animation's to simulate the operation of the equipment and is able to illustrate assembly and disassembly operations. The display is enhanced by the use of three dimensional images in a three dimensional space thus giving depth to the usual two dimensional plane. This provides a more visually realistic representation of the actual parts. Assembly and disassembly simulations are designed so that they could be undertaken under the direct control of the Service Engineer. Such a system allows the Engineer to select a sub-assembly from a machine view and explode the sub-assembly into its component parts and sub-assemblies. This process can be repeated at the next level of detail providing further simulations at greater levels of complexity. This enables the Service Engineer to access information at an appropriate level of detail required for the particular task in hand.

An additional feature of the system proposed is its ability to illustrate the overall conceptual operation of the equipment on a two dimensional plane enhanced by animated flows and part movements. For example, in the simple case of a coffee vending operation, this representation includes a 'brewer' boiling the water, the coffee dropping into the cup, the release mechanism of the cup to its despatch point, and the flow of the boiled water into the cup. This illustrates a complete cycle of the 'normal' operation of the machine, providing useful information for the Service Engineer prior to undertaking a detailed fault find on the composite parts.

# 4 Case Study

The EPSS approach has been applied in the development of an IMAS (Integrated Manual and Servicing) application. Development commenced by reviewing an existing service manual relating to maintenance of beer cooling, dispensing and retailing systems involving hydraulic, mechanical and electronic subsystems. Though by no means safety critical this application was felt to be both representative of service operations and in addition, it allowed detailed exploration of alternative ways in which the required service information could be delivered to the Engineer and used by Service Management in their monitoring of service operations. In particular, the application facilitated the testing of visually animated graphical simulations of assembly and disassembly of component parts.

It had been recognised that the majority of manuals are created on computer based DTP systems and therefore the information would be available in digital format. The methods used to elicit information to generate the content of current manuals and to update them were also examined. A parallel research activity sought the views of a small number of Service Engineers on the appropriateness of the content and the mode of presentation. This process highlighted a considerable dissatisfaction with existing paper manuals. In particular, the use of text illustrated by simple line drawings or photographs did not provide the information required to operate, assemble or dis-assemble the more complex equipment and systems encountered.

Not surprisingly, the more complex the equipment or system to be maintained the less satisfaction there was with existing paper based manuals. All the Engineers, questioned, however, insisted that textual information was still required, for example part numbers, model numbers, fault history etc. However, they did not believe that such information was very helpful in understanding how the equipment operated or could be assembled and disassembled.

To address this latter need a new electronic service manual was specified which incorporating some graphical simulation techniques.

The design of the manual and the underlying computer support was based on the following concepts:

- Encapsulation of the best knowledge available of the normal operation of the dispensing equipment and to make it available at the time of need.

- Use of appropriate digital media to maximise the effectiveness and efficiency of information and knowledge transfer.

- Automatic capture of fault and remedial information, which could be used to update and improve the manual content.

- Information and instruction linked to the item under investigation and retrieved via an image of the item, not via an index of part numbers.

- Items in the manual should appear as a virtual model which can be dis-assembled, assembled, calibrated, tested etc. alongside the real item.

- Information to be structured so that levels of 'how to' instruction provide a reminder for the experienced engineer and more detailed a step by step procedures for less experienced engineers.

The mode of presentation utilised a variety of complex animation's to simulate the operation of the equipment and its assembly and dis-assembly. The display would be enhanced by the use of three dimensional images in a three dimensional space thus giving depth to the usual two dimensional plane. It was thought that this would provide a provide a visually realistic representation of actual parts.

Assembly and disassembly simulations were designed so that they could be undertaken under the control of the Engineer. The system provides the ability to select a sub-assembly from a machine view and 'explode' it into its component parts and sub-assemblies. This process may is repeated providing further simulations at greater levels of detail. This hierarchical approach provided access to information at an appropriate level of detail as and when required.

An additional feature of the system specified is its ability to illustrate the overall conceptual operation of the equipment on a two dimensional plane enhanced by animated flows and part movements.

For example, in the case of a metered dispensing operation this representation includes an electro-mechanical pump operating a diaphragm to discharge an amount of beer, through a nozzle with a sensor operating to cut flow when the correct amount had been delivered. This illustrates a complete cycle of the 'normal' operation of the equipment and provides useful information to the Engineer on the process prior to undertaking detailed fault find on the composite parts.

Of course in the simple example above, a well trained Service Engineer would gain little benefit from such an obvious illustration of normal operation. However, if the illustration outlined above was linked to the operation of the complete system involving electronic point of sales terminals, stock monitoring and re-order systems then the value of the approach would increase significantly. In such circumstances, fault diagnosis of the whole system is inherently more complex.

## 5  EPSS and Management

The type of EPSS outlined above will also offer support to Service Management. It is able to accomplish this in several ways: First, EPSS's can support best practice by enforcing a well structured, systematic approach. The EPSS dictates both the pace and content of the diagnostic process in a manner analogous to computer aided co-operative working, [Voege et al, 1994].

Secondly, it facilitates planning and control of service operations by the improved accuracy of service records. From analysis of those records will come a better understanding of trends and problems. In addition, the ability to capture data at source will eliminate much of the drudgery and inefficiency involved in re-entering data centrally as information can be downloaded directly from the Engineers terminal to the computers in Service Management.

## 6  Training and Development

EPSS's offer considerable advantages in training and developing Service Engineers in the knowledge and skills associated with best practices in diagnostic engineering management. The system developed has been shown in another application area to be as effective as sending a Service Engineer on a training course for a particular piece of equipment. The widespread use of IMAS applications incorporating graphical simulation techniques will have considerable impact on the quality of the Service Engineers performance, this will be achieved by:

Enforcing and supporting best practice, or at least good practice in service operations, by taking the Service Engineer through a step-by-step diagnostic process.

Ensuring the efficient capture and processing of information at source and by facilitating its communication to others with a need to know.

# 7 Conclusions

The problems facing the Service Engineer in maintaining complex equipment and systems have been identified. A novel solution based on the concept of Electronic Performance Support Systems (EPSS's)incorporating animated graphical simulation techniques has been presented.

The solution proposed makes extensive use of graphical simulations to guide the Engineer through complex assembly and disassembly operations and in addition, it can appraise him or her of the potential causes of the problems, provide information on service histories, offer analytical tools to facilitate problem identification and solution generation and, not least, eliminate much of the drudgery and inefficiency associated with the maintenance and updating of service records.

Whilst only a simple application has been developed so far it is clear that the novel approach advocated would offer significant advantages in the very much more demanding task of maintaining safety critical equipment and systems. Further work is aimed at exploring such possibilities in greater detail.

# References

[Bentley, 1992], Bentley, T., *Training to Meet the Technological Challenge*, McGraw-Hill, Berkshire, UK.

[Geary, 1992], Geary, G., The Business of Training, in *Training to Meet the Technological Challenge*, Editor, Bentley, T., McGraw-Hill, Berkshire, UK.

[Moubray, 1991], Moubray, J., *Reliability-centred maintenance*, Oxford: Butterworth Heinemann.

[Voege et al, 1994], Voege, M Esser, M. and Hirsch, B. E., 1994, Computer Support for Co-operative Work on the Shop Floor, *Proceedings of the International Conference and Electronic Design Automation*, (Ed), Medhat, S., SCS International, San Diego, US.

# Invited Paper

# An industrial view of Requirements Engineering and Safety

Jean-Pierre Heckmann[1] & Stephen Shirlaw[2]
[1]Aerospatiale Avions, 31- Toulouse, France.
[2]GEC Alsthom Signalling Systems Division, 93- St. Ouen, France

## Abstract

Attempts to improve the Requirements Engineering (RE) process have so far been mainly concerned, either with proposing various specification formalisms, or with containing the problem by means of Requirements Management. The REAIMS project has gone further and developed a focused RE process Adaptation and Improvement Strategy. The proposed techniques have been evaluated on critical industrial developments in the Transport sector. The project results have then been developed into a generic improvement package to support various classes of complex system developments.

## 1. Overview

The activity of Requirements Engineering (RE) has often been shown to be a particular problem in the development of complex systems. Various approaches have been proposed to improve the process generally concentrating on System modelling tools. These only partially address the problem since they rarely deal with the properties of the system, but rather concentrate on functional or object modelling and are therefore most useful when requirements are already known. An alternative strategy is to accept the instability in requirements and to concentrate on requirements management, on systems validation and on making the development process receptive to changes. Similarly, a strategy emphasising prototyping is often proposed for dealing with human factors in systems that interact with operators and users.

However, none of these strategies propose a constructive and process approach able to generate complex requirements in a systematic and error-free manner. The objective of the REAIMS project is to propose such an approach. But, in doing this, it is necessary to take into consideration the fact that the RE process is a complex activity not yet amenable to a global approach. Therefore a focused improvement strategy is proposed based on an analysis of error causes that are particularly relevant to industry.

Furthermore, Requirements Engineering is an "integrating" or "link" activity between diverse corporate activities (such as safety, value, or market analyses), on the one hand, and the systems development process on the other. Hence, the process improvement strategy has to be as much concerned with adapting the RE process to particular corporate environments as with process innovations.

## 2. Process model and error cause analysis

The REAIMS process model [REAIMS 93] is based on the identification of four key activities that make up the Requirements Engineering process and which are significant sources of errors. These four activities are illustrated in Figure 1 and their failure modes described below :

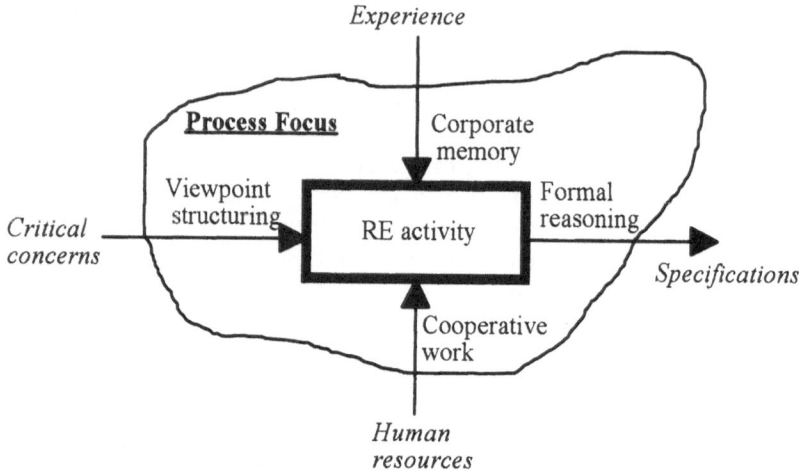

Figure 1 :  the REAIMS model

- The persons who formulate requirements are often domain experts. But new experience is also accumulated from past or similar systems developments and their expertise can still be incomplete or omit crucial practical information due to the difficulties in the transfer of experience within a company.

- The formulation of a set of consistent and valid requirements requires reasoning at an abstract level about the properties of the system. Any informal or incomplete reasoning about the system leads to errors as soon as the system becomes complex.

- The requirements specification for a system results from the integration of the results of different analyses from various sources. The triggers for this process are the critical concerns of the users, customers and supplier. In practice there are often problems in structuring this information and keeping track of all its sources and interrelations.

- Finally, even if the RE process can be broken down into the contribution of the viewpoints of experts and reasoning about the properties of the system, there still remains an important amount of co-operative work involving negotiations, knowledge transfer and confirmation of results. These activities are also important generators of errors.

# 3. An Adaptation and Improvement Strategy

The focusing on the REAIMS process model has driven the definition of a family of techniques whose development has been the main result of the project.

However, to create an improvement strategy, it is necessary to also give a definition of Good Practice, covering the Technical, Process Improvement, and Safety issues in Requirements Engineering. The REAIMS techniques can then be situated in the users problem context.

The REAIMS package, whose family members are PERE, MERE, FRERE and PET, is illustrated in Figure 2 below.

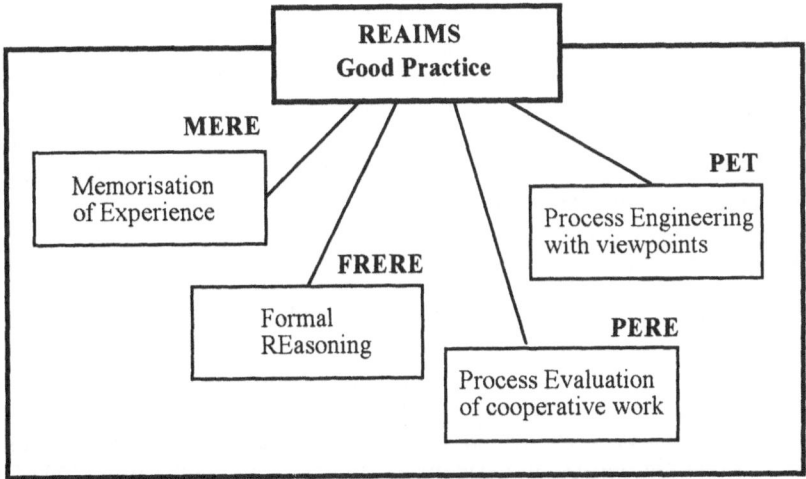

Figure 2. The RE Adaptation and IMprovement Strategy package

The four package members are closely related. Firstly, it is necessary to ensure that Requirements Engineering remains an integrating or link activity in Systems Development and leads to adaptation of the process to the problem domain context of users. The Process Engineering Technique based on the Viewpoints approach is therefore treated as a core technology used by the other members of the family.

Secondly, the model sees domain experts as the main contributors to viewpoints. They are persons who have a special intellectual skill or specific knowledge, but their involvement in the development of complex systems will still lead to two major failure modes that are particularly difficult to master, namely the inadequate memorisation of experience and incorrect reasoning.

However, co-operative work is still seen as the cement that holds the Requirements Engineering process together. Although there are wide variety of mechanisms that can be used for this activity, it is necessary, as a minimum verification, that these processes be systematically evaluated to ensure that they present dependable features.

# 4. The REAIMS Techniques

## 4.1 Memorisation of Experience

Making use of experience is particularly difficult since complex systems development involves a large number of stakeholders over a long period of time. The easiest way of handling such a problem is by techniques such as anomaly reports, suggestion schemes, and end of project experience reports. Such processes may be considered as short term memory based.

The importance and the value of corporate memory is well known in all industrial fields of activity. Nevertheless the organisation of the experience memorisation, the associated elaboration of Rules/Recommendations (R/R) and the systematic R/R application as Requirements in a particular system development, is not current practice in industry. The Memorisation of Experience for Requirements Elaboration (MERE) process defines means for these activities so that they can be an effective and efficient standard practice within a company.

The introduction of the MERE process enables not only an improvement in the quality and safety of products and of processes, but also an improvement in the adaptation between human behaviour and machine behaviour and in the adaptation to the users' needs. It also leads to the reduction of the number of modifications due to malfunctions that have to be carried out.

## 4.2 Formal Reasoning

The development of the FRERE process (Formal REasoning in Requirements Engineering) has been based on the results of the previous use on industrial applications at GEC Alsthom of the B method of formal specification, refinement and proof [Abrial 95]. The industrial application of formal methods, as opposed to case studies, requires a defined process. This includes clearly situating formal methods use within the system development life-cycle and defining the relation to all connected activities such as Quality Assurance, Verification and Validation. The first experiences showed that the method can be successfully used, but highlighted that the transition from an informal Requirements Specification to a formal specification also needs a defined process.

The FRERE process deals essentially with two aspects. The first is the redefinition of life-cycle concepts so that they deal with both informal and formal specification techniques and with the transition process. Informal methods are clear on objectives and on structuring, whereas formal methods are much more precise on the abstraction mechanisms that are used.

The second aspect is the issue of an error free transition from informal to formal. The act of formalisation is not in itself a guarantee that the right direction has been taken if there is any ambiguity in the informal specification. It is therefore necessary to propose effective mechanisms that can give the required confidence in the transition process.

The benefits of the FRERE process is that it fully develops the process level in formal methods use, up to now only formulated as a few commandments that should be followed. For the application, it allows a full return on the investment in formal methods use by ensuring a balanced error free process.

### 4.3 Process Engineering with Viewpoints

There is now wide recognition of the contribution of the process to the successful development of critical systems. However most models used for describing processes are essentially based on an activity driven, Document-In/Document-Out approach. Such approaches are particularly difficult to apply to the RE process since, by definition, there is no clear Document Input.

It is therefore natural to engineer the process using viewpoint structuring principles. The requirements are managed as discrete elements (Viewpoints) using explicit organisation principles. The sources of the elements, as well as the relationships between them, are recorded as attributes of the Viewpoint. The exact way the attributes of the viewpoints are organised depends on the criticality of the requirements and the process that is proposed for dealing with them. Hence a process model is constructed where sources are the input and a specification document the output. Such a way of working is a basic minimum constraint for a systematic approach to requirements engineering.

The use of System Viewpoints naturally leads to the consideration of Process Viewpoints. The development of other process oriented methods such as in the AMI project [AMI 92] have already shown that it is necessary to describe goals and viewpoints. In the case of the AMI measurement method, it was the need for a realistic process model, close enough to reality to enable the identification of relevant measures, that led to such considerations. Similarly, the need for practical modelling approaches of the RE process leads to the same techniques based on goal and viewpoint oriented process models.

### 4.4 Process Evaluation for human factors

The emphasis on Process Engineering with viewpoints needs in practice to be counter balanced by consideration of the "pipe work" of co-operative work that holds the process together. The REAIMS technique for this involves analysing the final process for dependability with special consideration of all the organisational issues that underlie such co-operative work.

The technique is based on an integration of an engineering oriented Failure modes or Hazops analysis with human factors methods of process capture and analysis.

The benefits of such a technique is that it transforms the traditional techniques of a quality or process audit to attain a much greater level of process detail and in a more systematic framework.

## 5. Conclusion

The REAIMS techniques have been applied to the development of critical applications in the transport sector, namely civil aircraft and train protection systems developments. The Aerospatiale aircraft development has been more concerned with the corporate memory aspects whereas the Train Protection systems application has involved more the formal reasoning issues.

Such activities of Process Improvement in Requirements Engineering are seen as fundamental ways for achieving the quality, cost and safety objectives for complex products especially when the products evolve towards product families.

In such a context, the REAIMS Adaptation and Improvement strategy can give industry a diverse approach to improvement that is compatible with existing methods and improvement techniques, but not dependent on any single assumption or global methodology.

## References

[ABRIAL 95] J-R Abrial, The B book, Assigning Programs to Meaning, to be published.

[AMI 92] AMI consortium, The ami Handbook : a quantitative approach to software management, South Bank University, London, 1992.

[REAIMS 93] REAIMS consortium (main partners : GEC Alsthom, Aerospatiale, Adelard, Lancaster University, RWTUV), ESPRIT Project EP8649, Project description, Technical Annex, 1993.

# Session 10
## Case Studies

# Safety Cases for Software Application Reuse

P Fenelon, T P Kelly, J A McDermid

High Integrity Systems Engineering Group,
University of York, Heslington,
York YO1 5DD, UK

e-mail: pete, tpk, jam @ minster.york.ac.uk

**Abstract**

In traditional engineering industries it is common to reuse tried and trusted components as one of the means of ensuring safety. Some low-level software components, e.g. libraries, are reused, but there are difficulties in justifying the reuse of software due to the complexity of interactions in a typical software system. This paper addresses the issue of reusing software applications by considering how to extend the safety case for the use of software in one application based on its use in another. It proposes an approach to analysing the change between two contexts of use of a software system, including analysing small changes in the software, and illustrates this through some examples based on an analysis of a reactor protection system.

## 1 Introduction

There is a long-established principle, in engineering safety-critical systems, to reuse tried and trusted components as a means of ensuring safety. These principles have been adopted in developing safety-critical software, but often only through the reuse of low level components, e.g. generic components of control laws. There is potential for considerable benefit in reusing larger software components, perhaps even a complete software 'application'. In practice it is unlikely that an item of software will be moved entirely unchanged from one domain to another but, even if it is unchanged, it is not obvious that the safety case can be preserved — the new environment may place new demands on the software, it may have a different distribution of demands thus rendering invalid existing statistical reliability data, and so on. Thus there is a challenge — to adapt the safety case to show how the reuse of an application (or part thereof) in a new domain may be justified. This amounts to adaptive reuse of the safety case, as well as reuse of the software.

We perceive re-use of safety cases as being valuable as the construction of a safety case is an expensive operation involving considerable time and effort on the part of engineers and managers from many disciplines [Ball89]. Almost inevitably however reuse involves change. Even if an application is well-suited to a new domain, small changes in the software will normally be needed. Thus we need a sound method for analysing the potential impact of change, both in the domain and in the software, identifying the aspects of the system design or operation which need to be re-assessed, then justifying the safety of the system in the new domain by reusing and extending the safety case. Such a method will have more than just economic benefits — it will avoid some classes of potential error in analysis and thus may contribute to safety.

We have developed some principles for structuring safety cases, for categorising imposed change (to the software, the system in which it is embedded and its operating environment), and for analysing the impact of change. These principles have been applied retrospectively to the safety case for a reactor protection system (the Stage 9 submission for the Dungeness B SCTS [NE91]), analysing several changes which were made late in the development of the system. In developing these principles, it became clear that they could also apply to the analysis of changes within the same environment — thus we believe that they are of general applicability in safety case evolution and maintenance (special cases of reuse).

Our aim in this paper is to illustrate the principles, and to show how they aid the analysis of change. For brevity we only describe a subset of the principles we have developed, and present a fragment of the example we have undertaken. The details of the example have been elided for ease of presentation, but the conclusions drawn accurately reflect the results of the study. We also include extensions of the analysis principles developed after the project was completed.

We first illustrate the problem in more detail, then present some of the key principles we have developed. The use of the principles is illustrated on a fragment of the safety case we analysed. The example is used to draw some conclusions about the utility of the approach, and to indicate what further developments would be needed to use the approach during development, rather than in retrospectively.

## 2 The Problem

Safety cases, like other products of complex engineering processes, are developed in an iterative manner, thus they are constantly subject to change during their development. Safety cases may also need to be reconsidered and changed after deployment of the system to which they refer. This may happen for a number of reasons, e.g. because of an unanticipated problem with the system, see for example [Hogberg94], because of a change in requirements or standards, e.g. the NUREGs [Queener94], or the desire to extend plant life [Clarke89] and the need to deal with operational history and changes in standards.

In supporting design iteration we need efficient ways of modifying the safety case, propagating the change and ensuring that we have re-established consistency of the case. We refer to this as *making changes*. In responding to external changes we are concerned with reasoning about the impact of change, showing that the system is still safe in the face of the change, or determining what consequential change is needed in order to preserve safety. We refer to this as *reasoning about change*.

In general we are concerned with an *imposed change*, i.e. one that is made or proposed outside the developer's control, and *consequential change*, i.e. one made by the developer in order to respond to the imposition, to produce an acceptable system and safety case. Our focus is on reasoning about imposed change, although we will need to discuss the analysis of consequential change to give a complete treatment of the situation. In order to handle imposed change efficiently we need to be able to preserve the initial safety case, so far as practical, adding the results of reasoning about the imposed and consequential changes to the safety case to show that the modified system is still safe.

We can illustrate the issues which arise using a simple scenario: a software-controlled reactor protection system, which trips on measured core temperature. Let us say that the reactor has been operating safely for some time with the trip level (referred to as the *set point*) at X degrees, and it is decided to raise the set point to X + 1 degrees (the reactor can be run more efficiently, hence more profitably, at a higher temperature, but the safety margin is reduced). Thus the change of 1 degree in the set point is the imposed change. What reasoning about change is required?

First, the reactor physics need to be considered to ensure that the change is safe in terms of the core operation. Second, evidence needs to be provided that the software implementing the trip function is still safe. The nature of the requisite evidence depends on the details of the change. It is instructive to consider some possibilities.

If the protection system software is recompiled with a change in a literal constant for the trip value, and the compiled code is bitwise identical to the previous version except for the constant, then we can argue that the previous testing evidence, etc. can be carried over to the new system as the change has not affected the program structure, the coverage of the testing programme, etc.

However, if the constant (set point) was changed from, say, 127 to 128 and, as a result, the compiler changed the code from an integer comparison to a shift followed by a 'branch if not zero' instruction (for speed), allowing further optimisation changing the executable code structure, then the testing results would not carry over directly. In this case more tests might be used to reconstruct the safety case. Thus the reasoning about change would be that the original test data, plus the additional tests and test results, were sufficient to show that the change was acceptable. This 'delta' on the safety case is needed to allow the original safety case to be reused.

In general it is a good policy to separate data such as trip points from the code, to simplify reasoning about change. In practice we can accommodate 'small changes' but there is a difficulty in determining when changes cease to be small and the software component needs to be treated as new. The conservative approach is to assume that all changes are 'large', and repeat the complete analysis. Our intention is to offer a more cost-effective way of dealing with small changes.

# 3    The Principles

The principles we developed were based around a particular approach to the structuring of safety cases, and some evolving techniques for software safety analysis. We amplify on the safety case approach and structuring principles below. We also developed a taxonomy of change for analysing both imposed and consequential change. Whilst this is important as part of an overall method for change analysis, we only need a small part of the taxonomy for discussing the examples, so we give a simple overview of the relevant ideas in section 3.4 below.

## 3.1    Types of Safety Case

In developing the principles, we identified two forms of software safety case — 'black box' and 'white box'. The 'black box' case places reliability figures on the software, and assumes no knowledge of the structure of the software. Thus, with a 'black box' approach it is necessary to use purely stochastic arguments to justify the deployment of the software (application) in the new domain, or to take into account the impact of an imposed change. Even an imposed change in the distribution or frequency of demands on the system could render the (relevant portion of the) safety case invalid — or require very subtle statistical analysis. Consequently we introduced the notion of a 'grey box' scenario, where we took an essentially 'black box' situation, and derived a small amount of pertinent structural information about the software, perhaps using reverse engineering — for example, so that we could show that test results still gave sufficient path coverage to sustain the safety case. However, none of our examples were based on a statistical analysis, so we do not consider the 'black box' approach any further.

The 'white box' approach constructs a deterministic safety case, at least so far as the programs are concerned. Thus the 'white box' case looks at the internal program structures, and employs software engineering and safety analysis techniques, e.g. static analysis and software fault-trees [Leveson83], to show that the proposed deployment of the software in the new domain is sound, or that the imposed change is benign. Probabilities of hazardous events are still derived, but the potential software contribution to a hazard is represented in the structure of the fault-trees, not as a probability of the software failing. Our examples were based on this 'white box' approach. The safety analysis techniques which we used in carrying out the examples are outlined in section 3.3.

## 3.2  Safety Case Structuring

Most safety cases are structured around a hazard log. We believe that this is appropriate to provide an index into a safety case, once it has been constructed, but inadequate to guide the construction of the safety case. In our approach we use the notion of *goal structuring* to assist in the construction of the safety case, to facilitate the management of scale and complexity in safety cases, and to assist in change management. Some of our early experience in using goal structuring in safety cases are reported in [McDermid94], and more recent applications to the nuclear domain are described in [Wilson95a]. The most fundamental concepts are:

- *goal* — is something that a stakeholder in the design and assessment process wishes to be achieved;

- *strategy* — a strategy is a (putative) means of achieving the goal or collection of goals, e.g. a system concept, or a sequence of activities.

Goals are decomposed through the strategies, and we will refer to sub-goals where this is helpful. Goals may refer to specific technical properties, or may be more general, e.g. to do with commercial objectives such as power plant profitability. Typically the high level goals in a safety case will reflect general principles, and the lower level goals will be interpretations of (derived requirements arising from) those principles for the particular system being considered. For example, the HSE set out some 333 safety principles [HSE92], a subset of which might form part of the top level goals of a safety case, see Figure 1. Here the term 'criticality based incident' refers to situations such as failing to trip which leave the power plant in a critical situation. Such incidents are distinct from, and less severe than, directly hazardous events such as a release of radioactive material to the environment.

Goal G1 states the intention to comply with the HSE safety assessment principles. The strategy is shown below the goal, and is to work to relevant principles, e.g. the principle (P126) to evaluate the probability of an aircraft crashing into the reactor may be deemed inappropriate for a reactor in a submarine. The sub-goals, G1.1 ... G1.N, are the specific principles which have been selected, and the ellipse is the justification for selecting these principles (see below). Our safety case support tool, SAM [Wilson95b], enables goals to be given simple names, and more details associated with the goals to be stored in a database. For example, a full description of a goal from figure 1 might include the full principle definition, e.g. G1.2 — 'A reactor should be provided with systems which can shut it down safely ...'.

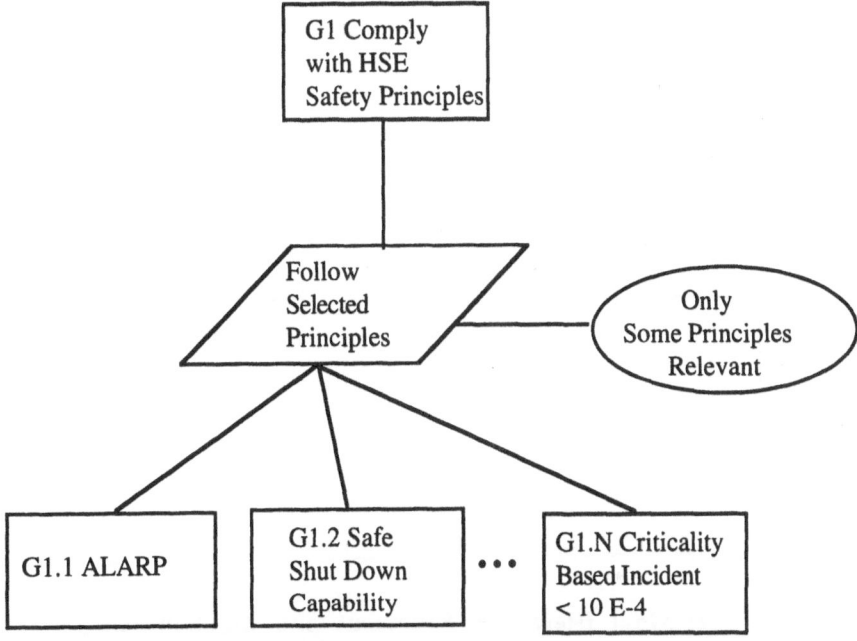

**Figure 1: Example Goal Hierarchy Fragment**

Some goals may be satisfied directly, e.g. by carrying out an action, or providing a product with the right properties. We use the term *solution* for the action or product which satisfies a goal. Goals with solutions are *leaves* of the goal structure, i.e. they have no strategy or sub-goals. Solutions are typically safety analyses, although they may be arguments (see section 4). Thus a lower level fragment of a safety case might be as shown in figure 2. The solution is represented as a circle and, in this case, is a system level fault-tree analysis.

The lower level goals are traceable to (from) the HSE principles. The terms Pfl and Pfh stand for probability of failing to trip on low temperature and high temperature, respectively. Here we have shown a common form of decomposition, where a particular criticality based incident — failure to trip on demand — is decomposed into lower level goals representing the contributory failure modes. The justification states that the budgeting (allocation) of rates between the different modes is based on historical evidence/experience.

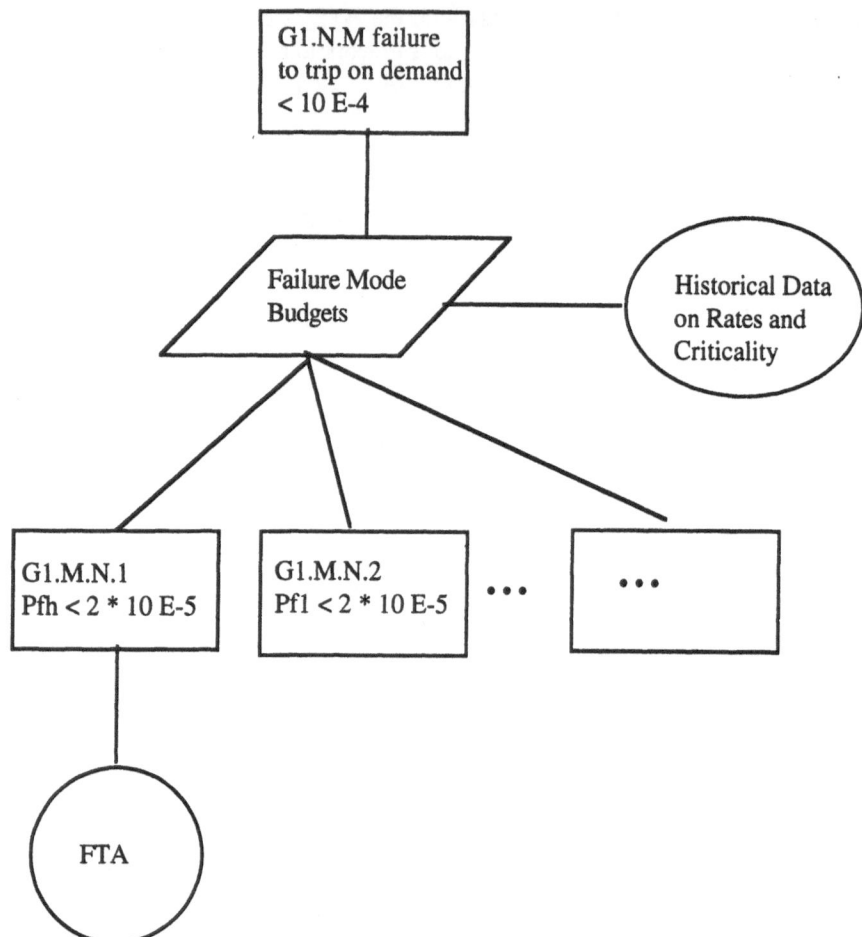

**Figure 2: Goal Hierarchy Fragment Showing Solution**

We use the term *constraints* to refer to those goals which are not solved directly, but which restrict the way in which other goals are solved, i.e. which restrict the set of allowable strategies (and models, see below). The satisfaction of constraints must be checked at multiple points in the goal hierarchy. Common safety requirements such as 'no single point of failure shall lead to a hazard' are representative of this class of goal. The HSE principles contain a number of constraints (in our terms). For example 'Defence in Depth' and 'Diversity in Detection and Control' are guidelines or constraints against which the design will be evaluated at many different stages in the design decomposition process.

There are other important facets of the structure which will be related to goals or strategies. These include:

- *models* — these represent part of the system of interest, its environment or the organisations associated with the system; goals will often be stated in terms of models, especially when they represent the system design;

- *justification* — a justification is an argument, or other information, e.g. the results of a safety analysis, presented to explain why a strategy is believed to be effective, i.e. that it meets the goals.

In our example, the models of interest are representations of the structure of the software in the trip system. In general, the goals and justifications give a basis for determining the impact of change, see section 3.4.

## 3.3  Safety Analysis Techniques

In general, we assume the use of standard safety analysis techniques, including the application of fault-trees to analyse software [Leveson83]. However, we also make use of one relatively new technique known as Failure Propagation and Transformation Notation (FPTN) [Fenelon93]. The purpose of this notation is to summarise the 'flow' of failures through complex integrated systems — hence the term propagation. As failures 'flow' they can be become transformed, e.g. the omission of a message may be detected by a watchdog timer, and an extrapolated value substituted, changing the omission failure to a value domain failure — hence the term transformation.

The notation is graphical, and it effectively acts as a summary for a set of fault-trees, enriched with the notion of failure types — omission. commission, etc. Thus FPTN can be thought of as being the functional analogue to FMES (Failure Modes and Effects Summary) — providing a succinct representation of the failure modes and effects, but structured around the system *functional* decomposition, not its *physical* decomposition. (FPTN is also intended to be used as a means of defining derived safety requirements, but we will not discuss this further, as we do not use the notation in this mode in our examples.) FPTN is represented as follows:

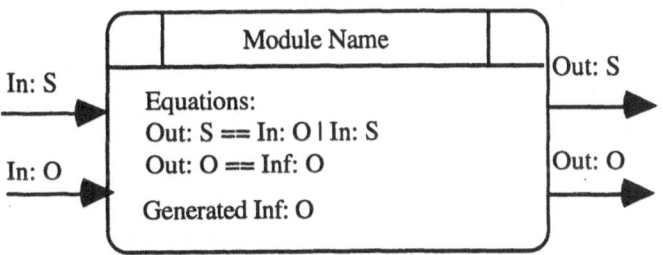

**Figure 3: Simple FPTN Module**

The notation is similar to Yourdon data-flows, but it is concerned with failures, not data flows. The round-cornered box corresponds to a software module, typically a process or a procedure. The inputs and outputs are failures, not data flows, but they are normally failures associated with the data flows, e.g. wrong data values. The labels before the colon are the flow name, and those after the colon are the 'types' of the failures (see [Bondavalli90] and [Pumfrey94] for more detail on failure types). The types used are O — Omission, C — Commission, S — Subtle, C — Coarse, E — Early and L — Late. The equations are representations of fault-tree cutsets. They say how the output failure modes are related to the input failure modes — in this case the subtle failure is propagated, and the omission failure is transformed into a subtle value failure. The generated failure arises from an omission failure of the computing infrastructure (Inf) which leads to an omission from the module.

In general we need a collection of FPTN modules to describe a system. We will get chains, or networks, of FPTN modules, and we need to allow module hierarchies, to describe a complex system. At the current state of development the method is largely qualitative, but we are investigating the issues in developing a quantitative calculus based on the FPTN structures. FPTN was developed to model properties of software systems, but can equally well be applied to other forms of discrete/digital system.

In analysing change, we need a further concept, that of the safety critical path (SCP). If a failure condition of a system, represented as an FPTN failure mode, is hazardous (critical) then all the modules which have failure modes that contribute to the hazardous failure mode are members of the SCP. We shall see that the notion of an SCP is important in sections 3.4 and 4, as the criticality of a change depends on whether or not it affects the SCP.

## 3.4 Taxonomy of Change and Analysis Approach

The analysis approach we developed has three main components:

- a taxonomy for identifying and classifying imposed changes in the environment and the system in which the software is embedded, and for classifying the consequential changes to the software or system;

- a form of analysis, based on the taxonomy and software safety analysis techniques, for identifying which changes are *benign*, and which pose a *challenge* to the system, software or safety case;

- a set of template arguments for justifying the use of the application in the new domain, based on the taxonomy and analysis.

The notions of changes which are benign or pose a challenge is central to the approach. Benign changes do not require any consequential change. An obvious category of benign change is one that either increases the ability of a system to meet its goals, e.g. increases its reliability, or reduces a requirement, e.g. a demand rate for a trip system. Other changes may also be benign, e.g. an increase in a trip demand rate might be benign if there is sufficient margin between the goal and the achieved reliability of the system. A challenge arises where a consequential change is needed to 're-substantiate' the overall safety case.

In general, we do not know *a priori* whether an imposed change is benign or poses a challenge, thus we need a process for analysing the impact of the imposed change:

1      categorise the imposed change — determine the source of the change, e.g. which system or software components are affected, and which goals are directly affected by the change, e.g. apply to the changed component;

2      categorise the change as benign or a challenge (by determining whether or not all the immediate goals are still met);

3      if the change is benign, present an argument why this is the case, e.g. reliability margins are reduced, but the design is still acceptable;

4        if the change is a challenge, determine and make a response (consequential change); note that this may not involve the system, but could be a change in operating procedures, or further testing to gain more evidence of safety;

5        repeat steps 1 to 4 for the consequential change, until all the consequences are shown to be benign, i.e. goals are reached which are now satisfied;

6        present the arguments why the system, incorporating/allowing for the consequential changes is safe (meets its goals).

FPTN can help to distinguish between benign and challenging software changes. An imposed change which is on an SCP is a potential challenge, as is a change which brings some module onto an SCP, when previously it was not on the SCP. A potential challenge is an actual challenge if it causes the goals immediately dependent on the changed module no longer to hold (or puts this in doubt). Changes which do not affect the SCP, or those which do but which do• not affect the immediate goals are benign. Thus, for software-based systems, we use FPTN as a key element in step 2 of our approach to change analysis.

Note that the fact that an imposed change represents a challenge does not mean that the system or software will need to be modified as a consequence. It may simply be that more evidence is needed to 'reconstruct' the safety case. Also note that the goals act as 'barriers' — change does not propagate any further if a goal is still met. This principle applies both to imposed and consequential changes.

# 4   An Example

The example is based on a safety case for a reactor trip system (the Stage 9 submission for the Dungeness B SCTS), and some changes that were imposed. We considered several different situations which affected this system. In all cases, there were new or modified requirements, but the aim was to leave the software unchanged, so far as practical, so as to avoid the cost of reverifying the modified software (see section 2). We briefly describe two of the changes then discuss their analysis in the terms introduced above. For presentational purposes, some details have been suppressed, but the structure of the arguments and the overall findings are unmodified.

## 4.1  Imposed Changes

The two imposed changes which we will consider are:

1        decreasing the maximum tolerable range of sensor values (i.e. the acceptable difference between the minimum and maximum recorded temperatures);

2        increasing the set point for high temperature trip.

Both changes present potential challenges, but the first is easy to justify as benign by a 'delta' to the safety case, as we shall show.

## 4.2  System Models

There are three system models of interest: the physical structure of the hardware, the logical structure of one lane of the software and relevant portions of the code. These are as shown in the following figures.

428

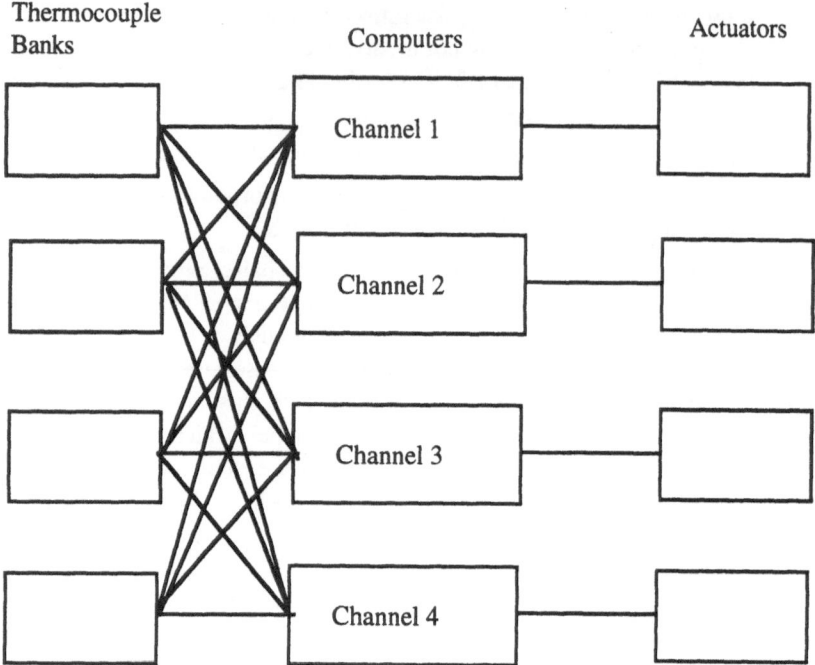

**Figure 4: Outline Hardware Structure**

The hardware is structured so that each channel needs to periodically send a signal to the actuator logic, otherwise the actuators will trip. This makes the design fail safe, in the event of power loss (another HSE principle).

The functional structure of a single channel is (in much simplified form):

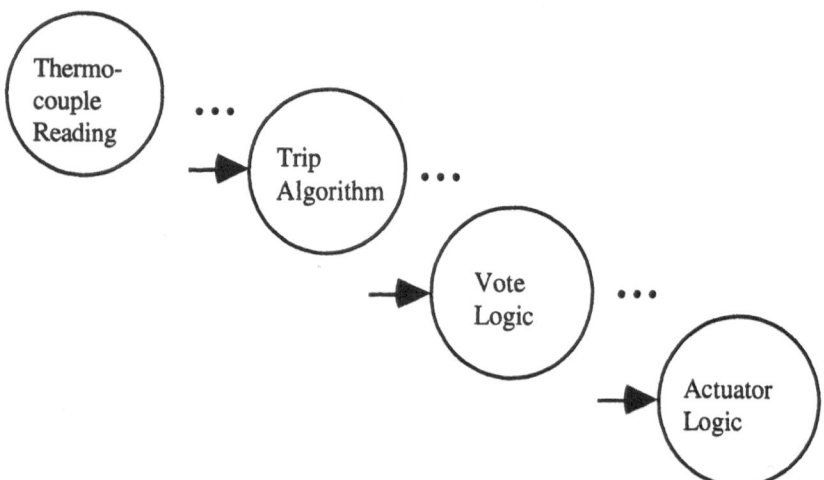

**Figure 5: Simplified DFD for Channel Software**

Each function is executed periodically. In the complete system there are a lot of functions in each channel, e.g. signal conditioning, but we show some of the key functions above. We will focus on the trip algorithm. A relevant fragment of the trip algorithm code is shown below. It was derived from published trip algorithm specifications, rather than being taken from the system, and is represented in Ada to enable us to use some of our analysis tools [Fenelon93], and simplified for the purposes of exposition. The fragment is:

```
procedure trip_alg is
    ...
if range_test
        -- $SSAP bg range_trip
        -- affected by change 1 (reduction in max_spread)
        t := range_calc;
        if t > max_spread
            then
                for s in signals loop
                if set(s)
                    then demand_trip (b, s);
                end if;
        end if;
        -- $SSAP eg range_trip
else
        for s in signals loop
            if set(s)
                then
                    -- $SSAP bg hot_trip
                    -- affected by change 2 (reduction in hot_set_point)
                    if temp(s) > hot_set_point
                        then demand_trip (b, s);
                    end if;
                    ...
                    -- $SSAP eg hot_trip
```

**Figure 6: Fragment of Trip Algorithm**

The variable signals is an array of augmented thermocouple readings pre-processed by the signal conditioning logic. Those signals which are set are from thermocouples which are deemed to be functioning correctly by the signal conditioning logic.

The range test logic determines the maximum spread of the temperature readings, and demands a trip for all the signals which are set. The temperature trip logic cycles through the pre-processed thermocouple signals, and demands a trip for any signal which exceeds the threshold. The function demand_trip passes a trip request to the voter logic which is the next function in the chain. It might be thought that an omission failure here could be hazardous, however such eventualities are addressed through the hardware redundancy, not the software structures.

The comments -- $SSAP are directives for a prototype fault-tree tool which we used to analyse the software. The terms bg and eg stand for 'begin group' and 'end group' respectively. These groups are 'components' of the program to be analysed, e.g. from which to produce a fault-tree.

These system 'models' serve as the basis for the analysis of change. They are far from sufficient to represent all the safety-relevant facets of the system, e.g. a real trip algorithm is considerably more complex, but are sufficient to illustrate the principles we have developed.

## 4.3   System Goals

The directly relevant goals for this analysis are G1.M.N.1 and G1.M.N.2 from Figure 2, dealing with the low and high temperature trip probability, respectively. In practice there are additional relevant goals to do with spurious trips (this is an availability issue, not a safety issue, but needs to be considered in developing the safety case), and concerned with operator procedures, e.g. vetoing thermocouples during maintenance operation. Some of the changes we considered had an impact on these goals, but they cannot be addressed in detail within the confines of this paper.

## 4.4   Safety Analysis

We are concerned with failure to trip on demand, e.g. satisfaction of goals G1.M.N.1 and G1.M.N.2. In considering the trip algorithm this corresponds to the omission of a call to demand_trip, and the consequent failure to request a trip from the voter. For the system overall to fail to trip on demand, we need to have a number of failures, but in change analysis we can focus on the trip algorithm.

The fault-tree for the hot set point logic is as follows. The fault-tree is generated in the spirit of the rules derived by Leveson et al [Leveson83]. The basic form of the fault-tree is produced automatically from the annotated Ada code. The top event is annotated with HighTrip: O to represent the fact that this is an omission failure. This label is used so that the correlation with the event in the FPTN is apparent. Similarly one of the leaves of the fault-tree is labelled to show the effect of failures of functions 'upstream' of the trip algorithm. The labelling with the FPTN failure types is currently a manual operation.

The fault-tree is the solution to the goal G1.M.N.1 in figure 2. In the complete safety case, there would be probabilities associated with some of the leaf events, representing the likelihood of particular hardware failures, e.g. thermocouple mis-readings. This would enable us to provide a top even probability, and show that goal G1.M.N.1 in figure 2 had been met. For our examples we do not need to make statistical arguments, so we do not include and failure rates or probabilities in the fault-trees.

A partial FPTN representation of the trip logic, sufficient for our analysis, is shown in figure 8. This shows possible input failures to the Trip Algorithm, specifically Omission and Subtle value failures of the conditioned thermocouple signals. The figure shows how the fault-tree of figure 7 contributes to the FPTN. Note that the FPTN for HighTrip includes a failure propagation not in the fault-tree: this arises because there is nothing in the logic of the software fault-trees to explain what happens if the software is not executed. Clearly, this is not a fault in the analysis approach; it merely points out that the FPTN needs to integrate analyses arising from a number of different sources. In a full analysis, we would also be interested in HighOut: C, etc. as these will represent spurious trips.

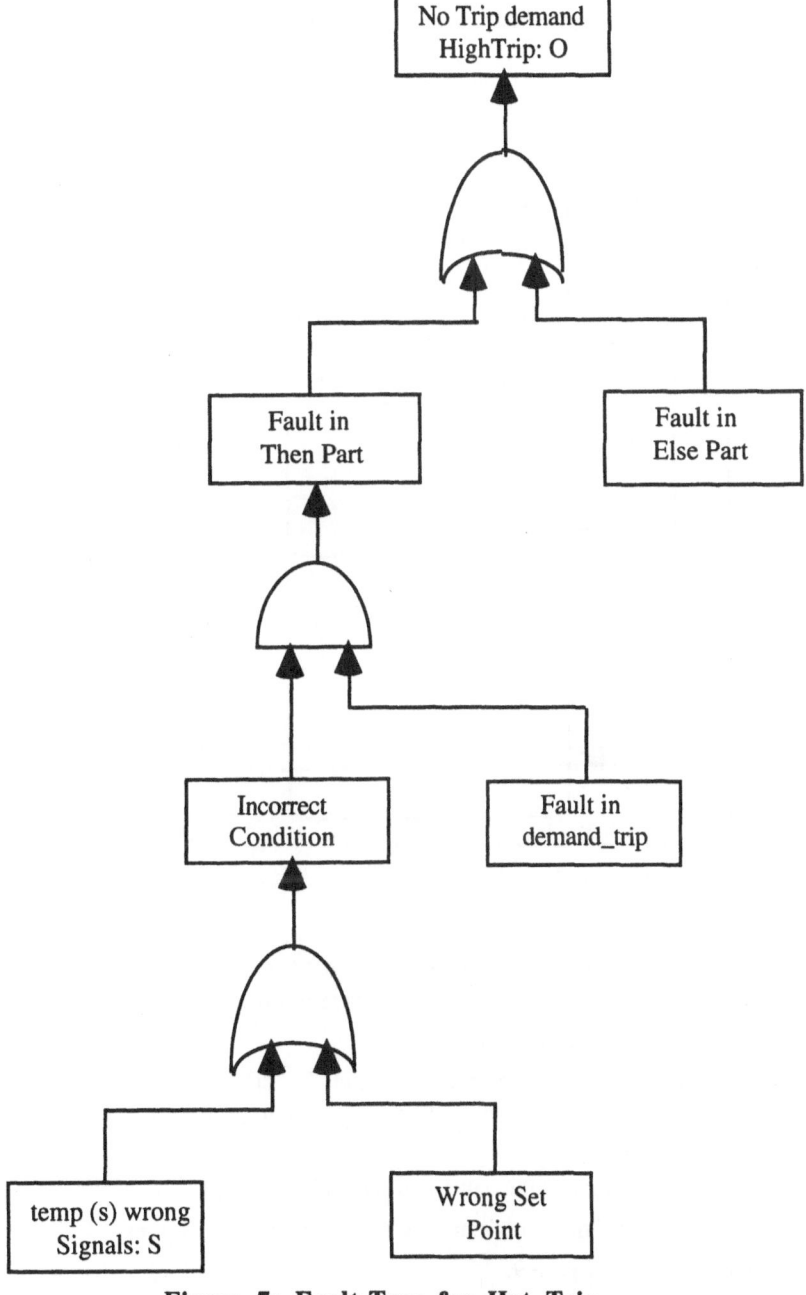

Figure 7: Fault-Tree for Hot Trip

The FPTN module Trip Algorithm is on the SCP. We now have sufficient information to discuss the change analysis.

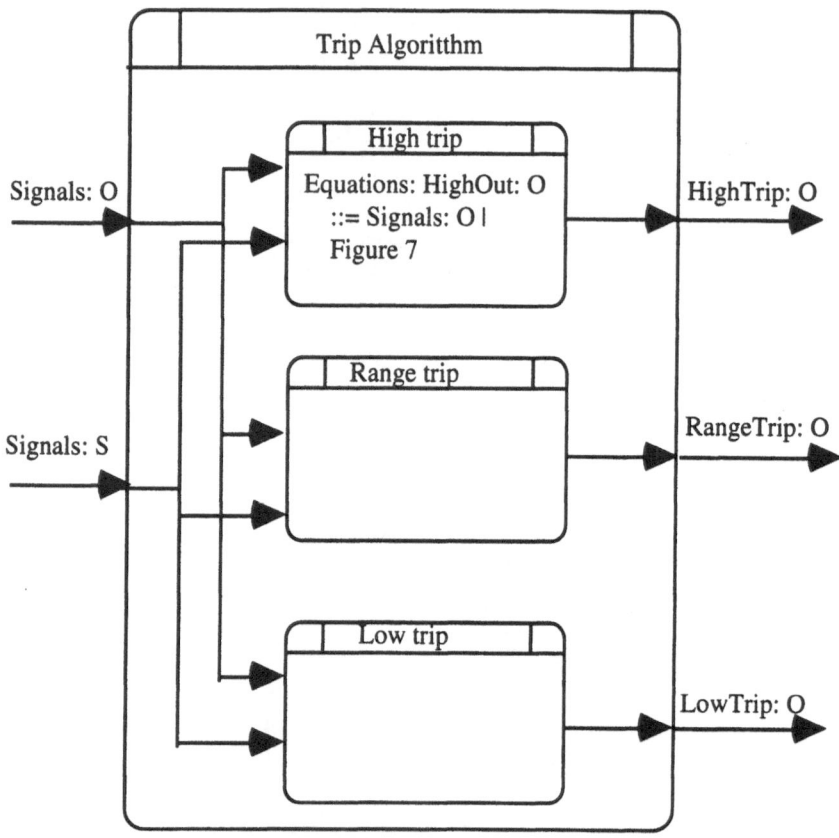

**Figure 8: FPTN for Trip Algorithm**

## 4.5 Change Analysis

We now consider the analysis of the changes identified in section 4.1, based on the information introduced above, and using the process outlined in section 3.4.

### 4.5.1 Decrease Range of Sensor Values

Strictly, the change is a challenge, but one which is 'discharged' immediately by analysis of the nature of the change. Thus we treat it as if it were benign for expository purposes:

1       source — TA module; the value max_spread is reduced.

2       category — benign as this can only lead to more trips, not less.

3       argument — see figure 9.

The argument why the change is benign is summarised in graphical form in figure 9. It is important to understand the context of the argument. The goal which is under challenge by the change is G1.M.N.1. The argument establishes that G1.M.N.1 still holds, even after the change, thus the analysis does not need to be propagated any further (the rest of the case is not challenged directly; G1.M.N.1 holds so the rest of the case is not challenged indirectly).

The argument form shown is very similar to a fault tree, except that the claim is shown on the right, and the data on which the claim rests is on the left. There are three data used in the argument. First, there is an assertion that narrowing the trip range can increase the trip rate, not decrease it. This assertion could be backed up by analysis of the trip algorithm, but we assume that it will be accepted as a correct informal analysis of the program. The second assertion is that only the trip algorithm on the SCP is affected, and so this is the only element which needs to be considered. This is clearly the case as the change does not affect the flow of data between any of the modules. However this is a crucial issue: the validity of reasoning in terms of goals and dealing with 'local' changes is contingent on the accuracy of the analysis of the scope of the impact of change. The third datum asserts that the testing of the code with the previous range value can be used to justify the current version of the software, as the change has not affected the code. These assertions seem to be 'obvious' but if they were challenged they can be 'backed up' by more detailed analysis of the programs/system, e.g. along the lines discussed in section 2 for the third datum.

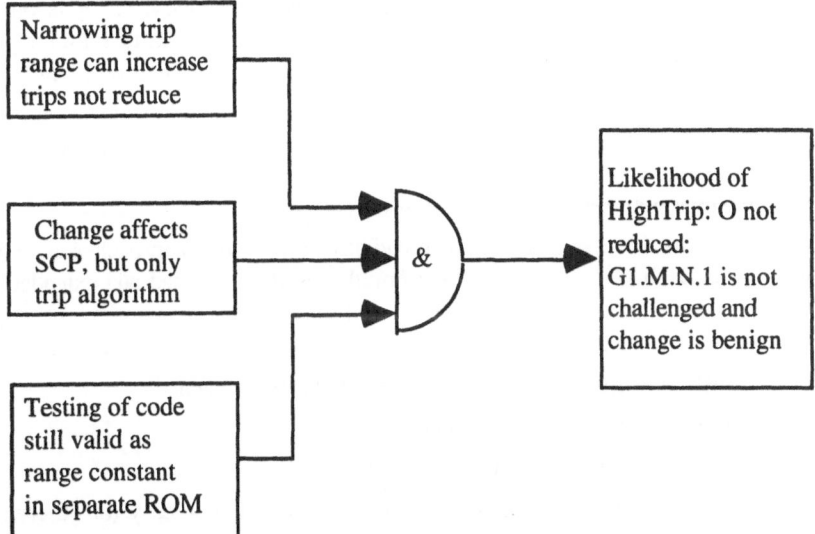

**Figure 9: Argument that Narrowing Trip Range is Benign**

In this case, the above argument and any supporting evidence would be the 'delta' on the safety case. This example illustrates the notion of argument template introduced in section 3.4. The general form of the argument is: the change is intrinsically benign; the scope of the impact of the change is limited to the item directly affected by the imposed change; the existing evidence for the affected item still holds — hence the claim of the safety case still holds. Many arguments justifying benign changes will be of this general form.

### 4.5.2 Increase Set Point

Increasing the set point is a challenge, and is analysed in the following way:

1        source — TA module; the value hot_set_point is increased.

2        category — challenge as there is a change on the SCP and it is not obvious that the immediate goal is still met as it could cause the 'wrong set point' event in the fault-tree of figure 7, and thus violate G1.M.N.1.

4        there are a number of possible responses, see below.

The challenge can only be met from the point of view of reactor physics, or evidence of successful operational experience. Clearly, there will have been no previous experience of successful and safe operation with temperatures between the old and new set points, so a direct appeal to history would be inadequate. Of the many possible solutions it is perhaps most likely that a modification of operational procedures enabling operation with the increased set point but more stringent monitoring would be required, before the plant could be operated with the new set point in a 'routine' manner. This experience might be used to alter the budgeting between G1.M.N.1 and G1.M.N.2 and to modify the justification, thus still satisfying G1.M.N. Here G1.M.N, not G1.M.N.1, would act as the 'barrier' in the goal structure beyond which no further challenges propagate.

## 5    Observations

The examples we have set out above are necessarily fragmentary, but it is our hope that we have illustrated enough of the approach we have adopted to reasoning about change to enable others to adopt the spirit of the approach on their problems.

We have analysed a number of other example changes including ones where there were consequential changes in the operating procedures. We suspect that this form of consequential change is quite common, and it makes clear that it is necessary to model the operating procedures as part of the safety case.

In carrying out the case studies, the broad structure of our approach has remained intact, although it is clear that the ideas could usefully be refined and extended. For example, the distinction between benign and challenging changes is not quite so clear cut as we once thought it was. We should perhaps distinguish three cases:

•        justifiable immediately as no safety goals are threatened;

•        justifiable after additional analysis;

•        justifiable after consequential change.

All changes need to be justified, but the above indicates how easy (or otherwise) it will be to come up with a justification.

Some of the examples we looked at exposed conflicts between goals, especially availability (spurious trip rate) and safety goals. One of the advantages of our approach (not illustrated here) is that the goals need not be confined to safety, and we can use the goal hierarchies to point out conflicts which arise as a result of change. One of the examples we considered involving the operating procedures highlighted a conflict between availability (avoidance of spurious trips) and safety.

The examples we have described have all been retrospective analyses, and clearly the method would be of much more value if it could also be applied in development. We can see no reason, in principle, why the method outlined here shouldn't work in development, although we would clearly also need procedures for folding the safety case 'delta' back into the main safety case, i.e. so we could connect the reasoning about change into the process of making change. We hope to have the opportunity to try out the ideas on, or in parallel with, a 'live' development in the near future, and this should enable us to put the hypothesis that our method will also apply in development to the test.

An issue which is still unclear is 'what is a small change?', or perhaps more subtly 'how do we know when the overall effect of a sequence of changes ceases to be small?'. We don't have a clear answer on this point. Our expectation is that the process we have described will break down when a change 'falls foul' of some unstated assumption or assumptions. This could occur after one change, or not occur even after several hundred changes. This is an issue which requires more study, but ultimately may rest on a judgement about the quality of the original safety case, in particular the extent to which all the salient information has been made explicit.

Finally, none of the examples we looked at required statistical treatment, although the cases were based around allowable failure rates. We do not know whether or not this is typical, however it was clear in many cases that a sub-text of the arguments presented was to show that the existing statistical data could be used unaltered in the new context. Even if this is not the typical case, there is a strong economic argument for trying to construct the 'delta' on a safety case to justify change in this manner.

# 6    Conclusions

We have developed a collection of taxonomies and analysis techniques which enabled us to analyse imposed changes, and to modify the system safety case in an appropriate way in response to those changes. These principles address general change control, as well as dealing with the initial problem which motivated the work — moving an application from one domain to another.

The examples we have addressed have enabled us to validate some of our principles, although not all of them could be tested on this small number of example changes we studied, e.g. the statistical ('black box') arguments were not tested. However the examples give us a measure of confidence in the overall approach adopted and it is clear that any method needs to identify the target and scope of change. We believe, therefore, that we have identified some useful principles for analysing and arguing about change, and for dealing with the movement of software from one domain to another. We hope to continue to develop and expand on these techniques as part of our overall programme of work on safety case development and management both in the University, and in co-operation with our industrial sponsors.

# 7    Acknowledgements

The bulk of the work reported here was supported by the HSE Nuclear Safety Research Programme, controlled by the nuclear Industry Management Committee

(IMC). Tim Kelly is supported by a CASE award funded by the EPSRC and Rolls-Royce and Associates. The SAM tool is being developed in the ASAM-II project, funded by the DTI and EPSRC, and involving BAe Airbus, BAe Military Aircraft, Lloyd's Register, Rolls-Royce Aerospace, Rolls-Royce and Associates, York Software Engineering Ltd and the University of York. Thanks go to all our colleagues in the ASAM-II project.

We are grateful to Gordon Hughes of Nuclear Electric and John Mitchell of the Nuclear Installations Inspectorate for the information and explanations which formed the basis of our case study.

# 8      References

[Ball89] Preparation of Fully Developed Safety Cases in Response to the NII Safety Audit, P W Ball, The Nuclear Engineer, Vol. 30, No. 2, pp34-40, 1989.

[Clarke89] Magnox Safety Review: Extending the Life of Britain's Work Horses, Nuclear Energy, Vol. 28, No. 4, pp215-220, 1989.

[Bondavalli90] Failure Classification with respect to Detection, A Bondavalli, L Simoncini, First Year Report: ESPRIT BRA Project 3092: Predictably Dependable Computing Systems, May 1990.

[Fenelon93] An Integrated Toolset for Software Safety Analysis, P Fenelon, J A McDermid, Journal of Systems and Software, Vol. 13, pp2-16, 1993.

[Hogberg94] Shutting down five reactors: reasons why and lessons learnt, L Hogberg, Nuclear Europe Worldscan, Vol. 1, No. 2, pp42-43, 1994.

[HSE92] Safety assessment principles for nuclear plants, Health and Safety Executive, 1992.

[Leveson83] Software Fault Tree Analysis, N Leveson, P R Harvey, Journal of Systems and Software, Vol. 3, pp173-181, 1983.

[McDermid94] Support for Safety Cases and Safety Arguments using SAM, J A McDermid, Reliability Engineering and System Safety, Vol. 43, No. 2, pp111-127, 1994.

[NE91] Stage 9 Submission, Dungeness 'B' Power Station, Single Channel trip System Reliability, Nuclear Electric 1991 (Private Communication).

[Pumfrey94] A Development of Hazard Analysis to Aid Software Design, D J Pumfrey, J A McDermid, In Proc. of COMPASS'94, IEEE, pp17-25, 1994.

[Queener94] Reports, Standards and Safety Guides, D S Queener, Nuclear Safety, Vol. 35, No. 2, pp339-344, 1994.

[Wilson95a] No more spineless safety cases: a structured method and comprehensive tool support, S P Wilson, J A McDermid, P Fenelon, P Kirkham, Proceedings of INEC'95: Second International Conference on Control and Instrumentation in Nuclear Installations, Institute of Nuclear Engineers, 1995.

[Wilson95b] ASAM II User Guide, S Wilson, ASAMII/UDOC/95.1, 1995. (Available from the authors.)

# The SHIP Safety Case Approach

P.G. Bishop and R.E. Bloomfield,
Adelard,
London, E3 2DA, England

### Abstract

This paper presents a safety case approach to the justification of safety-related systems. It combines methods used for handling software design faults with approaches used for hazardous plant. The general structure of the safety argument is presented together with the underlying models for system failure that can be used as the basis for quantified reliability estimates. The approach is illustrated using plant and computer based examples.

## Introduction

The SHIP project was sponsored under the EU Environment Programme (Major Industrial Hazards). The objective of the project was to assure plant safety in the presence of design faults but it was tackled from a novel standpoint. In software, all faults are design faults so techniques developed for software might well be applicable to the design of complete systems.

This paper describes a central element of this research—the SHIP safety case. The concept of a "safety case" grew out of work in the nuclear industry and is now a familiar term in many industries. For example, in the UK the CIMAH regulations implement the EC Directive on Major Hazards (the Seveso Directive) and require a report on the safety of the installation that addresses the dangerous substances involved; the installation itself; the management system; and the potential for major accidents. In SHIP, the safety case concept has been formalised and extended to cover both hardware and software systems.

## Elements of a safety case

We define a safety case as:

> "a documented body of evidence that provides a convincing and valid argument that a system is adequately safe for a given application in a given environment"

The development of the SHIP safety case is based on: a simple model of the system behaviour; the available evidence; and an argument which translates evidence into claims about behaviour.

## General Safety Case Structure

The safety case should be developed in parallel with the design. It will evolve and become more detailed as the system is developed. At each stage, the basis for the safety arguments should be clear. The safety case should:

- make an explicit set of claims about the system
- provide a systematic structure for marshalling the evidence
- provide a set of safety arguments that link the claims to the evidence
- make clear the assumptions and judgements underlying the arguments
- provide for different viewpoints and levels of detail

There is much work on the structuring and representation of arguments in mathematical logic [Gentzen69] and in formal methods [Hoare69, Jones90]. In the safety field we have the traditional representation such as fault trees [Vesely81] and the safety arguments in the Safety Argument Manager (SAM) [McDermid94]. The safety case structure developed in SHIP draws on this work. In addition we have formalised the basic argument structure, so that the safety argument could in principle be checked, supported and maintained by tools if the approach is further developed. Within this structure we consider the basic types of argument that can be deployed.

Within the SHIP model, a safety case consists of the following elements: a *claim* about a property of the system or some subsystem; *evidence* which is used as the basis of the safety argument; an *argument* linking the evidence to the claim, and an *inference* mechanism that provides the transformational rules for the argument. This is summarised in the figure below.

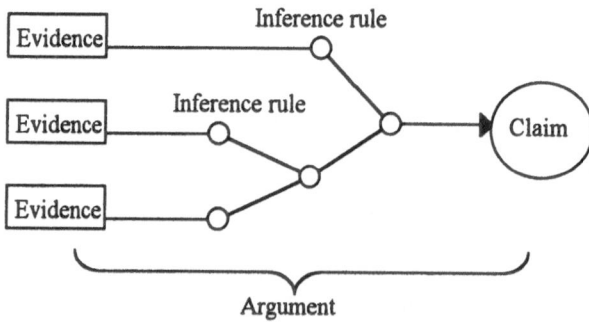

Figure 1: Argument Structure

The actual nature of the argument and the inference mechanism can vary depending on the system design and the safety case strategy. For example an argument could be:

- *Deterministic*, where the evidence can be axioms, the inference mechanism is the rules of predicate logic, and the safety argument is a proof using those rules.

- *Probabilistic*, where the evidence could be component failure rates and assumptions of independence, and the inference mechanism is statistical analysis.

- *Qualitative*, where the evidence might be adherence to standards, design rules, or guidance. The inference mechanism is some form of acceptance criterion based on this.

In addition the overall argument should be *robust*, i.e. the argument can be sound even if there are uncertainties or errors.

## Structuring a Safety Case

In practice it is unlikely that any safety case will be entirely deterministic or probabilistic. It may consist of a number of claims about specific properties of the system which may not necessarily be the same type. In addition it needs to be viewed at various levels of detail. It is proposed that a safety case can be structured as a hierarchy of claims as shown below:

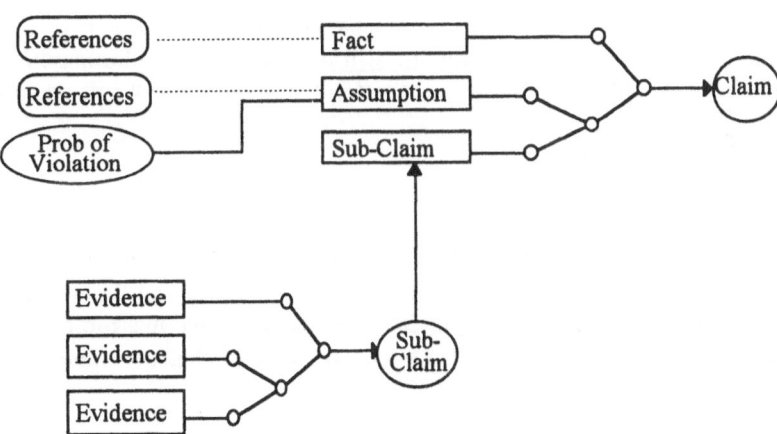

Figure 2: Hierarchic Argument Structure

In this model, the evidence used at one level of the argument can be:

- facts, e.g. based on established scientific principles and prior research

- assumptions, which are necessary to make the argument, but may not always apply in the "real world"

- sub-claims, derived from a lower-level sub-argument

This is a recursive structure which can represent arguments at successively finer levels of detail. This structure could evolve over the lifetime of the project. Initially some of the sub-claims might actually be *design targets*, but as the system develops the sub-claims might be replaced by facts or more detailed sub-arguments based on the real system. Deviations in implementation can be analysed to see how this affects a sub-claim, and how changes in sub-claim "ripple through" the safety argument.

In order to simplify the evaluation of this structure, it is proposed that:

- any argument tree must be consistent in type using a single consistent set of inference rules

- the evidence from sub-arguments must be consistent in type, or it must be possible to transform the type

For example, if the top-level argument is probabilistic, and there is a deterministic argument that some subsystem is free of design faults, the zero fault result is transformed into a zero failure rate in the top-level evidence. Equally if the top-level argument is deterministic, lower-level probability calculations can be linked to probabilities of violating the assumptions. For example it might be possible make a deterministic argument that a system is safe provided that "at least two feed pumps will always be available". A separate lower-level computation might calculate the chance that this assumption will be invalid. This information might then be used to qualify the claim (e.g. there is 0.999 chance/year of safe operation).

## Deterministic Argument

A deterministic argument supports a claim or sub-claim by showing that, given some assumptions and a model of the real world, certain hazardous behaviours are "incredible". A simple example for a chemical plant may be that the inventory of two chemicals is not sufficiently large for a critical explosive mass to be present. An example from the computer area would be a proof that a communications protocol cannot deadlock.

In addition to deterministic claims about the behaviour of the system, weaker deterministic arguments may be made about the faults in a system. These arguments require evidence of the complete absence of certain classes of faults for a particular system function or component. For example, the use of a CAD system may exclude some forms of translation fault or the typing mechanisms in a high level computer programming language may exclude certain errors.

Deterministic arguments would normally require a formal model of the system and a proof that the system is safe with respect to its safety requirements (the proof could be a rigorous style argument rather than a machine-checked proof). The supporting evidence could include:

- explicit validation of the model assumptions
- an independent check of the formal argument

In addition to arguments based on formal models, it may also be possible to claim "fault-freeness" on the basis of exhaustive test coverage of all required behaviours.

## Probabilistic Argument

Probabilistic argument combines parameter estimates to obtain an estimate of the probability of some top level property (e.g. dangerous failure). The inference structure might be some form of Bayesian combination, evaluation of some stochastic model (e.g. Markov model or a fault tree), or simple statistical combination. It should be noted that probabilistic arguments also make use of an underlying model of system behaviour and the relationships between the various forms of evidence.

## Qualitative Argument

Most safety cases will include important qualitative claims. In some ways they are similar to deterministic arguments in the sense that some particular property exists. While deterministic arguments might be binary, qualitative arguments might be more fuzzy (e.g. refer to a rating such as "good", "indifferent" or "bad") and the ratings may be assigned by expert judgement. Qualitative assessment can also be binary where some "tick-list" of criteria must be satisfied to demonstrate acceptability (e.g. conformity to standards, construction criteria, or design rules). This may be a valid approach if the design tick-list encapsulates past experience which has been shown to achieve safety.

## Dealing with Uncertainty

Any safety argument is susceptible to error (e.g. in the evidence, the argument or the assumptions) so there should be strategies for limiting the risks associated with such errors. One simple strategy is to design the overall argument so that can withstand a single flaw, i.e. we adopt a qualitative defence in depth approach to the argument. In this case it might be structured as follows.

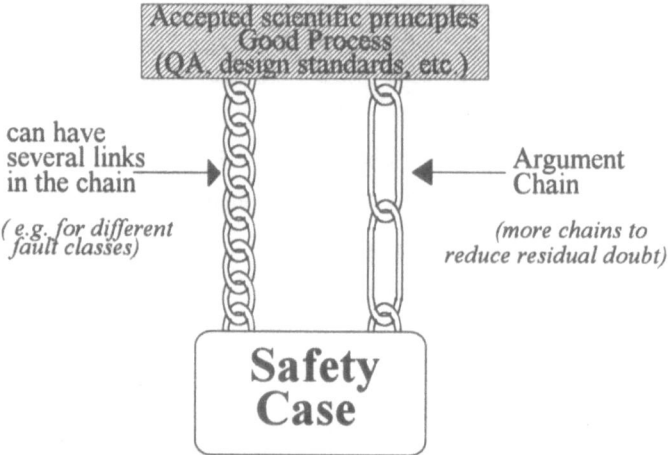

Figure 3: Example of a Safety Argument that Minimises Residual Doubts

Either chain would be sufficient to support the claim, and care should be taken to avoid any common links (e.g. common assumptions) between the two chains.

More sophisticated arguments could also be deployed which take a more quantified approach to such factors as: the confidence in specific assumptions, common mode failure probability, and numerical limits for claims made on any single leg

# Approach to Quantification

While the safety case structure can accommodate qualitative claims for system safety, the main objective of SHIP was to use a more quantified approach. In this section we outline the basic concepts that can support a quantified safety argument.

## Underlying Models

Our approach to reliability quantification in a safety case is based on two simple underlying models. The first is based on a standard model for software failure—and since software failures are due to design flaws, the same theories should be directly applicable to plant or computer hardware failures caused by design flaws. The second model considers the overall response of the system when a failure occurs, and demonstrates that the safety argument has to take into account the design methods, development process, and existing field experience.

*Failure model for software*

In trying to understand and predict the observed reliability of a software based system we need an underlying model for software failure. The reliability of a system is based on three factors:

1)    the number of faults

2)    the size and location of faults

3)    the input distribution (operational profile)

This is illustrated in the figure below.

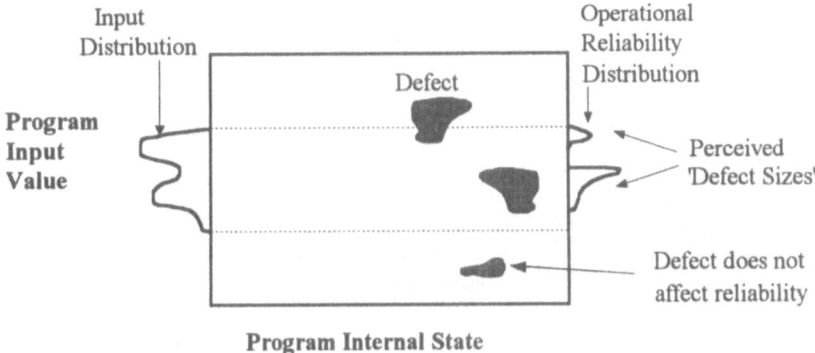

Figure 4: Illustration of the Software Failure Process

It is clear from the diagram that an alteration of the input distribution could radically alter the operational failure rate of the system. Where there is a single copy running in a fixed environment or where there are very many copies of the software running, the input distribution is likely to be effectively stable. Under such a stable input distribution, the faults are likely to have a fixed "perceived size" (which may be zero if a fault is not covered by input values).

In practice the number of faults within an item of software will not remain static. As operating experience is gained, faults will be revealed and corrected so the reliability of the software should grow with increasing execution time.

We considered that this model could be applied directly to hardware systems, so the related software reliability assessment methods could also be deployed. These methods include reliability growth modelling, testing and formal methods.

*System failure behaviour*

In safety related systems, we are not just concerned with reliability in general; we also need to distinguish between dangerous and safe failures. This leads to the underlying model of behaviour shown below.

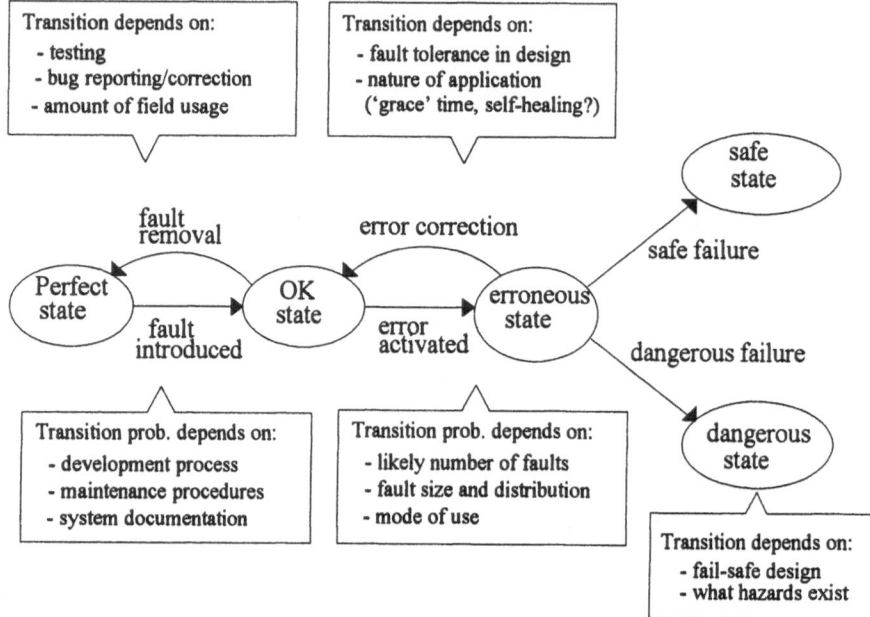

Figure 5: Model of System Failure Behaviour

This follows the standard fault-error-failure model for software. A *fault* is a defect in the program code and is the primary source of the failure. After development, the program could be perfect or faulty. However, even if it is faulty, the program may still operate correctly most of the time (i.e. stay in the OK state) until some triggering input condition is encountered. Once triggered, some of the computed values will deviate from the design intent (an *error*). However the deviation may not be large enough (or persist long enough) to be dangerous, so the system may recover naturally from the "glitch" in subsequent computations (*"self healing"*). Alternatively explicit design features (e.g. diversity, "firewalls", etc.) can be used to detect such deviations and either recover the correct value (*error recovery*) or override the value with a safe alternative (*fail-safety*).

If the failures are reported back then, in the longer term, the software can be corrected and a new version issued. Each version should (hopefully) contain fewer faults than the previous one and could potentially result in a "perfect" program. This is the principle underlying software reliability growth modelling.

These concepts should be equally applicable to faults in plant designs. The only difference in physical systems is that faults can occur spontaneously (e.g. random failures due to deterioration or stress) without any external intervention, and these faults can be fixed using a new part of the *same* design. However this aspect is already covered in conventional system reliability analyses.

Given that the basic concepts are valid, we still need to select which concepts will be deployed to support the specific safety argument. The strategy for developing a safety case is discussed below.

## Using the Models to Develop a Safety Case Strategy

The overall approach to generating the safety case involves:

- characterising the safety case arguments in terms of the transitions of the model

- ensuring the implementation strategy is compatible with the safety argument(s)

- determining and evaluating the evidence to support the claims made about the transition probabilities in the model

*Characterising the safety case*

As noted above one of the primary claims in developing the safety case is about the probability of dangerous failure. In developing the safety case arguments, it is useful to consider the mechanisms that determine the dangerous failure probability as indicated in the annotations of Figure 5. However, this is a general model, and a particular safety argument may focus on quantifying particular transition arcs. The main approaches are listed below:

1) A fault elimination and quantification argument can increase the chance of being in the "perfect" state and can also would reduce the probability of the $OK \rightarrow erroneous$ transition. An extreme example would be an argument of correctness. This would imply that the error transition rate was zero, and this would be sufficient to bound the dangerous failure rate.

2) A failure containment argument that would strengthen the $erroneous \rightarrow OK$ or $erroneous \rightarrow safe$ transition. An example would be a strongly fail-safe design which quantifies the fail-safe bias. This, coupled with test evidence bounding the error activation rate, would be sufficient to bound the dangerous failure rate.

3) A failure rate estimation argument that would estimate the $OK \rightarrow dangerous$ transition. The whole system is treated as a "black-box" and

probabilistic arguments are made about the observed failure rate based on past experience or extensive reliability testing.

It is also possible to apply the arguments selectively to particular components or fault classes, e.g.:

1) A design incorporates a safety barrier which can limit dangerous failures occurring in the remainder of the system. The safety argument would then focus on the reliability of the barrier rather than the whole system.

2) Different countermeasures might be utilised for different classes of fault. Each fault class then represents a separate "link" in the argument chain, and all fault classes would have to be covered to complete the argument chain. For example, design faults might be demonstrated to be absent by a deterministic argument, while random hardware failures are covered by hardware redundancy.

*Implementation supports the safety argument*

The previous discussion illustrates the key and closely coupled roles of the development processes and the design in formulating the safety case. Sometimes, the design and development approach is geared toward implementing the operational requirements; the need to *demonstrate* safety is only considered at a later stage. This can lead to considerable delays and additional assessment costs. The safety case should be an integral part of the design methodology and the feasibility and cost of the safety case should be evaluated in the initial design phase. This "design for assessment" approach should help exclude unsuitable designs and enable more realistic design trade-offs to be made.

*Sources of evidence and types of argument*

The arguments themselves may be either probabilistic or deterministic and utilise evidence from the following main sources:

- the design
- the development processes
- field experience

In considering the construction of a safety case, there are a range of options open to the designer at the preliminary design phase, as illustrated in Table 1 below. This is not a comprehensive list, but it serves to illustrate the basic approach to designing a system and safety case. Consideration of the readily-available evidence could have a strong influence on the economics of different design solutions.

| Type of Argument | Implementation Options/Evidence | | |
|---|---|---|---|
| | Development process | System design | Field experience |
| **Fault elimination and quantification**<br><br>Maximising the probability of a "perfect" state | Procedures, Standards, Documentation, Config. control, Testing, Reviews, Design tools Formal methods | Design simplification<br><br>Formal proof of system properties<br><br>Use of standard components | Prior operating history as evidence of correctness<br><br>Fault reporting, Design correction |
| **Error activation**<br><br>Minimising<br>OK→erroneous | Testing according to expected usage | | Avoid changes in the usage<br><br>Avoid known problem areas |
| **Failure containment**<br><br>Strengthening<br>erroneous → OK<br>erroneous → safe | | Fault Tolerant designs<br><br>Fail-safe designs | Fault injection tests |
| **Failure Estimation**<br><br>Estimating<br>OK → dangerous | Reliability testing | | Operational failure reports.<br><br>Reliability growth models |

Table 1: Arguments and Evidence

Given a list of possible implementation options, the designer then has to produce an overall system architecture which uses some cost-effective subset of these arguments. This may entail using different types of evidence for different components and may use different types of argument (e.g. fault elimination, failure containment and failure estimation). The safety case may also include diverse argument chains to allow for uncertainty. Some examples of different safety arguments are given in the following section.

## Illustrations of the Safety Case Approach

The work on incorporating some current software concepts within safety cases has also been beneficial in the reverse direction. The structuring concepts in safety cases (e.g. those for dealing with residual doubt) can be equally beneficial to software, especially where software is critical to the safety of the overall system. In SHIP the safety case approach was applied to both plant and software examples. Some examples of the different types of safety case for both plant and software-based systems are described below.

## Nuclear Pressure Vessel

We examined a pre-existing safety case for a nuclear pressure vessel [Hirsch87] and found that the arguments could be mapped on to the proposed argument structure as shown below.

| Transition | Cause | Safeguards |
|---|---|---|
| "Sound" → faulty | cracks grow due to normal ageing or abnormal transient | cracking minimised by: production processes, sound design, QA, avoidance of past problems<br><br>detected by: pre-service tests, on-line inspections. |
| faulty → erroneous | crack grows large enough to leak | minimised by periodic inspection of vessel |
| erroneous → safe | reactor trips before the vessel fails | on-line water leak detection initiates trip |
| erroneous → dangerous | catastrophic failure of vessel | judged incredible |

Table 2: Safety Case Arguments for a Nuclear Pressure Vessel

The safety case for the pressure vessel can be represented using the basic transitions of the model. The safeguards given in the final column show how the transition is minimised or eliminated. The top-level arguments are predicated on sub-claims that fast fractures cannot occur and that a vessel always leaks before it breaks. These sub-claims are supported by a large body of scientific evidence from fracture mechanics and metallurgy.

Errors in the argument can be tolerated because there are two forms of protection (periodic crack detection and on-line leak detection) either of which should be sufficient to maintain reactor safety. In addition there is a separate argument leg based on field experience where it is shown that the required level of reliability has been achieved on similar vessels in the past and that all known pressure vessel design flaws have been avoided.

## Boiler System Control

The Boiler System Control Specification study [Bishop93] looked at the use of formal specification methods in constructing a safety case. In this example the claim is essentially a deterministic argument that design faults are absent, coupled with probability estimates for random hardware failures. A top-down approach is used where the boiler dynamics are formally modelled using Temporal Logic Algebra [Lamport91]. Boiler safety constraints were identified (which were basically that the water level had to remain within upper and lower limits). The control and safety functions were then modelled and shown to preserve the safety constraints. Successive design iterations were made which identified additional component failure modes. The final software design specification was shown to satisfy the safety requirements provided certain assumptions were made about the failure behaviour of the components and the diagnostic capabilities of the software. By computing the likelihood of violating these assumptions the dangerous failure rate can be estimated.

This is quite an effective and systematic method of eliciting the underlying assumptions, and also for deriving an associated fault tree for random failure probability calculations. There is no independent "second chain" which could protect against flaws in the assumptions, so ideally an independent safety system would be needed to cater for residual doubt. This illustrates that there is a duality between safety arguments and system architectures. If diverse systems are used, a single argument can be used for each one. If only a single system is used, diverse arguments are needed to support the safety claim.

## Analysis of Industrial Controller Field Data

Field data can be used to provide supporting evidence of "perfection", or at least to give some lower bound on the expected level of reliability. This could be used either as the main safety case argument or as one leg in an overall safety argument. In SHIP we examined field data on industrial computer based control systems from both public and private sources. In one study of fault reports for a small industrial controller we observed the following pattern of fault discovery after a new industrial control system was released.

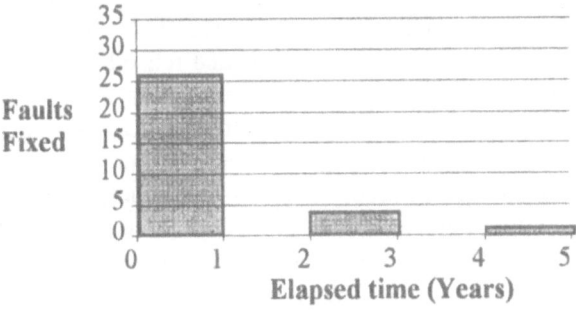

Figure 6: Fault Fixes over Time (Small Industrial Controller)

Faults reported by customers are recorded, and fault fixes are incorporated in later revisions of the design. This can happen several times in a year. It can be seen that in the second year no faults were fixed at all. Since these controllers are mass-produced with thousands being sold every year, this represents an extremely high reliability level. It is suspected that the subsidiary peaks in later years were actually due to *new* faults introduced when new features were added to the system. So it might be argued that after a year of fault fixing the design may be effectively "perfect".

When we looked at a large, complex, industrial controller, a different pattern emerged where there was little change in the fault fixes for successive years. We think this is due to a combination of factors. Firstly fewer units were sold so faults were not detected so rapidly in operational use. Secondly the system was more complex so it probably contained more faults. Finally there is more scope for introducing faults when new features are added.

A safety case argument based on observations of field reliability would therefore have to take into account the complexity of the system, the stability of the design (i.e. the rate of addition of new features) and the amount of field usage (which affects the rate of fault removal). A theory which models these effects is being developed in another research project [QUARC].

## Summary and Conclusions

The SHIP project set out to improve the state of the art in assessing the safety of systems containing design faults. We think that we have succeeded to the extent that we have identified an overall safety case structure, and the different forms of argument and evidence that can be used. We have also shown examples of different forms of safety case that can be constructed based on both probabilistic and deterministic design arguments, and evidence from field experience. Other examples were developed in SHIP using a range of approaches (including qualitative methods), but there is insufficient space to present them in this paper.

Currently most safety-related software standards focus on a development process which will minimise faults, and this does not lead to a quantified reliability estimate. The safety case approach proposed in SHIP takes a more global view and provides a framework for estimating the safety and reliability of the software and the associated systems. More work is needed to make this work routinely applicable to industry, and further research into the application of these concepts is being undertaken [QUARC]. We expect to see the SHIP work influencing relevant industrial standards. Indeed, some of the safety case concepts developed in SHIP have already been utilised by the authors in producing the forthcoming revision to the Ministry of Defence standard for safety-critical software [MOD91].

# Acknowledgements

The SHIP project (ref. EV5V 103) was carried out with financial support from the European Union in the framework of the Environment programme, sub-theme: Major Industrial Hazards.

We wish to acknowledge the contribution of the SHIP project partners to this work. The project partners are: Adelard, UK; the Centre for Software Reliability (CSR), UK; Objectif Technologie, France; Ente per le Nuove Tecnologie l'Energia e l'Ambiente (ENEA), Italy; Istituto di Elaborazione dell' Informazione (IEI-CNR), Italy; and the Finnish Technical Research Centre (VTT), Finland. Associate partners are: the Franco Polish School of New Information and Communication Technologies (Poland); and the Institute of Computer Systems (Bulgaria).

# References

[Bishop93] P.G. Bishop, G. Bruns, S.O. Anderson, "Stepwise Development and Verification of a Boiler System Specification.", *Int'l Workshop on the Design and Review of Software Controlled Safety-related Systems*, National Research Council, Ottawa, Canada, June 1993

[Gentzen69] G. Gentzen, *The Collected Papers of Gerhardt Gentzen*, North Holland, 1969

[Hirsch87] Hirsch Study Group, "An assessment of the integrity of PWR Pressure Vessels", Addendum to the Second Report of the Hirsch Study Group, HL/087, 1987

[Hoare69] C.A.R. Hoare, "An Axiomatic Basis for Computer Programming", *Communications of the ACM*, Vol. 12, No. 10, 1969

[Jones90] C.B. Jones, *Systematic Software Development using* VDM, Prentice-Hall International, London, UK, Second edition, 1990

[Lamport91] L. Lamport. "The temporal logic of actions", Technical Report 79, Digital Systems Research Center, 1991

[McDermid94] J.A. McDermid, "Support for safety cases and safety argument using SAM", *Reliability Engineering and Safety Systems*, Vol. 43, No. 2, 111-127, 1994

[MOD91] "The Procurement of Safety Critical Software in Defence Equipment", Ministry of Defence, Interim Defence Standard, IDS 00-55, 5 April 1991

[QUARC] Quantification of Reliability in Computer Based Systems, UK Health and Safety Executive Nuclear Safety Programme, Scottish Nuclear Contract: 70B/0000/006384

[Vesely81] W.E. Vesely et al., *Fault Tree Handbook*, NUREG 0942, US Nuclear Regulatory Commission, Washington DC 20555, 1981

# Safety Case: Structure and Role *

M.El Koursi, B.Letrung, H.Waeselynck and F.Baranowski

e-mail: miloudi@inrets.fr fax: (33) 20 43 83 59

INRETS: 20, rue Elisee Reclus, 59650 Villeneuve D'ascq.

## Abstract

Safety can be defined as a state in which the level of risk is acceptable for the user. The risk is an expression of the possibility of a mishap in terms of hazard severity and hazard probability. The aim of constructing safety is to prove that the system possesses the highest safety integrity level. The Safety Case forms the main proof that the system complies with the safety requirements. This paper gives an assessor point of view for the Safety Case structure and role. It presents the main properties that a Safety Case must have.

## 1  Introduction

The Safety Case is a documented safety evidence for a system, sub system or item of an equipment for a specific application, including the complete record of the Quality & Safety Management process followed and the technical evidence contained in the Technical Safety Case parts (system, subsystem, equipment, maintenance and operation documents).

The developer's task is to produce a Safety Case and thereby demonstrate that the system meets the safety requirements.The developer should list, analyse, classify the risks inherent in the activity and define the main principles of safety procurement together with the analysis methods and tools used. Whereas the assessor's task is to check the soundness, completeness and the traceability of the safety case argument. The assessor needs to confirm the adequacy of the hazard analysis, to check that the correct standards have been used and to verify that the correct measures have been selected and properly applied.

Various methods allow to proceed to a safety proof assessment according to the state of the art: the checklist, the examination, the reviews, the inspections and the audits by referring to relevant standards. Today, the situation is not clear about what the Safety Case should contain? All relevant standards recommend the construction of the Safety Case but they do not make explicit its content. They are safety process oriented. At best, as the case of the CENELEC pre-standards [CEN.94b], we are provided with a vague safety case plan. There are many reasons for this gap:

---

*This paper reflects work which is partially funded by the CEC under the ESPRIT III programme in the area of Information Processing Systems, Project number 9032:"Certification and Assessment of Safety-Critical Application Development.

- one reason is that the standard bodies come from developer companies and rarely from assessor companies. The developer proposes the minimum of safety case content for confidentiality and competitiveness reasons. this content is a mean to know about constructor's ability to produce a safe system. Assessor could suggest the safety case content which reflects the state of the art taking into account the technology progress and standards.

- another reason is that the content of the safety case is closely linked with its expected use (administrative or technical). The content of an administrative safety case could be a top level section.

This paper gives an assessor point of view for the safety case structure. Taking into account possible future progresses, it is advisable to leave constructors with a certain amount of freedom. Nevertheless, the constructor should refer to known standards, well established and recent, providing the appropriate terminology and rules.

# 2 Safety case role

The safety case provides the proof that the system complies with the specified safety requirements. The form of this proof, structure and content depends of the role that can be attached to it.

1) One primary objective of the safety case is to provide a technical evidence, by well documented demonstration, that the system complies with the safety requirements specifications. This compliance may be demonstrated by calculation, by appropriate experiments or by consensus of independent experts. The responsibility for specifying the overall safety requirement lies with the experts of the Safety Authority while the responsibility for demonstrating the safety of a system lies with the manufacturers. The manufacturers apply for the safety authority at the beginning of the project and supply well-structured, comprehensible safety case. The safety assessment is done by an assessor appointed by the Safety authority. Assessor approuves the constructor's validation and the safety works performed.

2) The safety case plays the role of the system memory. It is a reference for any change in the system. The users refer to the safety case for any improvement concerning operation, maintenance and extension of the system. For this role, throughout the operating life of a system, the safety case should be updated and modified parts submitted for a new approval by Safety Authority.

3) The safety case is a legal reference particularly when an accident occurs. It should contain the relevant administrative documents identifying step by step each activity, each author and each approval.

(4) The safety case is an official document which can be accepted by any European safety authority when it is accepted by one of them according to an European standard.

The central property a safety case must possess is traceability, that is, it must be structured in such a way that it is possible to trace the safety analyses throughout all development phases. Traceability requires the adoption of appropriate formalisms in the analyses, e.g. structured terminology, visual support of diagrams. The traceability being the possibility of ensuring that adequate precautions have been taken at each level, in order to be able to declare that all risks have been covered.

# 3   Safety case structure

We propose here a possible structure of the safety case taking into account its different roles. It is closely linked with its final use.
The final safety case should deal with the three evidences mentioned in the draft standard (CENELEC prEN 50129)[CEN.94b] and the results of the assessment work:

1) **Evidence of Quality Management :** The quality activities of the system shall be controlled by an appropriate management process, like EN 29001 ISO 9001 [ISO.91], throughout the life cycle;

2) **Evidence of Safety Management :** The safety activities of the system shall be managed by a well defined safety process (safety organisation, plans reviews, Hazard-log...);

3) **Evidence of Functional and Technical Safety:** This technical evidence for the safety of the design shall be demonstrated in the Technical Safety Report which forms a part of the safety case.

These evidences should be presented in proof of safety documents related to a generic system and in a safety case concerning a specific application (CENELEC 50129 fig 8.7, 8.8 and 8.9) [CEN.94b]. The overall documentary evidence for safety of specific application deals with the application design and installation.

4) **The result of the safety authority assessment:** including administrative and technical documents.

The safety case structure proposed here takes into account of a generic system and a specific application. It contains for instance:

- *Top level documents of the safety case*

- *System section*

- *Subsystem section*

- *Equipement section*

- *Safety critical software aspect*

- *Operation and maintenance section*

## 3.1  Top level documents of the safety case

This part of the safety case should describe the methodology used in the needed activities over the safety life cycle. It summarises the main results obtained to proof that the safety integrity level of the system is achieved. The role of this part is to allow to give a first indication about the effort allocated and the results of the safety activities in order to construct a preliminary opinion about the quality of the project safety organisation and assessment activities (e.g. safety review, inspection audit).

The main documents presented should be:

1. *A list of documents which form the safety case* (Title, author, delivery date).

2. *Methodology for safety construction:*

   This document describes the process to produce the safety case. It should describe the methods used to proof the safety of the system and should indicate the various standards followed.

   The safety construction methodology is a set of procedures that the constructor must comply with, in his presentation of the project. The safety authority adheres to this methodology, and verifies that it is well implemented all along the project .

3. *Documents related to the management, organisation and control of the safety construction:*

   - *Safety organisation document:* This document should define the safety organisation and the roles and responsibilities for key bodies.

   - *Quality Plan:* The Quality Plan defines the system development lifecycle. Its role is to define the rules and constraints to be complied with during the development stages, and the criteria for the passage from one stage to another. It defines the tasks and the products of each phase, enabling the visibility of the complete process.

- *Safety Plan :* It describes the way in which safety is built up throughout the project, and the way in which it is administered, in order to obtain the traceability of the work on safety. It should identify the milestones where approvals are required. It inludes plans for verifying that each phase of the life-cycle satisfies the specific safety requirements identified in the previous phase.

- *Verification Plan.* It describes the verification activities to be performed, documents all tools, techniques and methods used and lists the criteria for acceptance.

- *Integration Plan.* This document should describe the procedures to integrate properly the system including integration test data, test cases, tools, techniques and acceptance criteria.

- *Validation Plan:* The Validation Plan defines the last stage of the process "Safety Validation". Thus the Validation Plan has the role of defining the objective and the overall approach to be adopted for presenting the validation techniques and for determining the role of the validation team. It prepares the strategy and techniques of validation by taking into account safety criteria drawn from all the risk analyses such as a Risk Analysis of the Equipment, Risk Analysis of the hardware component, Risk Analysis of the Software component, and Maintenance and Operational Risk Analysis.

- *Management plan.*

4. **Documents related to the system approval:**

- *a list of documents submitted for an approval by Safety Authority.* This list may be different from the general list if the system under approval possesses an identical characteristic to an existing certified system. Each item of the list should present the approval date and should refer to the minutes of safety review by the safety authority.

- *independent assessment report done by different experts appointed by the safety authority,*

- *Minutes of the safety authority reviews,*

- *Final Safety Authority assessment report.*

## 3.2   System section

1. *General aspects:*

This section of the safety case consists of:

- *The safety requirement documents* describing the system mission, the general environment of the system and the limit of the safety studies.

- *The preliminary hazard list* that must be identified in this section.

- *The Hazard Identification and Analysis (HIA):* This part identifies and categorises actual and potential hazard. The hazard identification starts at the definition of the system and its environment with a Preliminary Hazard Identification (PHI). The Hazard Identification Analysis (HIA) identifies the system boundaries and the actual and potential hazard taking into account the limit of the system. These activities are carried out by an appropriate method (Check Lists, Hazops, FMEA, FT ..)

- *The Risk Assessment and Classification (RAC):* This part evaluates the probability of a hazard or sequence of hazardes leading to a potential accident or an accident.

2. *The safety objective assignement*

The safety objectives should be clearly defined.These principles govern the definition of the overall safety objective. The method of the allocation of safety objective is a top-down approach starting from the overall objective for the system and covering the subsystems and equipments.

3. *Description of the system:*

The general aspect of the system, its functionalities and the various types of operations should be described.

4. *System Safety studies :*

The constructor should list, analyse, classify the risks inherent in the activity and define the main principles of safety procurement together with the analysis methods and tools used. All safety analyses, Functional Risk Analysis (FRA), Product Risk Analysis (PRA) and Operation and Maintenance Risk Analysis (OMRA) described in the section 6.4 concerning the system, should be well documented. The manufacturer is free to adopt the appropriate method, but he should refer to known standards, both well established and recent, providing the appropriate terminology and rules. The allocation of a defined safety objective for each sub system should be clearly announced at this stage.

5. *System verification and validation.*

- verification report
- validation report
  This case presents all results of the safety tests performed on the system.

## 3.3  Subsystem section

1. *Description of the subsystem*

2. *SubSystem Safety studies :*

All safety analysis "FRA, PRA and OMRA" concerning the subsystem should be well documented. The constructor is free to adopt the appropriate method but he should refer to known standards (or recommendations), both well established and recent, providing the appropriate terminology and rules.

3. *Subsystem verification and validation*

- verification report
- validation report

  It presents different test results supplied on the subsystems.

## 3.4   Equipment section

1. *Description of the equipment*

2. *Equipment Safety studies :*

All safety analyses "FRA, PRA and OMRA" concerning the equipment should be well documented. This section contains the hardware and software component safety analysis.

3. *Equipment verification and validation.*

- verification report
- validation report

  It presents different test results performed on the equipment including hardware and software components.

## 3.5   Safety Critical Software aspect

Compliance with the requirements concerning the software component of the critical and safety related equipment is obtained by a set of specific procedures and rules such as those described in IEC and Cenelec draft standards. The software safety case proposed here is related to 4 and 3 integrity levels (safety critical software and safety related software). It comprises the following documents:

*a) Software Quality Assurance*

1. *Software Quality Assurance Plan.* This document describes all the phases of the lifecycle model and for each phase the quality requirements.

2. *Software Development Plan;*

3. *Software Safety Plan;*

4. *Software Configuration Management Plan;*

5. *Software Validation Plan* which describes the strategy and techniques to be used.

6. *Software Verification Plan.* It describes the verification activities to be performed, documents all tools, techniques and methods used and lists the criteria for acceptance.

*b) Software Requirement Specification and Validation* This part of the safety case describes a complete set of requirements for a software and the elements for the software validation at the intended integrity level. The safety critical functions should be derived from the equipment safety requirement specification.

1. *Software Requirement Specification;*

2. *Software Requirement Test Specification;*

3. *Software Requirement Verification Report;*

4. *Software Testing case.* It contains the required inputs, sequences and values, the expected results and the acceptance criteria.

5. *Software Validation Report.* This document identifies software version, the equipment used, the faults found and the corrective actions performed.

6. *Software/Hardware Integration Plan.* This document should describe the procedures to integrate properly the software and hardware including integration test data, test cases, tools, techniques and the acceptance criteria.

7. *Software/Hardware Integration Report.* It should contain test cases and their results, configuration and judgement according integration criteria.

8. *Software Assessment Report.*

*c) Software Design and Integration.* Software Architecture Specification. This document aims to present the preliminary design of software that achieves the software requirement specification according to the software integrity level.

1. *Software Design Specification;*

2. *Software Design Test Specification;*

3. *Software Integration Plan;*

4. *Software Integration Report;*

5. *Software Effect Errors Analysis Report.*

*d) Software Module Design and Testing* Software module design and
testing presents how the software is created and tested in order to achieve the
required integrity level.

1. *Software Module Design Specification;*

2. *Software Module Test Specification;*

3. *Software Module Test Report;*

4. *Software Module Verification Report.* This document contains the Software Module Specification test and the verification report;

5. *Software Effect Errors Analysis Report.*

*e) Systems Configured by Application Data (Invariant)* In railway
applications, the ATP "Automatic Train Protection" systems are configured
by data (Invariant) that allow type-approved generic software to be used. This
data is normally in the form of tabular information. The safety case for this
data base should contain:

1. *Installation Requirement Definitions;*

2. *Data Preparation Plan;*

3. *Data Test Plan;*

4. *Data Test report.*

*f) The Code*

1. *Software code Verification Plan;*

2. *Software Code Verification Report;*

3. *Software Source Code and supporting Documentation.*

*g) Software Maintenance documentation* These documentation are needed
for any correction or modification in order to preserve the required software integrity level. It comprises:

1. *Software maintenance Plan* which describes the procedures and rules to
   be followed;

2. *Software Maintenance log.* It should contain the software history, configuration, revalidation, testing data...

3. *Software Maintenance record.* This document contains the request modification, analysis of the impact of the maintenance activities and the
   required revalidation testing.

## 3.6 Operation and maintenance section

This section of the safety case consists of:

1. *Impact analysis of the operating instructions,*

2. *Validation of operation and maintenance procedures ,*

3. *Maintenance needed to maintain the safety level of the system.*

# 4 Functional, Product, Operation and Maintenance Risk Analyses

These analyses complete the Hazard Identification and Analysis (HIA) and the Risk Assessment and Classification (RAC). They are done at every level of the development process: system, subsystem, equipment, component, operation and maintenance.

## 4.1 Functional Risk Analysis (FRA)

Functional Risk Analysis (FRA): this analysis is performed on the specification of the item (system, subsystem and equipement). It is intended to demonstrate that the risk is covered by the function performed by the item and that the risk induced by the function is tolerable.

This analysis aims at validating the specification of the product, irrespective of its design and implementation. It is stated that the specified function is:

- complete, taking into account all the hazards that can occur in the environment of the function;

- consistent, as regards the processing of information and the calibration of parameters (e.g. time values, geometrical dimensions, band width). The item design cannot start before its specification has been demonstrated complete and consistent.

## 4.2 Product Risk Analysis(PRA)

Product Risk Analysis (PRA) : this form of analysis is performed on the product (system, subsystem and equipment) resulting from the specification. It is intended to demonstrate that the residual risk is in the permitted zone. It highlights the precautions taken in the interfaces with other items. This analysis aims at demonstrating that the design is in conformity with the specification, and that it covers the required properties even in case of internal hazard occurrence. The hazards to be taken into account depend on the nature of the internal components:

- the failure modes of hardware components can be found in catalogues;

462

- software failures may originate, for example, from computational errors, domain errors (the wrong path is exercised), or real-time errors.

  At a given design phase, two cases are possible:

- the identified risks are covered within the acceptable bounds;

- the identified risks are covered only if lower levels components satisfy some criteria : they must be further analysed in subsequent phases.

## 4.3  Operation and Maintenance Risk Analysis (OMRA)

Operation and Maintenance Risk Analysis (OMRA) : this form of analysis collects all the directives resulting from different Risk Analysis (FRA, PRA) and provides the safety operation and maintenance instructions. This analysis is intended to verify:

- that general and particular recommendations of operation and maintenance are consistent;

- the conformity between settlement, operation consigns and system functionalities.

This analysis validates the adequacy between the system and its use.

# 5  Conclusion

The Safety Case provides a proof, by well documented demonstration, that the system complies with the safety requirements specifications. The compliance with safety requirements is demonstrated by calculation, by appropriate experiments or by consensus of independent experts.

In this paper, we have proposed a possible structure of the Safety Case taking into account its different roles and we have discussed its final use. This paper also described the elements which form the main parts of the Safety Case.

The responsibility for specifying the safety requirements lies with the experts of the Safety Authority. The Safety Authority appoints out an assessor to perform the safety assessment. It approuves the validation of the system within the safety requirements. The responsibility of an assessor is to examine the safety work done by the developper and to report the result to the safety authority. The responsibility for demonstrating the safety of a system lies with the manufacturers.

The safety case produced by developer must exist in paper form printed by the supplier on its own paper. Electronic, magnetic and similar documents are only allowed for temporary use. The documents have to be kept in a book case. One copy of the safety case is deposit at the assessor. The Safety Case has to be preserved according to the legal requirements.

# References

[MLK.95] M.Elkoursi and B.Letrung " Current Assessment Approach Applied by INRETS for ATP Systems. M.Elkoursi and B.Letrung, safecomp'94, october 23-26, 1994, Anaheim, California, USA."

[CAS.94] CASCADE:"Provisional Generalised Assessment Method; 16th November 1994; CAS/LR/GP/D221/V0.4"

[CAS.95] CASCADE: "Generalised Assessment Method; to be published on june 1995, CAS/IC/MK/D231/V0.3"

[ISO.91] ISO9001: "Quality management and quality assurance standards - Part 3: Guidelines for the application of ISO9001 to the development, supply and maintenance of software, 1991."

[IEC.91a] IEC 65A(Secretariat)122: "Software for computers in the application of industrial safety-related systems; 26th September 1991; draft."

[IEC.91b] IEC 65A(Secretariat)123: "Functional Safety of Electrical/ Electronic/ Programmable Electronic Systems: General Aspects. Part 1. General Requirements; 26th September 1991; draft."

[CEN.93] Cenelec prEN 50126: "The Specification and Demonstration of Reliability, Availability, Maintainability and Safety (RAMS) of Railway Applications Part 0: Dependability (version 00, 06 June 1993)."

[CEN.94a] Cenelec prEN 50128: "Railway Applications -Software for railway Control and Protection Systems (draft February 1994)."

[CEN.94b] Cenelec prEN 50129:"Railway Applications -Safety-related Electronic Railway Control and Protection Systems (draft 1994)."

# Session 11
## Validation and Verification

Session 11

# Practical Approach for the Evaluation of Safety Related Programmable Electronics

Marita Hietikko
VTT Manufacturing Technology
Tampere, Finland

Risto Tiusanen
VTT Manufacturing Technology
Tampere, Finland

## 1 Introduction

At the same time as microprocessor-based systems spread to all consumption devices, the safety related parts of control systems contain more and more often programmable electronics (PE). The control systems of railways, lifts, cranes, or generally process or machine automation, are examples of the application areas. A common feature for systems in these areas is that a failure in the safety related part of a control system can lead into hazard, dangerous situation, accident or even loss of life, if the system is not designed safe enough. Therefore a high level of reliability and safety is required for the PE[1] to be accepted for use in this kind of safety related applications.

As we know, procedures for the evaluation of safety related programmable electronic systems (PES) have been widely studied from 1980's. Such institutes as, for example, ElectronikCentralen (EC), Health and Safety Executive (HSE) and The Swedish National Testing and Research Institute (SP) have published instructions and/or checklists for the evaluation of safety related PES. The European Workshop on Industrial Computer Systems Technical Committee 7 (EWICS TC7) has also published directions and methods for the design, development and analysis of critical computer systems, covering the whole system life cycle. In addition to this, the standardization work concerning the design instructions and evaluation methods for PES and safety related parts of control systems is being made in the working groups of both IEC and CEN/CENELEC.

In spite of the amount of research done within this area, there is a lack of generally accepted and consistent principles, methods, assessment criteria and especially practical instructions for ensuring the safety and reliability of PES in machine automation applications. Although the study of the evaluation of PES in process industry application area has been widely carried out from the middle of 1980's, this research area is quite new among the designers of machine automation.

---

[1] Here the PE include sensors and final elements.

The electronic designers in this area require cost effective, practical method and tool for safety design in order to fulfil the essential safety requirements of the machine directive in their new-technology applications.

The standardization work concerning safety related systems (SRS) takes several years. Simultaneously, from the designers' point of view, the design and verification of safety related functions including software have become a problem. Especially software safety evaluation and validation is problematic. The conventional analysis and test methods alone are no longer valid for the latest control system and microprocessor technology.

The goal of our study was to find a practical way for the identification and analysis of safety critical hardware and software faults and for the assessment of the safety measures related to these faults. In this paper we describe the identification of faults in three safety related PE systems by using a combination of analysis methods. In addition to these analysis methods we used a checklist for collecting information on the safety measures applied to the systems. We did not use any test equipment for the safety evaluation during the study.

The whole life cycle of an equipment under control (EUC), from system specification to use and maintenance, contains many separate safety related problems. In this paper we deal with hardware and software safety life cycle of safety related PE systems (Figure 1), except the validation planning and operation and maintenance procedures. The diagram of Figure 1 is sketched from the basis of models described in IEC 1508 standard draft [IEC 94] concerning safety related PES.

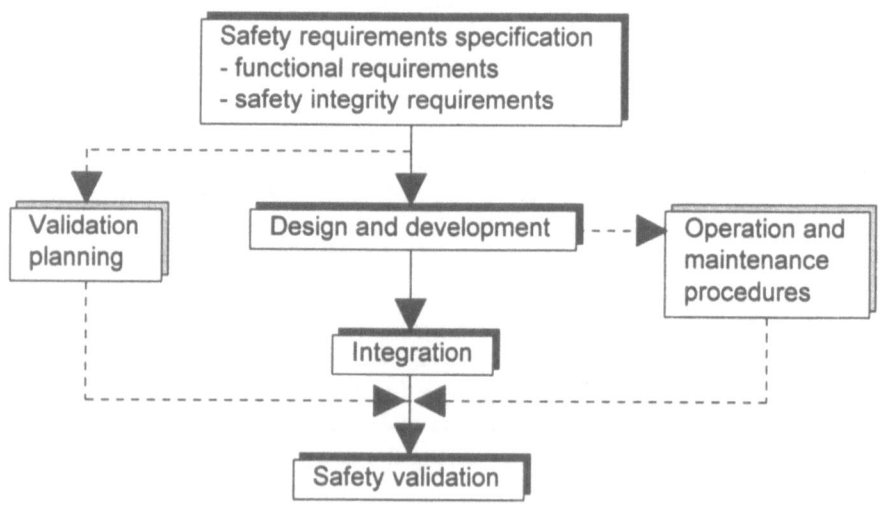

Figure 1. Hardware/software safety life cycle.

# 2 Methods used for the Safety Evaluation

The used analysis methods were Fault Tree Analysis (FTA), Software Fault Tree Analysis (SFTA) and Failure Mode and Effect Analysis (FMEA). We used these analysis methods qualitatively. The qualitative analysis is a non-mathematical method in which the factors having an effect on safety are studied.

The assessment of safety measures was carried out as a group work like design reviews by using a checklist for data collection and documentation. The checklist was developed in cooperation with VTT and the electronic design groups taking into account the design practices and control system technologies in machine automation. The checklist is used as a tool for estimating the safety integrity level of the control system under study.

## 2.1 Fault Tree Analysis and Software Fault Tree Analysis

Fault Tree Analysis (FTA) is a top-down (deductive) reliability analysis method intended to identify the causes or combinations of causes, which can lead to the defined undesired event, ie. top event, in the system under study [IEC 90]. FTA is mainly qualitative but it can be used quantitatively as well. FTA is actually a graphical representation of events or conditions contributing to the occurrence of the top event. Special symbols can be used in constructing a fault tree. FTA should be already started in the design phase of a system. Analysis is particularly suitable for complex systems.

Software Fault Tree Analysis (SFTA) is used in the same manner as FTA for ensuring that the logic in software design does not produce safety failures. The same symbols can be used both in FTA and SFTA. The SFTA is actually derived from FTA. The fault trees constructed separately to software and hardware can be linked together at their interfaces, and thus the whole safety related PES can be analysed [Clarke 93].

## 2.2 Failure Mode and Effect Analysis (FMEA)

Failure Mode and Effect Analysis (FMEA) is a bottom-up (inductive) reliability analysis method intended to identify failures and their effects on the system under study [IEC 85]. FMEA is mainly used as a qualitative analysis method for identifying failure modes of hardware. FMEA determines the relationship between element failures and system failures, malfunctions, operational restrictions and degradation of performance or integrity. FMEA usually does not include software errors.

FMEA may be used alone or to supplement other reliability analysis methods. FMEA is useful to carry out in the design phase of a system, in which case it supports design reviews. In order to ensure the quality of the analysis, a special training of personnel performing FMEA is required, and they must work in cooperation with system engineers and designers. By the end of a design project FMEA can be used to check the project design and may be essential for demonstration of conformity of a designed system to required standards, regulations, and user's requirements [Lehtelä 91].

## 2.3 Checklist

Within our project we created a checklist as a practical tool for the evaluation of safety related PES in cooperation with the electronic design groups. The checklist is based on the work done in this area by IEC [IEC 94], EWICS TC7 and CEN [CEN 94]. A goal of creating the checklist was to modify the heavy procedures of standard proposals to meet the design practices in Finnish design groups. The use of the checklist can be applied to the safety design of a product as well as to the validation of an existing system.

The designers of safety related PE systems use several software tools at their workstations. They may use windowing environment, and therefore the checklist should also be suitable to the same windowing environment as the other tools. One goal of our study was to create a demonstration version of the checklist based on Microsoft EXCEL. Our intention was to demonstrate the possibilities that the software tool of the checklist can offer to designers and inspectors. A subject specific help is included into the software tool. The main menu of the software tool with a help menu is presented in Figure 2.

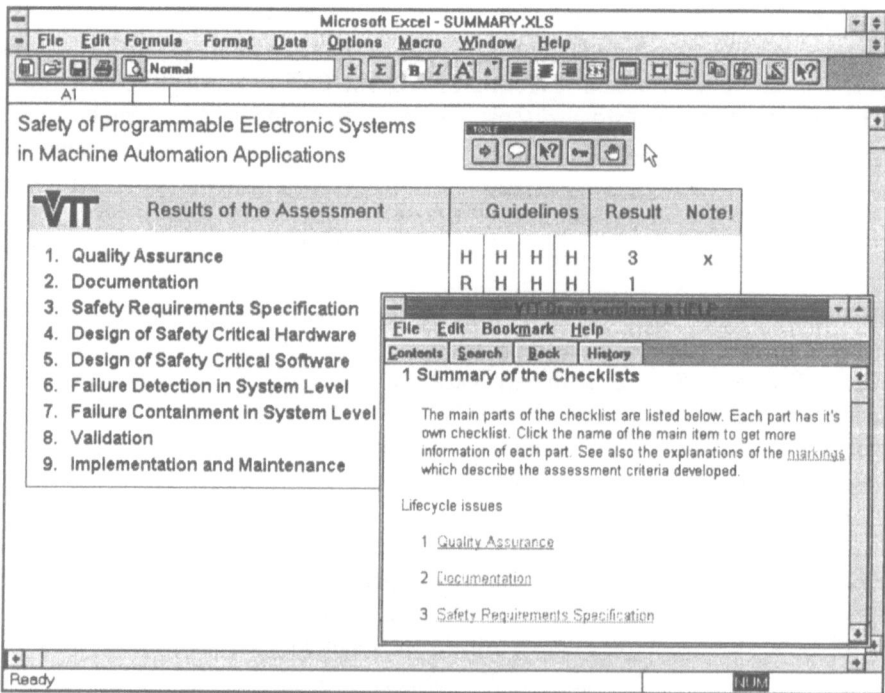

Figure 2. The main menu of the software demonstration tool for the checklist.

# 3 Results

From the basis of our study it is possible to create a suitable evaluation method for the system designers, taking into account their present projecting practice and recent regulations. The importance of using several safety evaluation methods to complement each other was clearly detected during the study.

## 3.1 An Electronic Actuator Controller

The electronic actuator controller is a microprocessor based device, which can be used in several safety critical applications. The design of the actuator controller was at a prototype stage, when we started the evaluation of the device. Four persons belong to the design group.

The hardware was analysed by using FMEA, and the software was analysed by using SFTA and the checklist. All these analyses were carried out as a group work. By this way the designers of the actuator controller obtained an important information about the functional safety of the device just before the following development stage of the product. All the possible failure modes were analysed by FMEA. Two severe failure modes were discovered, which led into changes in hardware implementation. The SFTA and the walk-through of the detected causes for the defined top events produced useful information on the logic and safety related checks of the software. However, by the aid of SFTA we could not analyse failures related to the timings of the parallel processes of the software. The safety integrity level of the device was defined by the checklist.

## 3.2 Distributed Control System of a Moving Machine

The machine includes controller area network (CAN) bus system with several intelligent modules for controlling different safety related functions. The number of persons in the design group was 8. They work in different departments of the company.

The hazards had been analysed before this study. We chose one safety critical event as a top event for FTA and looked for the causes of the top event in cooperation with the hardware, software and hydraulic designers. In this case we concluded that the other fault trees starting from the other top events are quite similar to the analysed one because of the CAN bus system. Thus the amount of work did not increase in direct relation to the number of fault trees. It was detected that the faults in cables and connectors are the most common failure modes in this system. The controlling of these failure modes requires safety measures related to the bus protocol and data transmission.

The checklist was used to complement FTA and to make sure that the applied techniques and measures are sufficient for the required safety integrity level. It was observed during this examination that the checklist as well as its reference standard IEC 1508 are not comprehensive enough for the usage of distributed systems. The recommendations of safety measures for ensuring data transmission in different safety integrity levels are required.

## 3.3 Remote Control System of a Machine

This equipment was a wireless remote control system intended to operate an earth moving machine in dangerous circumstances. The system consists of a portable transmitter and receiver mounted to the machine. The design group consisted of application engineers only.

The functional safety of the hardware was analysed by using FTA and FMEA. The result of the analyses was that a single fault can not lead to a hazard, but some failure modes may remain undetected.

The assessment of the remote control system was quite difficult because we could not interview the system designers. The functional safety requirements could be verified by FTA and FMEA. The software safety evaluation remained incomplete, because we could not get adequate information on the safety measures of the software.

# 4 Discussion

From the designers' point of view the approach of the IEC 1508 standard draft is laborious to apply to the present design practices in machine automation area. Within this study we created a practical approach for the identification and analysis of faults in safety related PE systems. The checklist modified to meet the designers' requirements gives a practical support to the verification of safety requirements. It also supports design reviews.

During the study we observed the importance of the cooperation between the hardware and software designers in evaluating safety related PES.

The possibility to evaluate distributed systems is required more and more frequently. There is an evident lack of safety measures and techniques in the IEC 1508 standard draft concerning distributed systems and the checks of data transfer and intercommunication related to these systems. There should also be recommendations of adequate measures for each safety integrity level.

The feedback that we received from the design groups concerning the safety evaluation was positive. The checklist will be utilized in several design groups in the design of new safety related systems.

## References

[CEN 94]     European Committee for Standardization, "Safety Related Parts of Control Systems, Part 1: General Principles for Design", Final Standard Draft No. prEN 954-1, November 1994.

[Clarke 93]   Clarke, S. and McDermid, J., "Software Fault Trees and Weakest Preconditions: a Comparison and Analysis", Software Engineering Journal, July 1993.

[IEC 85]     International Electrotechnical Commission, "Analysis Techniques for System Reliability - Procedure for Failure Mode and Effect Analysis (FMEA)", IEC 812, 1985.

[IEC 90]     International Electrotechnical Commission, "Fault Tree Analysis (FTA)", IEC 1025, October 1990.

[IEC 94]     International Electrotechnical Commission, "Functional Safety: Safety Related Systems", Standard Draft No. 1508, Technical Committee No. 65, Parts 1, 2 and 3, September 1994.

[Lehtelä 91]  Lehtelä, M., "Failure Mode and Effect Analysis of Electronic Circuits", Licentiate Thesis, Tampere University of Technology, July 1991.

# An Experience in Formal Verification of Safety Properties of a Railway Signalling Control System

A. Anselmi ◇, C. Bernardeschi ♣, A. Fantechi ♣ ♡, S. Gnesi ♡,
S. Larosa ♡, G. Mongardi ◇ and F. Torielli ◇

◇ Ansaldo Trasporti, Genova, Italy
♣ Dip. di Ingegneria dell'Informazione, Univ. di Pisa, Italy
♡ IEI - C.N.R., Pisa, Italy

## Abstract

An experience on the specification and verification of a railway inter-locking system produced in a joint project with Ansaldo and the Italian Railways is reported. In the project we have used the JACK environment both to build the algebraic and graphical specification of such a system and to perform the verification of logic formulae on the model of the system itself. JACK is an environment integrating a set of verification tools, supported by a graphical interface offering facilities to use these tools separately or in combination. The experiment carried on has shown that the methodology can be applied successfully in the verification of safety critical systems.

## 1  Introduction

This paper reports an experience on formal specification and verification carried on in a pilot project on the validation of the functional specifications of a railway computer-based interlocking system produced by Ansaldo Trasporti[1]. This system, described with more detail in [ABBM92], includes both hardware and software fault-tolerant features to meet very stringent safety requirements. The interlocking system has been validated with traditional methods (static analysis and simulation, Decision Tables, testing), and structured and rigorous methods have been used in its development, in order to achieve excellent software correctness and reliability [ABBM92].

The pilot project, we carried on, has addressed a re–validation of the interlocking logic, which is the part of the application software responsible for the interlocking of the equipments of a signalling system, in order to acquire experience in the application of formal methods to validation. Indeed, the use of formal methods is increasingly required by the international standards and guidelines (see, for example, [CEN94]). In this project the use of safety requirements validation was mainly devoted to examine the interactions between the

---

[1]Project developed under the supervision of the Italian Railways

controls of different entities: in fact, due to the structure of the specification, it is not difficult to be convinced of the correctness of the operations related to a particular entity, but is more difficult to trace the interactions between the different entities. The experiment has been done considering the control of two different entities, namely the level-crossing and the shunting route.

The operations related to these entities were specified as process algebra terms [Mil89] and some safety properties about the interactions between these objects have been expressed in ACTL logic [DNV90, DV92]. Then, we used the integrated verification environment JACK [BGL94] to check, by means of the included model checker [Fer94], the validity of the desired properties on the model of the global system obtained by composing the operations of the two entities. In the formal specification of the operations related to the entities we have maintained the operations/variables structure of the initial Ansaldo specification. The choice to stick to the original Ansaldo specification style has produced a specification with many processes. Since the dimension of the model, in general, grows exponentially with the number of component processes this has enhanced state explosion problems during the construction of the automaton on which the formulae are to be verified.
In order to cope with state explosion problems, which sometimes exhausted the memory available during the generation of the automaton, we have adopted some "abstraction" techniques that allow to cut parts of the specification which are not of interest for the required properties, in such a way that proving a property on the reduced model guarantees that the property is satisfied by the whole model.

In Section 2 we briefly introduce the formalisms and the verification tools used in the project; in Section 3 we introduce the interlocking system considered; in Section 4 we introduce the three main step of our work, namely the formal specification, the generation of the global model and the property verification, which are then presented in the three successive sections.

# 2   Background

Process algebras and their semantic models, i.e., Labelled Transition Systems (or, *state automata*) [Mil89], are generally recognized as a convenient mean for describing sequential or concurrent systems at different levels of abstraction. They rely on a small set of basic operators, which correspond to primitive notions of concurrent systems, and on one or more notions of behavioural equivalence or preorder. The operators are used to build complex systems from more elementary ones. The behavioural equivalences are used to study the relationships between descriptions of the same system at different levels of abstractions (e.g., specification and implementation). The concept of a Labelled Transition System it is formally defined below.

A *labelled transition system LTS* is a quadruple $(S, T, D, s_0)$, where $S$ is a set of states, $T$ is a set of transition labels, $s_0 \in S$ is the initial state, and $D \subseteq S \times T \times S$. An element of $D$ is denoted by $s \xrightarrow{\mu} s'$, that has to be read "from the state $s$ an action $\mu$ makes the system move to the state $s'$".

## 2.1 ACTL logic

Logic is a good candidate to provide abstract specifications, since it permits to describe system properties rather than system behaviours. Different types of temporal and modal logics have been proposed for the abstract specification of concurrent systems. In particular, specific logics modelled on Labelled Transition Systems have been recently defined, the so called *action-based logics*. Among them, we recall the action based version of CTL [EmH86, CES86], ACTL [DNV90]. This logic is suitable to express properties of systems defined by means of LTS's. Moreover, a set of ACTL formulae can be used to express those requirements that a safety-critical system must necessarily satisfy.

ACTL is a temporal logic of state formulae (denoted by $\phi$), in which a path quantifier prefixes an arbitrary path formula (denoted by $\gamma$). Also, it includes the logic for the definition of action formulae (denoted by $\chi$). The ACTL operators and their informal semantics over LTS's are reported in Figure 1, while the formal semantics is described in [DNV90, DV92].

| action | formulae | |
|---|---|---|
| $\chi ::=$ | *true* | "any action" |
| | *false* | "no action" |
| | *action* | |
| | $\neg \chi$ | "not $\chi$" |
| | $\chi \mid \chi'$ | "$\chi$ or $\chi'$" |
| **state** | **formulae** | |
| $\phi ::=$ | *true* | "any behaviour" |
| | *false* | "no behaviour" |
| | $\sim \phi$ | " not $\phi$" |
| | $\phi \& \phi'$ | "$\phi$ and $\phi'$" |
| | $E\gamma$ | "there exists a path in which $\gamma$" |
| | $A\gamma$ | "for all paths $\gamma$" |
| | $< action > \phi$ | "there exists a next state reachable with *action*, in which $\phi$" |
| | $[action]\phi$ | " for all next states reachable with *action*, $\phi$ is true" |
| **path** | **formulae** | |
| $\gamma ::=$ | $[\phi\{\chi\}U\{\chi'\}\phi']$ | "$\phi$ is true in each successive state reached by an action satisfying $\chi$, until a state in which $\phi'$ is true is reached by an action satisfying $\chi'$" |
| $\gamma ::=$ | $[\phi\{\chi\}U\phi']$ | "$\phi$ is true for each successive state reached by an action satisfying $\chi$, until a state in which $\phi'$ is true is reached |

Figure 1: ACTL operators.

Several logic operators can be defined starting from the basic ones. Let $\chi, \chi'$ range over action formulae, $E$ and $A$ be path quantifiers, $X$ and $U$ be the "next" and "until" operators, respectively. We will write:

- $EF\phi$ for $E[true\{true\}U\phi]$ and $AF\phi$ for $A[true\{true\}U\phi]$;
  these are called the *eventually* operators; $EF\phi$ $(AF\phi)$ means "on at least

a path (on all the paths, respectively) there exists a state in which $\phi$";

- $EG\phi$ for $\sim AF \sim \phi$ and $AG\phi$ for $\sim EF \sim \phi$;
  these are called the *always* operators; $EG\phi$ $(AG\phi)$ means "in all the states of at least a path (of all the paths, respectively) $\phi$";

The ACTL logic allows to simply express *safety* and *liveness* properties in terms of the actions a system can perform. Safety properties claim that anything bad does not happen; liveness properties claim that something good eventually happens.

When an ACTL formula is given and a system is described by means of a LTS, it is possible to verify by model checking algorithms if the LTS is a model for such a formula. In this case we say that the system satisfies the property the formula expresses.

## 2.2 The JACK environment

The integrated verification environment JACK [BGL94] is able to cover a large extent of the formal software development process, including the formalization of requirements, behavioural equivalence proofs, graph transformations and logic verification by model checking algorythms. The general structure of JACK is shown in Figure 2, while the subset of the tools used in this pilot project are listed below:

**MAUTO** - is a tool for both the specification and the verification for concurrent systems described by process algebras. MAUTO functionalities are: parsing of a specification; translation functions from the process written in the algebraic specification into the automaton representing it; automata analysis such as abstraction, minimisation and equivalence checking; translation of an automaton into either the standard FC2 format, suitable as input for other tools, or the ATG format to graphically view the specification.

**ATG** - This is a tool for a graphical representation of the specification that provides functionalities for a compositional development of a specification. As a general rule a window is a process specification. Process construction starts from automata which represent single sequential processes. Moreover, an automaton in FC2 format can be read and graphically represented. Editing can be done on the graphical representation and the final specification can be written in the customised ATG format or in FC2 format or as a standard postscript file. Processes surrounded by boxes are said to be network and are used to hide information and to represent parallel composition. Two networks can synchronise the signals they emit, thus representing communicating processes. There are MAUTO functionalities that allow a graphical representation of a specification to be automatically translated into an algebraic one and viceversa.

**FC2Link and Hoggar** - The first is a tool able to link together automata of a given network; the second one translates the network of automata in a single automaton which describes the behaviour of the whole set of entities of the specification.

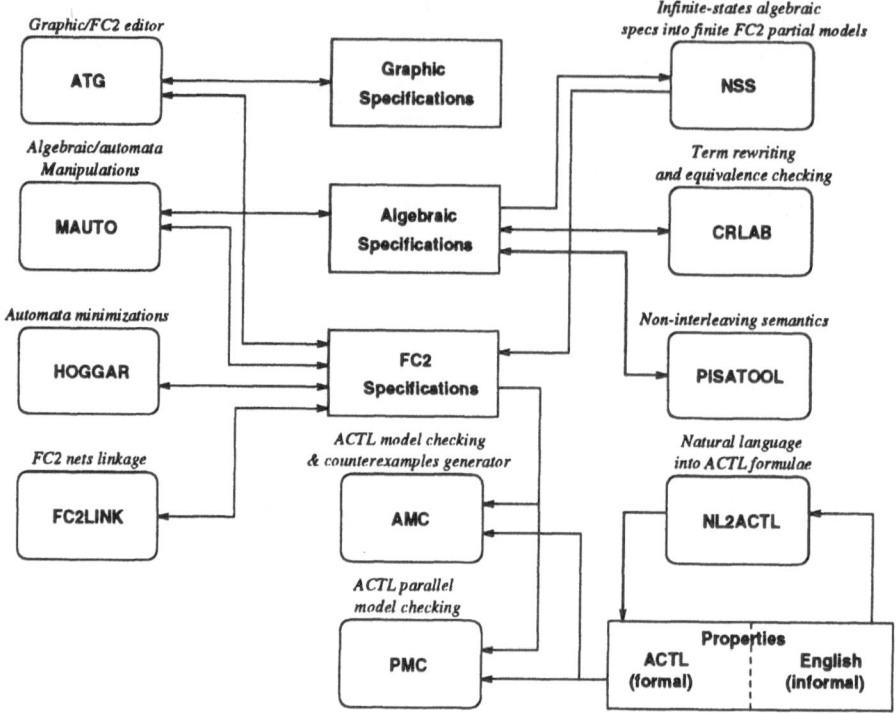

Figure 2: A general view of the JACK toolset.

**AMC** - AMC is the model checker for ACTL logic formulae and permits the validity of an ACTL formula to be verified on the automaton of the specification with linear time complexity. In the case in which a formula does not hold, the tool produces information on the part of the automata which falsifies the formula, thus reducing the time for the error detection on the specification.

JACK has a graphical interface to integrate the different tools; the interface provides a simple way to use the tools and to control the transmission of information between them.

# 3 The Computer Based Interlocking System - ACC

The Electronic Interlocking System, ACC (for "Apparato Centrale di Stazione a Calcolatore"), is based on a multiprocessor architecture and has been developed on the basis of a distributed architecture computer system with high modularity and configurability. The ACC is capable of carrying out, with very high reliability, all control, interlock and clearing functions performed by relay interlockings, both for mainline railway and for underground transport applications. Moreover the ACC carries out automatic diagnostic routines, data logging and connection with other systems that increase the operative efficiency of the Traffic Managers, as well as the Maintenance Engineers.

The ACC consists of two subsystems that independently perform vital and non vital functions respectively. The vital subsystem controls train movements and wayside equipment. It consists of a Central Logic Unit (Safety Nucleus - SN) and input/output Peripheral Units connected by data transmission links. The central SN is designed for the execution of safe operations and it is based on a Triple Modular Redundancy (TMR) configuration of computers implementing a two out of three voting scheme, with automatic exclusion of the unit in disagreement with the other two. It carries out safety logic and Operator Interface management functions. Peripheral Units, based on a vital two out of two microcomputer configuration, perform remote I/O functions, with direct control of the "field equipment" by special modules. The non vital subsystem (RDT - Recording, Diagnosis and data Transmission) is dedicated to auxiliary functions, like data recording, diagnostic management and remote control interface. To assure system availability, the said functions are carried out by two computers operating in a redundant stand-by configuration with automatic switch over.

Software reliability and correctness has been obtained by the application of structured and rigorous traditional methods in the development, verification and validation of the system. Software devoted to assure a safe behaviour of the system is separated by that for the operations required by a station. Programs for data communication, diagnostic tests, majority voting schemes, start up and automatic exclusion of a unit, belong to the first group. They are independent from the application and have been verified and validated for all applications. The second set includes the safety logic of the ACC, for which a structure based on common data has been adopted: it is composed by logic operations, each one with a list of verifications on variables associated to physical or logical entities and a list of assignments of values to other variables (the results of the operation).

In the system specification, the previous operations are grouped into processes, each one linked to the control of a physical entity (for example the level-crossing) or of a logical entity (for example the shunting route mechanism). Each of such processes is a collection of *operations*, that describe the behaviour of the process, *variables* that are used to store global information about the process and *attributes*, that are statical configuration parameters for the process itself. Below we show a fragment of the semi-formal description of the Level Crossing Control (LCC); such a fragment describes the operation related to the request of automatic closure, that is sent to the LCC by a shunting route process:

```
2.1.2 AUTOMATIC CLOSURE REQUEST
(Activated by the related command from a shunting route process).


I. VERIFY THAT
        a. the command state has the value "automatic";
        b. the LCC process has a value not equal to:
           - "request of closure"
           - "waiting for the timer".
```

II. ASSIGN
    a. the value "closed" to the command state
       JUST IF:
          the LCC process state has either the value
          "request of opening" or "opening is done"
          or "opening is doing";
    b. the value "waiting for the timer" to the LCC
       process state
       JUST IF:
          it has the value "idle".
EXCEPTIONS
    |a|  LCC command state has not the value "automatic".
        COMMAND IS LOST; NO RECOVERY ACTIONS.
    |b|  LCC process state has neither the value
        "request of opening" nor "opening is done"
        nor "opening is doing".
        COMMAND IS LOST; NO RECOVERY ACTIONS.

In the "VERIFY THAT" part some conditions on variables are tested that must be satisfied before continuing the operation. In the "ASSIGN" part the operation is really performed, and the "EXCEPTIONS" part specifies what should be done if any of the "VERIFY THAT" conditions is not satisfied (in the example, simply nothing).

# 4   Formal specification and verification

The pilot project analyses the subsystem composed by the level-crossing and the shunting route [PLG94], taking care in particular of the interaction between them: The interaction between entities turns out to be the most difficult to be validated with traditional verification and validation methods.

In the first part of the work, the formal specification of the level crossing and the shunting route has been developed in process algebra and a global model has been associated to the composition of these two entities. In the second part, some safety properties about the two entities and the interactions between them have been expressed in ACTL, in order to verify their satisfiability on the model of the specification.

The project was organized as follows:

1. Translation from the original semi-formal specification (referred to as Ansaldo specification in the following) to a specification in a formal language. The formal language we have used is the process algebra CCS (Calculus of Communicating Systems) [Mil89].

   In CCS a system consists of a set of communicating processes; each process executes actions and synchronises with other processes to execute its activities. An alternative way to specify the behaviour of a process is by means of a graphical formalism able to represent it as a finite state automaton. The graphical representation is equivalent to the algebraic one, but it has the advantage of being simpler for the user and more understandable. Moreover the possibility of using two different formats of a specification (the algebraic and the graphical one) is useful in the removal of errors from the specification.

2. Generation of a global model of the system (finite state automaton) which includes the level crossing and the shunting route and the interaction between them by using JACK. On this model we will perform the verification of the correctness of the system behaviour according to the selected properties. Some properties can be simply checked on the model of the shunting route subsystem, other properties instead need the whole model of the system to be verified. Since state explosion problems (that is, the exponential growth of the number of the states) arise during the generation of the whole model of the system, some "abstraction techniques" have been adopted. These help to reduce the state explosion problems and make the model tractable by the tools.

3. Application of the temporal logic ACTL to describe some safety properties of the system. The ACTL syntax allows a simple translation from the properties to their formal specification, since we can express directly the actions executed by the system components. The model checker AMC for the ACTL logic is used for the automatic verification of the properties on the finite state automaton which represent the behaviour of the system.

# 5 Formal specification model

In the Ansaldo specification, in which the main concern is on the states, there are *passive entities* (i.e., state variables, that do not execute operations) and *active entities* (i.e, those entities activated by state changes and which execute operations). On the contrary, in a process algebra specification, the whole set of entities are active.

The formalisation of the Ansaldo specification has been done by combining the process algebras formalism (CCS) and the graphical representation of automata. The use of these techniques permits to simply write specifications similar to that of Ansaldo (state machines) and specifications in which the main concern is the execution of "actions" instead of "state changes". Variables in the Ansaldo specification are mapped to processes in the process algebra specification. Operations are formalized using a graphical representation of the state machine they characterize. This formalization method allowed us to stick to the original Ansaldo specification style in the generation of the formal specification of the system; the Ansaldo specification model can be simply translated in process algebras specifications and into formal graphical specifications.

The proposed formalization method has the advantage that can be easily reused, since it is not tied to the particular considered subsystem; rather, it has been directly derived from the semi-formal specification style used by Ansaldo for all the entities of the interlocking logic.

## 5.1 Variables formalisation

Standard rules useful to formalise the concept of *discrete* variable by means of processes are:

- For each variable $X$, a process is associated to $X$. The process is called "process of $X$".

- For each value $v$ in the range of a variable $X$, a state $V$ is added to the set of states of the process of $X$, and the input action $s_v$? and the output action $v!$ are added to the set of its actions.

- If the process of the variable $X$, during an execution, is in the state $V$, it means that the value of the variable $X$ is $v$. When the process of $X$ is in the state $V$, it can synchronise with any other process, willing to read the value of the variable, by the action $v!$.

- If the values of the variable $X$ range over $a, b, \ldots$, in any state of the process $X$ the actions $s\_a?, s\_b?, \ldots$ may be executed, to synchronise with other processes which want to change the value of $X$. For example, if another process want to assign the value $b$ to the variable $X$, it must execute the $s\_b!$ action.

**Example: the "command state" variable used by the CLOSURE RE-QUEST operation of section 3.**

Possible values of the "command state" variable are "AUTOMATIC" (which is the initial value of the variable), "CLOSED" or "MANUAL". The CCS specification of the process associated to the variable "command state" is reported in the following:

```
parse command_state =
let rec {
      AUTOMATIC =  automatic! : AUTOMATIC +
                   s_manual? : MANUAL +
                   s_closed? : CLOSED +
                   s_automatic? : AUTOMATIC
and
         MANUAL =  manual! : MANUAL +
                   s_manual? : MANUAL +
                   s_closed? : CLOSED +
                   s_automatic? : AUTOMATIC
and
         CLOSED =  closed! : CLOSED +
                   s_manual? : MANUAL +
                   s_closed? : CLOSED +
                   s_automatic? : AUTOMATIC
} in AUTOMATIC;
```

## 5.2    Operations formalisation

The formalisation of an operation is a quite simple operation, because there is a strong binding between the Ansaldo semi–formal description and the automata formal specification. We just say that each operation is started by a process arbiter, by an explicit signal *?start* ...; such a signal is sent by the arbiter after either an explicit call of an interface operation, or a testing of the value of the state variable (i.e. a synchronization with the related process) for the internal operations. The initial state of an automaton is denoted by a double circle, and at the end of the operation the signal !end is sent back to the arbiter, so that it can schedule another operation.

**Example: Automatic Closure Request operation**

Let us consider the Automatic Closure Request operation: its formal specification is reported in graphical format, in Figure 3.

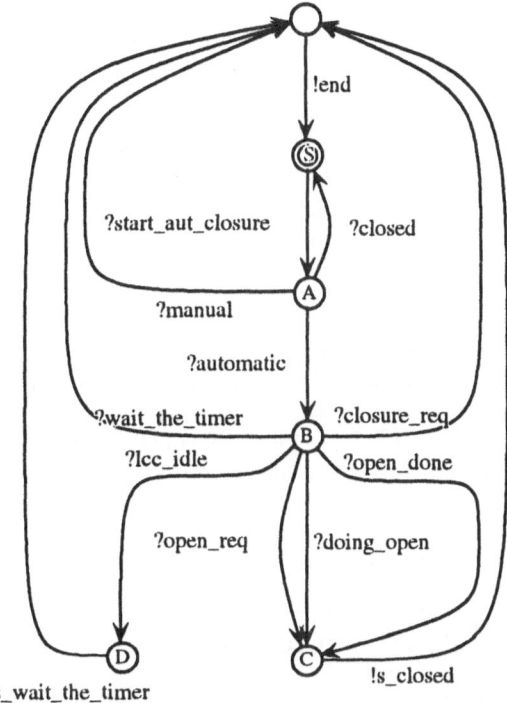

Figure 3: The Automatic Closure Request operation.

# 6 Generation of the global model

Once the formal specification has been defined for all variables and operations, it is necessary to generate the global automaton of the system. This is built composing together all the previous specifications. The composition is realized defining synchronization relations between the components. The main drawback of this composition process is in general related to the number of states of the obtained global automaton of the system, which grows exponentially with the number of processes composed in parallel. To avoid this problem it is useful to maintain the number processes small. Moreover, since the number of the states of the global model also depends on the level of synchronisation between the processes, a strict synchronisation between processes reduces the number of states of the model.

Unfortunately, in the interlocking logic specification it was not possible to follow these guidelines; in fact, we have defined a process for each variable and each operation, and often there is not a strict synchronisation between them. For these reasons, we adopt some preliminary abstraction criteria, to generate the global automaton of the system. These are necessary to cope with state explosion problems and limitations of the available automatic tools for the specification and the verification.

Essentially, we have used two kinds of abstractions:

- Abstract from the static configuration parameters, which were given the status of variables in the original specification. In practice, we have fixed some configuration parameters, in order to reduce the number of states. The verification made with this parameter value should then be repeated for the other values of the configuration parameter.

- Abstract from variables which do not interfere with the required property. This abstraction criteria requires the verification of each required property to be conducted on a possibly different reduced model.

The complete specification of the system composed by the LCC and by the shunting route process (SH_ROUT) is obtained by composing the specification of the different processes as explained before. The composition can be represented in a graphical way, by the use of the ATG tool, as shown in Figure 4. Each box represents a process (formally specified in a graphical form as an automaton or in the algebraic format), or in its turn, a net of processes.

The various nets and automata are linked together (by Fc2Link) to form a global network that is then transformed into a single global automaton (i.e. the model for the verification) by Hoggar; such an automaton has 77294 states.

# 7  Properties verification

In this section we introduce the properties that must be verified by the system and their ACTL formulae characterization; we will then show the verification of such formulae on the model of the system described in the previous sections.

## 7.1  Property 1

*The shunting route process does not send the proceed command to the shunting signal if the level crossing is not closed.*

Property 1 can be expressed on the model of the system in the following way: *if in* any *state of the model, it is true that the position of the LCC is* open or undefined, *then* there not exists *an execution of the processes* starting from such a state, *in which the position of the LCC never becomes* closed *and the shunting signal is sent the proceed command*

and then translated into the following ACTL formula

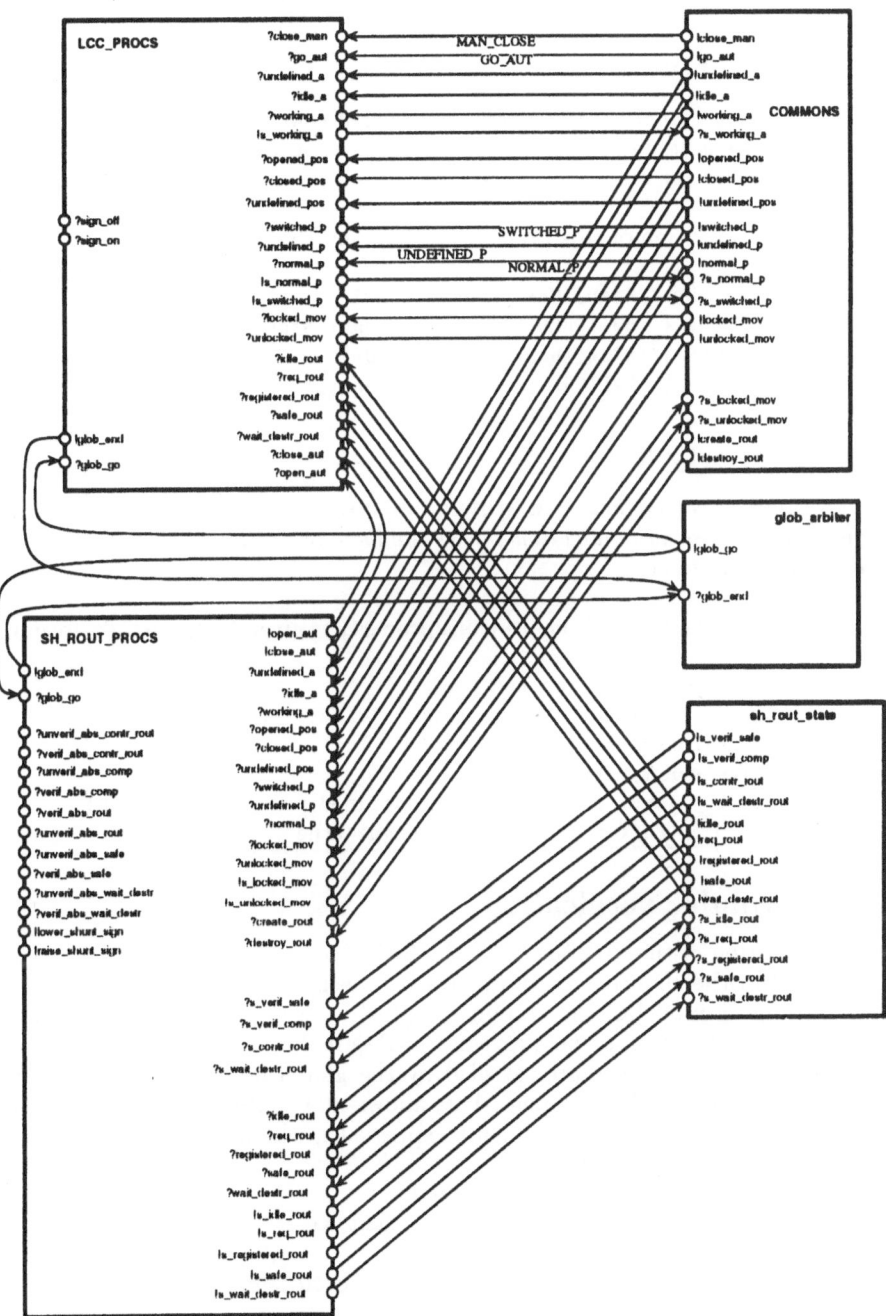

Figure 4: The complete specification.

```
AG ( [?undefined_pos | ?opened_pos]
 (~E[true {~ ?closed_pos}U (<!raise_shunt_sign> true)] ) )
```

## 7.2   Property 2

*If the shunting signal has been sent the proceed command and the level cross-
ing has been closed and locked by a single operation, the reset of the automatic
operation leaves the level crossing closed.*

The formalisation of this property is given by the ACTL formula

```
AG( [!raise_shunt_sign][MAN_CLOSE][GO_AUT]
        A[true {~ SWITCHED_P}U{!lower_shunt_sign} true])
& AG( [MAN_CLOSE][!raise_shunt_sign][GO_AUT]
        A[true {~ SWITCHED_P}U{!lower_shunt_sign} true])
```

## 7.3   Property 3

*If the shunting signal has been sent the proceed command, the loosing of the
"closed level crossing" indication causes the logic to switch off the shunting signal proceed command.*

The characterization of the previous property is given by the ACTL formula

```
AG ( [?undefined_pos | ?opened_pos]
     (~E[true {~ !lower_shunt_sign & ~ ?closed_pos}U
        (<!raise_shunt_sign> true)] ) )
& AG ( [!raise_shunt_sign]
    (~E[true {~ !lower_shunt_sign}U{?undefined_pos | ?opened_pos}
            (EG ( <~!lower_shunt_sign> true ) ) ] ) )
```

## 7.4   Formulae verification

The verification of the satisfaction of the formulae on the system composed by
the LCC and the shunting route process has been done by the model checker
AMC, using as model of the system the global automaton obtained by composing the specification of the processes in Figure 4.

The previous formulae are satisfied by the global model, as shown in the
following traces of AMC sessions. This allows us to affirm the correctness of
the model respect to the requirements expressed in the formulae.

```
|= AG ( [?undefined_pos | ?opened_pos]
    (~E[true {~ ?closed_pos}U (<!raise_shunt_sign> true)] ) )

The formula is TRUE in state 0 time: (user: 0.00 sec, sys: 0.00 sec)
...
|= AG( [!raise_shunt_sign][MAN_CLOSE][GO_AUT]
        A[true {~ SWITCHED_P}U{!lower_shunt_sign} true] )
 & AG( [MAN_CLOSE][!raise_shunt_sign][GO_AUT]
        A[true {~ SWITCHED_P}U{!lower_shunt_sign} true] )
```

```
The formula is TRUE in state 0 time: (user: 1.07 sec, sys: 0.06 sec)
...
|=   AG( [?undefined_pos | ?opened_pos]
         (~E[true {~ !lower_shunt_sign & ~ ?closed_pos}U
                                    (<!raise_shunt_sign> true)] ) )
   & AG( [!raise_shunt_sign]
         (~E[true {~ !lower_shunt_sign}U{?undefined_pos | ?opened_pos}
            (EG ( <~!lower_shunt_sign> true ) ) ] ) )


The formula is TRUE in state 0 time: (user: 0.01 sec, sys: 0.00 sec)
```

# 8   Conclusions

In this section some considerations about the project are reported. First we remind that the project was a re-validation of the interlocking logic, in order to acquire experience in the application to V&V of formal methods. The interlocking logic has been independently validated with traditional methods, therefore we expected the specification was correct. We proved, in fact, that the selected safety properties are satisfied by the specification.

Starting from the Ansaldo specification, a formal specification of the system has been done choosing a formalism useful to capture the original structure of the specification. We have defined a formalization method able to maintain the operations/variables structure of the initial Ansaldo specification. Alternatively, we could have directly expressed in process algebra the behaviour of the interlocking logic, but this would have meant to depart from the original specification developed by Ansaldo with well-established methodologies.

The passage from the Ansaldo level crossing control specification to its formal specification has been done initially with some difficulties, mainly due to the use of two different languages, the railway language and the formalists'one. These initial communication problems between the two groups involved in the project (the industrial staff and the academic group) disappeared once the academic group got acquainted with the railway language and the industrial team got acquainted with the used formal techniques. So, it has been possible to formalise the second process (the shunting route) in much less time.

We consider the outcomes of this pilot project to be successful; the major problem we had to face is the state explosion problem (it is easy to build automata reaching a million states), but we have experimented that formal verification by means of model checking of safety requirements on state models of the system is still feasible, when proper abstraction techniques are used and advanced tools able to handle a large number of states are available; in particular, verification tools based on the representation of the automata by means of Binary Decision Diagrams (BDDs) are able to deal with automata with a number of states several order of magnitude larger [BCM92].

# References

[ABBM92]  C. Abbaneo, G. Biondi, M. Ferrando, G. Mongardi. Testing of a Computer Based Interlocking Software: Methodology and Environment. SAFECOMP 1992, Zurich.

[BCM92]  J. R. Burch, E.M. Clarke, K. L. McMillan, D. L. Dill, L. J. Hwang. Symbolic model checking: $10^{20}$ states and beyond. *Information and Computation* 98(2), June 1992, pp. 142-270.

[BGL94]  A. Bouali, S. Gnesi, S. Larosa. The integration Project for the JACK Environment. *Bulletin of the EATCS*, n.54, October 1994, pp.207-223.

[CEN94]  CENELEC - pr EN 50128 - Railway Applications: Software fo Railway Control and Protection Systems.

[CES86]  E. M. Clarke, E. A. Emerson, A. P. Sistla. Automatic Verification of Finite–State Concurrent Systems Using Temporal Logic Specification. *ACM Transaction on Programming Languages and Systems*, Vol. 8, No. 2, April 1986, pp. 244 – 263.

[DNV90]  R. De Nicola, F. W. Vaandrager. Action versus State based Logics for Transition Systems. Proceedings Ecole de Printemps on Semantics of Concurrency. Lecture Notes in Computer Science 469, Springer-Verlag, 1990, pp. 407-419.

[DV92]  R. De Nicola and F.W. Vaandrager. Three Logics for Branching Bisimulation. Internal Report DSI-92-03, Dipartimento di Scienze dell'Informazione, Univ. di Roma "La Sapienza", 1992. To appear in *Journal of ACM*.

[EmH86]  E. A. Emerson, J. Y. Halpern. "Sometimes" and "Not Never" Revisited: on Branching Time versus Linear Time Temporal Logic. *Journal of ACM*, 33 (1),1986, pp. 151-178.

[Fer94]  G. Ferro. AMC: ACTL Model Checker. Reference Manual. IEI-Internal Report, B4-47 December 1994.

[Hoa85]  C.A.R. Hoare. *Communicating Sequential Processes*. Prentice Hall Int., London, 1985.

[Hem85]  M. Hennessy and R. Milner. Algebraic Laws for Nondeterminism and Concurrency. *Journal of ACM*, **32**, 1985, pp. 137-161.

[Mil89]  R. Milner. Communication and Concurrency. Prentice Hall,1989.

[PLG94]  Programma logica Gioul (vers.5.2) – Specifica funzionale – parte II, cap. 13, Ansaldo Trasporti, 1994.

# Dependability of Iterative Software: a Model for Evaluating the Effects of Input Correlation

A. Bondavalli[1], S. Chiaradonna[1], F. Di Giandomenico[2], S. La Torre[2]

[1] CNUCE-CNR, Via S. Maria 36, 56126 Pisa, Italy

[2] IEI-CNR, Via S. Maria 46, 56126 Pisa, Italy

### Abstract

This paper deals with the dependability evaluation of software programs of iterative nature. In this work we define a new model that is able to account for both dependencies between input values of successive iterations and the effects of sequences of consecutive failures. Differently from previously proposed models, it is based on steady state probabilities of an iteration outcome (success, benign failure or catastrophic failure) and those representing the correlation. As such it allows to analyse the effects of the correlation between successive inputs on the dependability attributes of iterative software and assumes, as starting knowledge, information usually obtainable by testing. Requiring the designers or users to determine just the steady state probabilities of an iteration outcome, rather than difficult and costly state transition probabilities, this model is more useful and more generally applicable.

## 1 Introduction

This paper deals with the dependability evaluation of software programs of iterative nature. The dependability of programs of iterative nature (as well as that of other software structures) is usually analysed using models [Arlat90, Chiaradonna94, Tai92, Tai93] that assume independence between the outcomes of successive executions of a program. This assumption, which is often false for many applications, strongly limits the realism of these models although make the associated mathematics simpler [Chiaradonna94]. Experiments and theoretical justifications show the existence of contiguous failure regions in the program input space and that, for many applications, such as real-time control systems, the inputs often follow a trajectory of contiguous points in the input space. For these reasons the inputs which originate failures of the software are very rarely isolated events but more likely grouped in *clusters* [Amman88, Bishop93, Bishop88]. For all the classes of applications to which these considerations apply, analyses of software dependability performed assuming independence between successive iterations seem to lead to results excessively diverging from the real behaviour of the analysed system.

Another important characteristic of these applications that should be captured by a realistic model (and usually is not) is the effect of clustering of failures on the system mission. Many (physical) systems can tolerate isolated or short bursts of benign failures, but long sequences of even benign failures prevent the system to get feed-back control often causing actual damage (from stopping a continuous production process to letting an airplane drift out of its safe flight envelope).

The problem of modelling and evaluating the effects of correlation between the outcomes of successive iterations has been addressed in the literature [Bondavalli95, Csenski89, Tomek93]. [Csenski89] models the behaviour of a recovery block structure [Randell75] composed of a primary version, an alternate version and a perfect acceptance test. Failures of the primary module are distinguished in :

i)   point failure: when the input sequence enters a failure region,

ii)  serial failure: a number of consecutive failures occurring after a point failure, i.e., after that the input trajectory enters a failure region.

The number of serial failures subsequent to any point failure is a random variable. From these modelling assumptions a simple Markov chain with discrete time is developed allowing an analytical evaluation of the reliability (MTTF) of the recovery blocks. [Tomek93] analyses the different forms of correlation of the recovery blocks structure, including correlation among the different alternates and among alternates and the acceptance test on the same inputs. While the previous two papers modelled correlation between inputs, we defined (in [Bondavalli95]) a dependability model for iterative software accounting for both dependencies between input values of successive iterations and the possibility that repeated, non fatal failures may together cause mission failure.

The effective utility of a model depends on many factors, that may also relate to its intended use. Among these we already recalled the realism, i.e. the plausibility of the assumptions made, but also very important are the ability to account for the relevant basic details, the robustness against inaccurate values assigned to some parameters and the possibility or easiness to obtain proper estimation of the parameters. A set of plausible values for the model parameters is normally derived from testing or from previous experience with similar software. Usually models kept simple by numerous assumptions, have a limited realism but use parameters which are easier to determine than those required by more realistic models. For example, models where the independence assumption holds allow to use constant probabilities of failure and success at each iteration for the entire mission duration. This has the advantage to allow dependability models with steady state probabilities as parameters. These steady state probabilities may be estimated through testing and then used for evaluating the system behaviour in different scenarios.

In this paper our aim is to improve the utility of modelling both dependencies between input values of successive iterations and the possibility that repeated, non fatal failures may together cause mission failure. The model in [Bondavalli95] is based on state transition probabilities. The estimation of these parameters, e.g. through testing, may be i) difficult to obtain and ii) in most cases useless since their estimation requires the same effort than estimating the dependability figures of interest. We develop a new model which is based on steady state probabilities easier to determine. In this way, an interesting compromise is reached between a fairly realistic model for obtaining predictions of dependability attributes of a system and difficulty (and costs) in obtaining the basic knowledge necessary to resolve the model.

The rest of the paper is organised as follows. Section 2 contains our assumptions, a brief recall to the model presented in [Bondavalli95] and a discussion on how values for model parameters may be obtained. In Section 3, a new, general model for iterative software based on steady state probabilities is developed. Examples of evaluations for specific classes of iterative software are presented in Section 4. Finally, Section 5 summarises our conclusions.

# 2 Background

## 2.1 Assumptions and Dependability Measures

Software applications (seen as a black box) of an iterative nature are assumed, where a *mission* is composed of a constant number $n$ of iterations of the execution of the program. Each iteration is started cyclically after a fixed time interval and is aborted by a watch-dog timer should it last more than a time threshold $\tau$ equal to the period. At each iteration, the program accepts an input and produces an output. Successive inputs form a trajectory, a random or deterministic walk with a step length that is small, compared to the size of the input space [Bishop88]. The outcomes of an individual iteration may be: i) *success*, i.e., the delivery of a correct result, ii) a *benign failure* of the program, i.e., an output that is not correct but does not, by itself, cause the entire mission to fail, or iii) a *catastrophic failure*, i.e., an output that causes the immediate failure of the entire mission. Failure regions are subsets of the program input space, consisting of contiguous points. They are separated by each other: to pass from one to another the input trajectories must cross at least one point for which the system executes successfully. "A priori", all points in the input space have the same probability of belonging to a failure region.

Any sequence of $n_c$ or more benign failures ($n_c > 0$) causes the failure of the entire mission. One success, after any series of less than $n_c$ benign failures, interrupts the sequence and the system looses the memory of the previous failure sequence. This is a plausible assumption if failure regions are small and widely dispersed in the input space. For instance, this is true in the case of convex failure regions and trajectories that change direction only slightly between any two consecutive inputs.

The two attributes of dependability considered are the probability of surviving a mission (reliability after a certain number of executions) and the performability [Meyer80]. The reward model used here as a basis for performability is the same used in [Chiaradonna94, Tai93]: successful executions add one unit to the value of the total reward accumulated over a mission; executions producing benign failures add zero; a catastrophic failure reduces the total reward to zero.

## 2.2 A Model Based on State Transition Probabilities

This model [Bondavalli95], described by the discrete-time Markov chain shown in Figure 1, is based on the same assumptions we have made here with just one difference: the input trajectories may directly pass from one failure region to another, that is contiguous failure regions are admitted. After an iteration with success (state S) there is a probability that the next execution will produce a benign failure (i.e., that the input trajectory enters a failure region). In this case the probability of staying there for i iterations is given by $p_i$, which models the variable "number of consecutive failures" conditioned on entering a failure region while $p_{nn}$ designates the probability of staying for at least $n_c$ iterations, i. e. $p_{nn} = 1 - \sum_{i=1}^{n_c-1} p_i$. The state representing the benign failure is split in $n_c-1$ states. So, each state $B_i$ designates a benign failure of the last execution. For instance, with probability $p_{sb}p_2$ the program enters a failure region, represented by state $B_2$ in our model, from which, unless a catastrophic failure occurs (arc from $B_2$ to C labelled $p_{bc}$) it will be compelled to move to $B_1$, after which it exits the failure region. Arcs exiting from $B_1$ and entering states $B_2$, .., $B_{nc-1}$, C, labelled with the same probabilities as those exiting from state S, model the case of input trajectories crossing two or more consecutive failure regions. The modelled behaviour (as it was intended) does not capture negative correlation.

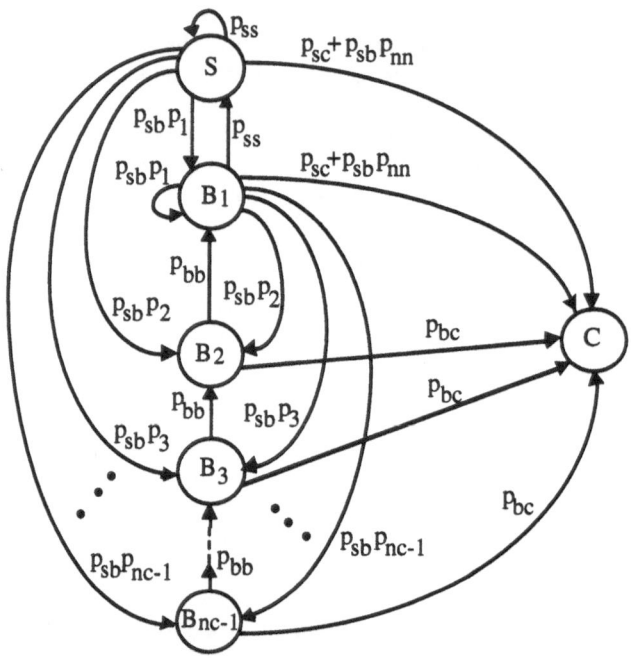

**Figure 1.** The model for iterative executions with failure clustering.

This model does not depend on any specific distribution of the length of stays in failure regions and has been used to evaluate the effects of several distributions on the dependability figures analysed. The resolution has been conducted by assuming specific values for the state transitions probabilities; then, the probability of mission failure and performability were evaluated as functions of two factors: 1) the probability of exceeding a sequence of $n_c$ -1 consecutive failure, $p_{nn}$ and 2) the mean stay in a failure region, once the input trajectory enters it. Moreover, two distributions representing the two extreme cases have been defined, and their figures were shown to bound those derived by all the analysed distributions.

## 2.3 Discussion

The model just described is, in principle, an interesting step forward in improving the predictions about system behaviour, but it assumes knowledge about the state transition probabilities (for example, the probability that the program enters in a failure region), which can be very difficult, or at least quite expensive, to obtain. Hence, our interest in developing an alternative process to obtain the same final evaluations, but starting from information which can be derived in a simpler and cheaper way, that is the steady state probabilities $p_s$, $p_b$ and $p_c$.

Usually, a set of plausible values for the model parameters is derived by testing. The accuracy of the measures so determined depends on how good the testing activity is. Obviously, the more the test is calibrated to the specific application and extends to a large number of experiments, the more the estimates of the measured quantities result to be accurate. Since it is impractical to test a program on every possible input, a variety of methods have been proposed in the testing field for selecting a subset of significant inputs on which to experiment the program. Testing on a randomly generated input profile could be satisfactory in finding bugs, but is generally much less interesting if the aim is to determine estimates of measures (such as reliability). In these cases, inputs more frequently encountered during the operational life of the program are normally privileged. Hence, the knowledge of the characteristics of the program under testing becomes crucial. For example, to get proper estimates for the state transition probabilities to be used in the model previously described, inputs for the testing should be selected in the form of realistic trajectories from a realistic distribution of trajectories. Moreover, to better represent the software in actual operation, the observation of a mission would stop at the moment of a mission failure. In the extreme case, when all the details about the operational environment of the program under testing are known, one could derive by testing directly estimates of the final quantities of interest (for example, the probability of mission failure, or the performability of a system). But, also in this extreme case, it could be prohibitive to derive by testing the final estimates due to the high costs involved (in terms of time required and skill in determining significant inputs).

Instead of providing reliability figures only through testing, i.e. considering the system as a black-box, its internal structure could be taken into account as well. For example, if the system has been designed according to some fault tolerant structure, the testing could be used to obtain an estimate of the probability of failure/success of the individual components of the software system. Then, models are widely available in the literature [Arlat90], to derive the probability of success, benign and catastrophic failure for each single iteration. Last, these probabilities of the outcome at each individual execution can be interpreted as the steady state probabilities of the system, and estimates of the performability or the probability of surviving missions of a certain duration can be found using models as that proposed in Section 3.

Here, we are not much interested in discussing all the possible ways of obtaining the steady state probabilities $p_s$, $p_b$ and $p_c$ of a software system, but we simply point out that this knowledge can be reasonably provided. In particular, we assume to know them from testing. To simplify the testing process, we consider that the behaviour of the system under testing differs from the behaviour of the system in its real operational environment for two aspects: 1) the effects of sequences of benign failures are not taken into account; and 2) missions are not made of a fixed number of iterations but they are terminated only by the occurrence of a pointwise catastrophic failure (after a catastrophic failure the system is reset to the initial state). A practical use of this technique is possible when a realistic distribution of inputs can be generated by synthesising the population of trajectories expected for the controlled system. This makes testing much cheaper than testing the system on realistic trajectories. The steady state probabilities thus determined are obviously not exactly those of our original system due to the simplifications at point 1) and 2) above. Clearly, the quality of the measured probabilities heavily depends on how properly

494

the inputs used for testing represent the trajectories that will be encountered by the system during its operational lifetime.

# 3 The new Developed Model

In our scenario, the steady state probabilities $p_s$, $p_b$ and $p_c$ obtained through testing are the steady state probabilities of the system described by the Markov chain in Figure 2.

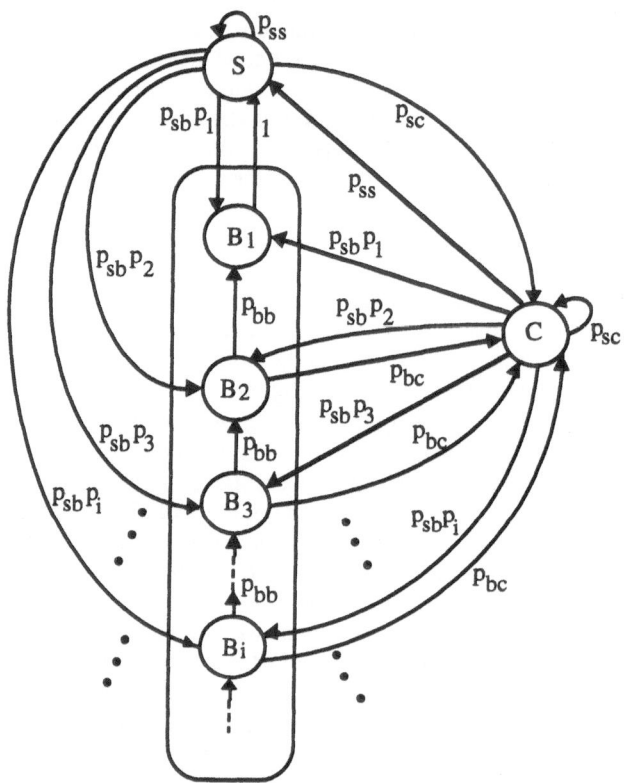

**Figure 2.** The model of the system under testing.

In this model, the state representing the benign failure state is split in an infinite number of states $B_i$, where the state $B_i$ is reached from S when the input trajectory enters a failure region and remains in it for i iterations (if no catastrophic failures are encountered). Then, the steady state probability of benign failure must be seen as the sum of the probabilities $p_{bi}$ of being in state $B_i$, that is $p_b = \sum_{i=1}^{\infty} p_{b_i}$. We consider that each trajectory, after crossing a failure region, experiences at least one successful execution. $B_1$ is the state corresponding to the last failure in the crossed regions and we assign the value 1 to the arc from the state $B_1$ to S.

This general model takes into account the correlation between successive inputs through the parameters $p_i$ on the arcs from S to the states $B_i$, without being tied to any specific distribution representing the permanence of the input trajectory in failure regions. At this stage, the possibility for repeated benign failures to cause a catastrophic failure is not considered (only pointwise catastrophic failures are considered). In addition, the assumption that after a catastrophic failure the system is reset to the initial state is represented by transitions from state C to states S, $B_i$ and C, all with the same probabilities as transitions from S.

Let $(p_{b1},.....,p_{bi},.....,p_s, p_c)$ be the steady-state distribution of the probability of staying respectively in the states $B_1,........, B_i,......,$ S and C and $X_i$ be the random variable representing the state of the system at the i-th iteration, i.e.:

$$\lim_{i \to \infty} P(X_i = B_j) = p_{b_j} \quad \forall j; \qquad \lim_{i \to \infty} P(X_i = S) = p_s; \qquad \lim_{i \to \infty} P(X_i = C) = p_c.$$

For all the distributions p, such that the Markov chain is irreducible, aperiodic and with all recurrent non-null states, the vector $(p_{b1},.....,p_{bi},....., p_s, p_c)$ is the unique solution to the equation system:

$$\begin{cases} p_s = p_{ss}p_s + p_{b_1} + p_{ss}p_c \\ p_{b_1} = p_{sb}p_1p_s + p_{bb}p_{b_2} + p_{sb}p_1p_c \\ \qquad \vdots \\ p_{b_i} = p_{sb}p_ip_s + p_{bb}p_{b_{i+1}} + p_{sb}p_ip_c \quad i = 2,3,... \\ \qquad \vdots \\ p_c = p_{sc}p_s + (p_b - p_{b_1})p_{bc} + p_{sc}p_c \end{cases} \qquad 3.1$$

The steady state probabilities and transition probabilities in this model are interdependent and our aim is to derive the transition probabilities ($p_{ss}$, $p_{sc}$, $p_{sb}$, $p_{bb}$ and $p_{bc}$) from the steady state ones ($p_b$, $p_s$ and $p_c$) plus the $p_i$'s representing the distribution of the length of staying in failure regions. Unfortunately, this knowledge is not sufficient to solve system 3.1 and to determine all the transition probabilities in the Markov chain in Figure 2. More information must be collected. We assume that further information may be obtained from analysing the system; for example such analysis could suggest special relationships between transition probabilities. Some reasonable examples are the following:

a) a system for which catastrophic failures occur independently from the trajectory being crossing failure regions, thus the probability of a catastrophic failure is the same if the last execution produced a benign failure or a success. Setting pbc = psc in our model allows to represent this case;

b) a system for which the probability of the next iteration producing a catastrophic failure increases if the last iteration produced a benign failure. This is modelled by setting $p_{bc} > p_{sc}$. This looks like a realistic assumption in many cases: for instance, one may assume that a benign failure implies that the program has entered a region of its input space where failure in general is especially likely, and that a fixed proportion of such failures happens to be immediately catastrophic;

c) the case of a controlled system where most erroneous control signals are immediately "catastrophic", but the control system is engineered to detect its own internal errors and then issue a safe output and reset itself to a known state from which the program is likely to proceed correctly. One may then assume that most benign failures are due to this mechanism, and likely to be followed by successes: $0 < p_{bc} < p_{sc}$;

d) the extreme of case c) above: the system issues a safe output but after the reset takes place pointwise catastrophic failures cannot immediately happen ($pbc = 0$).

The different situations illustrated can be summarised with the relation $pbc = k \, p_{sc}$, where k is a non negative real number. We can thus determine the values of the transition probabilities as a function of k and of the distribution of $p_i$. The resulting expressions are shown in Table 1 where two cases ($k = 0$ and $k > 0$) have been distinguished for mathematical reasons. In Section 4 we will discuss the effects of these scenarios (assigning proper values to k) on the probability of mission failure and the performability.

|  | $k = 0$ | $k > 0$ |
|---|---|---|
| **Pbb** | $1$ | $\dfrac{(1-p_{bb})}{k} = \dfrac{p_c - p_b(1-p_{bb})}{(1-p_b)} + $ $+ \dfrac{p_b(1-p_{bb})^2 \sum_{i=1}^{\infty} p_i p_{bb}^{i-1}}{(1-p_b)\left(1-p_{bb}\sum_{i=1}^{\infty} p_i p_{bb}^{i-1}\right)}$ |
| **Pbc** | $0$ | $1-p_{bb}$ |
| **Psb** | $\dfrac{p_b}{(1-p_b)\sum_{i=1}^{\infty} i p_i}$ | $\dfrac{p_b(1-p_{bb})}{(1-p_b)\left(1-p_{bb}\sum_{i=1}^{\infty} p_i p_{bb}^{i-1}\right)}$ |
| **Psc** | $\dfrac{p_c}{1-p_b}$ | $\dfrac{1-p_{bb}}{k}$ |
| **Pss** | $1 - \dfrac{p_b + p_c \sum_{i=1}^{\infty} i p_i}{(1-p_b)\sum_{i=1}^{\infty} i p_i}$ | $\dfrac{1}{\left(1-p_{bb}\sum_{i=1}^{\infty} p_i p_{bb}^{i-1}\right)} - \dfrac{p_c}{1-p_b} + $ $+ \dfrac{(2p_b p_{bb} - p_b - p_{bb})\sum_{i=1}^{\infty} p_i p_{bb}^{i-1}}{(1-p_b)\left(1-p_{bb}\sum_{i=1}^{\infty} p_i p_{bb}^{i-1}\right)}$ |

**Table 1.** Expressions for the transition probabilities in case of $k = 0$ and $k > 0$.

Note that in the case of $k > 0$ an implicit expression for $p_{bb}$ is given. So we have to solve numerically the equation for $p_{bb}$ to obtain values for the other transition probabilities. The solution must refer to specific distributions of probability and is restricted to those distributions such that the expression $\sum_{i=1}^{\infty} p_i p_{bb}^{i-1}$ admits a solution. For a number of distributions this solution is known [Trivedi82] and we will show the numerical results for some of them.

Having obtained the state transition probabilities we now model the system in its operational context where sequences of $n_c$ or more benign failures cause the system to fail and missions last at most n iterations. Furthermore, a mission terminates successfully if no pointwise catastrophic failure or sequences of at least $n_c$ benign failures are experienced otherwise terminates with failure as soon as one of these two events is observed. This scenario is represented by the model of Figure 3 to which the values for the transition probabilities just obtained are applied. All states $B_i$, for $i \geq n_c$, disappeared and the term $p_{sb}p_{nn}$, where $p_{nn} = \sum_{i=n_c}^{\infty} p_i$, has been added to the arc from S to C to capture that sequences of benign failures longer than $n_c - 1$ now lead to a mission failure. A very insignificant approximation has been introduced, this is described in [Bondavalli94] where an upper bound to the error introduced in the probability of surviving a mission and the performability is also given.

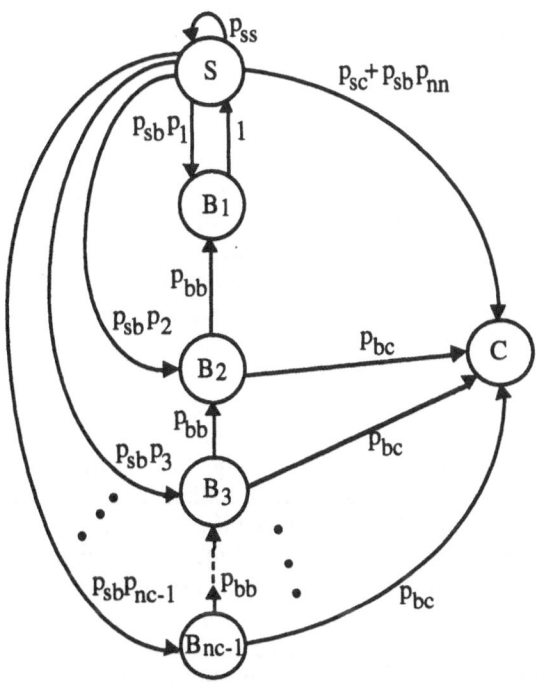

**Figure 3.** The new model for iterative executions with failure clustering.

As stated in Subsection 2.1 the dependability figures we are interested in are the probability of surviving a mission of a constant number of iterations n and the performability. The expressions for them are the following.

$$P(\text{mission success}) = \prod_{i=1}^{n} P(X_i \neq C \mid X_{i-1} \neq C), \text{ where } P(X_i \neq C \mid X_{i-1} \neq C) =$$

$$= (1 - p_{sc}) +$$
$$+ (p_{sc} - p_{bc})p(X_{i-1} = B_2 \text{ or..or } X_{i-1} = B_{n_c-1} \mid X_{i-1} \neq C) +$$
$$- p_{nn}p_{sb}\left(1 - p(X_{i-1} = B_2 \text{ or..or } X_{i-1} = B_{n_c-1} \mid X_{i-1} \neq C)\right) +$$
$$+ (p_{sc} + p_{nn}p_{sb})P(X_{i-1} = B_1 \mid X_{i-1} \neq C).$$

$E[M_n] = E[\text{number of successes}|\text{mission success}]P(\text{mission success})$

$= (n - E[\text{number of benign failures}|\text{mission success}])P(\text{mission success}) =$

$$= P(\text{mission success})n - \left( \sum_{h=1}^{n_c-1} h p_h P_{bb}{}^{h-1} \right) \Big/ \left( \sum_{h=1}^{n_c-1} P_h P_{bb}{}^{h-1} \right)$$

$$\sum_{i=1}^{\lfloor (n-1)/2 \rfloor} i \sum_{\substack{1 \le j_1,\dots,j_i \le n_c-1 \\ j=j_1+\dots+j_i \le n-i+1}} \prod_{k=1}^{i} p_{j_k} \left[ \binom{n-j}{i} p_{sb}{}^i P_{bb}{}^{j-i} (1 - p_{sc} - p_{sb})^{n-j-i} + \right.$$

$$\left. + \sum_{h=1}^{j_i} \binom{n-j+j_i-h}{i} p_{sb}{}^i P_{bb}{}^{j-i+h-j_i} (1 - p_{sc} - p_{sb})^{n-j-h-i+j_i+1} \right].$$

# 4 Evaluations

The model proposed in Section 3 is not tied to any specific distribution of the length of stays in failure regions. The two main factors characterising the effects on dependability figures of any distribution function are 1) its mean, and 2) the probability of sequences of benign failures equal or longer than the critical threshold $n_c$. We concentrate on the latter one in the following reliability and performability evaluation (while keeping all other parameters constant) showing the variation obtained by applying distribution functions belonging to different families (with the same values of the probability $p_{nn}$). Then we show that the figures obtained are approximately the same for all the types of system considered in the previous section: heavy variations of the factor k (determining the ratio between catastrophic failures obtained after a success or after a benign failure) from 0 to $10^4$ do not imply significant changes in reliability and performability. Last, we perform an analysis of the variations in the considered dependability measures as a function of the probability $p_s$ obtained by testing, fixing all the other parameters and considering two different values for the ratio between $p_b$ and $p_c$. The families of distributions used include some common distributions from the literature and some special limiting distributions, one of which can be shown to provide lower bounds on the dependability figures while the others are useful to explain some tendency. These distributions are:

- geometric distribution, $p_i = q(1-q)^{i-1}$, $i = 1,2,3,\dots$, with $q \in (0,1]$;

- modified Poisson distribution, $p_i = \dfrac{e^{-\alpha}\alpha^{i-1}}{(i-1)!}$, $i = 1,2,3,\dots$, $\alpha > 0$;

- modified negative binomial, $p_i = \binom{i+r-2}{r-1} q^r (1-q)^{i-1}$, $i = 1,2,3,\dots$, $q \in (0,1]$ and $r=1,2,3,\dots$ (we use r=5);

and, once a value for $p_{nn}$ has been fixed,

- a distribution d1 defined as: $p(1) = (1 - p_{nn})$ and $p(n_c) = p_{nn}$;

- a distribution d2 defined as: $p(1) = \dfrac{(1 - p_{nn})}{2}$, $p(n_c - 1) = \dfrac{(1 - p_{nn})}{2}$ and $p(n_c) = p_{nn}$;

- a distribution d3 defined as: $p(n_c - 1) = (1 - p_{nn})$, $p(n_c) = \dfrac{p_{nn}}{2}$ and $p(2n_c) = \dfrac{p_{nn}}{2}$.

Figures 4 and 5 show, respectively, the results for the probability of mission failure and the performability measure obtained from the model as a function of the variation of the probability of exceeding a sequence of $n_c$ -1 consecutive failures, $p_{nn}$. We use the six distributions previously described to model the correlation between successive inputs and a set of plausible values for the model parameters, as shown in Table 2.

| Parameter values | |
|---|---|
| $P_b = 10^{-4}$ | $n_c = 10$ |
| $P_c = 10^{-9}$ | $n = 10^6$ |
| $P_s = 1 - P_{sb} - P_{sc}$ | $k = 100$ |
| | $p_{nn}$ variable from 0 to $10^{-3}$ |

**Table 2.** Parameter values used in the evaluation conducted as a function of $p_{nn}$.

The number of iterations in a mission, n, is $10^6$ (a realistic number, e.g., for civil avionics where the average duration of one iteration could be 20-50 milliseconds and the mission duration could be around 10 hours). The value for k is fixed to 100, modelling case b) in the previous section (pbc = 100 $p_{sc}$), which represents a quite common scenario where the probability of the next iteration producing a catastrophic failure increases if the last iteration produced a benign failure.

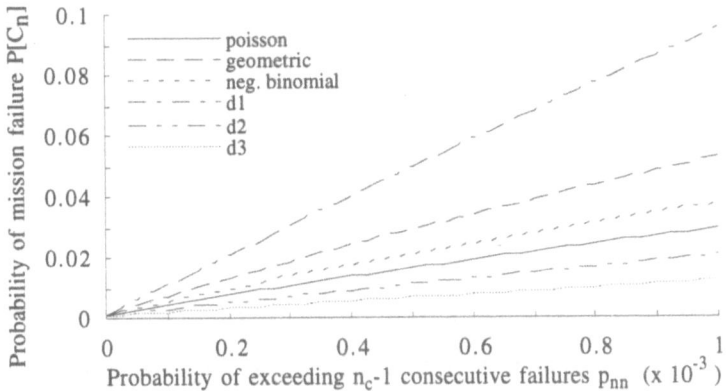

**Figure 4.** Probability of mission failure as a function of $p_{nn}$.

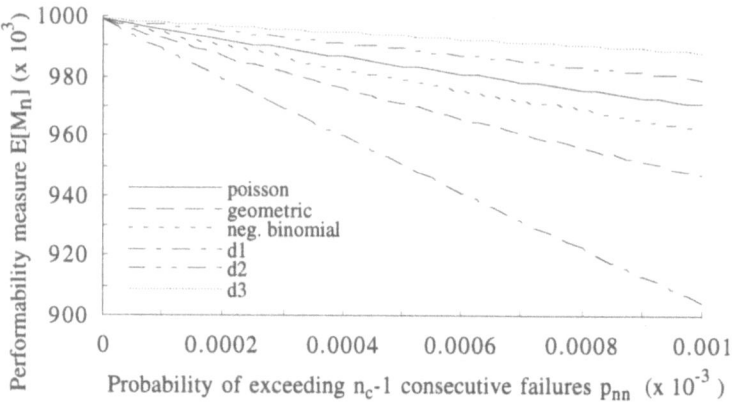

**Figure 5.** Performability as a function of $p_{nn}$.

In our scenario, where we have fixed the steady state probabilities $p_s$, $p_b$ and $p_c$, distributions with low mean represent input trajectories crossing failure regions more often but with a shorter permanence than those represented by distributions with higher mean. For these low-mean distributions we obtain higher values of the probability $p_{sb}$. Among these distributions, it is easy to derive that, for a given value of $p_{nn}$, d1 is the one having the lowest mean. Since the dominant factor responsible for increasing the probability of mission failure due to exceeding a sequence of $n_c$-1 failures is the term $p_{sb}p_{nn}$ on the arc from S to C in Figure 3, it can be argued that distributions functions having low mean show worse behaviour. In fact, in Figures 4 and 5 we observe that d1 has the worst behaviour and is a lower bound for all the other distributions. Moreover, from the evaluation performed for $p_{nn}$ in the range $(0, 10^{-3})$, we observed that the dependability figures obtained for the various distributions are ranked in the same order as the mean. Distributions d2 and d3 have the best behaviour among the distribution considered and, having the highest mean given $p_{nn}$, exemplify the behaviour just described.

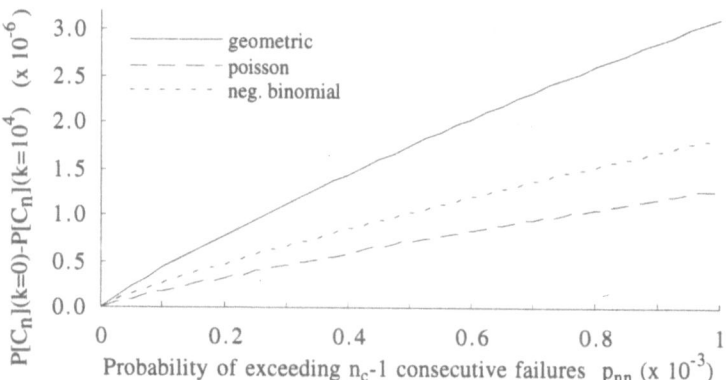

**Figure 6.** Differences in the probability of mission failure for $k = 0$ and $k = 10^4$.

To extend our analysis to the other scenarios described in Section 3, we computed the probability of mission failure for different values of k $(0, 10^{-4}, 10^{-2}, 1, 10^2$ and $10^4)$ using the values in Table 2 for the other parameters. The distributions selected in this case have been the geometric, the negative binomial and the modified Poisson. The figures obtained for the various k were so close to each other that plotting them would have been useless. So, we decided to show a plot of the variations on the probability of mission failure between the lowest (k= $10^4$) and the highest (k=0) case. Figure 6 shows the plot of these differences. It can be observed from the figure that for all the three distributions the maximum difference is extremely low, of the order $10^{-6}$ against absolute values of $10^{-2}$ as it can be observed in Figure 4. More details can be found in [La Torre95].

| Parameter values | |
|---|---|
| $p_b$ variable | $n_c = 10$ |
| $p_c$ variable | $n = 10^6$ |
| $p_s$ variable from 1-2.5 $10^{-4}$ to 1- 5 $10^{-5}$ | k=100 |
| | $p_{nn} =3\ 10^{-4}$ |

**Table 3.** Parameter values used in the evaluation conducted as a function of $p_s$.

Last the probability of mission failure has been computed varying the probability $p_s$. The aim is to investigate how sensitive is the model to such a parameter, since the values provided by testing could be not extremely precise and therefore it becomes useful to study an interval around the value derived by testing. In doing this we wanted also to take into account different ratios between $p_b$ and $p_c$. Using the values in Table 3 we produced two such plots (Figures 7 and 8) for $p_b = 10^5\ p_c$ and $p_b = 10^3\ p_c$ respectively.

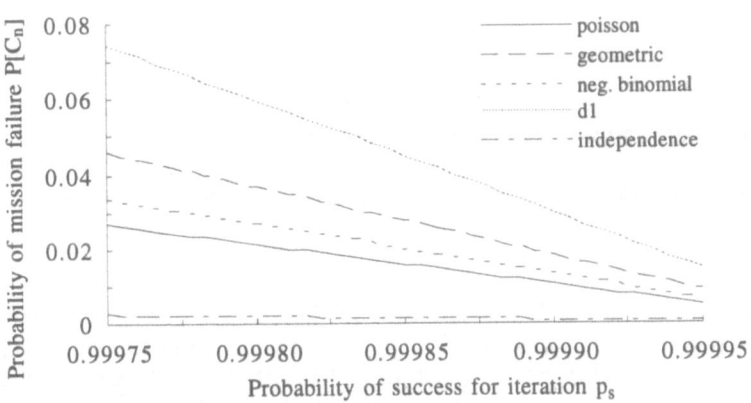

**Figure 7.** Probability of mission failure as a function of ps, with pb = $10^5$ pc.

From Figures 7 and 8 it can be observed that, as expected, passing from $p_b = 10^5\ p_c$ to $p_b = 10^3\ p_c$ the probability of pointwise catastrophic failures increases thus increasing (in the same proportion) the probability of mission failure. The other fact that can be immediately observed is that the curves in Figure 8 are much more close to each other and are ranked in the same order than in Figure 7. More precisely we observed that the absolute distance between any two curves for the same value of $p_s$ is exactly the same. Our explanation is that we have left $p_{nn}$ unchanged, and that the value of $p_{nn}$ determines the contribution to mission failure due to sequences of benign failures. In the Figures we included also the probability of mission failure in case of independence between successive inputs. In such an hypothesis the value of $p_{nn}$ is not fixed but depends on $p_b$ and is usually extremely low ($p_b{}^{nc}$). This case can be therefore considered a lower bound for the probability of mission failure that can be expected given $p_s$, $p_b$ and $p_c$ thus helping to appreciate the effects of different distributions and different degrees of input correlation on the expected behaviour of the system.

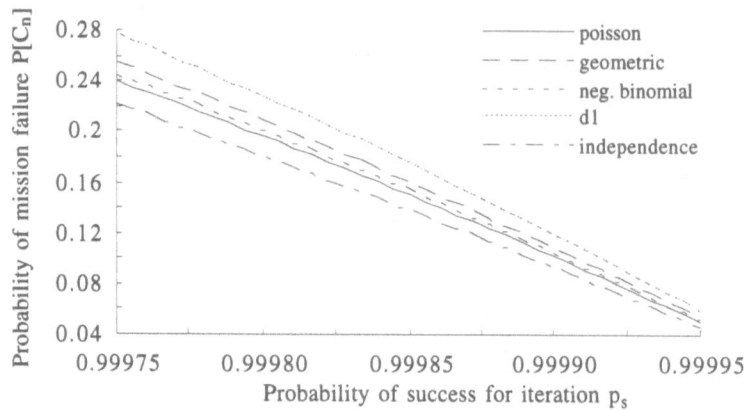

**Figure 8.** Probability of mission failure as a function of ps, with $p_b = 10^3 p_c$.

# 5 Conclusions

In this paper, we addressed one of the main causes of the lack of realism of most structural models for predicting the dependability of iterative software. We defined a new model that is able to account for both dependencies between input values of successive iterations and the effects of sequences of consecutive failures. To improve utility with respect to previously proposed models, ours is based on steady state probabilities of an iteration outcome (success, benign failure or catastrophic failure) and those representing the correlation. It requires the designers or users to determine just the steady state probabilities of an iteration outcome, which are relatively easy to obtain, rather than the difficult and costly state transition probabilities on which previous proposals are based. The proposed model can accommodate different distributions of the length of stays in failure regions; a number of distributions have been taken into consideration and their effects on the dependability figures analysed. Analyses have been made also to investigate the effects of different structural characteristics of the system and the variations due to different ratios of benign versus catastrophic failures on the dependability attributes considered.

# Acknowledgements

This research was supported by the CEC in the framework of the ESPRIT Basic Research Action 6362 "PDCS2" ("Predictably Dependable Computing Systems").

# References

[Amman88]        Amman P E, Knight J C. Data Diversity: An Approach to Software Fault Tolerance. IEEE TC 1988; C-37:418-425

[Arlat90]        Arlat J, Kanoun K, Laprie J C. Dependability Modelling and Evaluation of Sofware Fault-Tolerant Systems. IEEE TC 1990; C-39:504-512

[Bishop93]        Bishop P G. The Variation of Software Survival Time for Different Operational Input Profiles (or why you can wait a long time for a big bug to fail). Proc. of FTCS-23, Toulouse, France, 1993, pp. 98-107

[Bishop88]        Bishop P G, Pullen F D. PODS Revisited - A Study of Software Failure Behaviour. Proc. of FTCS-18, Tokyo, Japan, 1988, pp. 1-8.

[Bondavalli95]        Bondavalli A, Chiaradonna S, Di Giandomenico F, Strigini L. Dependability Models for Iterative Software Considering Correlation between Successive Inputs. Proc. of IEEE Int. Conference on Performance and Dependability, Erlangen, Germany, 1995, pp. 13-21

[Bondavalli94]        Bondavalli A, Chiaradonna S, Di Giandomenico F, Strigini L. Modelling correlation among successive inputs in software dependability analyses. C94-20, CNUCE/CNR, 1994

[Chiaradonna94]        Chiaradonna S, Bondavalli A, Strigini L. On Performability Modeling and Evaluation of Software Fault Tolerance Structures. Proc. of EDCC1, Berlin, Germany, 1994, pp. 97-114

[Csenski89]        Csenski A. Recovery block reliability analysis with failure clustering. Proc. of DCCA-1 (Preprints), Santa Barbara, California, 1989, pp. 33-42.

[La Torre95]        La Torre S, Chiaradonna S, Di Giandomenico F, Bondavalli A. The Effects of Input Correlation on the Dependability of Iterative Software. B4-24, IEI-CNR, Pisa, Italy, 1995

[Meyer80]        Meyer J F. On evaluating the performability of degradable computing systems. IEEE TC 1980; C-29:720-731

[Randell75]        Randell B. System Structure for Software Fault Tolerance. IEEE TSE 1975; SE-1:220-232

[Tai92]        Tai A T, Avizienis A, Meyer J F. Evaluation of fault tolerant software: a performability modeling approach. In: C. E. Landwher, B. Randell and L. Simoncini (ed) Dependable Computing for Critical Applications 3. Springer-Verlag, 1992, pp 113-135,

[Tai93]        Tai A T, Avizienis A, Meyer J F. Performability Enhancement of Fault-Tolerant Software. IEEE TR 1993; R-42:227-237

[Tomek93]        Tomek L A, Muppala J K, Trivedi K S. Modeling Correlation in Software Recovery Blocks. IEEE TSE 1993; SE-19:1071-1085

[Trivedi82]        Trivedi K S. Probability & Statistics with Reliability, Queuing, and Computer Science Applications. Prentice-Hall, London, 1982

# The Verification of Compiled Code

T. Jennings and P. Taylor
EDS
Milton Keynes U.K.

## 1 Introduction

For safety critical applications it is desirable to have a procedure for positive validation of the correctness of the compiler employed to generate object code from high level source code. Confidence in a compiler is usually based upon comprehensive testing and experience of its use but more formal approaches are possible. Two approaches are:

- systematic testing of the object code exploiting knowledge of the source code and the transformations employed by the compiler.

- a mathematical proof of correctness of the compiler transformation.

Both of these have advantages and disadvantages. Systematic testing has the advantage that the code actually generated by the compiler is being examined and that the task is bounded by the individual module under consideration. Its disadvantage is that it is difficult to ensure that testing is comprehensive, so that the test data exercises all paths through the code and is representative of all the positive behaviours of the code. A mathematical proof of correctness has the advantage that it is static and so in principle gives comprehensive verification regardless of the data inputs to the code. Its disadvantage is that such a proof must address all possible samples of source code that could be presented to the compiler, and so for any actual compiler a proof of correctness is at a level of complexity beyond the current state of the art for automatic proof checking.

In this paper a third approach to compiler verification is discussed, which has been investigated in the context of the Ada language. The verification approach is static and focuses upon individual code modules. By combining aspects of both of the approaches described above it exploits the advantages of each.

# 2 An Approach to Verification

## 2.1 The Verification Problem

A compiler performs a complicated transformation to generate, from source code in a high level language, a sequence of machine code statements. The process of compilation is organised via the structure of the source language so that a compiler can be explained in terms of the code mappings associated with each of the different constructs identified by the syntax of the source language. Often some form of documentation of these code mappings will provide the major part of the available information as to the mechanism of a compiler. Based upon this fact, compiler verification approaches have been suggested that involve the decompilation of the object code using knowledge of these mappings. However in practice such an approach is of limited value since:

- The information on the mappings is partial; often they will be presented via examples.

- The computation of the inverse becomes computationally intractable; often a choice of the inverse mapping cannot be easily determined from the object code.

- Even if decompilation could be achieved, only the accuracy of the compilation transformation would be checked. That the various code mappings correctly implement the semantics of the source language would still not have been addressed.

These problems are compounded when code optimisations are introduced by the compiler. However, knowledge of the code mappings can provide valuable information when employed as part of a more general, sounder, approach to verification.

To discuss the verification problem clearly it is useful to have a more formal understanding of the aim of the compilation process, to this end a mathematical formulation of compiler correctness is sketched below.

```
Correct compiler
   = ∀source : AdaCode
      • let object  =  compiler source
        in
        ∀ env : AdaEnvironment, mem : MachineEnvironment
        •
        let new_env  =  ada_exec sourceenv;
            new_mem  =  machine_exec object mem
        in
        mem represents env  ⇒  new_mem represents new_env
```

Where the various functions have the interpretations:

- compiler : AdaCode → AssemblyCode,
  is a compilation function;

- represents : AdaEnvironment ↔ MachineEnvironment,
  is a relation which is valid only when the machine environment accurately
  represents the Ada environment

- ada_exec : AdaCode → AdaEnvironment → AdaEnvironment,
  defines the effect of executing the Ada code upon the Ada environment;

- machine_exec :
  MachineCode → MachineEnvironment → MachineEnvironment,
  defines the effect of executing the machine code.

Whilst this formula is a drastic oversimplification it does highlight a number
of points:

1. Simply to state the correctness problem for a compiler various semantic
   concepts are required, namely "represents" to get some handle upon the
   static organisation of the memory and "ada_exec" and "machine_exec" to
   describe the dynamic aspects of code execution.

2. To verify a compiler one has to prove its accuracy for all source code text,
   thus presenting a very extensive task. However if a compiler is correct
   then the correctness property must hold for any source text. Thus the
   accuracy of a compiler could be verified for particular compilation units;
   this being a more circumscribed and hence more feasible task.

3. The correctness property relates to the effects of programs, thus if a way
   of approaching it directly could be devised then complexities introduced
   by the details of compilation, such as optimisations, could be bypassed.

Based upon these insights an approach to compiler verification at the level
of individual code units has been investigated at EDS. The approach builds
upon work carried out to perform static analysis of code and employs: the
parallelisms between Ada environments and machine environments discovered
in translation to an intermediate language to identify the "represents" property; path expressions computed from the intermediate language representations to present the dynamic aspects of the semantics i.e. "ada_exec" and
"machine_exec"

## 2.2 Static Analysis

During the past decade, tools, derived from code analysis techniques developed
for compilation, have been successfully employed to examine large quantities of
code to identify the presence or confirm the absence of certain types of anomaly
known to lead to errors at run time (e.g. uninitialized variables). These static
analysis techniques have an advantage over testing in that, because they inspect
the code directly and do not just rely upon exercising the code with a necessarily

restricted sample of input data, they can guarantee the absence of certain types of error.

SPADE and MALPAS [CNCD86, TACS92] are two of the more widely used static analysis tool sets. Both of these require that the source code under analysis be translated into an intermediate language form, FDL for SPADE and IL for MALPAS. Tools that take this intermediate language as input can then check the source for specific properties and carry out other analysis processes in a source language independent and mathematically tractable setting. Thus the SPADE FDL provides a notation for describing graph structures to model the flow of control in a program, and uses mathematical functions and mapping statements labelling the edges of the flow graph to model the effects of program statements upon the execution state.

From an intermediate language representation, both the SPADE and MAL-PAS tools allow for the computation of path functions. These are mathematical expressions that capture the overall effect of the various statements in a program and hence give a mathematical presentation of the semantics of the program. The calculation of path functions makes use of the flow graph of the program. Specifically path functions are calculated between pairs of "cut points" in the flow graph, i.e. a set of points where the graph can be divided into a collection of loop-free subgraphs. Each path function summarises the effect of program between two points in terms of a finite number of cases of the form:

$$\text{TraversalCondition} \implies \text{Action}$$

The traversal condition is a boolean expression involving the initial values of the abstract program variables at the start cutpoint, and the action is a set of equalities defining the final values of the program variables, at the end cutpoint, in terms of their initial values. Each of these cases is computed by forming the composition of the mapping statements along each of the various paths between the pair of cut points; different choices at the conditional tests give rise to the the different paths and hence the separate cases.

## 2.3 The Strategy for Verification

The strategy for verification of an Ada compiler at the level of compilation units, building upon the foundations prepared by static analysis is thus the following:

1. Via translators for Ada and the compiler's target assembler language, transform the source and its compiled image into a common semantic setting, the intermediate language, where the interpretation of the high level and low level variables is aligned as much as possible.

2. Generate the two sets of path functions from the intermediate language representations of the source and object code.

3. Using logical simplification demonstrate that equivalent paths in the source and the object have equivalent effects.

Whilst simple in concept a number of obstacles must be overcome to make the approach practical.

# 3   Making it Work

## 3.1   Comparing Path Functions

Reducing the problem of verifying compiler correctness over a module to that of proving the logical equivalence of sets of path functions depends upon being able to make a correspondence between the path functions arising from the Ada code and its compiled image. However in general this is a non-trivial problem since, as noted previously, the computation of the path functions associated with a code fragment exploits knowledge of the looping structure of that code via the identification of cut points in the control flow graph. Whilst compilers do not usually change the algorithms specified by the source code and hence will not drastically affect the topology of the control flow graph from the source to the object even non-optimising compilation can employ transformation that will increase or decrease the amount of looping in a program. Examples of this occur where:

1.  Additional looping is introduced to initialize the elements of a compound data structure.

2.  "FOR" loops over small data ranges are unfolded to deal explicitly with each case, thus reducing the number of loops.

3.  The original looping structure may be unfolded to an equivalent form more suitable for the target language, thus changing the distribution of the cut points whilst preserving the number of loops e.g. `while C do S` becomes `S;if C then repeat S until not C`

It is in dealing with these situations that a knowledge of the code mappings employed by the compiler can be of value, since an obvious and effective solution is to transform the discrepancy out of existence before translation to the intermediate language. Moreover one has a choice of employing translations of the assembler or of source depending on which is most convenient. (A third option would be to transform one of the intermediate language representations).

A similar problem occurs with respect to the inline expansion of library functions (or user functions identified by the INLINE pragma for inline expansion). Whilst these might not always introduce extra looping they usually will lead to additional tests and hence expand the number of paths requiring comparison, a task one would wish to minimise as much as possible. Thus, again, it will sometimes be advisable to perform transformations upon one or other of the representations to compress (undo in some sense) such expansions. For example, an assembler will not have its own "ABS" function and so will inline expand such calls, however such expansions can be identified and folded back into a pseudo assembler statement, representing such an "ABS" function.

As well as aligning the cut points, it is also useful to align the translation into the intermediate language of local variables of the Ada source and the assembler object. In practice there are no general strategies for achieving this but again knowledge of the memory mapping approach employed by the compiler can be exploited for this.

A further technique that can be of benefit in the comparison of path functions is less pragmatic in its approach. One cause of the difficulty in comparing path functions arising from the source and object code, is that the computation of path functions, relying as it does on cut points, by its very nature fragments the unity of a code module into several parts. Comparison would be facilitated if these parts could be put back together again. In fact reconstructing a single function is not too difficult since, formally, each loop may be replaced by a call to a hypothetical auxiliary function. Since the function represents a loop, it will be a recursive function whose transfer is computed by the path functions associated with the loop. Loops within loops will give rise to further auxiliary functions, so continuing the process enables a set of mutually recursive functions to be identified that define the overall behaviour of the code. Note, essentially we are making use of the fact that, with a continuation semantics, arbitrary looping may be modelled by a set of mutually recursive functions. Of course this has neither added nor subtracted to our information about the program semantics, it has however transformed our representations and as such can sometimes make comparisons more apparent. The chief advantage of this technique is in providing an expression of the effect of the module as a transfer function effecting only the input and output variables, whilst isolating the effects of any local variables to the auxiliary functions. By supporting this focus, comparison of the source and object is made easier, since the alignment between the interface variables is usually readily discovered.

## 3.2 Coping with a Path Explosion

Whilst path functions provide a means for encoding within a mathematical formula the semantics of a sequence of program statements (that may include conditionals but excludes loops), the mathematical formula produced is usually far from being compact and will involve substantial repetitions and redundancy. The reason for this is that computation of the bundle of condition-action cases presented in a path function is equivalent to unfolding the decision tree associated with the code fragment. At each conditional, the tree bifurcates so that that the number of paths may increase exponentially with the number of conditions.

However in practice when comparing the semantics of two program fragments one is often most interested in the sets of values taken by a restricted set of the program variables. In such situations simplifications can often be discovered, since whilst the specific values are set for the variables in a limited number of places, these same values have been scattered across many irrelevant cases by the path function generation process. By what is effectively undoing the path function generation process the constraints that are really relevant to the values acquired by the variables may be made apparent.

A tool that achieves such an inversion of path function generation can be devised that: by pattern matching groups together cases according to desired criteria for the end values of variables; and by exploiting properties of the disjunctive normal forms simplifies the resulting logical expression to an equivalent form that is often much more compact.

## 3.3  Algebraic Simplification

In establishing the logical equivalence of semantic effects of the source and its compiled image, it is necessary at some stage to carry out algebraic simplifications of the mathematical expressions presented in the intermediate language mapping statements or in the path functions that result from these. Two major classes of simplifications are significant:

1. simplifications that realise the definitions of the arithmetic and boolean operations, such as addition, subtraction, negation etc.

2. simplifications that model the semantics of the machine operations such as memory and register transfers.

Whilst the stage at which simplification is carried out is usually theoretically irrelevant, advancing or delaying simplification can often have a substantial influence on the tractability of a process. Thus the ability, not just to apply simplification but to control where and when it is applied, is significant to the feasibility of the verification of compilation.

## 3.4  Translation

A further point of significance to feasibility of comparison of path expressions is the alignment of variable names in the translation to the intermediate language of both the Ada source and the assembler image. Ideally the same mathematical representation should be given for an Ada variable and its assembler image. In a sense at this level and for variables one is trying to achieve decompilation. The major difficulties arise with the translation of the assembler data representations since these cannot usually be straightforwardly mapped into simple mathematical structures. This is a problem that always arises when defining translators for assembler. However, in the case of a compiler, the situation is more tractable than in the case of hand generated code. A compiler must follow rules and will not, at random points, decide to depart from some coding guideline. The memory map the compiler specifies is either documented or can be experimentally discovered.

In preparing translators two major related areas present problems, namely in:

1. identifying when components of compound structures such as arrays or records are being referenced.

2. tracking the intended target of indirect addressing, where a value is located by the address referenced by an address.

At EDS, techniques have been developed using the data environment information, interface definitions for subroutine calls, and the computation of referencing relations to allow a translator to deduce the intended references of the assembler code.

# 4    A Trial Investigation

The strategy and tactics discussed in this paper arose as the result of a short study that investigated the compiler verification problem in the context of the XD-Ada compiler. A full account of this work was reported in [JT94]. The compiler takes Ada 83 as source and produces 68020 assembler as target.

Whilst a full trial could not be achieved within the confines of the study period, experiments were carried out to confirm the viability of the ideas presented. The experiments were based upon a sample of Ada units conforming to the SPARK [CJMFG90] Ada subset. In the experiments:

1. Ada code was checked for conformance to the SPARK subset and then the associated path functions generated, using the SPARK support tools.

2. The same Ada code was compiled by the XD-Ada compiler and then disassembled into 68020 assembler. By a combination of automatic and hand translation the assembler was converted into SPADE FDL and its associated path functions generated.

3. The comparison of the two sets of path functions was achieved by means of the techniques previously described. In particular prototypes were used of tools that:

   (a) reformulated the path expression characterisation of the module's effect into a set of inter-related recursive functions.

   (b) exploited the decision tree structure of the code to organise the path functions.

   (c) selected cases in the path functions on the bases of their action and then exploited properties of the disjunctive normal form to simplify the resulting logical expressions.

The results of these experiments were very promising indicating that with a suitably optimised sequence of transformations a largely automatic (of the order of 90%) verification of the logical equivalence of the source and object codes should be possible.

# 5   Related Work

In [PW93] an approach to the verification of compiled code similarly based at the module level is described. This was employed to verify the accuracy the PLM-86 compilation used upon the code for the protection software of a nuclear reactor. The technique is similar in that it exploits the machinery developed for a related static code analysis exercise to support compilation verification. The strategy used path expressions computed from the source module to act as the specification of the compiled object module. From there on standard program verification techniques were followed. Whilst effectively applied to a large volume of code it appears that substantial manual guidance of proof was required. In this paper a more flexible strategy is advocated. This encompasses transformations of both source and object, based upon an understanding of the compiler mappings, and exploits some general tools to recover the structure of the code. It is believed that only by building such an informed approach into the tools can an automatic strategy for compiler verification be successful and the need for manual intervention substantially reduced.

# 6   Conclusions

In this paper an approach to achieving verification of compilation at the level of individual compilation units has been described. The approach is rigorous since many transformations must be employed in order to render the source and object codes capable of comparison. It is however a verification strategy and not a testing strategy. The transformations employed are in principle sound and may themselves be verified independently of the actual compilation process.

Moreover the approach is feasible, in that it builds upon the experience static code analysis and uses well understood transformations. One would anticipate that, once optimum orderings of transformations had been identified, much of the process could be achieved automatically without human intervention.

It should also be noted that many of the transformations required might well play a part in providing a mathematical proof of correctness of a compiler. Consequently a pragmatic program for developing proofs of compilation correctness at the module level could provide an arena in which more broadly applicable expertise in practical proof techniques could be developed, that ultimately would be sufficient for the static verification of a complete compiler.

Finally it should be emphasised that the compiler verification strategy outlined above is reliant upon the availability of a toolkit - one that contains a flexible and configurable collection of tools which may be connected together in various ways to provide the means for examining different aspects of the organisation and semantics of program texts. This is similar in concept to that adopted in the LCF/HOL family of theorem provers. It is also the approach that has been evolved at EDS, over a number of substantial code analysis projects, to the construction of translators to intermediate languages and for code verification. Indeed processes required to achieve translation to IL or FDL often duplicate some of the analyses undertaken by MALPAS and SPADE.

# 7 Acknowledgement

The authors would like to thank Eurofigher 2000 Project Office in the UK Ministry of Defence and the UK Aircraft and Armaments Evaluation Establishment for their support, both moral and financial, in developing the techniques discussed in this paper.

# References

[CJMFG90] Carré B. A, Jennings T. J, Maclennan F. J., Farrow, P. F, and Garnsworthy J. R: SPARK – The SPADE Ada Kernel, Program Validation Limited Report, Third Edition, 1990.

[JT94] Jennings T. J, and Taylor P.M.: Verification of Compiled Code, EDS Report 1994.

[TACS92] TA Associates: User Guide for MALPAS Version 6.0, TA Associates 1992

[CNCD86] Carré B. A, O'Neil I.M., Clutterbuck D.L., and Debney C.W.: SPADE - The Southampton Program Analysis and Development Environment, In Sommerville I. (editor) Software Engineering Environments, Peter Peregrinus 1986

[PW93] Pavey D.J. and Winsborrow L.A.: Demonstrating Equivalence of Source Code and PROM Content, The Computer Journal 1993.

# Author Index